TARGETS IN HETEROCYCLIC SYSTEMS

Chemistry and Properties

Volume 1 (1997)

Reviews and Accounts on Heterocyclic Chemistry

Editors

Prof. Orazio A. Attanasi
Institute of Organic Chemistry, University of Urbino
Urbino, Italy

and

Prof. Domenico Spinelli
Department of Organic Chemistry, University of Bologna
Bologna, Italy

Italian Society of Chemistry
Division of Organic Chemistry
Division of Medicinal Chemistry

Published by:
Società Chimica Italiana
Viale Liegi, 48
00198 Roma
Italy

First edition 1997

Printed and bound in Italy by:
Centro Grafico, Università di Parma, Parma

Editorial Advisory Board Members

Preface

This book is dedicated to the memory of Prof. Luigi Minale (Ω May 11, 1997)

The Editors are indebted both to the Division of Organic Chemistry and the Division of Medicinal Chemistry of the Italian Society of Chemistry who encouraged them to continue a previous editorial venture on this subject. The most sincere gratitude is due to the late President of the Division of Organic Chemistry, Prof. Luigi Minale, who supported this project with enthusiasm, although not directly involved in this chemistry.

The Editors would like to thank the Members of the Editorial Advisory Board who believed in the scientific value of this new undertaking. Their suggestions and improvements were very much appreciated and some of these will be included in the next edition of this series. In this way, the Editors hope very soon to reach an international standard in all respects.

It is well known that heterocyclic derivatives represent nearly half of the some thirteen million chemical species discovered up to date. Owing to their singular properties, several of these are important in organic chemistry, both from the mechanistic and synthetic points of view, as products (including natural) and/or useful tools in the construction of more complicated molecular entities. Their utilization in polymeric, medicinal and agricultural chemistry is widely documented. Both dyestuff structures and life molecules frequently involve heterocyclic rings that play an important role in several biochemical processes. Therefore, it is readily predictable that investigations into this field will receive growing attention in the near future.

Among the various publications dealing with the chemistry of heterocycles, the main objective of the series ***TARGETS IN HETEROCYCLIC SYSTEMS - Chemistry and Properties*** is that of hitting targets able to stimulate creative hypotheses and determine innovative progress by researchers working in this area. Thus, comprehensive Reviews reporting the overall state of the art and personal Accounts highlighting interesting developments by research groups on their specific themes have been solicited from leading authors. Hence, the submission of articles having the above-mentioned aims and concerning highly specialistic topics is strongly recommended. A brief layout of the contribution should be sent to a Member of the Editorial Advisory Board.

The Authors, who contributed most competently to the realization of this volume, and the Referees, who cooperated unselfishly (often with great patience) spending valuable attention and time in the review of the manuscripts, are both gratefully acknowledged.

Orazio A. Attanasi and Domenico Spinelli
Editors

Contents

Polyazolylbenzenes and related compounds: propellene-like aromatic compounds

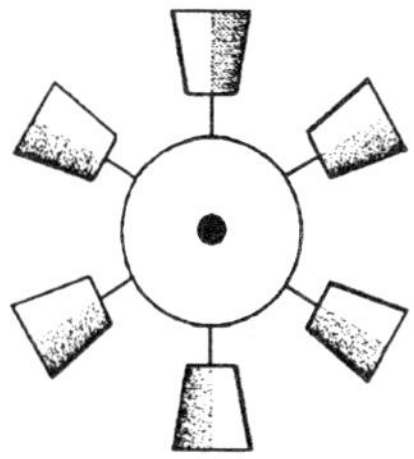

Rosa María Claramunt,[a] José Elguero,[b] Consuelo Escolástico,[a] Cristina Fernández-Castaño,[c] Concepción Foces-Foces,[c] Antonio L. Llamas-Saiz[c] and María Dolores Santa María[a]

[a] *Departamento de Química Orgánica y Biología, Facultad de Ciencias, Universidad Nacional de Educación a Distancia, Senda del Rey, s/n, E-28040 Madrid (Spain)*

[b] *Instituto de Química Médica, CSIC, Juan de la Cierva, 3, E-28006 Madrid (Spain)*

[c] *Departamento de Cristalografía, Instituto de Química-Física "Rocasolano", C.S.I.C., Serrano, 119, 28006 Madrid (Spain)*

Abstract. *Our contribution to the synthesis and structural studies of polyazolylbenzenes and polyazolylpyridines "propellenes" is summarized. After an introduction reporting related systems, the synthesis of propellenes from fluoroaromatic and heterocyclic derivatives is described. The main part reports the crystallographic and NMR studies of propellenes in relation to their conformation. Both, the static (more stable conformers in solution and in the solid state) and dynamic aspects (barriers to interconversion) have been studied experimentally. Semi-empirical (AM1) calculations prove invaluable for rationalizing the experimental observations.*

1. Introduction

This review deals with compounds for which we have coined the trivial name 'Propellenes' by analogy with the term 'Propellanes' which was used by David Ginsburg for compounds like **1** while Jean-Marie Conia employed the term 'Rotanes' for compounds such as **2**.[1a,1b,2]

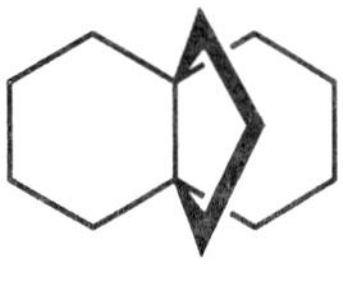

1

2

3

There are compounds like **3**,[1c] which resemble propellers and, consequently, when we first prepared the compounds subject of this review, we called them 'Propellenes'. Our compounds are related to MacNicol's 'Hexa-Hosts',[3] Vögtle and Weber's 'Octopus'[4] and to Toda's 'Hexa-Pedals'.[5]

We have summarized in Table 1 the three situations we will discuss although our work is only related to the part framed in black. The MacNicol, Weber and Toda compounds belong to R_6-benzenes with R = others (often a CC triple bond or a OR/SR substituent).

Table 1. R_6-Benzenes, R_5-pyridines and R_8-naphthalenes

	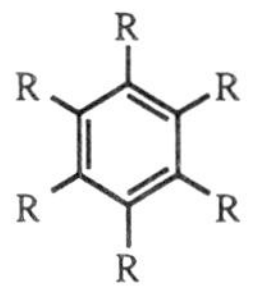	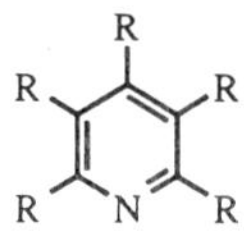	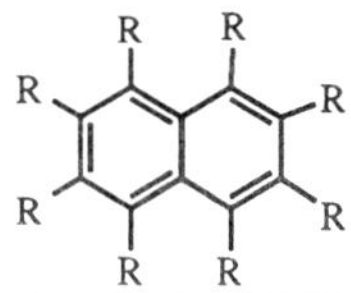
R = aromatic	Gielen[6] Gielen[7] Müllen[8]		
R = heterocyclic	Henrie[9] Weiss[10] Meijer[11]		Meijer[11]
R = others	Biali[12] Astruc[13] Vollhardt[14]		MacNicol[15]

Gielen and Willen have studied the conformational behaviour of benzene derivatives with R = *o*-methyl and *o*-methoxybenzene,[6,7] while Müllen has prepared terphenylyl and quaterphenylyl derivatives.[8] Henrie was the first to describe hexapyrazolylbenzenes[9] and Weiss succeeded in preparing the first-ever per(onio) substituted benzene **4**;[10] both used hexafluorobenzene as starting material as did Meijer for hexakispyrrolylbenzene.[11]

Biali discussed the case of hexakisisopropylbenzene **5** prepared by other authors,[12] and Astruc described hexakis(but-3-ynyl)benzene (R = $CH_2CH_2C{\equiv}CH$) and the hexacid R = $CH_2CH_2C{\equiv}C\text{-}CO_2H$;[13] other hexakis-alkynylbenzenes have been prepared by Vollhardt.[14]

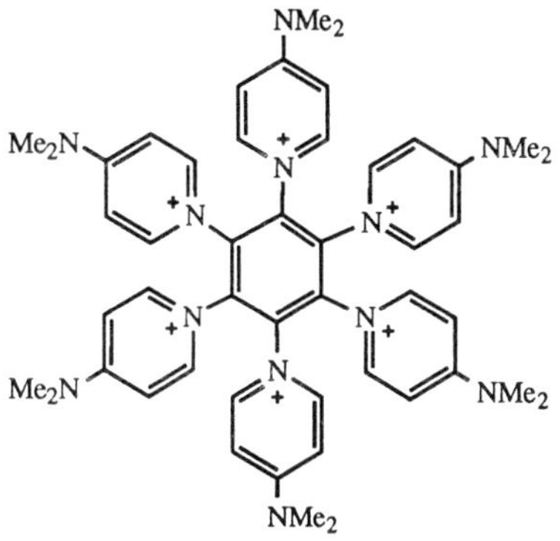

4

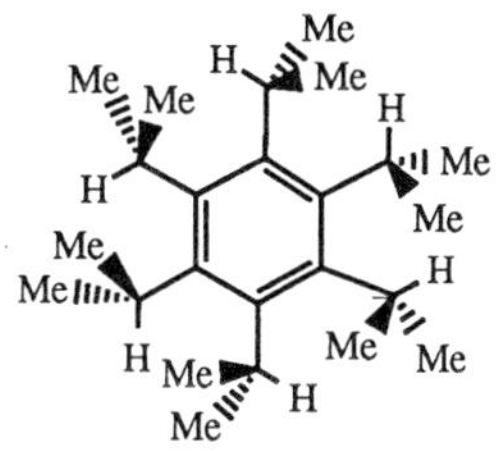

5

In the series of R_8-naphthalenes, MacNicol has recently prepared the case R = SAr with a variety of Ar substituents; these compounds are described as hosts of unparalleled versatility.[15] The naphthalene ring is the central core of octa(pyrrol-1-yl) derivatives.[11]

Clearly, the main interest of these compounds lies in their host-guest properties. We will summarize our contribution to this field, the published part,[16-20] the part that is in press,[21-23] and some unpublished results.[24]

Together with these compounds, in which the substituent is directly linked to a benzene (or a heterocycle), the structures recently described by Steel are worth reporting. Poly(pyrazol-1-yl)benzenes **6** and **7** ("Coelen-terands") are extremely interesting ligands in coordination chemistry.[25,26]

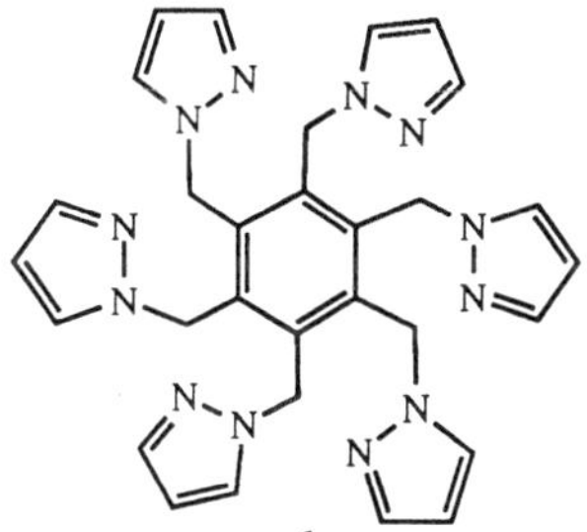

6

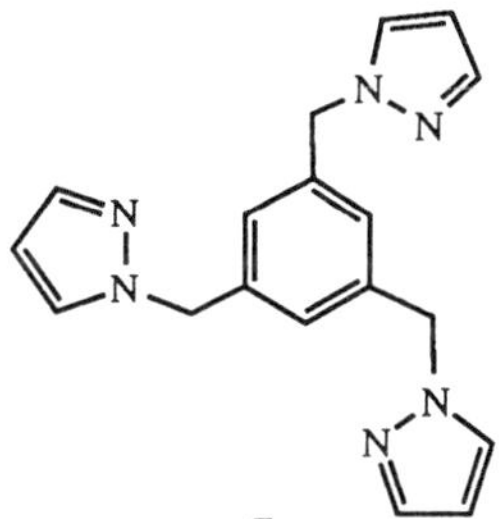

7

2. Chemistry: Synthesis

We have represented in Scheme 1 the compounds we have synthesized and studied belonging to pyrazolylfluorobenzenes, hexakispyrazolylbenzenes, pentakispyrazolylpyridines and their inclusion compounds.

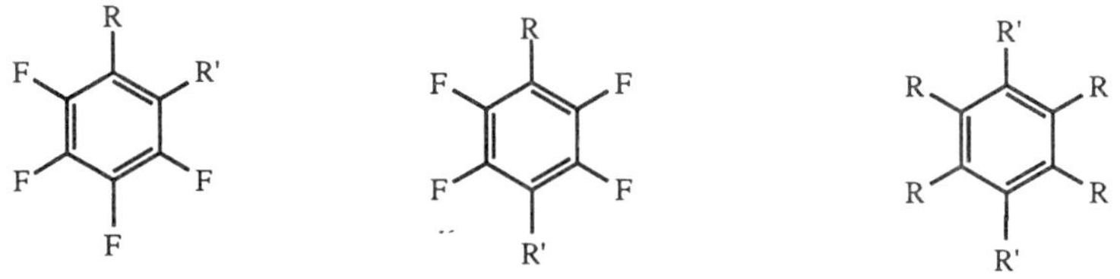

8, R = R' = pz
9, R = R' = dmpz
10, R = pz, R' = dmpz

11, R = R' = pz
12, R = R' = dmpz
12a, **12**.2 CH_3CO_2H
12b, **12**, 2 $CH_3CH_2CO_2H$
13, R = pz, R' = dmpz

14a, **14b**, R = R' = pz
15, R = R' = dmpz
16a, **16b**, R = pz, R' =dmpz
16c, **16**.2 CH_3CO_2H
17, R = dmpz, R' = pz
17a, **17**.H_2O
17b, **17**.2 CH_3CO_2H
17c, **17**.2 $CH_3CH_2CO_2H$
17d, **17**.2 $CH_3(CH_2)_3CO_2H$
17e, **17**.2 $CH_3CH_2CH(CH_3)CO_2H$
17f, $\mathbf{17}^{2+}$.2 $(C_6H_2N_3O_7)^-$

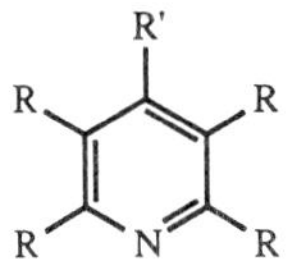

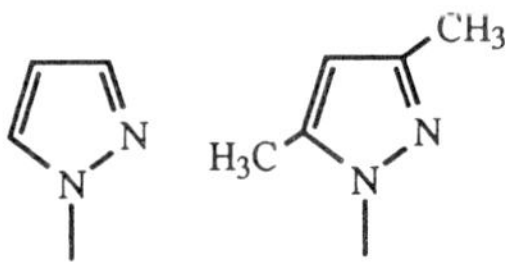

pz dmpz

18, R = R' = pz
19, R = R' = dmpz
20, R = pz, R' = dmpz
21, R = dmpz, R' = pz

Scheme 1

2.1. Pyrazolylfluorobenzenes 8-13.

In Scheme 2 we have summarized the synthetic procedure starting always from commercial pentafluorophenylhydrazine and the corresponding β-dicarbonyl compounds, malonaldehyde (in the form of its tetraacetal) and acetylacetone. The separation of isomers was carried out on final pyrazoles (**8/11**, **9/12**, **10/13**) and not on the arylhydrazine intermediates [**24** (20%) - **25** (80%); **26** (15%) - **27** (85%)]. The pentafluoro derivatives **22** and **23** are liquid compounds and their study was not pursued further.

2.2. Hexakispyrazolylbenzenes 14-17.

This series of compounds was prepared by reaction between pyrazolate anions (pz⁻ or dmpz⁻) and polyflurobenzene derivatives in THF. In this way, following the Henrie-Yeager's procedure,[9] compounds **14** and **15** were obtained from hexafluorobenzene and the mixed compounds **16** and **17** from the tetrafluoro derivatives **12** and **13**.

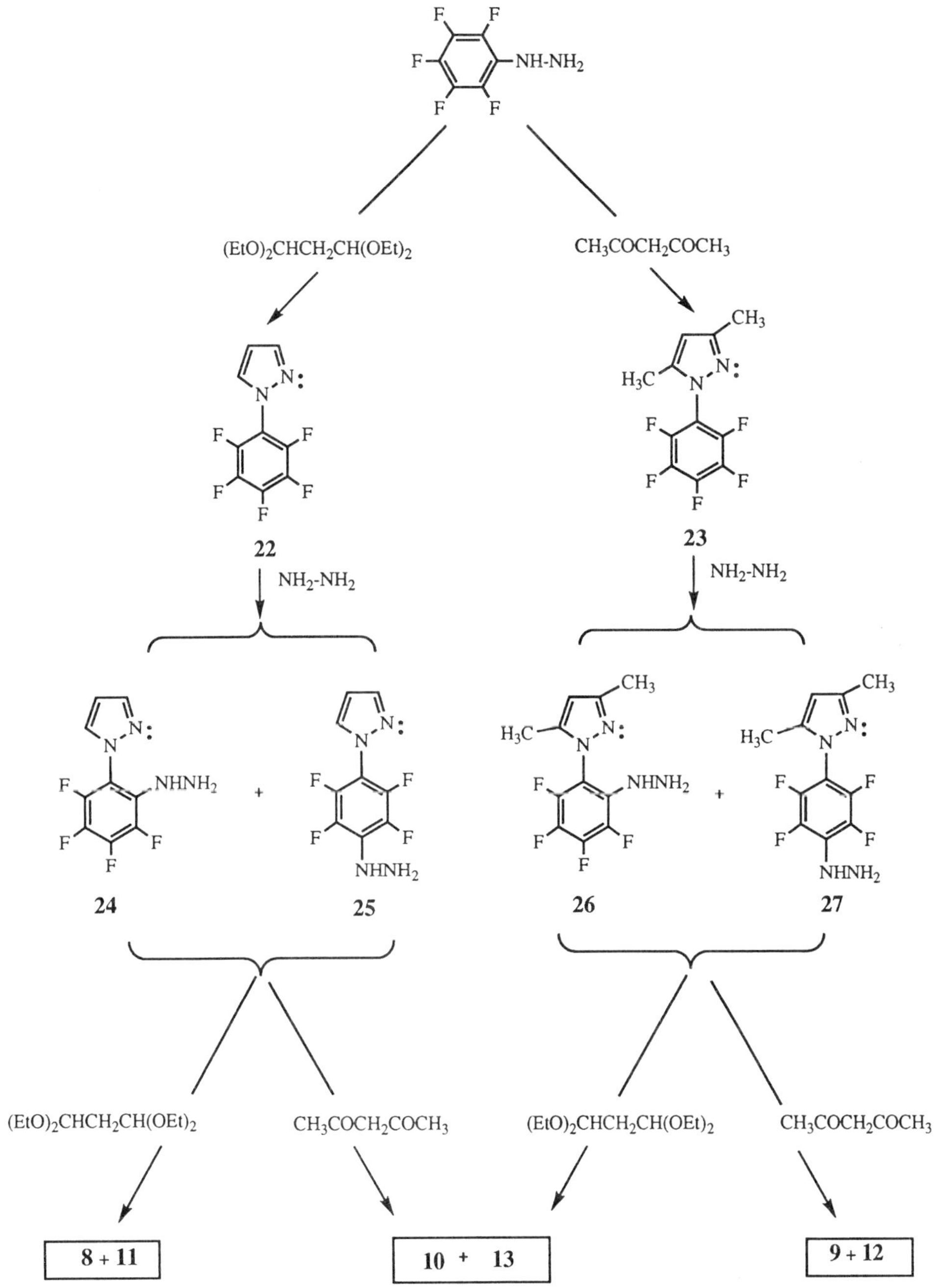
NH-NH2
(EtO)2CHCH2CH(OEt)2
CH3COCH2COCH3
22
23
NH2-NH2
NHNH2
24
25
26
27
8 + 11
10 + 13
9 + 12

Scheme 2

2.3. Pentakispyrazolylpyridines 18-21.

When the molar ratio pyrazole/pentafluoropyridine was 1:1, five compounds were formed: the mono- **28a**, **28b**, bis- **29a, 29b**, tris- **30a, 30b**, tetrakis- **31a, 31b**, and pentakis-pyrazolyl substituted pyridines **18** and **19**. The **pz** and **dmpz** series correspond respectively to R = pyrazol-1'-yl and 3,5-dimethylpyrazol-1'-yl (Scheme 3). By using a molar ratio pyrazole/pentafluoro-pyridine 5:1 only the 2,3,4,5,6-pentakis(pyrazol-1'-yl)pyridine (**18**) and 2,3,4,5,6-pentakis(3',5'-dimethyl-pyrazol-1'-yl)pyridine (**19**) were isolated (Scheme 4). Starting from 4-(pyrazol-1'-yl)-2,3,5,6-tetrafluoropyridine (**28pz**) and from 4-(3',5'-dimethylpyrazol-1'-yl)-2,3,5,6-tetrafluoropyridine (**28dmpz**) the mixed derivatives 2,3,5,6-tetrakis(3',5'-dimethylpyrazol-1'-yl)-4-(pyrazol-1'-yl)pyridine (**21**) and 2,3,5,6-tetrakis(pyrazol-1'-yl)-4-(3',5'-dimethylpyrazol-1'-yl)pyridine (**20**) were obtained (Scheme 5).

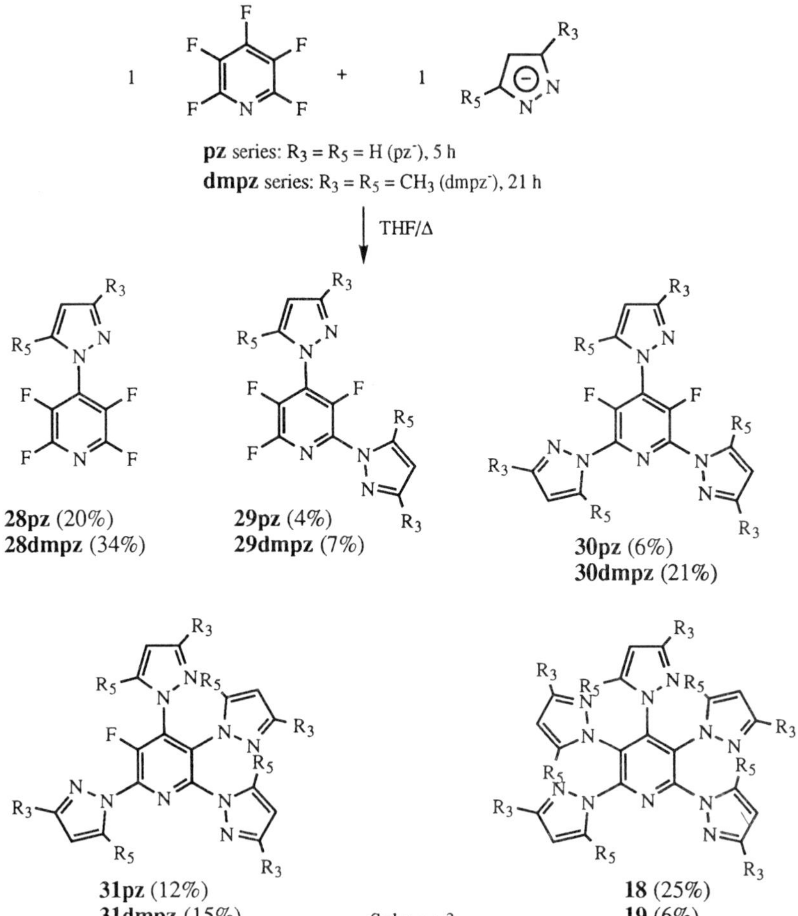

Scheme 3

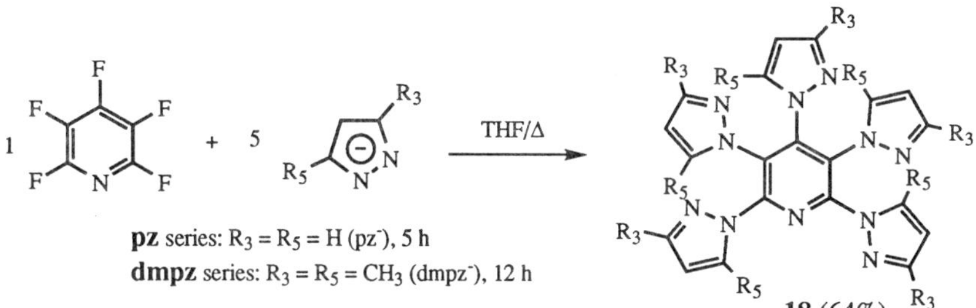

pz series: $R_3 = R_5 = H$ (pz$^-$), 5 h
dmpz series: $R_3 = R_5 = CH_3$ (dmpz$^-$), 12 h

18 (64%)
19 (56%)

Scheme 4

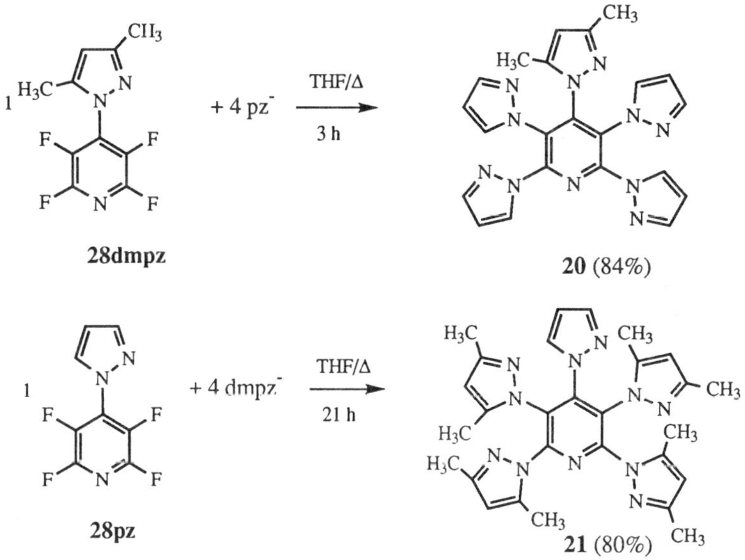

28dmpz

20 (84%)

28pz

21 (80%)

Scheme 5

Considering the structure of the products of partial substitution (Scheme 3), the reactivity of the fluorine atoms in pentafluoropyridine follows the order 4-F > 2-F/6-F > 3-F/5-F. This is the order described for other nucleophiles although tetra substituted derivatives like **31** were not reported.[27,28]

2.4. Hexakisimidazolylbenzenes and octakisazolylnaphthalenes

As an extension of the reactivity of polyfluorinated benzenes and naphthalenes with azolates, we have reacted the imidazole, 2-methylimidazole and benzimidazole anions with hexafluorobenzene in tetrahydrofurane (THF) to obtain the hexakisimidazolylbenzenes **32-34** in quantitative yields. In the same conditions, anions of 1-hydroxypyrazole and 1-aminopyrazole also afforded the hexakis-substituted derivatives **35** and **36**, but in lower yields (Scheme 6).[24]

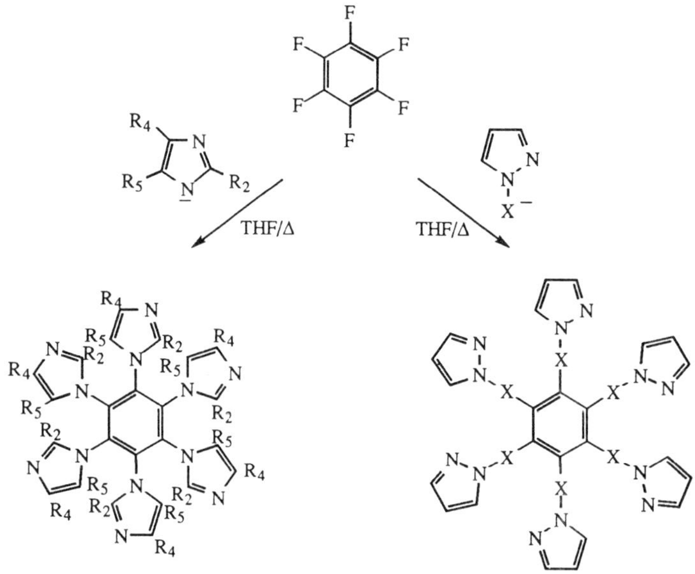

$R_2 = R_4 = R_5 = H$	**32**
$R_2 = Me,\ R_4 = R_5 = H$	**33**
$R_2 = H, R_4, R_5 =$	**34**

X = O,	**35**
NH	**36**

Scheme 6

Starting from octafluoronaphthalene it was also possible to get octakis(pyrazol-1'-yl)- and octakis(3',5'-dimethylpyrazol-1'-yl)-naphthalenes. Better yields were obtained when changing the THF solvent by *N,N'*-dimethylformamide (DMF).[24]

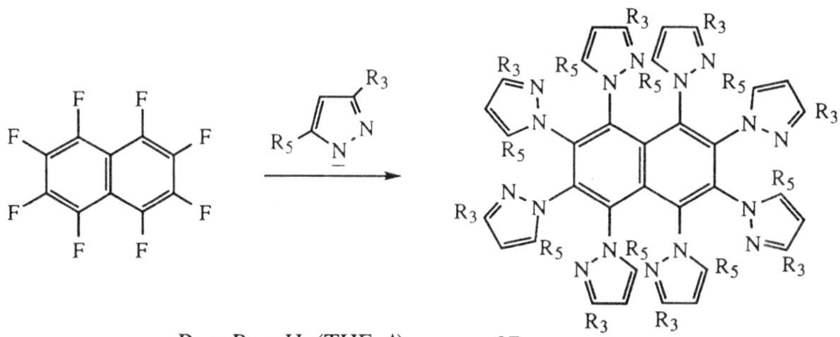

$R_3 = R_5 = H$, (THF, Δ)	**37**
$R_3 = R_5 = CH_3$, (DMF, Δ)	**38**

Scheme 7

The reactivity of propellenes is largely unexplored. There are indications (from NMR experiments) that protonation of one pz/dmpz substituent induces the rotation of this substituent in such a way that the N^{+}-H becomes close to the lone pair of a neighbour pz/dmpz substituent.

3. X-Ray Crystallography.

3.1. Introduction

The structural work reported here has been performed on a set of related molecules referred to as aromatic propellenes[16,17,18] and some intermediate products[20] obtained in their synthesis, Scheme 1. Using this information, we will try to correlate the molecular structure with the crystal packing, paying special attention to the description of host-guest complexes, polymorphism, pseudoisomorphism and, in general, to the characterization of packing modes. In order to complement the description of the intermolecular interactions that govern their packing, several statistical studies have been performed using the data compiled in the Cambridge Structural Database[29] [CSD hereinafter]. Finally, semiempirical calculations have been carried out to study the molecular structure and conformation of these compounds. The relative stabilities of all conformers will be described in section 6, together with their interconversion paths and the energy barriers between the most stable ones.

3.2. Molecular structure

All molecular structures described in this section (Scheme 1) are consist of a central core (benzene or pyridine) and fluorine atoms, partial or totally substituted by pyrazole (pz) or 3,5-dimethylpyrazole (dmpz) rings (Figure 1 and 2). Due to the different possibilities in the disposition of the nitrogen atom lone pairs with regard to the central ring, several conformations can be obtained and some of them have been observed experimentally in the solid state.

3.2.1. Angular deformation in the propeller core

3.2.1.1. Benzene ring

The geometry presented by the benzene ring is sensitive to the electronic nature of the substituents.[30-32] It should be possible to explain the differences with regard to the standard values (120°, 1.39 Å)[33] due to the substitution of one or more hydrogen atoms and, as a consequence, of the redistribution of the electron density in the system. The variations shown by the endocyclic angles in terms of the influence of the substituent are usually more significant than those presented by the bond distances, because they are less influenced by the thermal motion.

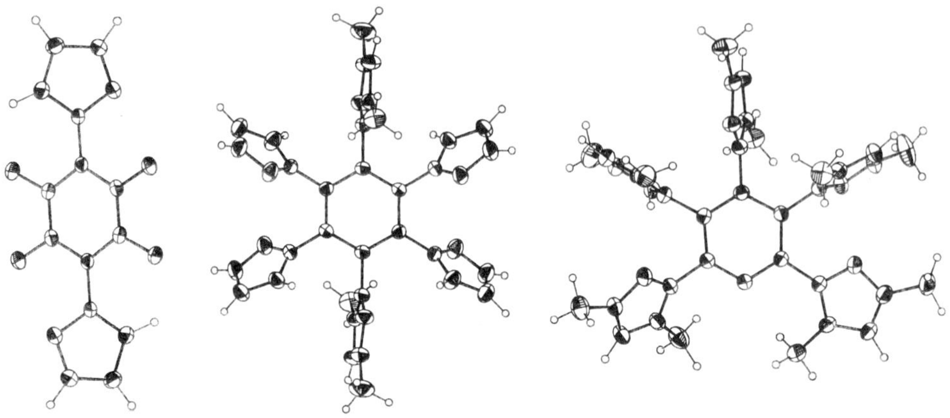

Figure 1: Molecular structure of **11**, **16c** host and **19** compounds. Ellipsoids are drawn at the 30% probability level.

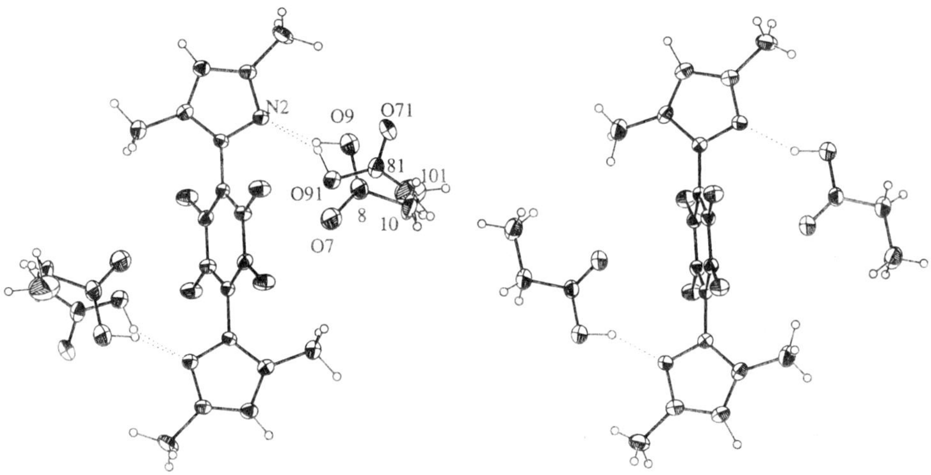

Figure 2: Molecular structure of the **12a** and **12b** inclusion compounds showing the disorder of the acetic acid (O7, C8, O9, C10 and O71, C81, O91, C101).

The X-ray molecular structure of *p*-phenylenediamine[34] was taken from the CSD[29] [Refcode: PENDAM] to be used as model of ring deformation induced by the substitution of H atoms by pyrazole rings. The average angular values for the three independent molecules in the asymmetric unit cause the angular deformations $\Delta\alpha$ and $\Delta\beta$ shown in Figure 3a. When comparing the calculated bond angles using these parameters with the experimental ones, we can conclude that the pyrazole rings induce similar deformations in the benzene as in the amino groups. The main differences were observed in the inclusion complexes which present a strong O-H···N hydrogen bond interaction between the guest and host moieties (Figure 2).

In summary, the influence of the substituents over the benzene rings is exhibited in the corresponding endocyclic angles, mainly in the *ipso* one (α), that is increased by effect of the F atoms[32] and diminished by the pz and dmpz substituents.

3.2.1.2. Pyridine ring

The pyridine moiety in the derivatives (**18**, **19** and **20**) presents a similar pattern of bond distances and angles to that of pentachloropyridine[35] (Figure 3b). In both cases, it is observed an enlargement of the C-C bond distances, an opening of the bond angles at atoms N(1) and C(4) and a lowering at C(3) and C(5) with regard to the X-ray structure determination of pyridine.[36] The nitrogen atoms of pyrazole rings, which are directly bonded to the pyridine, deviate significantly from its least-square planes, showing values up to 0.096(3), 0.205(3) and 0.174(4) in **18**, **19** and **20** respectively. These kind of distortion have been also observed in other bulky pentasubstituted pyridines[37] reported in the CSD.

3.2.2. Peripheral rings

The influence of the methyl group at positions 3 and 5 of the pyrazole ring is mainly noticed in the opening of the exocyclic bond angle at N(1) [the nitrogen atom bonded to the benzene] which is closer to the methyl substituent, reflecting the steric hindrance associated to this group. Moreover, a lengthening of the bond distances marked in Figure 3c and an interchange of the angular values at C(4) and C(5) are observed when the ring is either pyrazole or 3,5-dimethylpyrazole.

The 3,5-dimethylpyrazole protonated ring that is present in the picrate salt, shows variations in bond distances and angles with regard to the neutral pyrazolyl group making the five-membered ring to display an almost 2-fold symmetry along the line connecting the C(4) atom and the mid point of the N(1)-N(2) bond.

3.3. Crystal structure

In general, the polyazolyl derivatives gathered in this review form inclusion compounds and not salts with several carboxylic acids containing aliphatic chains ranging from 1 to 4 carbon atoms. With stronger acids, like picric, organic salts are formed, **17f**. Polymorphism (**14a**, **14b**; **16a**, **16b**) and isomorphism (or pseudoisomorphism) (**14a**, **15**) have been found in the case of single host structures; however, only pseudo-isomorphism (**12a**, **12b**; **17b**, **17c**) has been detected in host-guest complexes. Furthermore, some crystal

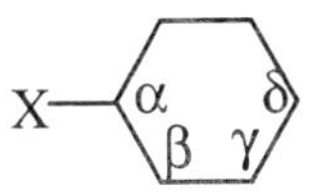

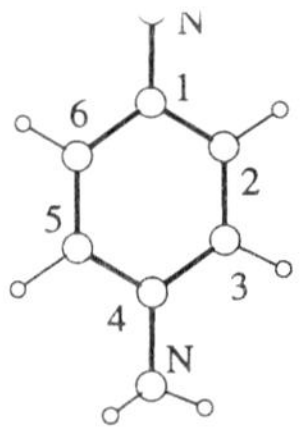

Three independent molecules:

$\Delta\alpha = 117.3(5) - 120.0 = -2.7(5)°$

$\Delta\beta = 121.4(8) - 120.0 = 1.4(8)°$

Substituents at:	*para*		*ortho*	
	Experimental	Calculated	Experimental	Calculated
C6-C1-C2	117.5(4)	117.3	118.8(3)	118.7
C1-C2-C3	120.9(3)	121.4	118.5(4)	118.7
C2-C3-C4	121.6(3)	121.4	121.6(4)	121.4
C3-C4-C5	117.5(4)	117.3	120.1(4)	120.0
C4-C5-C6	120.9(3)	121.4	119.5(4)	120.0
C5-C6-C1	121.6(3)	121.4	121.6(3)	121.4

(b)	Experimental*	py	Cl_5py
N1-C2	1.329(8)	1.337	1.323
C2-C3	**1.393(4)**	**1.378**	**1.411**
C3-C4	**1.393(2)**	**1.376**	**1.410**
C4-C5	**1.386(4)**	**1.376**	**1.410**
C5-C6	**1.387(11)**	**1.378**	**1.411**
C6-N1	1.335(1)	1.337	1.323
N1-C2-C3	122.8(7)	123.8	123.9
C2-C3-C4	**117.3(4)**	**118.6**	**116.8**
C3-C4-C5	**120.4(5)**	**118.8**	**119.8**
C4-C5-C6	**117.4(2)**	**118.6**	**116.8**
N1-C6-C5	123.2(7)	123.8	123.9
C2-N1-C6	**118.8(3)**	**116.5**	**118.8**

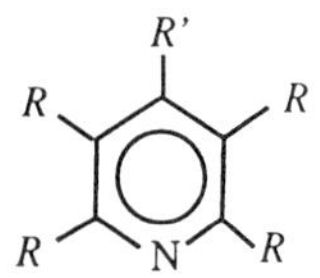

*Estimated standard deviations of the sample in brackets.

(c)

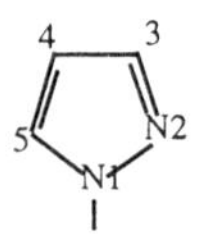

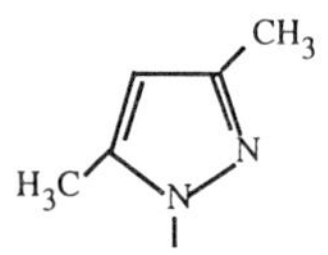

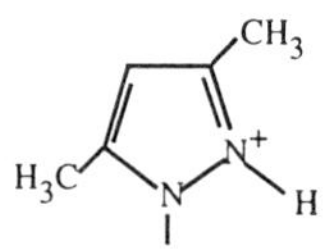

	Pz	Dmpz	$Dmpz^+$
N1-N2	**1.357(7)**	**1.371(6)**	**1.350(5)**
N2-C3	1.324(7)	1.327(6)	1.341(7)
C3-C4	**1.380(13)**	**1.397(6)**	**1.374(7)**
C4-C5	1.357(6)	1.368(6)	**1.377(7)**
C5-N1	1.350(9)	1.359(7)	1.358(6)
C5-N1-N2	112.1(5)	112.8(4)	**109.9(4)**
N1-N2-C3	103.5(5)	103.9(4)	**108.2(4)**
N2-C3-C4	112.5(4)	111.4(4)	**107.6(4)**
C3-C4-C5	**105.2(4)**	**106.6(4)**	108.5(5)
C4-C5-N1	**106.6(6)**	**105.3(3)**	105.7(4)

Figure 3: (a) Angular deformations of benzene ring in the 8-13 compounds due to the azole substituents influence. (b) Weighted mean values of the geometrical parameters (Å, °) for the pyridine rings in 18-20 compounds; py and Cl5py represent the pyridine and pentacloropyridine compounds (CSD refcodes PYRDNA01, PCLPYR respectively); (c) peripheral rings.

packing similarities can be established between pairs of compounds (**16a** *vs* **16c**; **17** *vs* **17a**; and hexafluorobenzene/hexadeuterobenzene *vs* **12**). A comparison between selected crystal structures is made below in order to illustrate all these phenomena.

3.3.1. Host-guest complexes

All compounds display a similar pattern of hydrogen bonds that relates the host and guest moieties through a strong O-H···N interaction. Furthermore, in all complexes of compound **17**, the guest is located interconnecting both 3,5-dimethylpyrazole rings through the O-H···N hydrogen bond previously mentioned in such a way that the distance between the methyl group of the adjacent pyrazole ring and the keto group of the carboxylic acid are in the 3.558(5)-3.875(7) Å range (Figures 4a,b), giving rise to what Desiraju has called a supramolecular synton.[38] Generally, the fragment of the guest that is close to the acid group is almost coplanar with the pyrazole ring to which it is strongly bonded. The value for the dihedral angle between both planes ranges from 4.1(2) to 21.6(2)° (Figure 4a).

The picrate salt, **17f**, presents a similar model of intramolecular interactions (Figure 4d), since two protons have been transferred to compound **17**, protonating two 3,5-dimethylpyrazole groups and producing N^{+}-H···O^{-} hydrogen bonds instead of the O-H···N ones. Furthermore, the presence of three nitro groups capable to act as acceptors of hydrogen interactions modifies the crystal packing.

The only exception encountered to the general rule of host-guest interactions has been found in **16c** where the guest molecules are hydrogen bonded with each other forming dimers that only interact weakly with the host (Figure 4c and 5a).

It is already known that pure acetic acid form dimers in gas and liquid phase,[39] while in solid state it presents chains of molecules where the keto oxygen is involved in two hydrogen bonds, the strong O-H···O one and the weak C-H···O one, that take place along the directions of the oxygen electron lone pairs [O···O = 2.631(8), H···O = 1.64(1) Å, O-H···O = 165(1)°; C···O = 3.429(8), H···O = 2.41(2) Å, C-H···O = 158(1)°].[40] When it is cocrystallized with other organic molecules, only in 4 out of 30 crystal structures reported in the CSD, the acetic acid molecules form dimers, even when the host molecule contains groups able to accept hydrogen bonds.

It is worth to mention that none of the hexakisazolylbenzenes equally substituted, neither pyrazolyl[17] nor 3,5-dimethylpyrazolyl,[16] form inclusion complexes with acetic acid, while both compounds with mixed substituents, **16** and **17**, cocrystallize with it. Even more, the crystal samples used to determine the structures of **14a** and **15** (Figure 6b,c) were obtained from recrystallization in acetic acid.

3.3.2. Polymorphism

Compounds **14** and **16** present the polymorphic structures **14a**, **14b** and **16a**, **16b** respectively (Figures 6b,a and 5b,c). In the first pair of structures each polymorph corresponds to a different conformer, *ududud* for **14a** and *udduud* for **14b**, while in the second pair, the molecules present the same *up-down* conformation (*ududud*) (See section 7 for *u-d* definitions).

(a)

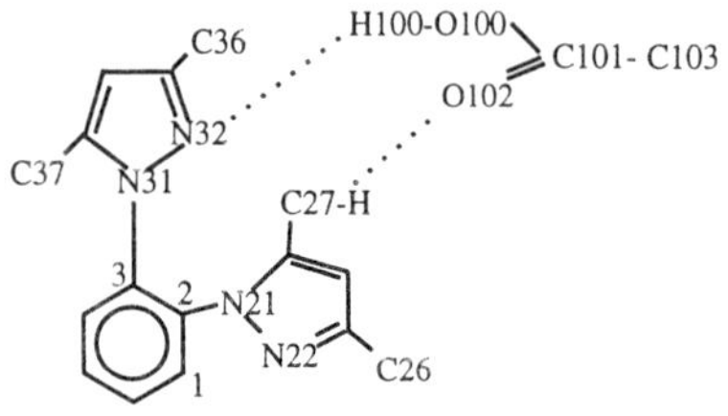

	C3-N31-N32···O100	N31-N32···O100-C101	N32···O100-C101-O102
17a	41.0(2)	-	-
17b: Mol.1	-6.2(4)	22.5(5)	3.8(8)
17b: Mol.2	5.2(5)	-0.2(6)	-2.5(7)
17c	-3.7(6)	16.9(9)	4.0(9)
17d: Mol.1	18.0(5)	-16.6(5)	0.0(6)
17d: Mol.2	7.6(4)	-0.5(4)	-2.9(5)
17e	7.3(5)	-8.9(7)	2.0(9)

	Π1-Π2	O100-H100···N32 O···N	O100-H100···N32 O-H···N	C27-H273···O102 C···O
17a	-	2.997(4)	162(11)	3.558(5)
17b: Mol.1	21.6(2)	2.753(4)	151(7)	3.875(7)
17b: Mol.2	12.8(2)	2.697(4)	163(5)	3.862(7)
17c	17.1(2)	2.733(6)	180(-)	3.761(12)
17d: Mol.1	14.1(1)	2.711(4)	165(5)	3.714(7)
17d: Mol.2	13.4(1)	2.741(5)	174(6)	3.739(7)
17e	4.1(2)	2.761(5)	176(-)	3.603(9)

(b) (c) (d)

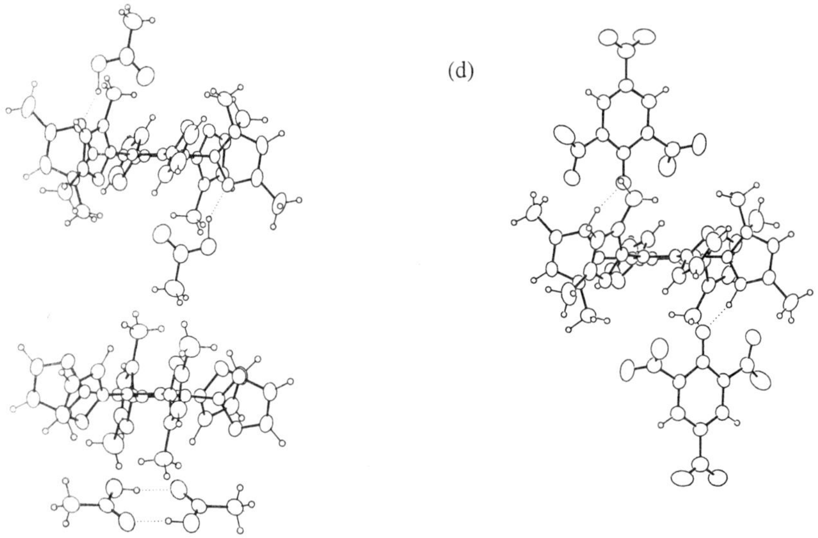

Figure 4: (a) Geometrical parameters (Å, °) characterizing the guest disposition with respect to the host (in **17** complexes). Π1, Π2 represent the least-square planes through the N31···C35 pyrazole ring and through O(100) C(101), O(102) and C(103) atoms respectively. (b, c, d) Parallel views to the benzene rings showing the conformation of the host and the guest disposition in **17b**, **16c** and the **17f** picric salt, respectively.

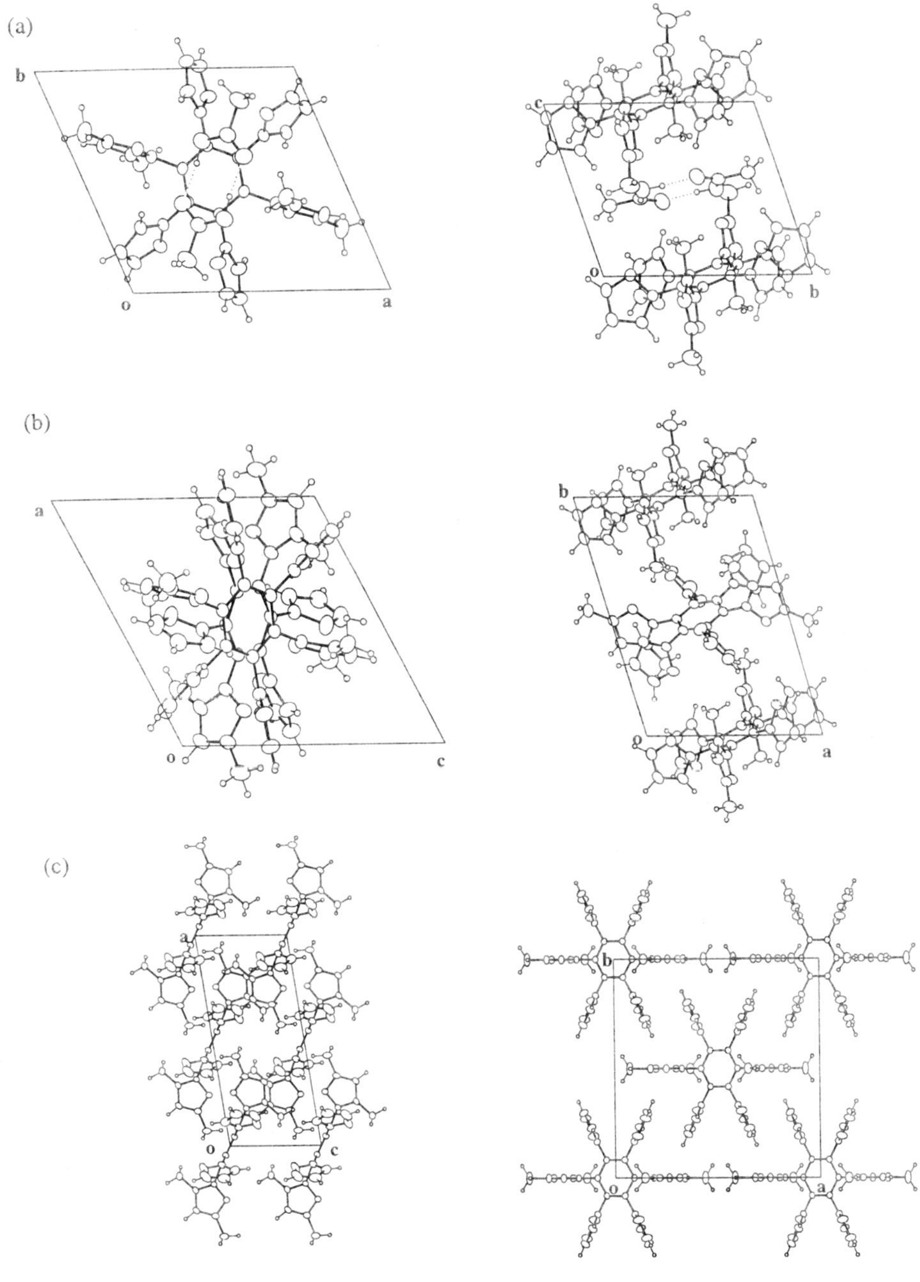

Figure 5: Crystal packings of (a) **16c**, (b, c) the two polymorphic forms (**16a**, **16b**).

In **14a** and **14b** the molecular conformations and the crystal packing are very different and, therefore, it is quite difficult to establish any structural relationship between them, Figure 6a,b. The higher conformational energy that presents the molecule in **14b** is probably compensated by its more dense packing, according to their total packing coefficients[41] of 0.69 and 0.65 respectively.

In **16a**, there are two crystallographically independent molecules stacked along the **b** axis (Figure 5b). The only interactions between them are weak C-H···N contacts that stabilize these columns of molecules. The angle between the central benzene rings is 13.1(2)° and the distance between their centroids is half the **b** axis, 7.1706(24) Å. In **16b**, identical stacked molecules constitute the rows that in this case are oriented along the crystallographic **c** axis. Only the 3,5-dimethylpyrazole groups are involved in the C-H···N interactions that stabilize the chains. The central benzene rings are parallel and the distance between their centroids is equal to the length of the **c** axis, 6.4408(2) Å, that is smaller than the value found in **16a**.

3.3.3. Pseudoisomorphism

Pseudoisomorphic structures are observed in the packing of poliazolylbenzene molecules (hosts) as well as between inclusion complexes with different guest molecules.

In **14a** and **15**, both structures (Figure 6b,c) present the same space group symmetry and analogous cell dimensions except for the length of the **c** axis [5.5469(2) and 7.6743(2) Å respectively] which is enlarged by the presence of the methyl groups at positions 3 and 5 of the pyrazole rings. The packing is formed by columns of molecules along the **c** axis, being the separation between the central rings the same as the unit cell dimension.

In the crystal packing of the inclusion complexes of **12** with acetic and propanoic acids, **12a** and **12b**, Figure 7, the host molecules produce a three-dimensional structure leaving elliptical channels parallel to the **c** axis[41] that are occupied by the guest molecules giving rise to pseudoisomorphic structures. As it can be seen in Figure 7, the acetic acid is disordered over two positions with population parameters of 0.69 and 0.31 respectively. The only differences between both structures are the intermolecular contacts along the **b** axis and the interactions that join these chains through the guest. There is an overlapping of the 3,5-dimethylpyrazole rings that is only partial in the complex with propanoic acid.

In the case of the inclusion complexes formed by **17** with acetic (**17b**) and propanoic (**17c**) acids, a similar comparison to the previous one can be established, Figure 8a,b. Complex **17b** presents an asymmetric unit formed by two half molecules located at crystallographic symmetry centers [(0, 1/2, 1/2) and (1/2, 1/2, 0) and the symmetry equivalent ones], in a way that the crystal is built by alternated sheets of crystallographically independent molecules along **a** and shifted by **c**/2. Both host molecules are related by a pseudo two-fold screw axis parallel to **a** and located at (x, 1/2, 1/4) approximately. The only significant intermolecular interaction that holds together the host and guest molecules is a O-H···N hydrogen bond. In **17c**, there is a crystallographic two-fold screw axis relating the molecules. The cell constants are analogous after the following transformation of the axes in **17b**: $\underline{a}'=\underline{b}$, $\underline{b}'=-a$ and $\underline{c}'=\underline{c}$.

17b:	a' =b = 10.5951(9) Å	α' = 81.588(7)°	**17c**:	a = 10.8856(6) Å	α = 90°
	b' = a = 20.8260(20) Å	β' = 109.432(5)°		b = 20.2867(17) Å	β = 112.296(3)°
	c' = c = 9.2370(5) Å	γ' = 88.281(11)°		c = 9.8256(5) Å	γ = 90°

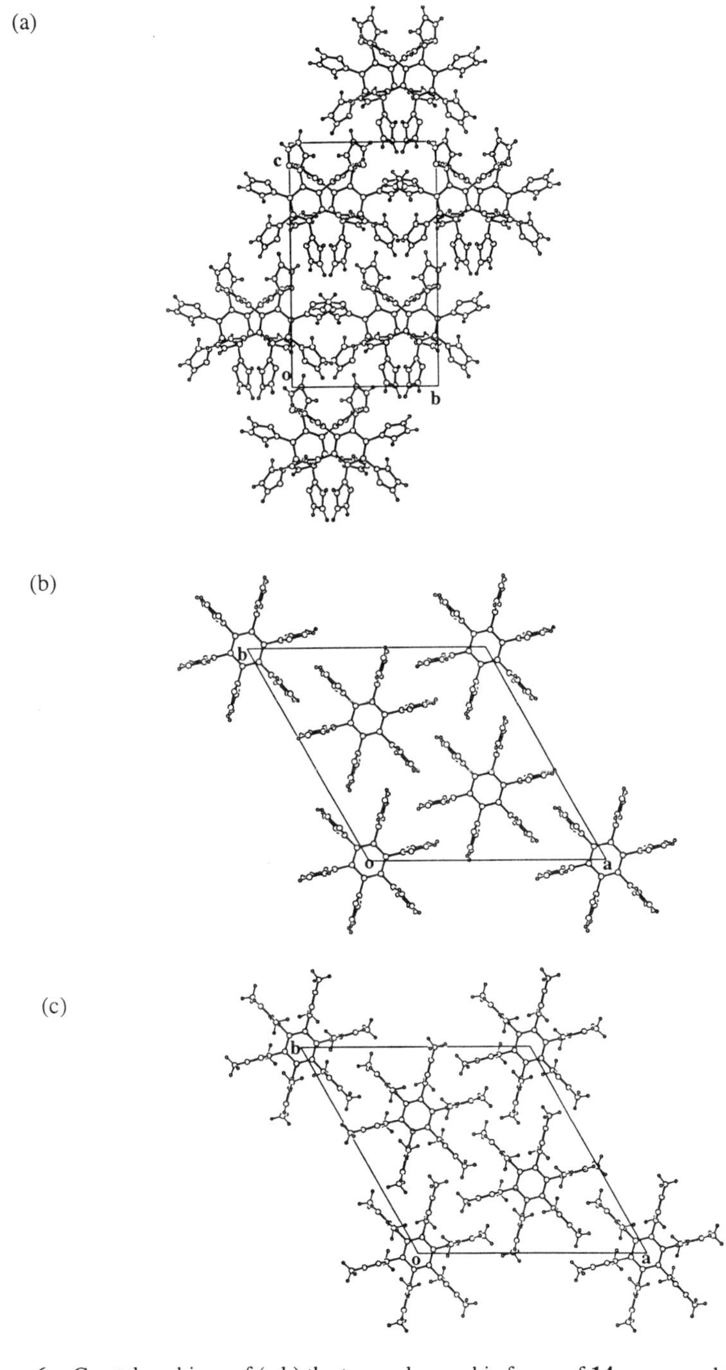

Figure 6: Crystal packings of (a,b) the two polymorphic forms of **14** compound (**14b**, **14a**), and (c) **15**.

(a)

(b)

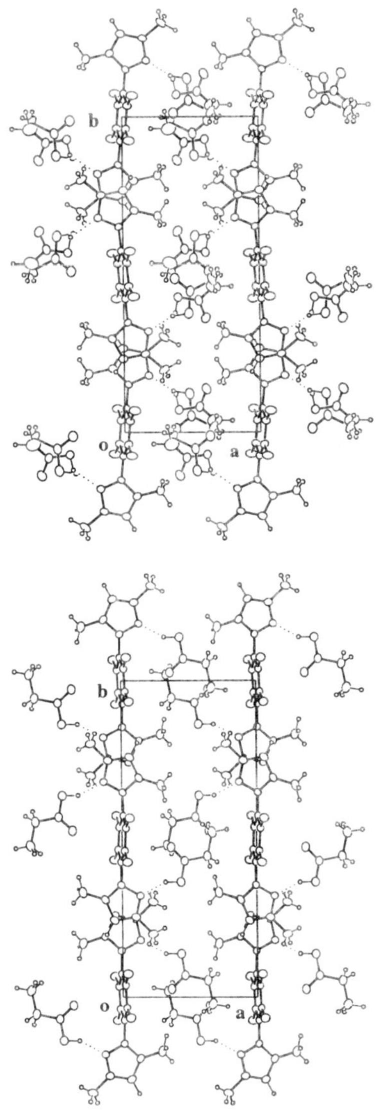

Figure 7: Crystal packings of the two pseudoisomorphic forms: (a) **12a** and (b) **12b**.

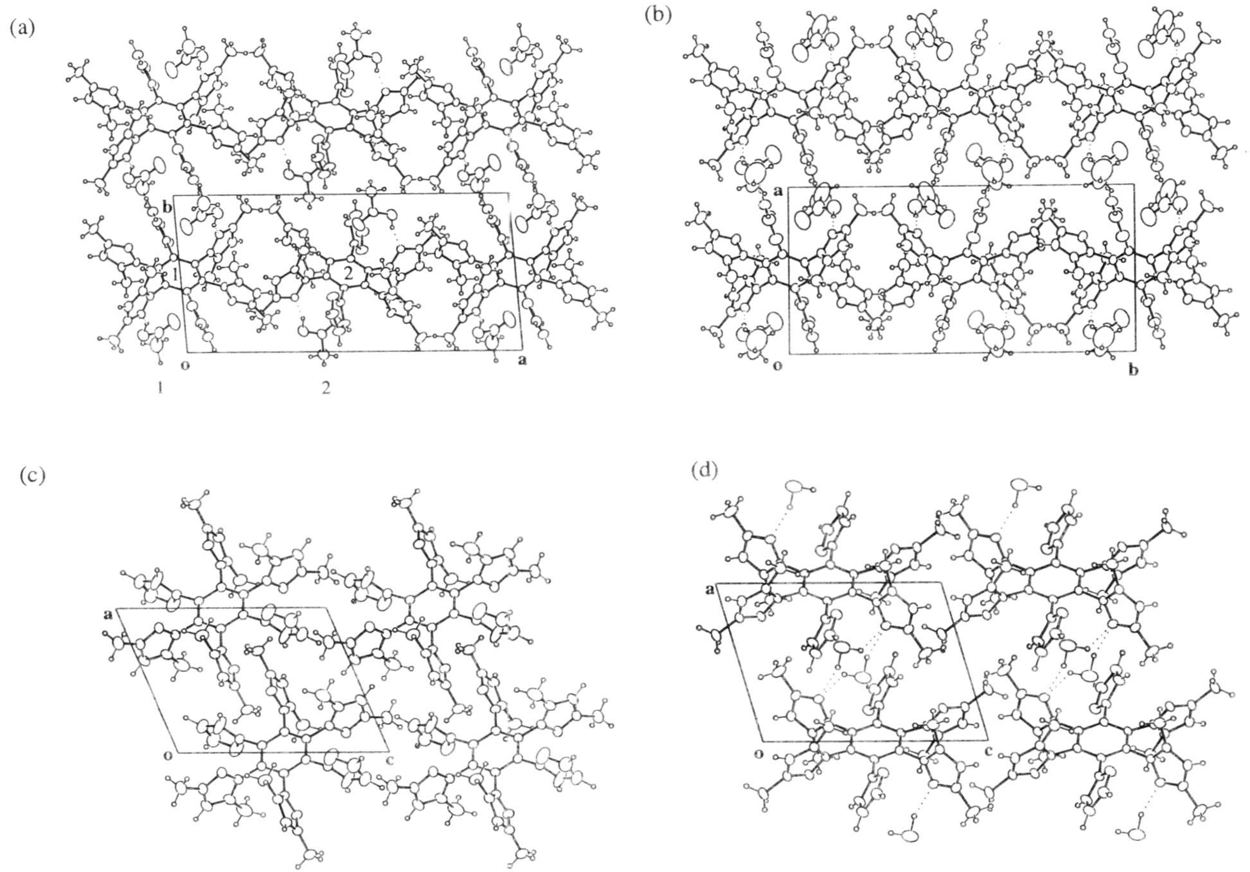

Figure 8: Crystal structures for **17b**, **17c** (a, b) and **17**, **17a** (c, d).

Both crystal structures are very similar; however, the different disposition of one acetic acid guest molecule (labelled as 2, Figure 8a) with regard to the host results in a decreasing of the crystal symmetry, that, in contrast, does not change the overall packing mode.

3.3.4. Other analogous packings

In complex **16c**, the host molecules and the acetic acid dimer are packed in columns parallel to the **c** axis that are analogous to those found in one of the polymorphic forms of the host compound, **16a**, but in this case one of the two crystallographically independent molecules of the host has been replaced by the dimer. There is a shrinkage of the crystallographic **c** axis in **16c** compared to the **b** axis in **16a** [14.3411(24) Å versus 8.3867(7) Å], Figure 5a,b.

In both structures, **17** and **17a**, (Figure 8c,d) the molecules present a symmetry center and the distance between the centroids of the benzene rings in the three space directions is the length of the **a**, **b** and **c** crystallographic axes. In **17a** the host molecules are joined, through water molecules, by means of the O-H···N hydrogen bonds forming chains that do not bear other interactions among them than the van der Waals ones.

The similarity between the packings of hexafluorobenzene-hexadeuterobenzene complex and **12** is shown in Figure 9a,b. In the first one,[42] [CSD Refcode: BICVUE01], it can be observed how the molecules stack forming columns in the **c** directions. The disposition of the benzene rings of the two species is intermediate between eclipsed and alternated [being the pseudotorsion angle C···Centroid···Centroid···C= 16.2(3)°], the molecules are almost parallel being 4.7(1)° the angle between planes and the distance between centroids is 3.763(2)Å. The columns are packed through weak C-D···F interactions: C···F = 3.272(6) Å and C-D···F = 118.2(3)°. On the other hand, the two independent molecules in **12** form chains along the directions [1 1 0] and [1 -1 0] through C-H···F interactions that connect alternatively both molecules. The central benzene rings are also stacked along the **c** crystallographic axis, but the substitution of D or F for pyrazoles makes bigger the angle between their planes and the distance between their centroids, 16.7(1)° and 4.310(1) Å respectively.

3.3.5. Crystal packing prediction

The general problem of crystal packing prediction using as starting point the knowledge of the molecular structure has been treated by several authors that have used different approaches.[43,44] The application of one of these methods[41] to the compounds studied in this review has been successful only in some cases. Figure 10 and 11 show the crystal packing of compounds **13** and **18** obtained experimentally and theoretically using the experimental space group and several available possibilities for the coordination of the original molecule as starting point. In **13** the agreement between experiment and theory of one type of environment is quite good (Figure 10a,b), while the second predicted structure is completely different (Figure 10a,c). In **18**, both coordinations afford the same crystal structure as the experimental one, Figure 11. The computed one, in both compounds, corresponds to a closer packing structure than the experimental one, as measured by the calculated atomic volume ratios (V/N. atoms).

(a)

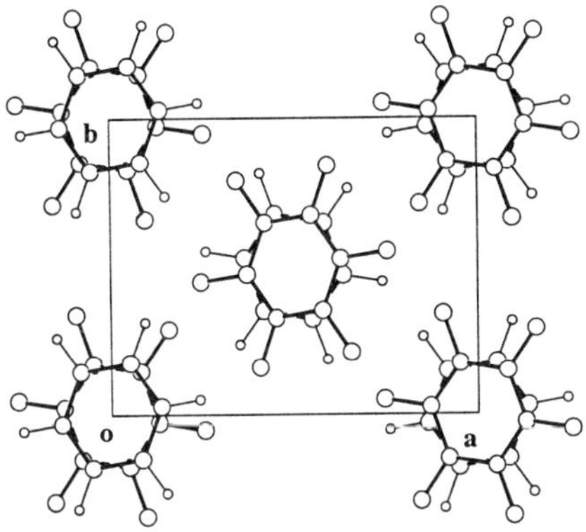

Interplanar angle = 4.7(1)°

Centroid1···Centroid2 = 3.763(2)Å

(b)

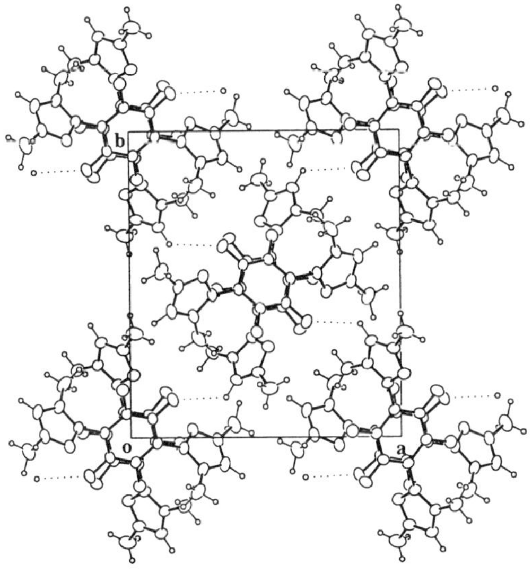

Interplanar angle = 16.7(1)°

Centroid1···Centroid2 = 4.310(1)Å

Figure 9: Crystal packings for compounds: Hexafluorobenzene-hexadeuterobenzene complex (CSD refcode: BICVUE01), (a) and **12**, showing the same packing mode and the differences due to the fluorine substitution by 3,5-dimethylpyrazole rings (b).

(a)

Observed

13

a=18.5739(12)

b=17.2372(12)

c=8.4961(3)

90, 90, 90

V= 2720.1(3)

V/N. atoms= 15.5

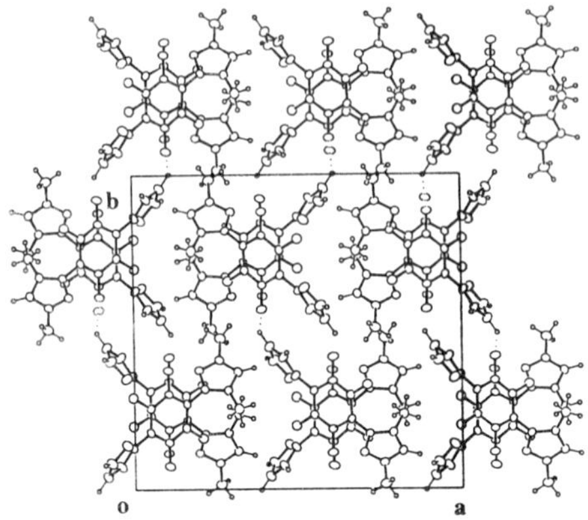

(b)

Computed

13

a=17.9462

b=17.3456

c=8.3377

90, 90, 90

V= 2595.4

V/N. atoms= 14.8

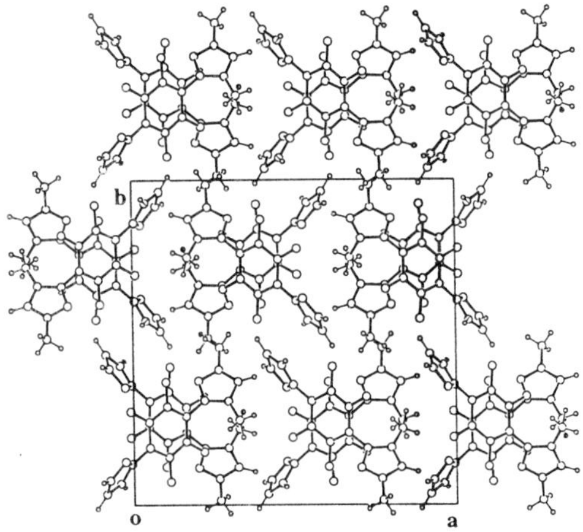

(c)

Computed

13

c=18.3025

b=17.6960

a=8.6397

90, 90, 90

V= 2798.2

V/N. atoms= 15.9

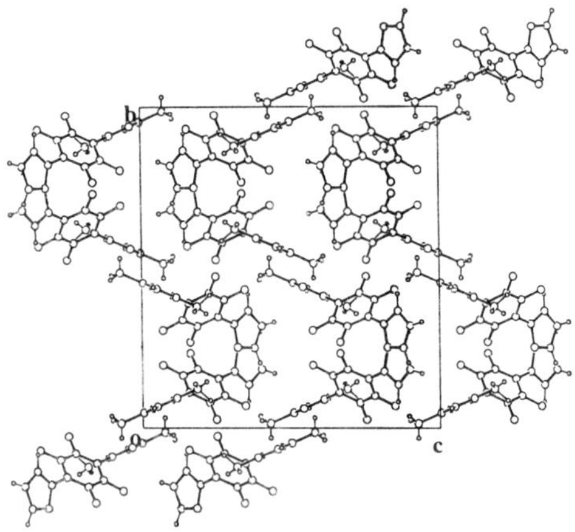

Figure 10: (a) Observed and (b, c) computed crystal packing of **13** using different molecular environments (CB, CC respectively, space group Pbca).

(a)

Observed

18

a=11.7056(6)

b=9.3228(8)

c=10.0245(13)

α=101.07(1)

β=115.45(1)

γ=78.87(1)

V= 961.9(2)

V/N. atoms= 15.5

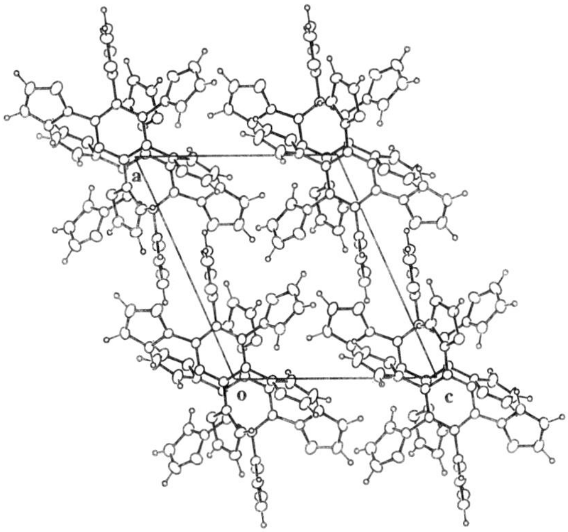

(b)

Computed

18

b=11.5705

c=8.9989

a=9.9210

β=100.23

γ=113.85

α=79.96

V= 923.7

V/N. atoms= 14.9

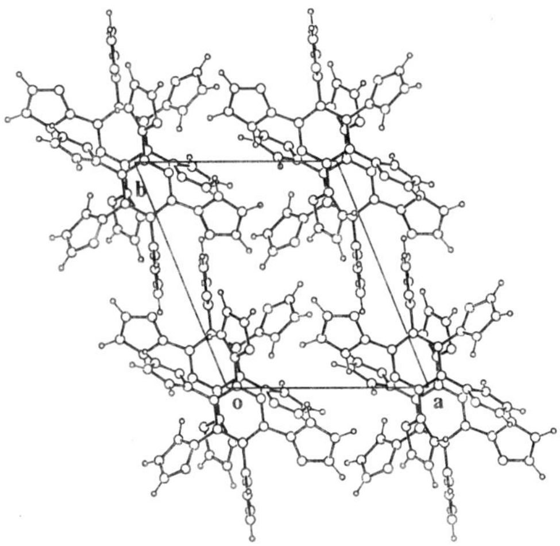

Figure 11: (a) Observed and (b) computed crystal packing of **18** (molecular environment AB, CA, space group P-1).

3.4. Intermolecular contacts

A summary of the geometries shown by the different types of intermolecular interactions present in the crystal structures is gathered in Figure 12. Using the CSD[29] as analytical tool, the C-F···π, F···F and C-H···F contacts have been analyzed as described below.

3.4.1. C-F···π(arene) systems

The crystal structure of hexaflurobenzene presents distances between fluorine atoms and phenyl rings centroids shorter than 3.5Å (Figure 13) that could show, in spite of the electronegative nature of fluorine, a tendency to adopt similar geometrical dispositions to that which the hydrogen atom adopts with respect to the electronic π cloud of aromatic rings in the crystal packing of one polymorphic form of benzene (CSD Refcode: BENZEN) (Figure 13).

The F···Centroid intermolecular contacts contained in the CSD satisfying the search conditions listed in Figure 14a have been retrieved and analyzed only in those cases where the fluorine atom is bonded to a benzene ring. There are two limiting situations for these interactions; "T-type" and parallel (schemes b and c of Figure 14) that can be characterized by the van der Waals radii[45] of carbon and fluorine atoms and considering the last one placed in the normal line to the benzene ring constructed through its centroid. The first histogram (Figure 14d) shows that a small number of the structures (4%) present F···Centroid distances shorter than the sum of van der Waals radii, being 3.67(24) Å the mean value of the distribution.

The histogram of the C···Centroid distances (Figure 14e) displays several maxima, the lowest at 3.50 Å approximately and a second one close to the mean value of the distribution, 4.26(51) Å. The third histogram (Figure 14f) represents the angle between the C-F vector and the normal to the plane of the aromatic ring with which it is interacting. It shows three different regions: the angular range 0-20° contains 3% of the interactions (38) corresponding to the "T-type" situation; the range 80-90° comprises most of the contacts (53%, 664 interactions) that cause a pronounced maximum enclosing the structures forming the first maximum in the C···Centroid distribution (Figure 14e), therefore, the geometrical arrangement corresponds to two parallel but shifted aromatic rings; and finally, the remaining interactions (44%, 542 interactions) that correspond to intermediate situations.

The distribution of the minima of the F···C distances (Figure 14g) between the fluoride atom and any of the carbon atoms of the benzene ring has a quasi-gaussian behaviour with little asymmetry. The mean value is 3.25(23) Å and the maximum of the distribution is at 3.15 Å approximately, that corresponds to the sum of van der Waals radii.

It can be concluded that although only 4% of the cases present F···Centroid distances shorter than the sum of van der Waals radii, we have found a considerable number of structures where the aromatic rings tend to arrange parallel but displaced (situation c in Figure 14). This observation is in good agreement with the recent published results where the energetic and geometry of benzene-hexafluorobenzene dimers has been studied by means of *ab initio* calculations.[46] According to them, the perpendicular situation (Figure 14b) is repulsive while the parallel one (Figure 14c) is attractive, at least without the shifting of the rings.

•Hydrogen bonds:

Contact	Angle
O-H···O: 2.671(3)Å	172(8)°
O-H···N: 2.697(4)-3.084(4)Å	130(8)-180(-)°
N^{+}-H···O^{-}: 2.633(5)Å	166(5)°

• Weak interactions:

Contact	Angle
π···π: 3.731(3)-4.965(1)Å	
C-H···π: 3.554(3)-4.252(5)Å	116(3)-153(5)°
C-H···N: 3.191(4)-3.657(9)Å	107(3)-180(4)°
C-H···O: 3.261(9)-3.875(7)Å	106(5)-173(4)°
F···F: 2.886(6)-3.329(4)	
C-H···F: 3.215(6)-3.458(6)Å	123(5)-148(4)°

Figure 12: Summary of the geometry characterizing the interactions in the studied structures.

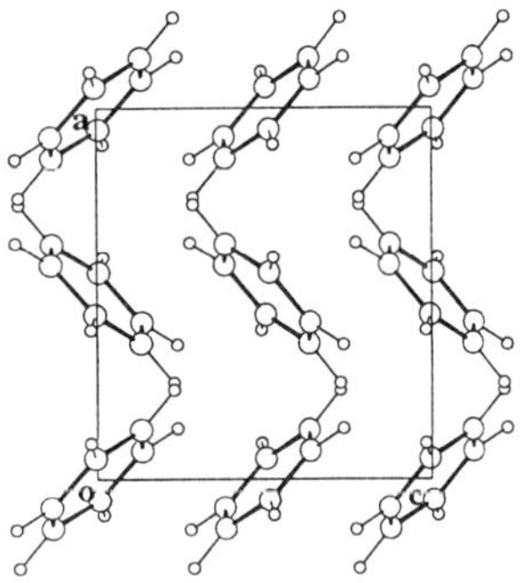

Benzene

(BENZEN)

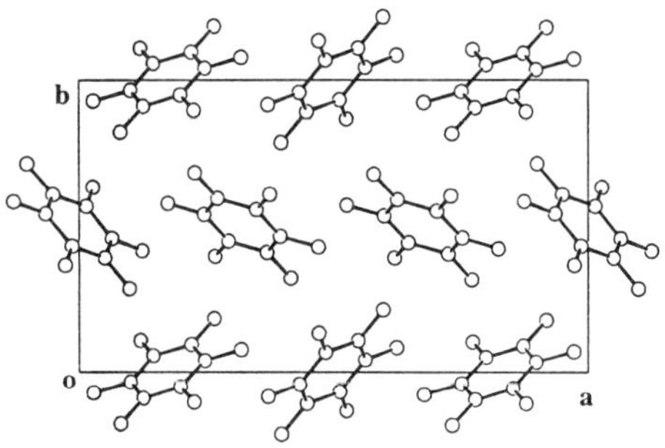

Hexafluorobenzene

(HFBENZ)

	C-F	F···Centr.	C···Centr.	C-F···Centr.
C-F···C(A)	1.331(-)	3.174(-)	4.237(-)	136(-)
C-F···C(A)	1.328(-)	3.306(-)	4.542(-)	155(-)
C-F···C(B)	1.339(-)	3.395(-)	4.326(-)	126(-)

(-) Means no standard deviations available.

Figure 13: Crystal structures of the orthorhombic benzene polymorph and the hexafluorobenzene, and the geometry of the C-F···Centroid contacts.

(a) **CSD search conditions**

- Organic compounds, R < 0.10

d(F···Centroid) < 4Å ⇨ 686 structures and 2003 contacts

d[C(benzene ring)-F···Centroid] < 4Å ⇨ 343 structures and 1244 contacts

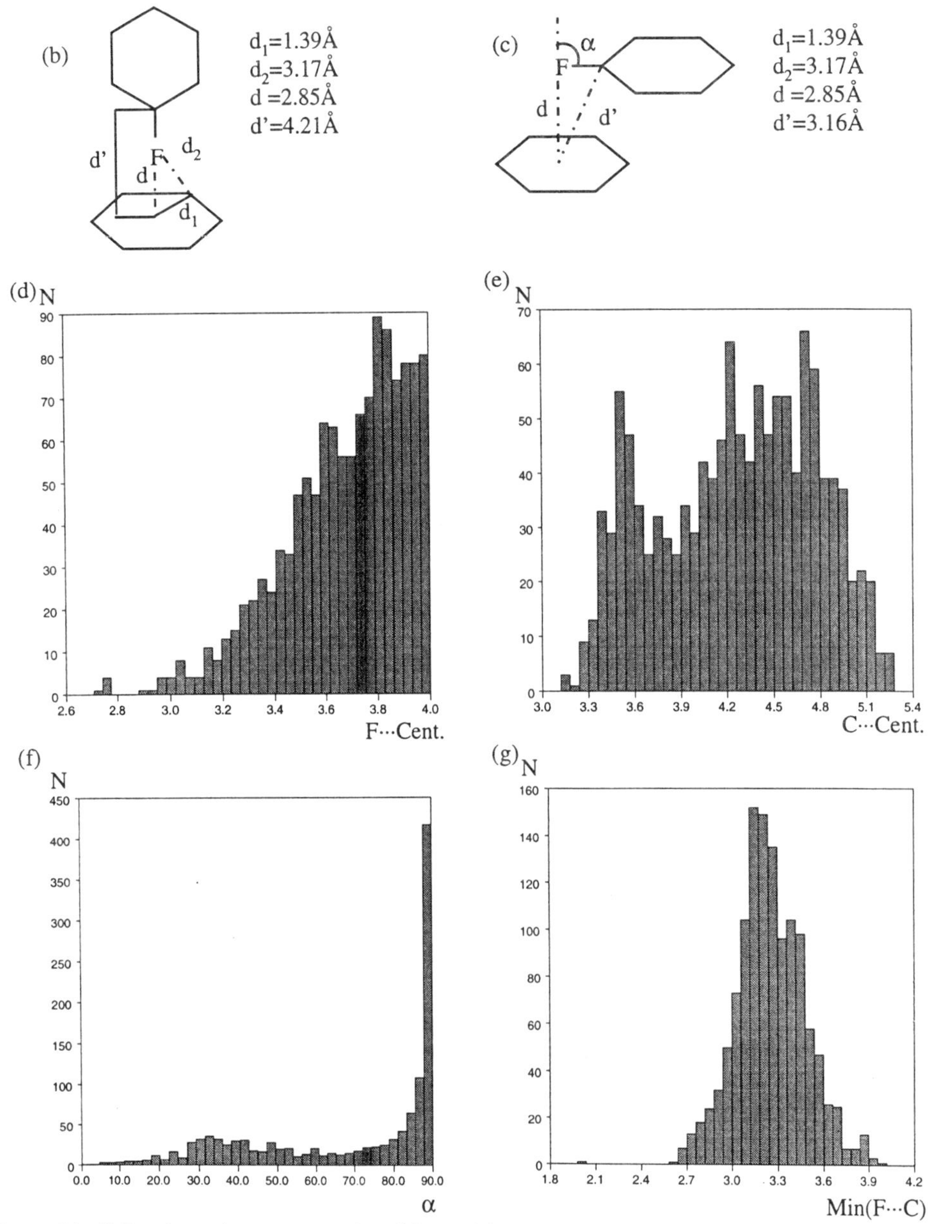

Figure 14: C-F···π(arene) systems. (a) Conditions of the search. (b, c) The perpendicular and parallel types. (d-g) Histograms characterizing the system geometry.

3.4.2. F···F short contacts

The halogen···halogen contacts seem to play a significant role determining the crystal structure of some compounds, and in order to clarify the controversy risen among different authors who have studied the problem,[47-49] we have performed a new statistical analysis of the F···F contacts using the CSD and fulfilling the search conditions shown in Figure 15a. Two cases were considered, taking into account values of the F···F distances below or above 2.95 Å.

The distribution of the pseudotorsion angle C-F···F-C for F···F distances below 2.95 Å is shown in Figure 15b. The highest frequency value is 180° approximately, and this maximum arises from a quasi uniform background having about half frequency value.

Figure 15c displays the histogram for all F···F distances up to 4.5 Å found on CSD. The distribution shows that there are a considerable number of structures with F···F distances shorter than the sum of van der Waals radii, 2.94 Å.[45] It is possible to define a variable, d_{FF}, as the interaction distance where the histogram presents half the height of its first peak or shoulder and its physical meaning corresponds to the sum of van der Waals radii. In our study the value for d_{FF} is located between 2.80 and 2.90 Å, and it is coincident with the value recently obtained by other authors[50] and it is closer to that given by Bondi (1.47 Å),[51,52] although the accordance is worse with the radius reported by Pauling (1.35 Å).

The representation of the minimum values of (μ_1,μ_2) versus the maximum ones for all distances shows a big number of structures located in the central zone, Figure 15d. There is also an important set of interactions having $\mu_1 = \mu_2$ that could be due to the presence of a crystallographic symmetry center that relates both C-F fragments.[48]

Figure 15e represents the polar diagram for all retrieved F···F distances which has been plotted in order to investigate how the F···F interaction distributes around one of the intervening atoms. This analysis allowed us to define a limit or frontier for the region where the intermolecular contacts were observed. The lower limit indicates the van der Waals distance between F atoms, therefore, the effective shape of the F atoms in crystals is ellipsoidal (mayor axis = 2.84 Å and minor axis = 2.66 Å), the shortest axis being aligned with the C-F bond vector ("polar shrinkage"). A previous study[47] reached the same conclusion with similar values (2.76 and 2.60 Å).

3.4.3. C-H···F interactions

Figure 16a displays the search conditions used to retrieve all structures containing C-H···F interactions from the CSD. Only 9 interactions corresponding to 7 crystalline structures present C···F distances up to 0.2 Å shorter than the sum of van der Waals radii.[45] All of them, except one, are intramolecular interactions.

As it can be seen in Figure 16b, no significant differences were found when the hybridization state of the donor fragment was taken into account. The distribution of the C-H bond distance (Figure 16c) is bimodal; the second maximum, centered around 1.08 Å, is due to the contribution of data obtained by neutron diffraction and mainly when the H atoms positions have been positioned and fixed at the distance obtained by neutron studies. Only 4% of the interactions (57 fragments) present C···F distances (Figure 16d) shorter than the sum of van der

(a) **CSD** search conditions:

- Organic compounds, R < 0.075, 90°< (C-F···F) < 180°
- Neither error nor disorder
- d (F···F) < 2.95Å ⇨ 385 contacts
- d (F···F) > 2.95Å ⇨ 1908 contacts

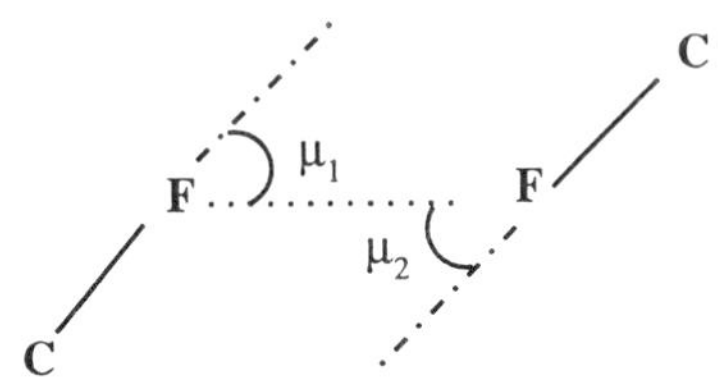

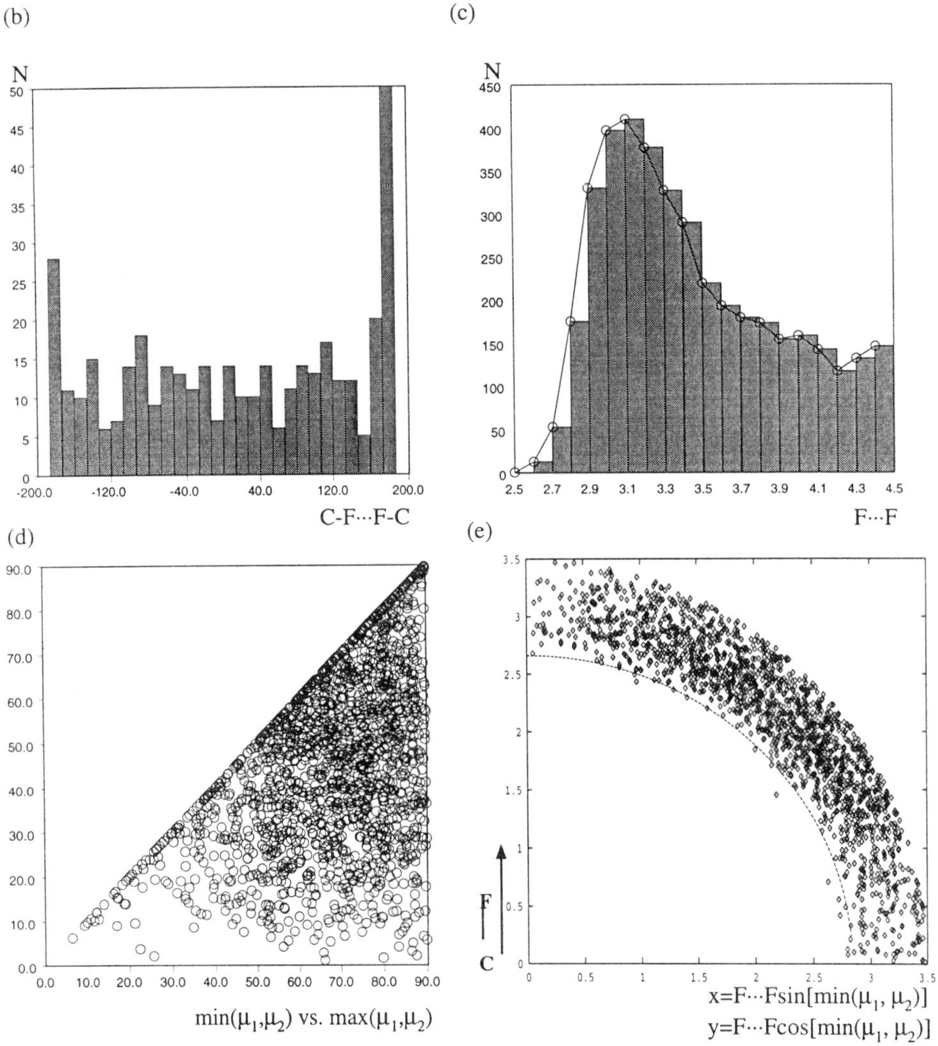

Figure 15: F···F contacts. (a) Conditions of the search and contact scheme. (b-e) Histograms of the most relevant parameters characterizing the geometry of the contact (see scheme a).

(a) **CSD** search conditions:

- Organic compounds, R < 0.075, neither error nor disorder
- d (H···F) < 2.64Å
- 130°< (C-H···F) < 180°

⇨ 700 structures and 1630 interactions

(b)

	Nfrag	Nent	C-H	H···F	C···F	C-H···F
Donor						
Csp^3	985	455	1.01(7)	2.52(10)	3.41(12)	150(12)
Csp^2	644	350	0.99(7)	2.52(11)	3.40(12)	150(12)
Acceptor						
C-F	626	405	1.00(7)	2.53(10)	3.41(12)	149(12)
Interaction						
Inter	1525	669	1.00(7)	2.52(10)	3.41(11)	150(12)
Intra	105	72	0.98(7)	2.45(13)	3.28(16)	144(12)

(c)

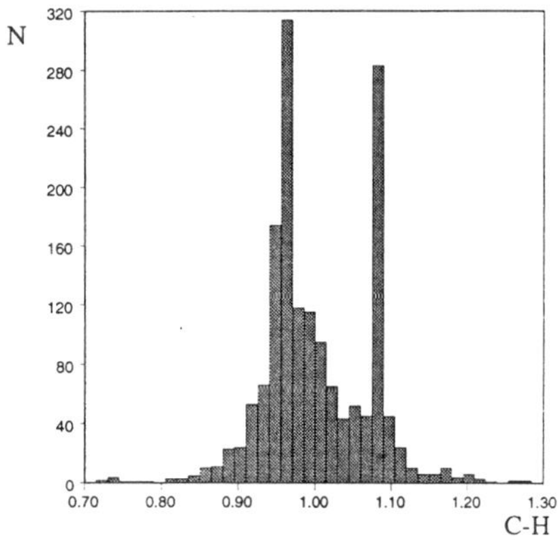

(d)

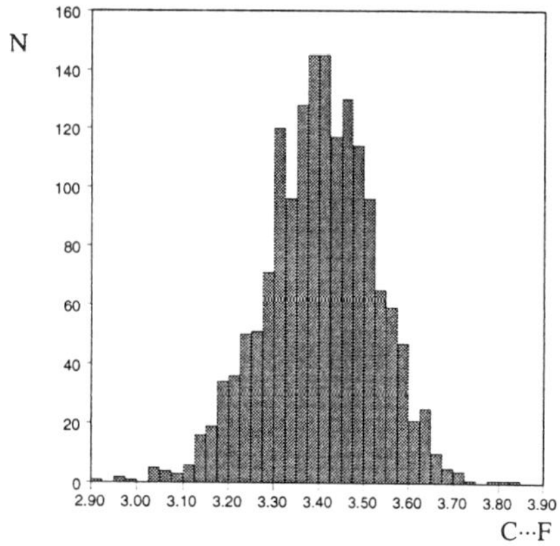

(e)

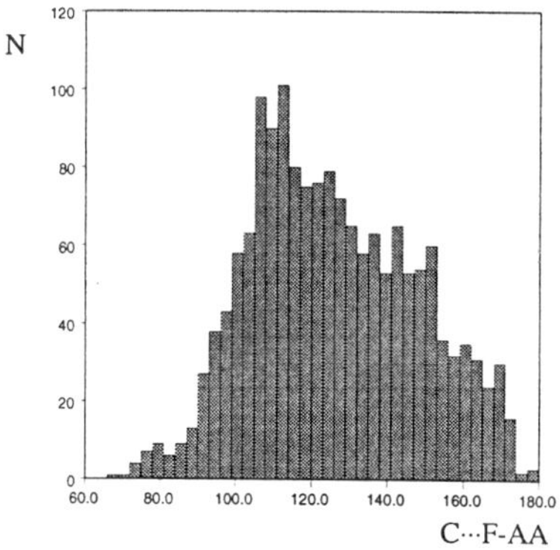

Figure 16: C-H···F interactions. (a) Conditions of the search. (b) Brief summary of the geometrical parameters (Å, °) according to the nature of the donor or acceptor and the intra or intermolecular contacts. (c-d) Histograms characterizing the geometry of the contact.

Waals radii (3.17Å),[45] versus to the most probable value of the distribution located at 3.40(2) Å. The interaction is quite linear (Figure 16b), presenting a mean value for the C-H···F angle of 150(12)°. The last histogram (Figure 16e) displays the information concerning the direction where the contact takes place: the most probable value for the angle at the F atom is 110.8°, but there is a significant asymmetry towards higher values of the angle.

Finally, it can be concluded that the intramolecular interactions seem to be stronger than the intermolecular ones and that, although these interactions are weak in terms of the C···F distance, we have found an important number of structures which present the C-H bond pointing to fluorine atoms.

3.5. Crystallographic Conclusions

Regarding the analysis of the molecular structures, it can be concluded that the most stable conformation seems to be the one presenting the electron lone pairs of the nitrogen atoms alternating at both sides of the benzene ring. In the inclusion complexes the host-guest interactions take place in a similar way in all cases and always by means of a strong O-H···N and a weak C-H···N hydrogen bonds with two contiguous 3,5-dimethylpyrazole rings.

Small modifications in the crystal packing have been observed when either variations in the molecular structure, or in the size of the guest have been introduced.

From the analysis of the intermolecular interactions present in the studied compounds it can be concluded that the hydrogen bonds are the driving forces in the crystal packing, although there are other weaker interactions, like C-H···F, C-H···N, C-H···π, F···F, C-H···O, π···π, etc, that are also detectable.

Finally, and considering the efforts made to predict the packing modes from the molecular structure of these compounds, it is worth to notice the big percentage of failure which denotes the inherent difficulty of these kinds of studies and the pressing necessity of developing a methodology development found in this area, that seems to be only in its commencements.

4. NMR: SOLUTION AND SOLID STATE.

4.1. Pyrazolylfluorobenzenes 8-13.

The ^{1}H NMR results are reported on Table 2 and 3. The assignment of the hydrogen signals was straightforward because the coupling constants between H-4' and the substituent at position 5' (H or CH_3) are always longer than those with the substituent at position 3' (H or CH_3). Since in a benzene derivative bearing two 3,5-dimethylpyrazol-1-yl substituents surrounded by two OH groups, we had observed a *meso-d,l* isomerism,[53] we have studied the behaviour of compound **9** at low temperature (in $CDCl_3$, Table 3). At room temperature only one type of signals is observed, not due to a very rapid interconversion of both isomers (average signals), but to the fact that the signals corresponding to the minor isomer are too weak. At low temperatures, the signals of the minor isomer appear and grow in intensity; moreover, they are broad showing that the barrier minor to major isomer is not very high. Recording the spectra several times it is possible to have accurate relative intensities

Table 2. ^{1}H-NMR Data (Solvent $CDCl_3$). δ in ppm, J in Hz.

Comp.	H-3'	H-4'	H-5'	H-4"	CH_3-3'	CH_3-5'
8	7.66 (dd)	6.33 (dd) $J_{3'5'}$=0.6	7.30 (dd) $J_{3'4'}$=1.9	---- $J_{4'5'}$=2.6	----	----
9	----	5.82 (s)	----	----	2.13 (s)	2.09 (s)
10	7.65 (dd)	6.29 (dd) $J_{3'5'}$=0.6	7.26 (dd) $J_{3'4'}$=1.9	5.85 (m) $J_{4'5'}$=2.6	2.20 (m)	1.86 (m) $J_{4'Me5'}$=0.8 $J_{4'Me3'}$=0.4
11	7.88 (dd)	6.58 (dd) $J_{3'5'}$=0.6	7.77 (dd) $J_{3'4'}$=1.9	---- $J_{4'5'}$=2.6	----	----
12	----	6.07 (s)	----	----	2.30 (s)	2.19 (s)
13	7.88 (dd)	6.58 (dd) $J_{3'5'}$=0.6	7.77 (dd) $J_{3'4'}$=1.9	6.08 (m) $J_{4'5'}$=2.6	2.31 (m)	2.21 (m) $J_{4'Me5'}$=0.8 $J_{4'Me3'}$=0.4

Table 3. ^{1}H-NMR Chemical Shifts and Relative Intensities of 1,2-Bis(3',5'-dimethylpyrazol-1'-yl) -3,4,5,6-tetrafluorobenzene **9** at different temperatures (Solvent $CDCl_3$). δ in ppm, T in K.

T (K)	δ H-4'	Intensity	δ CH_3-3'	Intensity	δ CH_3-5'	Intensity
308	5.81	4.00	2.12	20.10	2.09	11.42
290	5.82	3.98	2.12	20.00	2.08	8.82
283	5.82	4.10	2.13	20.08	2.08	10.73
275	5.82	4.38	2.13	20.17	2.07	13.61
263	5.83	4.44	2.14	20.45	2.04 1.96	16.09 0.72
253	5.83 5.87	4.56 0.24	2.14	20.08	2.04 1.96	16.07 0.84
243	5.84 5.87	4.40 0.28	2.14	20.01	2.03 1.95	15.37 1.00
233	5.84 5.88	4.78 0.37	2.15	20.01	2.02 1.95	16.02 1.25
233	5.85 5.88	5.19 0.49	2.16	20.11	2.00 1.95	17.00 1.60

(Table 3); they correspond (n = 9, r^2 = 0.999) to ΔH = 2.2±0.5 kJ mol^{-1} and ΔS = 14.5±0.2 cal mol^{-1} K^{-1}. Since we have not succeeded in obtaining crystals of **9** we are compelled to assume that the major isomer is the *d,l.* (at 233 K there is 93% of this isomer). The numbering system used in the NMR Tables is represented in Scheme 8.

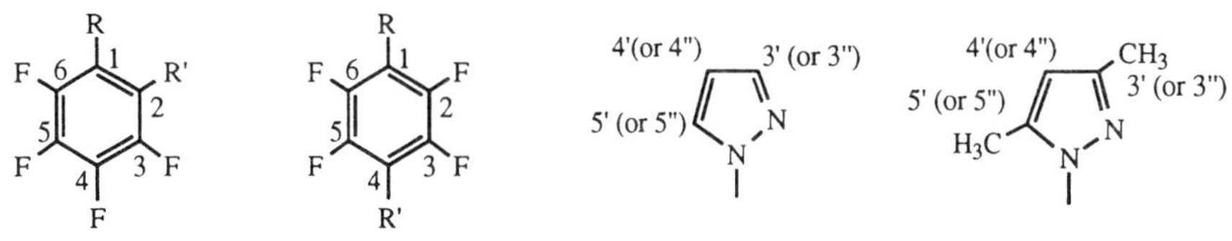

Scheme 8

The ^{13}C NMR results are reported in Tables 4-6. The chemical shifts have been assigned considering the wealth of information about ^{13}C NMR spectroscopy of pyrazoles as well as the ^{13}C-^{19}F coupling constants.[20] Taking into account the ^{13}C chemical shift of hexafluorobenzene (138.0 ppm) and assuming that the replacement of a fluorine atom by a pyrazol-1-yl ring has no effect on the *meta*-carbon atom, it is possible to calculate the effect of replacing a C-F by a C-pz (or C-dmpz) on the *ortho*-carbon atom bearing a fluorine (the same approach can be used for *para* effects, the goodness of the model proving that *meta* effects are negligible). The results of the multiregression in ppm (n = 13, r^2 = 0.995) are represented in Scheme 9.

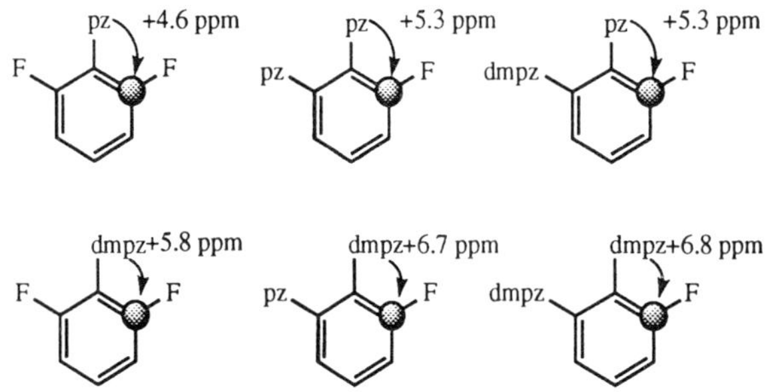

Scheme 9

The effect of dmpz is larger than that of pz and it depends on the torsion angle: when ϕ increases (due to another *ortho'* pyrazole) the effect also increases. The chemical shifts of benzene carbon atoms *ortho* and *meta* to an azole ring have been used to estimate the torsion angle between both rings.

The solid-state ^{13}C CPMAS NMR spectra of compounds **9**, **12** and **13** have been recorded (the values of the chemical shifts are in the corresponding Tables). The similarity of the solution and solid state values is remarkable, the only difference being observed for the methyl groups. We have already noted that these groups

Table 4. ^{13}C-NMR Data in Solution (Solvent $CDCl_3$) of pz derivatives **8** and **11**. δ in ppm, J in Hz.

Comp	C-1	C-2	C-3	C-4	C-5	C-6	C-3'	C-4'	C-5'
8	123.1	123.1	143.3 $^{1}J_{F3}$=256.8	141.1	141.1	143.3	142.2 $^{1}J_{H3'}$=187.5 $^{2}J_{H4'}$=5.8 $^{3}J_{H5'}$=8.5	107.7 $^{1}J_{H4'}$=179.0 $^{2}J_{H3'}$=10.3 $^{2}J_{H5'}$=8.5	132.1 $^{1}J_{H5'}$=190.8 $^{2}J_{H4'}$=9.3 $^{3}J_{H3'}$=4.6
11	119.8	142.2 $^{1}J_{F2}$=258.5	142.2	119.8	142.2	142.2	142.7 $^{1}J_{H3'}$=188.0 $^{2}J_{H4'}$=5.6 $^{3}J_{H5'}$=8.5	108.1 $^{1}J_{H4'}$=179.5 $^{2}J_{H3'}$=10.3 $^{2}J_{H5'}$=8.5	132.1 $^{1}J_{H5'}$=191.4 $^{2}J_{H4'}$=9.3 $^{3}J_{H3'}$=4.5

Table 5. ^{13}C-NMR Data in Solution (Solvent $CDCl_3$) and in the Solid State (CPMAS) of mixed (pz, dmpz) derivatives **10** and **13**. δ in ppm, J in Hz.

Comp	C-1	C-2	C-3	C-4	C-5	C-6	C-3'	C-4'	C-5'	C-3"	C-4"	C-5"	CH_3-3"	CH_3-5"
10	124.8	121.1	144.7 $^{1}J_{F3}$=253.8	140.6 $^{1}J_{F4}$=257.6	141.7 $^{1}J_{F5}$=257.8	143.3 $^{1}J_{F6}$=256.1	142.2 $^{1}J_{H3'}$=187.3 $^{2}J_{H4'}$=5.7 $^{3}J_{H5'}$=8.5	107.5 $^{1}J_{H4'}$=179.0 $^{2}J_{H3'}$=10.3 $^{2}J_{H5'}$=8.5	131.7 $^{1}J_{H5'}$=191.7 $^{2}J_{H4'}$=9.3 $^{3}J_{H3'}$=4.6	151.4 $^{2}J_{H4'}$=6.3 $^{2}J_{Me3''}$=6.3	106.5 $^{1}J_{H4'''}$=174.4 $^{3}J_{Me3''}$=3.4 $^{3}J_{Me5''}$=3.4	142.6	13.5 ^{1}J =127.6	10.4 ^{1}J =129.1
13	120.7	142.0 $^{1}J_{F2}$=255.4	143.8 $^{1}J_{F3}$=254.6	118.7	143.8	142.0	142.7 $^{1}J_{H3'}$=188.0 $^{2}J_{H4'}$=5.7 $^{3}J_{H5'}$=8.4	108.1 $^{1}J_{H4'}$=179.5 $^{2}J_{H3'}$=10.0 $^{2}J_{H5'}$=8.7	132.1 $^{1}J_{H5'}$=191.4	151.9 $^{2}J_{H4''}$=6.1 $^{2}J_{Me3''}$=6.1	107.1 $^{1}J_{H4''}$=174.8 $^{3}J_{Me3''}$=3.3 $^{3}J_{Me5''}$=3.3	142.3 $^{2}J_{H4''}$=6.1 $^{2}J_{Me5''}$=6.1	13.6 ^{1}J =127.6	10.7 ^{1}J =129.1
13[a]	119.8	143.2	143.2	121.7	143.2	143.2	143.2	109.9	130.9 129.0	152.3	109.9	143.2	11.7	8.8

[a] CPMAS.

Table 6. ^{13}C-NMR Data in Solution (Solvent $CDCl_3$) and in the Solid State (CPMAS) of dmpz derivatives **9**, **12**, **12a** and **12b**. δ in ppm, J in Hz.

Comp	C-1	C-2	C-3	C-4	C-5	C-6	C-3'	C-4'	C-5'	CH_3-3'	CH_3-5'
9	123.8	123.8 $^1J_{F3}$=256.3	144.8 $^1J_{F4}$=260.0	141.2	141.2	144.8	150.9 $^2J_{H4'}$=6.1 $^2J_{Me3'}$=6.1 $^3J_{Me5'}$=3.0	106.1 $^1J_{H4'}$=174.1 $^3J_{Me3'}$=3.0	143.1 $^2J_{H4'}$=6.5 $^2J_{Me5'}$=6.5	13.4 ^1J=127.5	10.8 ^1J=128.1
9[a]	124.1	124.1	146.0	139.9	139.9	146.0	150.8	108.3	144.3	12.1	9.8
12	119.6 $^2J_{F2}$=16.9	143.8 $^1J_{F2}$=257.0	143.8	119.6	143.8	143.8	151.9 $^2J_{H4'}$=6.5 $^2J_{Me3'}$=6.5	107.1 $^1J_{H4'}$=174.7 $^3J_{Me3'}$=3.3 $^3J_{Me5'}$=3.3	142.3 $^2J_{H4'}$=6.7 $^2J_{Me5'}$=6.7	13.5 ^1J=127.7	10.7 ^1J=129.3
12[a]	120.0	143.0	143.0	120.0	143.0	143.0	149.9	107.1	143.0	13.2	8.5
12[b]	118.7	143.4	143.4	118.7	143.4	143.4	149.6	107.2	142.3	13.5	8.7
12a[c]	118.7	143.0	143.0	118.7	143.0	143.0	151.9	106.7	143.0	11.9	9.1
12b[d]	118.6	142.7	142.7	118.6	142.7	142.7	152.7	106.4	142.7	11.0	9.0

[a] CPMAS (50 MHz); [b] CPMAS (75 MHz); [c] CPMAS (75 MHz), $2CH_3CO_2H$: 19.6 (CH_3) and 172.2 ppm (C=O); [d] CPMAS (75 MHz), $2CH_3CH_2CO_2H$: 8.3 (CH_3), 26.8 (CH_2), 175.6 ppm (C=O).

are much more sensitive than sp^2 carbons to phase effects. Since these shielding effects are characteristic of the solid state they must arise from intermolecular interactions and most probably from the perpendicular proximity of a methyl group to the π cloud of an aromatic ring.

Compound **9** presents only a series of signals indicating that in the solid state there is a unique structure, probably the *d,l* isomer. The same happens for compound **12** although there are two independent molecules in the unit cell, but very similar in geometry. Compound **13**, with only one independent molecule in the unit cell, shows a splitting of C-5' of the pyrazole ring, may be due to a quadrupolar effect of N-1'.

The ^{13}C CPMAS NMR spectra of **12a** and **12b** (Table 6) show, besides the signals of the guests (acetic and propionic acid in a 1:2 host-guest ratio), those of the host which are similar to the free host save for C-3' (from 149.6 to 151.9 or 152.7 ppm) and CH_3-3' (from 13.5 to 11.9 or 11.0 ppm). These variations are due to the hydrogen bonds between the host and the guests (see crystallographic discussion in Section 4) being similar to the effects observed in solution between $CDCl_3$ and CF_3CO_2H.

The ^{19}F NMR data of compound **9** in [$^2H_6$]acetone was recorded at low temperature in a series of experiments similar to those reported for ^1H-NMR. At 294 K the signals of the *d,l* isomer [reference C_6F_6, F-3(6) 18.54, F-4(5) 9.82] are narrow but those of the *meso* isomer are very broad [reference C_6F_6, F-3(6) 18.86, F-4(5) 8.99]; they become well resolved at 267 K (94% *d,l* - 6% *meso*).

4.2. Hexakispyrazolylbenzenes 14-17.

The ^1H and ^{13}C NMR solution spectra of the two polymorphs of the hexakis(pyrazol-1'-yl)benzene, **14a** and **14b**, are identical and show only a narrow set of signals. The pyrazole ring protons in $CDCl_3$ at 200 MHz appeared at δ 7.41 (dd, H-3'), 6.13 (dd, H-4') and 7.25 (dd, H-5'), the ^{13}C NMR data (50 MHz) of the $(pz)_6bz$ in the same solvent being: δ 136.6 (C-1, benzene carbons), 141.5 (C-3'), 107.2 (C-4') and 132.3 (C-5'), with coupling constant values within the normal ranges for substituted pyrazoles.[17]

In the case of the mixed compound **16**, $pz_4(dmpz)_2bz$, the ^1H NMR spectra of a freshly prepared solution and its evolution with time in solvents such as $CDCl_3$ and methanol-d_4 were recorded (see Table 7). The spectra of the freshly prepared solution show signals which correspond to the single isomer **16**(*1, ududud*) (see Figure 20 for the 14 conformational isomers). After 24 h the presence of a minor new rotational isomer was detected and the proportion did not change with time, the initial conformer being the major component with a ratio of 9:1. When an enriched mixture in the minor isomer was left to evolve for 24 h the same equilibrium mixture was obtained.

We have assigned structure (*7, duuduu*) to the minor isomer according to the following criteria: i) its NMR spectra with only signals for one pz and one dmpz (ratio 4:2) and ii) the δΔH value of 3.0 kcal mol^{-1} with respect to the most stable conformer (*1, ududud*), obtained from the semiempirical calculations.

The ^{13}C NMR spectra are reported in Table 8. The ^{13}C CPMAS NMR spectra of polymorphs **16a** and **16b** and its acetic acid complex **16c** show data which agree with the existence of $pz_4(dmpz)_2bz$ in the solid state as only one conformer, but with two different structures: in **16a** there are two kinds of dmpz whereas in **16b** there is only one kind. The corresponding inclusion compound **16**·2 CH_3CO_2H shows the signals of acetic acid in a 2:1 ratio and only one kind of dmpz.

Table 7. ^{1}H-NMR chemical shifts (δ, ppm) and coupling constants (*J* , Hz) of 1,2,4,5-tetrakis(pyrazol-1'-yl)-3,6-bis(3",5"-dimethylpyrazol-1"-yl)benzene **16**

Compound	H-3'	H-4'	H-5'	H-4"	CH_3-3'	CH_3-5'	Solvent
16(*1*)	7.35(dd)	6.11(dd) $J_{H3',H4'}$=1.8 $J_{H4',H5'}$=2.5	7.36(dd) $J_{H3',H5'}$=0.5	5.61(s)	2.10(s)	1.82(s)	$CDCl_3$
16(*7*)	7.41(dd) $J_{H3',H4'}$=1.9 $J_{H3',H5'}$=0.5	6.11(dd)	7.11(dd) $J_{H4',H5'}$=2.5	5.64(s)	2.03(s)	1.97(s)	$CDCl_3$
16(*1*)	7.36(dd) $J_{H3',H5'}$=0.4	6.17(dd) $J_{H3',H4'}$=1.9 $J_{H4',H5'}$=2.6	7.41(dd)	5.71(s)	2.05(s)	1.86(s)	Metanol-d_4
16(*7*)	7.38(dd) $J_{H3',H4'}$=1.9 $J_{H3',H5'}$=0.5	6.17(dd)	7.29(dd) $J_{H4',H5'}$=2.4	5.74(s)	1.99(s)	1.99(s)	Metanol-d_4

Table 8. ^{13}C-NMR chemical shifts (δ, ppm) and coupling constants (*J*, Hz) of 1,2,4,5-tetrakis(pyrazol-1'-yl)-3,6-bis(3",5"-dimethylpyrazol-1"-yl)benzene **16**

Compound	C-1	C-2	C-3	C-3'	C-4'	C-5'	C-3"	C-4"	C-5"	CH_3-3'	CH_3-5'
16(*1*)[a]	137.2	137.2	135.1	141.4 $^{1}J_{H3'}$=186.6 $^{2}J_{H4'}$=5.9 $^{3}J_{H5'}$=8.4	106.7 $^{1}J_{H4'}$=178.5 $^{2}J_{H3'}$=10.2 $^{2}J_{H5'}$=8.9	132.0 $^{1}J_{H5'}$=191.9 $^{2}J_{H4'}$=9.4 $^{3}J_{H3'}$=4.6	150.4	105.5 $^{1}J_{H4''}$=174.1	143.1	13.4 ^{1}J=127.4	10.5 ^{1}J=129.2
16(*7*)[a]	137.2	137.2	135.4	141.4	107.1	131.1	150.5	105.5	143.3	13.4	10.5
16a[b]	138.7	138.7	136.7	140.8	108.3	131.2 131.7	148.3 151.4	107.1	142.2	13.0 14.1	9.7 10.8
16b[b]	138.1	138.1	138.1	138.1	107.0	130.9	148.1	105.0	142.6	15.0	11.5
16c[b,c]	137.5	137.5	133.7	141.5	107.0	131.8	148.9	106.3	141.5	13.3	10.3

[a]In $CDCl_3$; [b]CP MAS; [c] Acetic acid: 21.7 (Me), 179.7 (CO)

Our initial NMR solution studies on the hexakis(3',5'-dimethylpyrazol-1'-yl)benzene **15** were performed, due to its low solubility in organic solvents, in a mixture of $CDCl_3$ containing 5% of CF_3CO_2H. In these conditions, when the equilibrium was reached there was 20% of isomer **15** (*1*, *ududud* in Figure 19) and 80% of **15** (*2*, *uduuud*).[16]

Because there are indications that the presence of CF_3CO_2H affects the distribution populations, protonation will change repulsive *uu* interaction by attractive *uu*+ one,[23] we studied again the behaviour of **15** in pure $CDCl_3$. We observed that the initial amount of isomer **15** (*1*, only one dmpz, Table 9) diminishes leading to conformer **15** (*2*, *uduuud*)[16] which presents four dmpzs and the latter in turn, disappears to yield an isomer showing the 1:1:1 pattern for dmpzs to which we have assigned the structure **15** (*3*, *udduud*). At the equilibrium, the relative amounts are 44% **15**(*1*), 12% **15**(*2*), 44% **15**(*3*).[23] The ^{13}C NMR results corroborate that the two major isomers in a 1:1 relative ratio are **15** (*1*) and **15** (*3*) (see Table 9).

The free host 1,4-bis(pyrazol-1'-yl)-2,3,5,6-tetrakis(3",5"-dimethylpyrazol-1"-yl)benzene **17** exists in $CDCl_3$ solution (see Table 9) as the most stable conformation **17**(*1, ududud*) that remains unaltered in all the inclusion complexes: **17a-17e** (water, acetic, propionic, pentanoic, and (±)-2-methylbutyric acids).[21] The CPMAS spectra of compounds **17-17e** (see Table 10) show the splitting of several signals (C-4', C-3", C-4" and the methyl groups); these splittings result in differences in magnetic environments in the solid state. The signals of the different guests are clearly observed and the corresponding chemical shifts agree with literature data in solution available for all acids save (±)-2-methylbutyric acid **17e**. The splitting of signals also affects the guests reflecting the 1:2 host-guest stoechiometry.

In solution, compound **17**(*1,ududud*) evolves to a mixture of three conformations 63% **17**(*1*), 23% **17**(*5* or *6*), 14% **17**(*3* or *4*), the equilibration taking place through the minor isomer **17** (*3* or *4*) (see Figure 20a). Moreover the data gathered in Tables 9 and 10, the complete assignment of all the ^{1}H and ^{13}C NMR signals is described in reference 23.

4.3. Pentakispyrazolylpyridines 18-20.

The ^{1}H NMR results for pentakispyrazolylpyiridines **18-21** are reported in Table 11, while the ^{13}C NMR results are reported in Tables 12 and 13. The assignments are based on the extensive use of 2D-NMR spectroscopy (^{1}H-^{1}H and ^{1}H-^{13}C). We have checked the additivity of the substituents effects (SCS) both in ^{1}H and in ^{13}C NMR.

Two kinds of experiments deserve a longer comment. First, in ^{1}H NMR some compounds appear as mixtures of isomers and not others, fact related to the $u \rightarrow d$ rotational barriers (see Section 7.5).[19] Thus, pz derivatives, like **18** and **20**, have low rotational barriers and only average spectra are observed in solution while dmpz derivatives, like **19** and **21**, have high rotational barriers and two isomers are observed in a 80:20 ratio for **19** and 50:50 ratio for **21**. We assign these isomers to conformations *1* and *4* of Figure 21. We have checked that these relative proportions are time independent by recording the ^{1}H NMR spectra with very few scans (very short time) and after a week in $CDCl_3$ solution.

The other experiment which deserves to be commented is the series of ^{13}C CPMAS NMR spectra. Compounds **18** and **19** show solid state spectra consistent with the X-ray structures (*1:C* and *4:A* respectively).

Table 9 ^{1}H NMR (between brackets) and ^{13}C NMR chemical shifts (δ in ppm) of $(dmpz)_6bz$ **15** isomers and $(pz)_2(dmpz)_4bz$ **17** isomers in $CDCl_3$

Comp.	C-3' (H)	C-4' (H)	C-5' (H)	CH_3-3' (H)	CH_3-5' (H)	C-3'	C-4' (H)	C-5'	CH_3-3' (H)	CH_3-5' (H)	C-3'	C-4' (H)	C-5'	CH_3-3' (H)	CH_3-5' (H)	C(bz.)
	Positions 1,2,3,4,5,6															
15(*1*)	149.3	104.5	143.8	13.2	11.3											138.8
		(5.52)		(1.94)	(2.13)											
	Positions 1,6					Positions 2,5					Positions 3,4					
15(*3*)	149.6[b]	104.6	144.2[c]	13.2	11.4	149.4[b]	104.9	143.8[c]	13.3	11.0	149.2[b]	105.4	143.8[c]	13.3	11.2	138.3, 138.8,
		(5.51)		(1.91)	(2.09)		(5.55)		(2.03)	(1.86)		(5.58)		(2.02)	(1.91)	138.9
	Positions 1,4					Positions 2,3,5,6										
17(*1*)	141.1	106.3	131.9	-	-	149.8	105.0	143.7	13.3	11.1						138.7 (C-1,4)
	(7.29)	(6.03)	(7.32)				(5.57)		(1.98)	(2.02)						136.8 (C-2,3,5,6)
17(*5*)	141.5	106.9	130.6	-	-	150.1	105.1	143.0	13.3	10.9						137.6 (C-1,4)
17(*6*)	(7.42)	(6.10)	(7.13)				(5.57)		(2.01)	(1.92)						136.9 (C-2,3,5,6)
	Position 1					Position 2					Position 3					
17(*3*)	141.4	107.0	130.7	-	-	n.o.	105.8	n.o.	13.3	10.8	n.o.	105.0	n.o.	13.3	10.8	n.o.
17(*4*)	(7.41)	(6.07)	(7.09)				(5.60)		(2.07)	(1.79)		(5.57)		(2.048)	(1.86)	
	Position 4					Position 5					Position 6					
	141.0	106.4	132.1	-	-	n.o.	104.7	n.o.	13.3	11.4	n.o.	105.3	n.o.	13.3	10.7	
	(7.30)	(6.02)	(7.22)				(5.54)		(1.92)	(2.16)		(5.53)		(2.052)	(1.81)	

n.o. = not observed

Table 10. ^{13}C-NMR data in $CDCl_3$ solution[a] and in the solid state CPMAS[b] of [$(pz)_2(dmpz)_4bz$] **17** and its inclusion compounds **17a-17e**.δ in ppm, *J* in Hz. The guest signals are also given.

Comp	Benzene ring C-1/C-4	Benzene ring C-2/C-3	Pyrazole ring C-3'	Pyrazole ring C-4'	Pyrazole ring C-5'	3,5-Dimethylpyrazole ring C-3"	3,5-Dimethylpyrazole ring C-4"	3,5-Dimethylpyrazole ring C-5"	CH_3-3'	CH_3-5'
17[a]	138.7	136.8	141.1 $^1J_{H3'}$=186.2 $^2J_{H4'}$=6.0 $^3J_{H5'}$=8.5	106.3 $^1J_{H4'}$=178.0 $^2J_{H3'}$=10.5 $^2J_{H5'}$=8.7	131.9 $^1J_{H5'}$=192.5 $^2J_{H4'}$=9.5 $^3J_{H3'}$=4.3	149.8	105.0 $^1J_{H4''}$=173.2 $^3J_{Me3''}$=3.2 $^3J_{Me5''}$=3.2	143.7	13.3 ^1J=127.2	11.1 ^1J=129.3
17a[b]	139.4	137.1	141.6	103.8 106.4	132.3	147.1 149.5 150.9	106.4	142.8	13.3 14.1	11.2 12.2 12.6
17b[b]	140.7	136.8br	140.7	104.9 106.8	131.4	147.4	107.3 108.1	142.4	13.0 13.9	9.7 12.1
17c[b]	140.6	138.0br	141.1	105.3	132.3	147.1 148.7 149.7	106.9 107.9	142.5	12.6 14.2	10.6 12.6
17g[b]	139.7	139.7	141.7	105.0 105.1	131.3 132.4	147.2 148.3 149.1 149.8	106.5 107.6	141.7	12.9 13.4	12.1
17d[b]	139.2	137.0	141.6	104.4 105.3	132.0	147.8 149.0	106.2 108.7	140.0	12.9	9.4 11.6
17e[b]	140.3	137.0	141.5	104.5	130.3	148.8 148.9	106.1	142.5	13.0 13.6	9.3 12.4

Guests	C=O	C-2	C-3	C-4	C-5	C-2'
17b[b]	171.8 172.5	20.6 22.1				
17c[b]	175.4	27.8	9.4			
17g[b]	175.0	35.0	17.8	13.9		
17d[b]	173.4 174.2	34.4	27.4	23.0 23.4	14.7 15.4	
17e[b]	177.5	40.5 39.7	27.3	18.4br		12.4
c	183.6	40.9	26.5	16.2		11.4

[c] Guest (±)-2-methylbutyric acid in $CDCl_3$

Table 11. ^{1}H NMR data (chemical shifts, δ in ppm, and coupling constants, J in Hz) in $CDCl_3$ for compounds **18-21**

Comp	H-4'	$H(CH_3)$-3'	$H(CH_3)$-5'	H-4'	$H(CH_3)$-3'	$H(CH_3)$-5'	H-4'	$H(CH_3)$-3'	$H(CH_3)$-5'
	Positions 2, 6			Positions 3, 5			Position 4		
18	6.29	7.58	7.38	6.40	7.60	8.02	6.12	7.43	7.21
	$J_{H\text{-}5'}$=2.4	$J_{H\text{-}4'}$=1.9		$J_{H\text{-}5'}$=2.6	$J_{H\text{-}4'}$=1.8		$J_{H\text{-}5'}$=2.5	$J_{H\text{-}4'}$=1.8	
19[a]	5.63	2.00	1.97	5.84	2.04	2.35	5.58	1.94	2.07
19[b]	5.63	1.99	1.97	5.81	2.04	2.39	5.67	1.84	2.21
20	6.26	7.53	7.40	6.40	7.57	8.03	5.61	2.09	1.88
	$J_{H\text{-}5'}$=2.5	$J_{H\text{-}4'}$=1.8	$J_{H\text{-}3'}$=0.5	$J_{H\text{-}5'}$=2.8	$J_{H\text{-}4'}$=1.8	$J_{H\text{-}3'}$=0.5			
21[c]	5.70	2.10	1.93	5.85	2.06	2.40	7.47	6.15	7.05
							$J_{H\text{-}4'}$=1.8	$J_{H\text{-}5'}$=2.6	
21[d]	5.68	2.00	1.82	5.84	2.03	2.31	7.35	6.10	7.19
							$J_{H\text{-}4'}$=1.8	$J_{H\text{-}5'}$=2.5	

[a] Major isomer (80%); [b] Minor isomer (20%); [c,d] Two isomers in a 50:50 ratio

Table 12. ^{13}C NMR data (chemical shifts, δ in ppm, and coupling constants, *J* in Hz) of compounds **18** and **19**in $CDCl_3$

Comp	C-3'	C-4'	C-5'	CH_3-3'	CH_3-5'	C-3'	C-4'	C-5'	CH_3-3'	CH_3-5'	C-3'	C-4'	C-5'	CH_3-3'	CH_3-5'
	Positions 2, 6					Positions 3, 5					Position 4				
18	141.4 ^1J=186.7 $^2J_{H-4}$=5.6 $^3J_{H-5}$=8.5	107.4 ^1J=179.1 $^2J_{H-3}$=10.5 $^2J_{H-5}$=8.4	133.0 ^1J=190.6 $^2J_{H-4}$=9.3 $^3J_{H-3}$=4.7	-	-	143.6 ^1J=187.9 $^2J_{H-4}$=5.8 $^3J_{H-5}$=8.8	108.4 ^1J=179.1 $^2J_{H-3}$=10.7 $^2J_{H-5}$=8.4	130.2 ^1J=193.1 $^2J_{H-4}$=9.6 $^3J_{H-3}$=4.3	-	-	141.8 ^1J=187.4 $^2J_{H-4}$=5.9 $^3J_{H-5}$=8.5	107.4 ^1J=179.1 $^2J_{H-3}$=10.5 $^2J_{H-5}$=8.5	131.8 ^1J=192.4 $^2J_{H-4}$=9.4 $^3J_{H-3}$=4.6	-	-
18[a]	143.0	106.8	134.2	-	-	144.3	110.1	132.1	-	-	142.4	108.8	134.2	-	-
19[b]	150.0	104.9 ^1J=173.2 ^3J=3.3	143.5	13.2[d] ^1J=127.1	10.9 ^1J=128.8	149.5	107.2 ^1J=173.6 ^3J=3 4	141.7	13.3[d] ^1J=127.3	12.3 ^1J=129.4	149.9	105.0	143.9	13.0[d] ^1J=127.1	11.1 ^1J=129.2
19[c]	150.2	105.3	142.0	13.3[e] ^1J=127.3	10.6	149.7	106.9	141.3	13.3[e] ^1J=127.3	12.4	149.9	105.8	144.4	13.0[e] ^1J=127.1	10.8
19[a]	150.5	105.8	142.2	14.1	11.4	150.5	107.8 109.1	141.7	14.7	12.6	147.7	105.8	142.2	12.6	12.2

[a] This row corresponds to the CPMAS; [b] Major isomer; [c] Minor isomer; Signals marked [d], [e] can be reversed;

Table 13. ^{13}C NMR data (chemical shifts, δ in ppm, and coupling constants, J in Hz) of compounds **20** and **21**in $CDCl_3$

Comp	C-2(6) [C-4] C-3(5)	C-3' Pos. 2,6	C-4' Pos. 2,6	C-5' Pos 2.6	CH_3-3' CH_3-5' Pos. 26	C-3' Pos. 3,5	C-4' Pos. 3,5	C-5' Pos. 3,5	CH_3-3' CH_3-5' Pos. 3,5	C-3' Pos. 4	C-4' Pos. 4	C-5' Pos. 4	CH_3-3' CH_3-5' Pos. 4
20	146.4 [147.5] 125.1	141.3 ^1J=186.6 $^3J_{H-5'}$=8.4 $^2J_{H-4'}$=5.9	107.1 ^1J=178.3 $^2J_{H-3'}$=10.4 $^2J_{H-5'}$=8.7	132.1 ^1J=191.0 $^2J_{H-4'}$=9.3 $^3J_{H-3'}$=4.5	-	143.4 ^1J=187.6 $^3J_{H-5'}$=8.9 $^2J_{H-4'}$=5.9	108.3 ^1J=179.1 $^2J_{H-3'}$=10.4 $^2J_{H-5'}$=8.7	130.1 ^1J=192.9 $^2J_{H-4'}$=9.5 $^3J_{H-3'}$=4.2	-	150.9	105.8 ^1J=174.3 $^3J_{Me3'}$=3.3 $^3J_{Me5'}$=3.3	142.7	13.3 ^1J=127.5 10.6 ^1J=129.4
20[a]	145.6 [145.6] 124.3 125.3	138.8	107.3	132.1	-	143.0	110.2	130.9	-	150.3	106.0	143.0	12.6 11.4
21[b]	149.4 [146.9] 126.4	150.0	105.3 ^1J=173.3 ^3J=3.2	143.5	13.4 ^1J=127.5 10.8	150.2	107.2 ^1J=173.4 ^3J=3.5	141.9	13.2 ^1J=127.3 12.3 ^1J=129.5	141.7 ^1J=186.6 $^3J_{H-5'}$=8.5 $^2J_{H-4'}$=5.9	107.4	130.6 ^1J=192.1 $^2J_{H-4'}$=9.6 $^3J_{H-3'}$=4.4	-
21[c]	149.0 [148.4 125.3	150.1	105.3 ^1J=173.3 ^3J=3.2	143.7	13.3 ^1J=127.3 10.6 ^1J=129.2	150.3	107.2 ^1J=173.4 ^3J=3.5	142.0	13.3 ^1J=127.3 12.1 ^1J=129.4	141.7 ^1J=186.6 $^3J_{H-5'}$=8.5 $^2J_{H-4'}$=5.9	107.0	131.8 ^1J=193.0 $^2J_{H-4'}$=9.5 $^3J_{H-3'}$=4.6	-
21[a]	149.0 [148.4] 127.5 127.8	150.6 151.5	105.1 104.8	143.5	13.6 11.0 9.5 9.1	150.1	110.0 108.6	142.5 142.9	13.6 12.1 10.4	141.5	107.5 107.1 106.7	131.2	-

[a] This row corresponds to the CPMAS spectrum; [b, c] Two isomers

Compound **20** presents the signals of only one isomer which according to the X-ray structure has the conformation *4:A*, only the pyridine carbon C-3(6) is split owing to a loss of symmetry in the solid state. The case of compound **21** is more complex. After many attempts (different solvents, different conditions) we failed to obtain crystals of **21**, the apparently most beautiful crystals were always twins. The ^{13}C CPMAS NMR spectrum (Table 13) proves without doubt that this compound in the solid state is a mixture of isomers *1* and *4*, the ***A-C*** conformation cannot be ascertained.

In summary, i) when there is only one isomer in solution (**18** and **20**) there is only one isomer in the solid state; ii) when there are two isomers in solution but one is largely predominant (**19**) only the most stable is found in the solid state, and iii) when there are two isomers of the same stability in solution (**21**), both are present in the solid state.

5. OTHER PHYSICAL PROPERTIES (DSC, UV, MS, DIPOLE MOMENTS).

The Differential Scanning Calorimetry (DSC) analysis confirmed, in the case of 1,2-bis(3',5'-dimethyl-pyrazol-1'-yl)-3,4,5,6-tetrafluorobenzene **9**, hexakispyrazolylbenzenes **15-17**, pentakis(3',5'-dimethylpyrazol-1'-yl)pyridine **19** and 2,3,5,6-tetrakis(3',5'-dimethylpyrazol-1'-yl)-4-(pyrazol-1'-yl)pyridine **21**, the existence of several conformational isomers in solid state.[24] For the host-guest complexes of 1,4-bis(pyrazol-1'-yl)-2,3,5,6-tetrakis(3",5"-dimethylpyrazol-1"-yl)benzene **17a-17e**, the guest loss occurs in several steps caused by the desorption of the solvent molecules. These steps correspond to the endotherms seen in the DSC before a sharp melting point for the host compound at onset temperature 281 °C.[21]

The HPL chromatography of the hexakispyrazolylbenzene derivatives allowed us to observe different peaks for the various conformers in compounds: $(dmpz)_6bz$ **15**, $pz_4(dmpz)_2$ **16** and $pz_2(dmpz)_6$ **17** in $MeOH:H_2O$, for several proportions and using two reverse phase columns (Waters Spherisorb ODS2 150x4 mm, 5µm and Merck Lichrospher 100 RP18, 125x4 mm, 5µm). Only one peak was detected in hexakis(pyrazol-1'-yl)benzene **14**.[22]

In the hexakis(3',5'-dimethylpyrazol-1'-yl)benzene the use of different mixtures of $MeOH:H_2O$ and different temperatures permitted us, by applying the Lebl and Gut method,[22] to estimate the energy barrier which corresponds to the rotation of a dmpz (see Section 7) in approximately 26 kcal mol^{-1}.

The UV spectra of all hexakispyrazolylbenzenes **14-17** were very similar with only one maximum at around 262-266 nm. Dipole moment measurements were achieved for polypyrazolylpyridines of Schemes 3, 4 and 5, and the values obtained contain indications about the conformations of the molecules.[19]

FAB mass spectra of all polyazolylbenzenes and pyridines of this review were recorded in glycerol (G) and *meta*-nitrobenzyl alcohol (NBA), clearly showing the corresponding $[M+H]^+$ ions. A F/H reduction process in pyrazolylfluorobenzenes was also observed.[54]

6. SEMIEMPIRICAL CALCULATIONS: GROUND STATES AND ROTATIONAL BARRIERS.

In a propellene system, the peripheral rings can be considered as blades that project out the central unit or rotation axis (propellene axis) and that can be rotated in the same way to produce a helical molecular conformation that is chiral.[55,56]

The possible conformation of these molecules can be described (in a first level of approximation and considering the peripheral rings perpendicular to the central one) by the situation of the electron lone pair of the nitrogen atoms with regard to each side of the benzene central ring, *up* or *down* (*u* or *d* hereinafter). The interconversion between them requires to cross considerable high energy barriers (See section 7.6), passing through a situation where both rings, central and peripheral, are coplanar. If the substituents are not perpendicular to the central ring, several helical conformations appear for each *up-down* arrangement and their interconversion (from *P* to *M* conformation) goes through the perpendicular situation of the rings which involves a much lower energy barrier than the *up-down* transition described above (See reference 17 for a more detailed discussion).

The relative stabilities of the conformers have been studied by means of semiempirical calculations using the AM1 parametrization for the Hamiltonian of the system.[55] The only constrain was the planarity of the aromatic rings (benzene and pyrazole).

In this study, a black circle or a square (Figure 17 and the following ones) represents the electronic lone pair of a pyrazole or 3,5-dimethylpyrazole ring on one side of the central ring and the white one on the opposite side. The conformation *2*, according to the scheme, is obtained twisting one ring of pyrazole or 3,5-dimethylpyrazole 180° from the situation in *1*.

6.1. *Ortho* and *para* bis-pyrazolyl-tetrafluorobenzene derivatives

All the tetrafluoro derivatives and their inclusion complexes present a *ud* conformation (Figure 17c). In all cases, the most stable conformation presents the nitrogen lone pairs oriented to opposite sides of the central ring, conformer *1* in the scheme. The relative twist of a pyrazole ring joined to the aromatic central ring is the result of several factors, mainly the steric hindrance, the degree of delocalization of the electronic π clouds and the nature (attractive or repulsive) of the interactions N(lone pair)···F/H-C(central ring). As the counterbalance of these facts the 3,5-dimethylpyrazole substituents in the tetrafluoro derivatives are oriented more perpendicular to the benzene central ring than the corresponding pyrazole moieties (Figure 17d) presenting differences of up to 20° in the torsion angles.

The potential energy surfaces (Figure 17b) have been computed for the *para* substituted bis-pyrazole and bis-3,5-dimethylpyrazole derivatives as a function of the torsion angles between the azole groups and the central ring and using as starting points the fully optimized geometries of the absolute minima (Figure 17d). Both surfaces present mm2 planar group symmetry and the position for the global and the first local minima have been annotated with and asterisk (*) in the Figure. In both cases, there are two kinds of global minima with the same energy and the corresponding symmetry equivalent ones. The first one corresponds to a molecule with Ci point group symmetry [*(40, 320°), (140, 220°), (320, 40°) and (220, 140°) or due to the periodicity of torsion angles (40, -40°), (140, -140°), and so on*] in good agreement with the crystallographic results (Figure 17d). The second one corresponds to a molecule with C_2 symmetry [*(40, 220°), (140, 320°), (220, 40°) and (320, 140°) or*

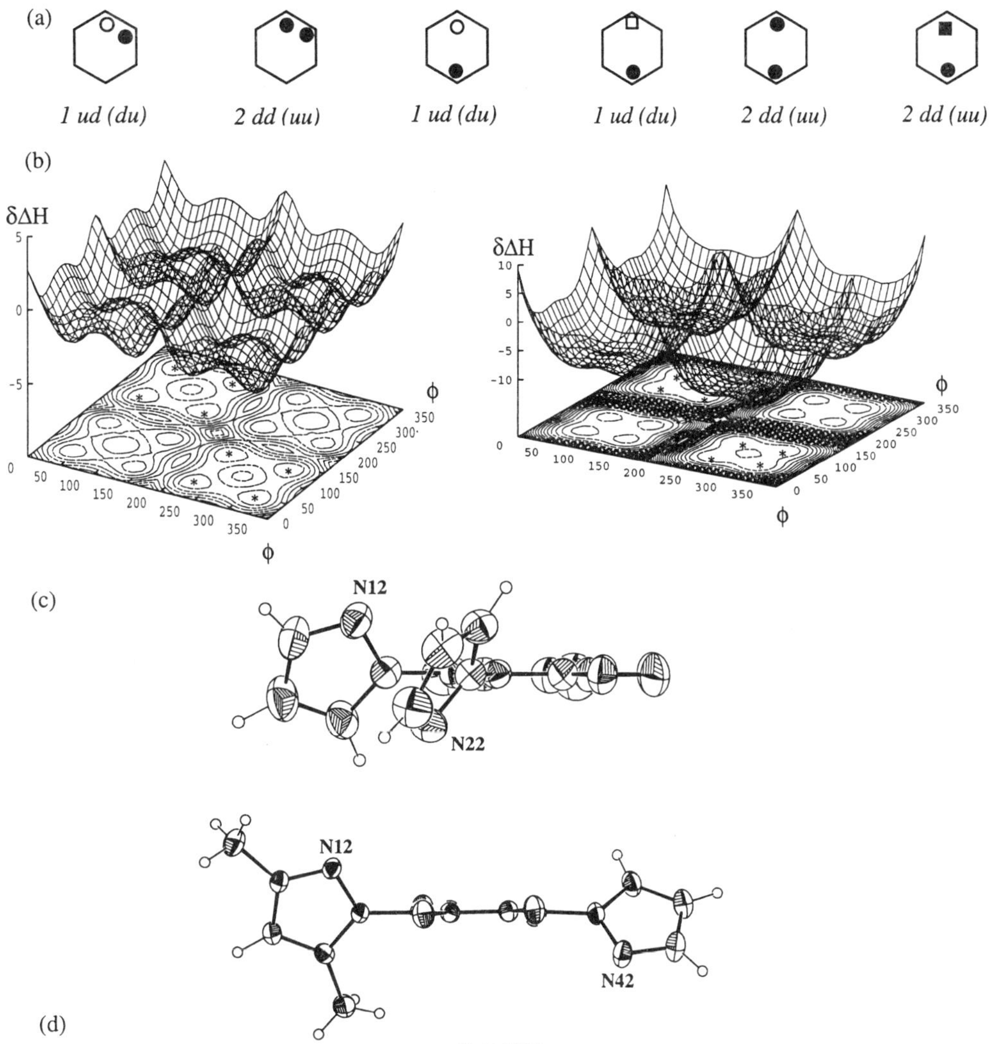

(d)

		C-C-N-N	
		Exp.	Cal.
8: Mol.1	x=1	-122.3(4)	-131.5
	x=2	57.0(5)	47.4
8: Mol.2	x=1	-119.7(4)	
	x=2	51.5(5)	
11	x=1,4	-136.4(2)	-136.1
12: Mol.1	x=1,4	-111.8(2)	-119.4
12: Mol.2	x=1,4	-112.1(2)	

Figure 17: Semiempirical calculations. (a) Possible isomers in the **8**-**13** compounds (see text). (b) Potential energy surface as a function of the C-C-N-N torsion angle for the **11** and **12** compounds. (c) Molecular structure of **8** (molecule 1) and **13** showing the conformation. (d) Experimental and computed torsion angles corresponding to the global minimum.

(40, -140°), (140, -40°), and so on]. A similar set of points is obtained for the 3,5-dimethylpyrazole derivative the only difference is that the peripheral rings are 20° more perpendicular to the central one for each minimum [*(60, 300°),... for the Ci symmetry and (60, 240°),... for the C_2 one*]. However, the maxima in both surfaces are in the same positions, (0, 0°) and (0, 180°), that correspond to a situation where all rings, peripheral and central, are coplanar. The differences in energy with respect to the minimum are 6 and 18 kcal mol^{-1} for the first and second derivative respectively.

6.2. Analysis of the nature of the ortho-ortho interactions

Due to the scarce number of molecular structures experimentally determined (Figure 18), and in order to understand the nature of the interactions between the C-H/C-F substituents of the benzene ring and the electronic lone pair of the nitrogen or the C-H/C-Me groups of the peripheral rings, the geometries of the molecules shown in Figure 18 have been optimized. A multiregression analysis[20] has been performed assuming an additive model for the different types of interactions (correlation coefficient = 0.993). The most repulsive interactions take place between the methyl and C-F groups and the least between N lone pair and C-H. The substitution of hydrogen atoms by fluorine ones in the benzene ring always produces an increment in the dihedral angle with the peripheral ring, that is dependent of the nature of the azol substituent and that presents values ranging from 6° for a hydrogen and 17° for a methyl group. This means, that the *ortho-ortho* interaction between the N lone pair and the C-F has a repulsive component of electronic origin that is added to the steric effect, while for C-H/C-F that component is attractive. Pascal *et al.*[58] have studied the donor-acceptor properties of the N(lone pair)···Cl-C interactions, however, the fluorine seems to present an opposite behaviour to the chlorine, in spite of being a more electronegative atom.

6.3. Conformational isomerism of hexakispyrazolylbenzene and hexakis(3,5-dimethyl-pyrazolyl)benzene

When the six peripheral rings are identical (types a and b) there are only 8 possible *up-down* conformations (Figure 19a), *3* being the only one that is chiral with independence of the values of the torsion angles. The variation of the relative energies for the 8 *up-down* conformers has the same sequence in **14** and **15** (Figure 19b), although it is clear the bigger influence of the *uuu* interaction in making conformers more unstable in the second type of derivatives, mainly for *6*. In the solid state, two conformers (Figure 19c,d) have been observed *ududud* and *uuddud* respectively.

6.4. Conformational isomerism of bispyrazolyl-tetrakis(3,5-dimethylpyrazolyl)benzene and tetrakispyrazolyl-bis(3,5-dimethylpyrazolyl)benzene

There are two types of mixed derivatives shown in Figure 20: type c, with two 3,5-dimethylpyrazoles (dmpzs) in positions 1 and 4 and four pyrazoles (pz) in the remaining ones and type d, with two pzs in positions 1 and 4 and four dmpzs in the remaining positions.

(a)

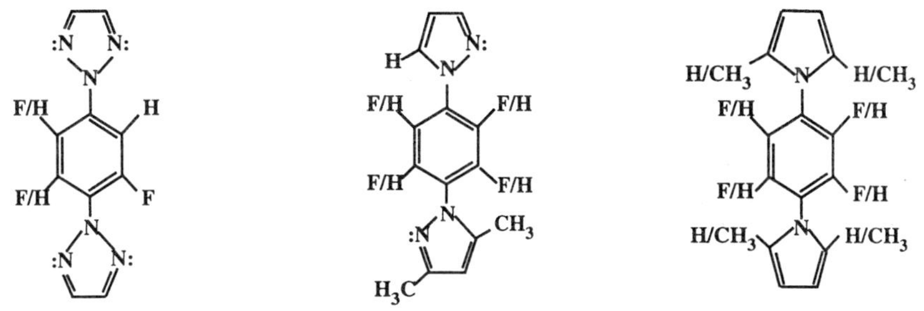

(b)

Interaction	N/CH	CH/CH	Me/CH	N/CF	CH/CF	Me/CF
Coefficient (°)	10.2(21)	14.3(21)	23.0(21)	26.1(21)	19.7(21)	40.1(21)

Interaction								
N/CH	CH/CH	Me/CH	N/CF	CH/CF	Me/CF	φ(AM1)	φ(X ray)	φ(model)
1	-	-	1	-	-	43.3	-	36.3
2	-	-	-	-	-	11.5	1.8[a)]	20.4
-	-	-	2	-	-	54.1	-	52.2
-	-	-	1	1	-	43.6	43.6(4)	45.8
-	-	-	1	-	1	59.0	59.5(8)	66.2
1	1	-	-	-	-	25.3	27.4[b)]	24.5
1	-	1	-	-	-	38.4	41.8[c)]	33.1
1	-	-	-	1	-	29.7	-	29.9
-	-	1	1	-	-	46.7	-	49.1
-	1	-	1	-	-	41.5	-	40.4
1	-	-	-	-	1	55.2	-	50.3
-	2	-	-	-	-	23.2	-	28.7
-	-	-	-	2	-	39.1	-	39.5
-	1	-	-	1	-	33.1	-	34.1
-	-	1	-	-	1	69.2	-	63.1
-	-	2	-	-	-	42.9	89.5[d)]	45.9
-	-	-	-	-	2	69.9	-	80.3
-	1	1	-	-	-	37.5	-	37.3
-	-	-	-	1	1	67.1	-	59.9
-	1	-	-	-	1	64.3	-	54.5
-	-	1	-	1	-	39.6	-	42.7

(c)

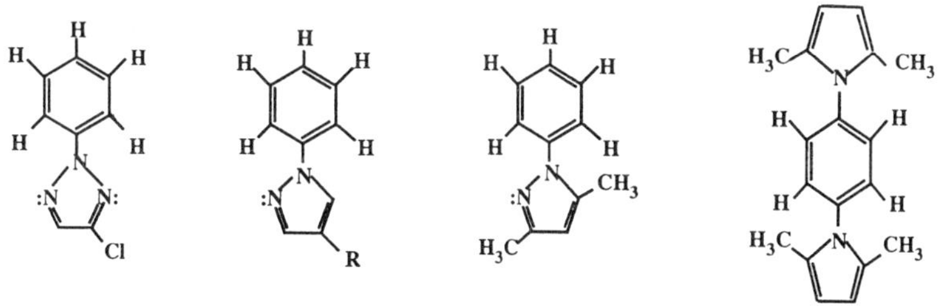

[a)]**COGYAY** [b)]**JEPPEZ** [c)]**COGREV** [d)]**KIXHUU**

Figure 18: *Ortho-Ortho* interactions. (a) Types of studied interactions. (b) Coefficients of the regression equation, the AM1, experimental and computed φ torsion angles (°).

(a)

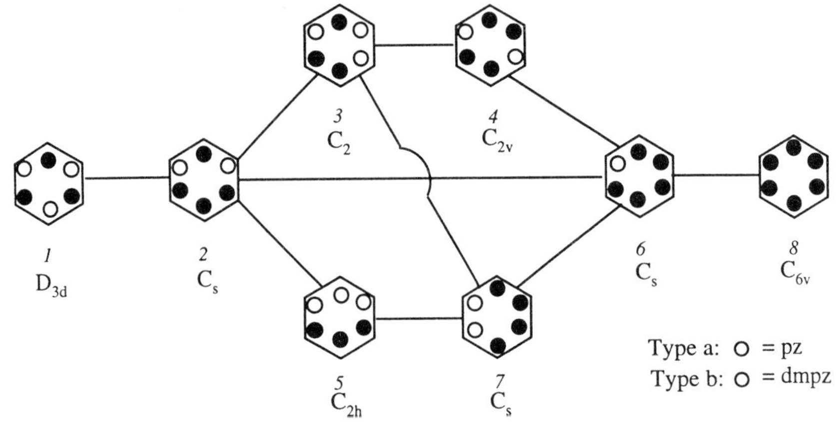

(b)

		1	*2*	*3*	*4*	*5*	*6*	*7*	*8*
δΔH	Type a:	0.0	2.2	2.0	2.1	4.2	4.9	4.3	8.1
	Type b:	0.0	2.8	2.0	2.5	4.7	7.1	5.3	11.3
Interactions:	*uu*	0	2	2	2	4	4	4	6
	uuu	0	1	0	0	2	3	2	6

(c) (d)

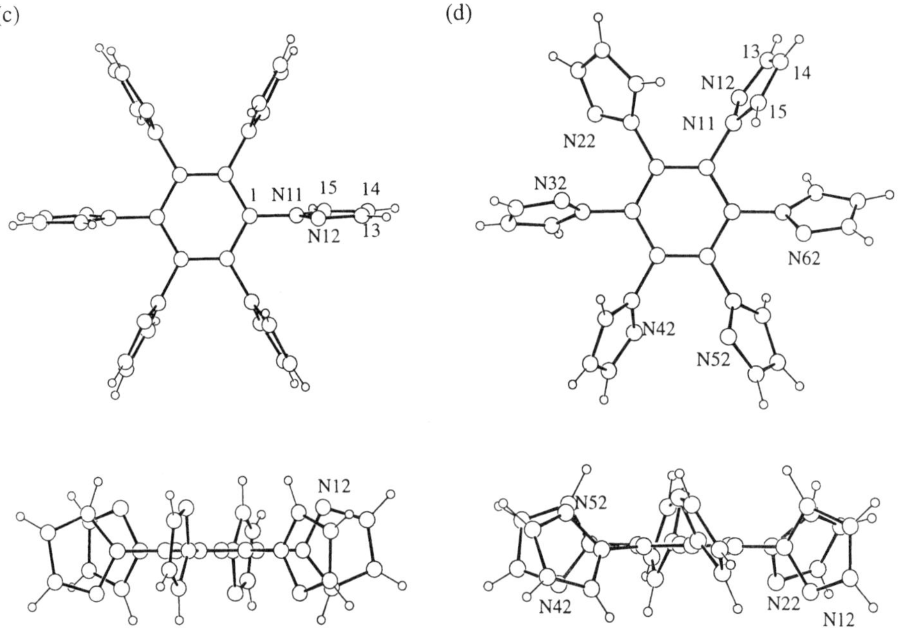

Figure 19: Semiempirical calculations. (a) Possible isomers in compounds **14** and **15**. (b) Differences in heats of formation, δΔH, in kcal mol^{-1} and number of interactions (*uu*, *uuu*), corresponding to the local minima. (c-d) Two views of the molecular structure of **14a** and **14b**.

In all the crystal structures of these two series the nitrogen lone pair is located alternatively in one side (*u*) and the other (*d*) of the benzene ring showing torsion angles close to ±90° in the methyl substituted rings and close to ±60° in the non-methyl substituted ones. The lower torsion angles presented by the pyrazole rings are sustained by (C(x5)-H(x5)···N(22) intramolecular interactions with distances shorter than the sum of van der Waals radii, 3.28Å, in **16a**, 3.191(4) and 3.238(4) Å, for x = 1,3 in molecule *1* and 3.196(4) Å for x = 1 in molecule *2*, while in **16c** this value is 3.203(3) Å). In type d (labelled as **17**), in general, the most perpendicular 3,5-dimethylpyrazole rings are those involved in the host-guest interaction.

There are 14 conformers (Figure 20a), plus one additional enantiomer for each situation that presents a chiral symmetry point group, either C_1 (*3, 4, 10, 12*) or C_2 (*5, 6*) regardless of the pyrazole ring orientation.[18,23] The differences in energy relative to the absolute minimum together with the number of interactions (*uu* or *dd* and *uuu* or *ddd*, that means two or three adjacent pyrazole rings with the lone pair on the same side of the central ring) are gathered in Figure 20b. The results show that the energy increases with the number of interactions. For type c, all conformers with 2 interactions are approximately 2.6 kcal mol^{-1} less stable than *1*, and among those which present 4 interactions, the ones with 3 contiguous and equally oriented pyrazoles are the most unstable. For type d, the most stable conformers, with energies just above the absolute minimum, are *5* and *6*, that present the fewer number of interactions. They are the only ones that have one pyrazole and one dimethylpyrazole ring (dmpzs) contiguous and with the lone pairs on the same side of the benzene ring and that do not have two contiguous dmpzs. The remaining 4 conformers with 2 interactions have 3 contiguous rings with the same orientation (*2, 3*) or only 2, but in this case both are 3,5-dimethylpyrazoles (*4, 7*), *7* being the least stable of all of them, having two pairs of equally oriented dimethylpyrazoles.

Two different conformations were obtained for the situation *1* of type c (*1* and *1'*), Figure 20c. They are only 0.7 kcal mol^{-1} different in energy and present the same point group symmetry C_i. The most stable displays torsion angles similar to those found in the crystal structures. For type d there are also 2 minima with different point groups (C_i and C_2), but with the same energy, being the first one the conformation observed in the solid state. Taking as starting points the geometry of the two minima in type d, two opposite 3,5-dimethylpyrazole rings have been protonated and the geometry of the resulting cation, present in the picrate salt, has been optimized. Both cases converge to the same centrosymmetric conformation which present torsion angles similar to those found in the crystal structure, being the protonated rings almost perpendicular to the central one.

6.5. Conformational isomerism of pentakispyrazolylpyridines

Analogously to the previous compounds, the possible conformers are shown in Figure 21a together with their interconversion pathways. If the torsion angles are different from 90°, it is possible to consider three situations (***A-C*** in Figure 21b) for each of the conformations.

In all cases, the most stable conformer is *1A* followed by *4A* that implicates a change of *up-down* disposition of the pyrazole ring in position 2. In the solid state we have observed two conformations, labelled as *1A* (compounds **19** and **20**) and *4C* (compound **18**) that is 2.1 kcal mol^{-1} less stable than the absolute minimum. In most caes we have observed good agreement between the observed and calculated torsion angles, although, the calculated values indicate that the pyrazole rings are more perpendicular to the central ring than the experimental ones.

(a)

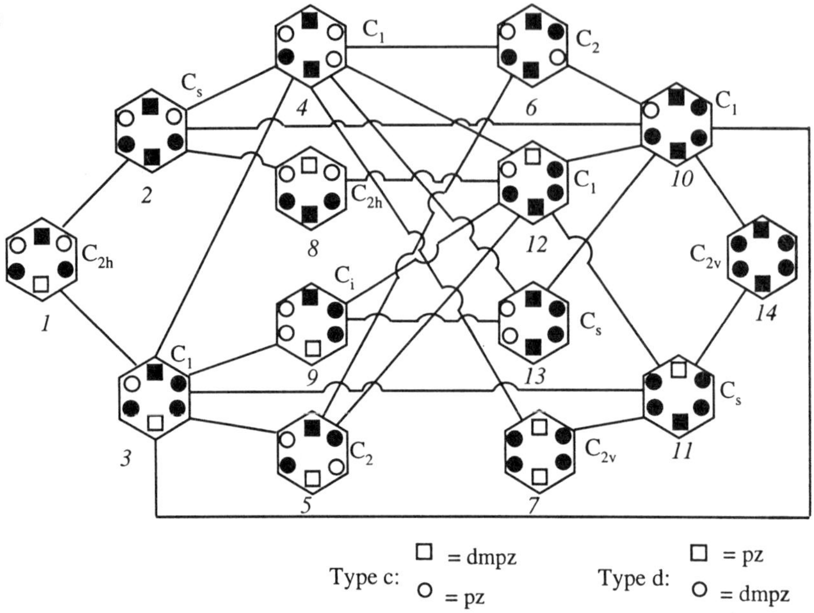

(b)

		1	*2*	*3*	*4*	*5*	*6*	*7*	*8*	*9*	*10*	*11*	*12*	*13*	*14*
δΔH	Type c:	0.0	2.8	2.5	2.6	2.7	3.0	2.9	4.7	4.5	5.7	5.3	5.3	5.3	9.2
	Type d:	0.0	2.3	2.3	1.8	1.2	1.3	3.0	4.6	3.8	5.1	6.2	4.5	4.1	9.5
Interactions:	*uu*	0	2	2	2	2	2	2	4	4	4	4	4	4	6
	uuu	0	1	1	0	0	0	0	2	2	3	3	2	2	6

(c)

	Type c		Type d		Type d
	1	*1'*	*1*	*1'*	1^{2+}
ΔH	508.7	509.4	480.1	480.1	820.1
δΔH	0.0	0.7	0.0	0.0	-
P. g.	C_i	C_i	C_2	C_i	C_i

Figure 20: Semiempirical calculations. (a) Possible isomers in compounds **16** and **17**. (b) Differences in heats of formation, δΔH, in kcal mol^{-1} and number of interactions (*uu*, *uuu*), corresponding to the local minima. (c) Heats of formation ΔH, differences in heats of formation, δΔH, in kcal mol^{-1} and point groups.

6.6. Transition states

The energies of the transition states, TS in Figure 22a, for the first transition step between conformers have been computed for all benzene derivatives. This implies to arrange one ring of pyrazole or 3,5-dimethylpyrazole coplanar with the central benzene ring. The transition states that involve 3,5-dimethylpyrazole rings are higher in energy and present larger deformations in the geometry than those where the pyrazole ring is rotated, probably due to the higher degree of steric hindrance.

6.7. AM1 calculations and NMR studies in solution

The stability of all possible conformers of hexakis(pyrazol-1'-yl)benzene [$(pz)_6bz$] **14**, hexakis(3',5'-dimethylpyrazol-1'-yl)benzene [$(dmpz)_6bz$] **15**, 1,2,4,5-tetrakis(pyrazol-1'-yl)-3,6-bis(3",5"-dimethylpyrazol-1"-yl) benzene [$(pz)_4(dmpz)_2bz$] **16** and 1,4-bis(pyrazol-1'-yl)-2,3,5,6-tetrakis(3",5"-dimethyl-pyrazol-1"-yl) benzene [$(pz)_2(dmpz)_4bz$] **17** has been determined by semiempirical AM1 calculations. In all series the absolute minimum corresponds to the conformation where the nitrogen lone pair of consecutive pyrazole rings is alternatively oriented to opposite sides of the benzene plane. The case of $(pz)_2(dmpz)_4bz$ **17** has been studied by ^1H NMR in solution. Three isomers have been identified and all the signals assigned. The slow evolution of the isomer found in the solid state to other isomers has been followed by ^1H NMR (classical kinetic) thus allowing to determine the barrier to rotation of a (dmpz) substituent ($\Delta G^\ddagger = 21\text{-}22$ kcal mol^{-1}).

The most important conclusion of the AM1 calculations for the NMR studies in solution is that these compounds present two types of barriers (see the Tables in Figs. 19 and 20 as well as Figs. 22 and 23). The first one corresponds to the rotation of a pz ring (represented in Fig. 23 with a grey circle) and has an activation barrier of 11.2-12.6 kcal mol^{-1}. The second one corresponds to the rotation of a dmpz (represented in Fig. 23 with a square, black for *up* and white for *down*) to which corresponds a barrier of 21.2-23.1 kcal mol^{-1}. The first barrier is low enough to observe in NMR narrow average signals in the range of usual temperatures; for this reason we have used a grey representation since by NMR it cannot be determined if the pz rings are *up* or *down*. The second one is large enough to observe in NMR, narrow signals for each isomer and this for the usual NMR temperatures. Thus, in the case of compound **17**, the actual conformational space involves only the "average" conformations of lower energy: (**1**, **2**, **8**)/(**3**, **4**, **10**, **12**)/(**5**, **6**) which are found at 63-14-23 % at the equilibrium (several hours at room temperature in $CDCl_3$). The AM1 calculations reproduce fairly well this result (70-5-25 %) considering that solvent effects have been neglected.

In a first approximation, the six barriers reported in the Tables of Figs. 19 and 20 can be decomposed into a sum of two terms, one for each adjacent pyrazole. For pz rotation the contributions are pz 6.1, dmpz 5.4 kcal mol^{-1} and for a dmpz rotation, they are 11.6 and 10.6 kcal mol^{-1}, roughly twice. Thus, an adjacent pz appears to produce more steric hindrance than a dmpz, which may seem paradoxical, but one has to consider the structure of the transition states (see Fig. 22) in which the 3,5-substituents of the adjacent rings are far away from the contact zone in the TSs.

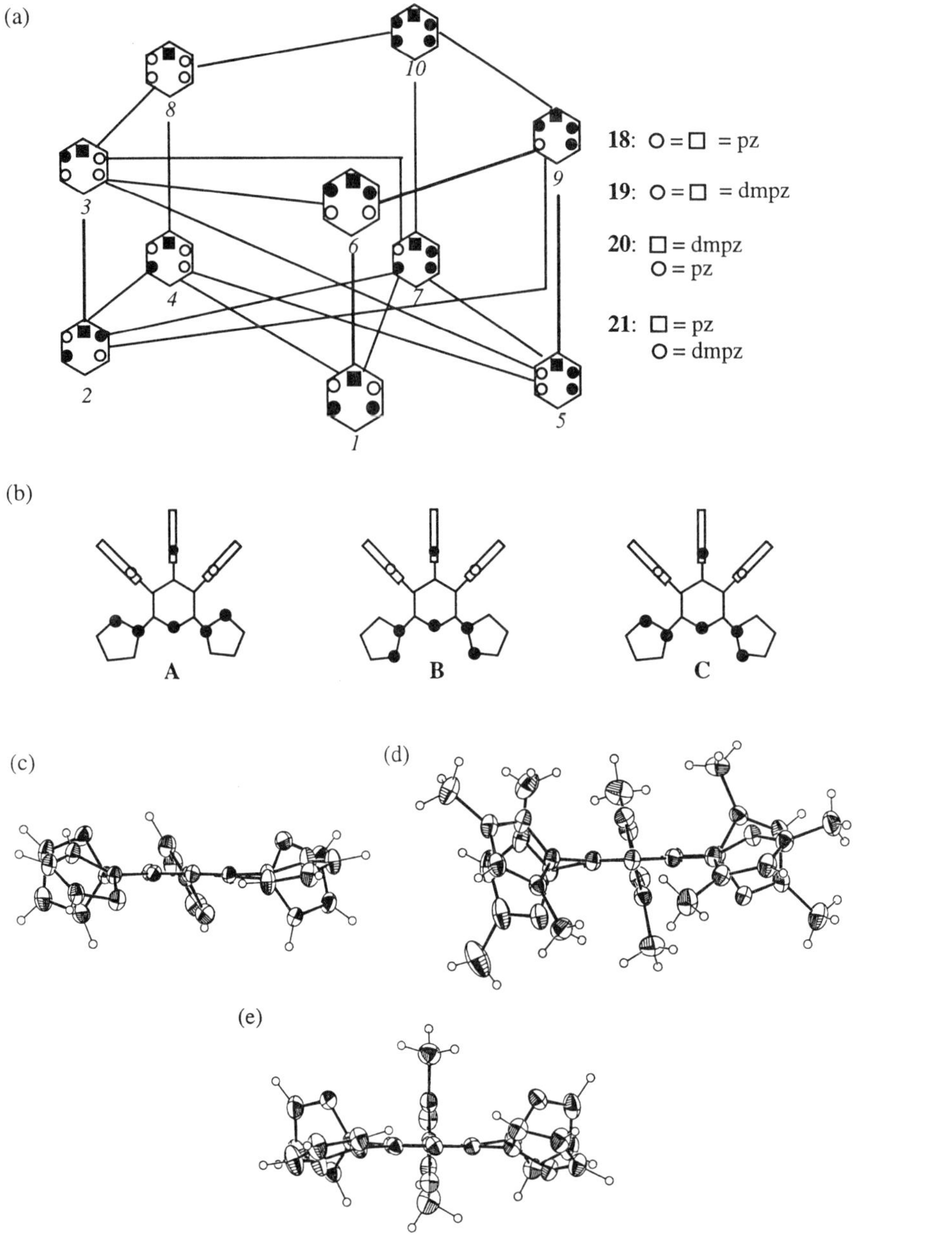

Figure 21: Semiempirical calculations. (a-b) Possible isomers in the pyridine derivatives. (c-e) Parallel view to the pyridine ring of **18**, **19** and **20** showing the molecular conformation.

(a)

TS	δΔH	TS	δΔH	TS	δΔH
1a to *2a*	12.6	*1c* to *2c*	23.1	*1d* to *2d*	11.2
1b to *2b*	21.2	*1c* to *3c*	11.2	*1d* to *3d*	22.1

(b)

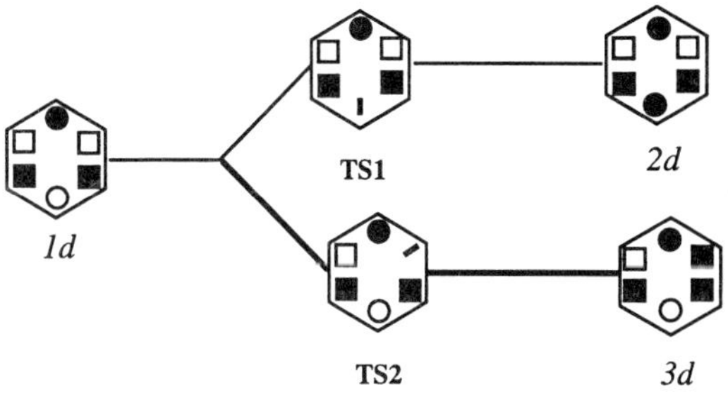

(c) (d)

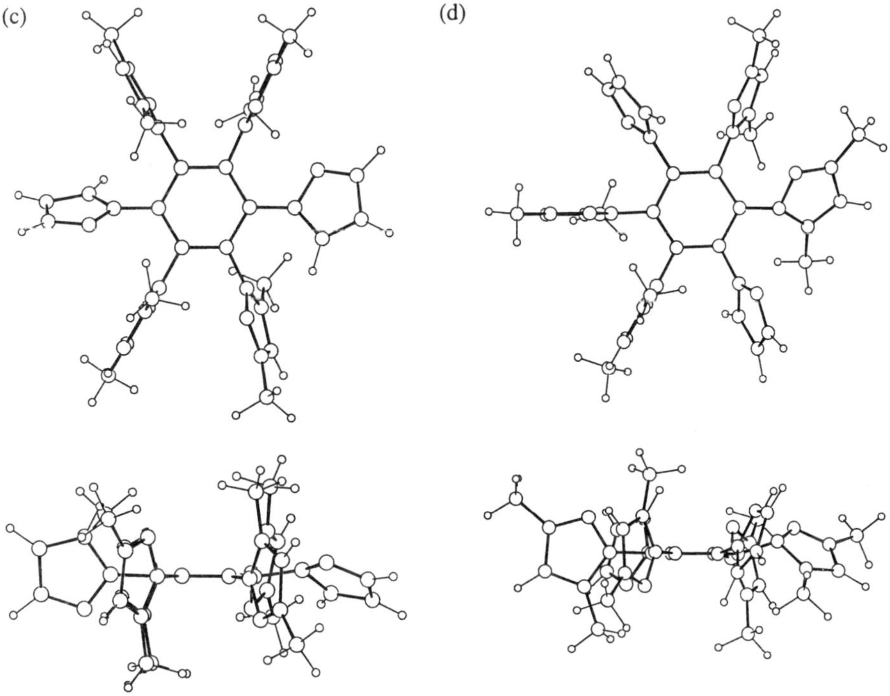

Figure 22: Transition states. (a) δΔH (kcal mol^{-1}) in the four series of compounds. (b) Schematic representation. (b-c) Two views of **17** in the transition states from *1d* to *2d* and from *1d* to *3d*, corresponding to a rotation of a 3,5-dimethylpyrazole and a pyrazole ring respectively.

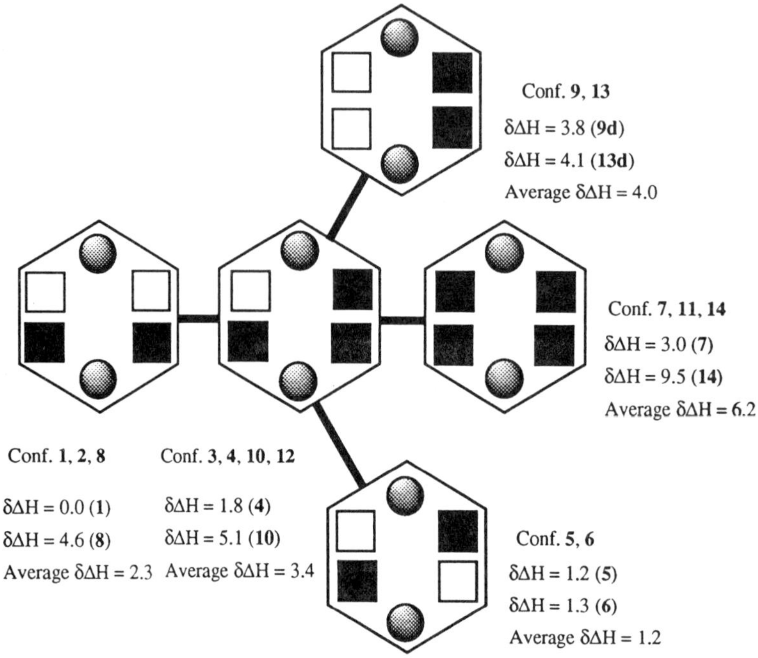

$[(pz)_2(dmpz)_4bz]$ **17**
(see, Fig. 20)

Figure 23

Acknowledgments

Financial support was provided by the Spanish DGICYT (PB93-0125 and PB93-0197-C02). C. F-C. and C. E. thank to the Comunidad de Madrid for a fellowship (Acción Especial Nº AE00067/95). We are grateful to Bruker Analytische Messtechnik GmbH for some ^{13}C CPMAS NMR spectra and to Dr. Pilar Goya for carefully reading this manuscript.

References

1. Nickon, A.; Silversmith, E. F. *Organic Chemistry: The Name Game*, Pergamon Press: New York, **1987**, a) p. 30, b) p. 33, c) p. 35.
2. Eliel, E. L.; Wilen, S. H. *Stereochemistry of Organic Compounds*, John Wiley & Sons: New York, **1994**, p. 794.
3. MacNicol, D. D. *Inclusion Compounds* (Atwood, J. L., Davies, J. E.; MacNicol, D. D. Eds.), Academic Press: London, **1984**, Vol. *2*, p. 123. MacNicol, D. D.; Downing, G. A. in *Comprehensive Supramolecular Chemistry*, Pergamon Press: New York, **1996**, Vol. *6*, Chapter 14, p. 421.

4. Vögtle F.; Weber, E. *Angew. Chem. Int. Ed. Engl.* **1974**,*13*, 814. Weber, E. in *Comprehensive Supramolecular Chemistry*, Pergamon Press: New York, **1996**, Vol. *6*, Chapter 17, p. 535.
5. Bourne, S. A.; Caira, M. R.; Nassimbeni, L. R.; Sakamoto, M.; Tanaka, K.; Toda, F. *J. Chem. Soc. Perkin Trans. 2* **1994**, 1899.
6. Pepermans, H.; Gielen, M.; Hoogzand, C.; Willem, R. *Bull. Soc. Chim. Belg.* **1983**,*92*, 465.
7. Pepermans, H.; Willem, R.; Gielen, M.; Hoogzand, C. *Magn. Reson. Chem.* **1988**, *26*, 311.
8. Keegstra, M. A.; De Feyter, S.; De Schryver, F. C.; Müllen, K. *Angew. Chem. Int. Ed. Engl.* **1996**, *35*, 774.
9. Henrie II, R. N.; Yeager, W. H. *Heterocycles* **1993**, *35*, 415.
10. Weiss, R.; Pomrehn, B.; Hampel, F.; Bauer, W. *Angew. Chem. Int. Ed. Engl.* **1995**, *34*, 1319.
11. Biemans, H. A. M.; Zhang, C.; Smith, P.; Kooijman, H.; Smeets, W. J. J.; Spek, A. L.; Meijer, E. W. *J. Org. Chem.* **1996**, *61*, 9012.
12. Columbus, I.; Biali, S. E. *J. Org. Chem.* **1994**, *59*, 3402.
13. Marx, H.-W.; Moulines, F.; Wagner, T.; Astruc, D. *Angew. Chem. Int. Ed. Engl.* **1996**, *35*, 1701.
14. Diercks, R.; Armstrong, J. C.; Boese, R.; Vollhardt, K. P. C. *Angew. Chem. Int. Ed. Engl.* **1986**, *25*, 268. Boese, R.; Green, J. R.; Mittendorf, J.; Mohler, D. L.; Vollhardt, K. P. C. *Angew. Chem. Int. Ed. Engl.* **1992**, *31*, 1643.
15. Downing, G. A.; Frampton, C. S.; Gall, J. H.; MacNicol, D. D. *Angew. Chem. Int. Ed. Engl.* **1996**, *35*, 1547.
16. Foces-Foces, C.; Llamas-Saiz, A. L.; Claramunt, R. M.; Jagerovic, N.; Jimeno, M. L.; Elguero, J. *J. Chem. Soc. Perkin Trans.* **1995**, *2* 1359.
17. Foces-Foces, C.; Llamas-Saiz, A. L.; Escolástico, C.; Claramunt, R. M.; Elguero, J. *J. Phys. Org. Chem.* **1996**, *9*, 137.
18. Foces-Foces, C.; Fernández-Castaño, C.; Claramunt, R. M.; Escolástico, C.; Elguero, J. *J. Phys. Org. Chem.* **1996**, *9*, 717.
19. Cornago, P.; Escolástico, C.; Santa María, M. D.; Claramunt, R. M.; Fernández-Castaño, C.; Foces-Foces-C.; Fayet, J.-P.; Elguero, J. *Tetrahedron* **1996**, *52*, 11075.
20. Fernández-Castaño, C.; Foces-Foces, C.; Cano, F. H.; Claramunt, R. M.; Escolástico, C.; Fruchier, A.; Elguero, J. *New J. Chem.* **1997**, *21*, 195.
21. Fernández-Castaño, C.; Foces-Foces, C.; Claramunt, R. M.; Escolástico, C.; Elguero, J. *Tetrahedron*, **1997**, submitted.
22. Lavandera, J.-L.; Claramunt, R. M.; Escolástico, C.; Santa María, M. D.; Cornago, P.; Elguero, J. *Bull. Soc. Chim. Belg.* **1997**, submitted.
23. Foces-Foces, C.; Llamas-Saiz, A. L.; Fernández-Castaño, C.; Claramunt, R. M.; Escolástico, C.; Lavandera, J. L.; Jimeno, M. L.; Elguero, J. *J. Chem. Soc. Perkin Trans. 2*, **1997**, submitted.
24. Claramunt, R. M.; Escolástico, C.; Santa María, M. D.; Cornago, P.; Elguero, J. unpublished results.
25. Hartshorn, C. M.; Steel, P. J. *Aust. J. Chem.* **1995**, *48*, 1587.
26. Hartshorn, C. M.; Steel, P. J. *Angew. Chem. Int. Ed. Engl.* **1996**, *35*, 2655.
27. Chambers, R. D.; Sargent, C. R. *Adv. Heterocycl. Chem.* **1981**, *28*, 1.

28. Silvester, M. J. *Adv. Heterocycl. Chem.* **1994**, *59*, 1.
29. Allen, F. H.; Davies, J. E.; Galloy, J. J.; Johnson, O.; Kennard, O.; Macrae, C. F.; Mitchell, E. M.; Smith, J. M.; Watson, D. G. *J. Chem. Info. Comp. Sci.* **1991**, *31*, 187.
30. Domenicano, A.; Vaciago, A. *Acta Crystallogr. Sect. B* **1979**, *35*, 1382.
31. Norrestam, R.; Schepper, L. *Acta Chem. Scand.* **1981**, *35*, 91.
32. Domenicano, A.; Murray-Rust, P. *Tetrahedron Lett.* **1979**, *24*, 2283.
33. Domenicano, A.; Murray-Rust, P.; Vaciago, A. *Acta Crystallogr. Sect. B* **1983**, *39*, 457.
34. Povet`eva, Z. P.; Zvonkova, Z. V. *Kristallografiya* **1975**, *20*, 69.
35. Rossell, H. J.; Scott, H. G. *J. Cryst. Mol. Struct.* **1973**, *3*, 259.
36. Mootz, D.; Wussow, H.-G. *J. Chem. Phys.* **1981**, *75*, 1517.
37. Maas, G.; Fink, J.; Wingert, H.; Blatter, K.; Regitz, R. *Chem. Ber.* **1987**, *120*, 819.
38. Desiraju, G. *Angew. Chem. Int. Ed. Engl.* **1995**, *34*, 2311.
39. Perlstein, J.; Steppe, K.; Vaday, S.; Moses, E.; Ndip, N. *J. Am. Chem. Soc.* **1996**, *118*, 8433.
40. Jonsson, P.-G. *Acta Crystallogr. Sect. B* **1971**, *27*, 893.
41. Cano, F.H.; Martínez-Ripoll, M. *J. Mol. Struct. (Theochem)* **1992**, *258*, 139.
42. Williams, J. H; Cockcroft, J. K; Fitch, A. N. *Angew. Chem. Int. Ed. Engl.* **1992**, *31*, 1655.
43. Holden, J. R.; Du, Z.; Ammon, H. L. *J. Comp. Chem.* **1993**, *14*, 422 and references included.
44. Williams, D. E. *Acta Crystallogr. Sect. A* **1996**, *52*, 326 and references included.
45. Vainshtein, B. K.; Fridkin, V. M.; Indenbom, V. L. *Modern Crystallography II*, Springer-Verlag: Berlin, 1982; p. 87.
46. Hernández-Trujillo, J.; Colmenares, F.; Cuevas, G.; Costas, M. *Chem. Phys. Lett.* **1997**, *265*, 503.
47. Nyburg, S. C.; Faerman, C. H. *Acta Crystallogr. Sect B* **1985**, *41*, 274.
48. Ramasubbu, N.; Parthasarathy, R.; Murray-Rust, P. *J. Am. Chem. Soc.* **1986**, *108*, 4308.
49. Kirin, D. *Acta Crystallogr. Sect B* **1987**, *43*, 405.
50. Rowland, R. S.; Taylor, R. *J. Phys. Chem.* **1996**, *100*, 7384.
51. Bondi, A. *J. Phys. Chem.* **1964**, *68*, 441.
52. Bondi, A. *Physical Properties of Molecular Crystals, Liquids and Glasses.* John Wiley: New York, 1968; p. 450.
53. Escolástico, C.; Santa María, M. D.;, Claramunt, R. M.; Jimeno, M. L.; Alkorta, I.; Foces-Foces, C.; Hernández-Cano, F.; Elguero, J. *Tetrahedron* **1994**, *43*, 12489.
54. Aubagnac, J. L.; Gilles, I.; Claramunt, R. M.; Escolástico, C.; Sanz, D.; Elguero, J. *Rapid Commun. Mass Spectrom.* **1995**, *9*, 156 and unpublished results.
55. Mislow, K. *Acc. Chem. Res.* **1976**, *9*, 26.
56. Willem, R.; Gielen, M.; Hoogzand, C.; Pepermans, H. *The Internal Rotation Dynamics of Polyaryl Compounds in Solution* in *Advances in Dynamic Stereochemistry*; Gielen, M. F. Ed.; Freund Publishing House: London, 1985; Vol. 1, p. 207. Brocas, J.; Gielen, M.; Willem, R. *The Permutational Approach to Dynamic Stereochemistry*, McGraw-Hill, New York, 1983.
57. Stewart, J. J. P. *J. Comp.-Aided Mol. Des.* **1990**, *4*, 1.
58. Xu, K.; Ho, D. M.; Pascal Jr, R. A. *J. Org. Chem.* **1995**, *60*, 7186.

HETERO DIELS-ALDER REACTIONS IN AQUEOUS MEDIUM

Francesco Fringuelli, Oriana Piermatti and Ferdinando Pizzo
Dipartimento di Chimica, Università di Perugia, Via Elce di Sotto, 8
06123 Perugia, Italy

Abstract. *Hetero Diels-Alder reactions in aqueous medium between dienophiles having nitroso, carbonyl and iminium functionalities with various carbon dienes and between etherodienes, such as 2,3-diazadienes and oxa-aza-dienes, with carbon dienophiles are illustrated. Examples of retro Diels-Alder reactions in water are also presented and the role of the water is examined.*

1. Introduction

Since the pioneer work of Breslow [1] which showed that the Diels-Alder reaction between the poorly water-soluble cyclopentadiene and buten-2-one is greatly accelerated in aqueous medium, there has been an increasing interest in water as a reaction medium for organic reactions.[2]

Many organic reactions have been investigated [2] in aqueous medium and, to date, the low solubility of the reactants and the incompatibility of the intermediates with water are no longer considered as *a priori* obstacles for reactions carried out in water. On the contrary, the aqueous medium offers some of the following advantages with respect to organic solvent (a) inflammable and anhydrous solvents are not handled, (b) the protection-deprotection of functional groups such as reactive hydroxy or acidic groups may not be necessary, (c) the pH can be accurately controlled (d) additives, such as salts, surfactants and cyclodextrins, can be used, (e) the reactivity and selectivity of the reaction can be dramatically influenced, and (f) the process is more economical and environmental friendly.

The aqueous medium has proven to be especially advantageous for the [4π+2π] cycloadditions involving heterodienophiles and/or heterodienes (hetero Diels-Alder reaction). This review illustrates some significant and recent results obtained in this field. The inverse reaction (retro hetero Diels-Alder reaction) will also be reviewed.

2. Heterodienophiles

2.1. Nitroso Function

In principle, the water should not be a good reaction medium for reactions involving compounds having a nitroso function directly linked to an electron-withdrawing group such as in N-acylnitroso compounds **2**, because they are short-lived, extremely reactive and rapidly solvolysed. The nitroso moiety of the N-acylnitroso function is, however, a powerful dienophile, in according to very low LUMO energy and weak π bond,[3] and therefore N-acylnitroso compounds are trapped rapidly, especially in an intramolecular reaction, with a diene, allowing the Diels-Alder reaction to occur also in water.

N-Acylnitroso compounds **2** are generated *in situ* as transient intermediates by periodate oxidation of hydroxamic acids **1** and react thermally with 1,3 dienes to give N-acyl derivatives of 3,6-dihydro-1,2-oxazine **3**.

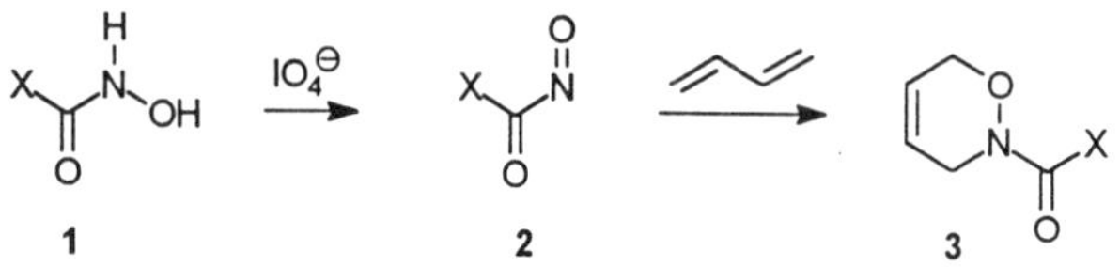

X = R, RO, RR'N

Hetero Diels-Alder cycloaddition with nitroso dienophiles is a subject of great interest [4] and Kibayashi and co-workers used the intramolecular nitroso Diels-Alder approach to prepare key chiral intermediates in the synthesis of naturally occurring alkaloids.[4-8]

When the reaction was carried out in $CHCl_3$ solution, the cycloaddition occured with low diastereoselectivity (Table 1). A significant enhancement in favour of *trans* adduct was observed when the Diels-Alder reaction was executed in water.

Table 1. Intramolecular Diels-Alder Reactions of N-Acylnitroso Compounds

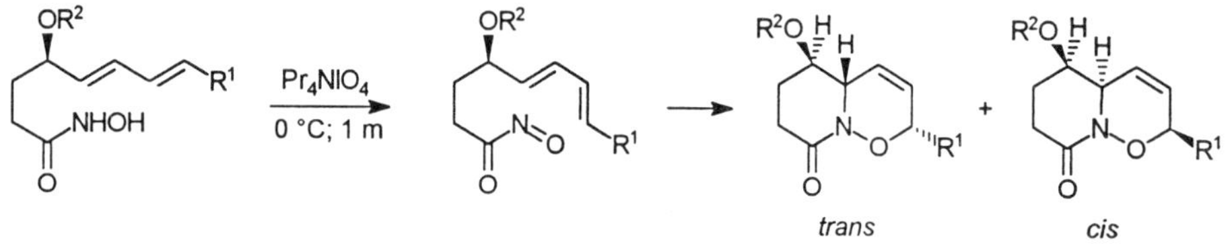

R^1	$R^{2\,a}$	Medium	trans/cis	Yield (%)
H	Bn	$CHCl_3$	54/46	76
H	Bn	H_2O	80/20	87
H	MOM	$CHCl_3$	63/37	75
H	MOM	H_2O	82/18	93
Et	MOM	$CHCl_3$	66/34	86
Et[b]	MOM	H_2O	83/17	83

[a] Bn = CH_2Ph; MOM = CH_2OMe. [b] Oxidation carried out with Bu_4NIO_4.

The periodate oxidation of 4-O-protected chiral hydroxamic acids occurs in heterogeneous phase at 0°C and the generated N-acylnitroso compounds immediately cyclize to 1,2-oxazinolactams (Table 1). The presence of cyclodextrins in the aqueous medium sometimes improves the reaction yield but does not enhance the diastereoselectivity of the cycloaddition. Mixtures of H_2O-DMSO and H_2O-MeOH were used with hydroxamic acids sparingly soluble in water alone.

The preference for the *endo-trans* adduct in aqueous medium was attributed to the hydrophobic packing effect [9] which stabilizes the compact boat-like transition state **5** more than the chair-like one **4**.

4 **5**

Scheme 1 illustrates the enantioselective approach to (–)-swainsonine [6] and (–)-pumiliotoxin C [8] in which the aqueous intramolecular Diels-Alder reactions of chiral N-acylnitroso dienes have been employed as key reactions.

Scheme 1

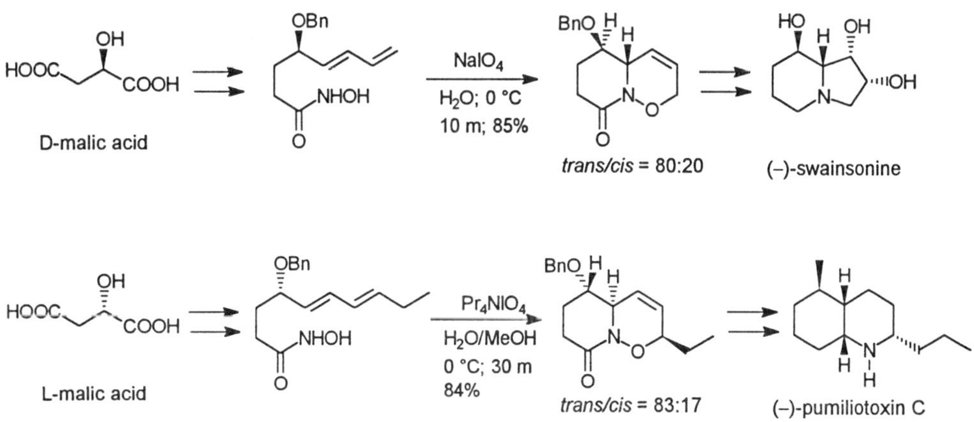

2.2. Carbonyl Function

The carbonyl function, used as dienophile, allows dihydropiran structures, present in many natural compounds, to be sinthesized by Diels-Alder reaction. Generally, [4+2] cycloadditions in organic solvent must be catalysed by Lewis acids or require high pressure or high temperature and the carbonyl function must be activated by electron-withdrawing groups in α-position.[10]

α-Activated carbonyl compounds, such as glyoxal **6** ($R^1 = R^2 = H$), glyoxylic acid **6** ($R^1 = H$; $R^2 = OH$), pyruvic acid **6** ($R^1 = Me$; $R^2 = OH$) and pyruvaldehyde **6** ($R^1 = H$; $R^2 = Me$) are for the most part in the hydrated form in water and the Diels-Alder cycloaddition should, in principle, be compromised because of the very low concentration of the reactive specie. In contrast, Lubineau and co-workers [11,12] have shown that these compounds give [4+2] heterocycloadditions in water with reactive and less reactive dienes. The results of reactions with 2-methyl pentadiene are reported in Table 2. The yields of the reactions are good with the exception of the reaction with glyoxal and the ratio of the adducts **7** and **8** reflects the thermodynamic control of the reaction. In organic solvent the cycloadditions occur under kinetic control with reversed diastereoselectivity.

Table 2. Synthesis of Dihydropyrans by Aqueous Diels-Alder Reactions

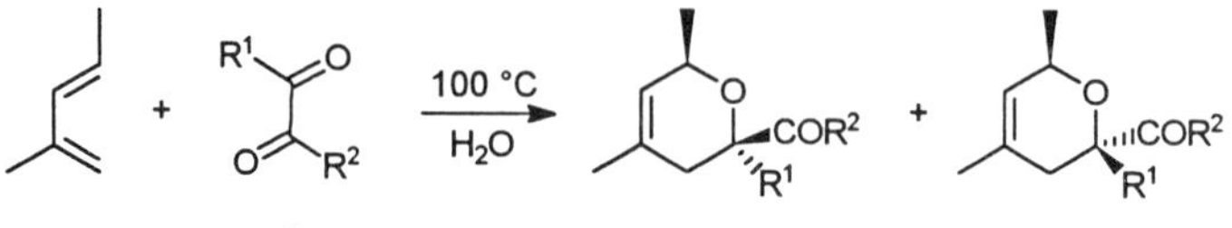

6 7 8

R^1	R^2	t(h)	**7**/**8**	Yield (%)
H	OH	1.5	64/36	97
H	Me	48	53/47	96
H	H	60	50/50	36
Me	OH	48	33/67	74

The reaction of cheap, commercially available glyoxylic acid aqueous solution with the reactive cyclopentadiene gives a mixture of α-hydroxy-γ-lactones **9** and **10** (n = 1, Scheme 2). The reaction yield and the ratio of lactones depend greatly on the pH of the medium. The best result is obtained at 40 °C and pH = 0.9. It is assumed that the reaction occurs through the Diels-Alder adduct which spontaneously rearranges to **9** and **10** (n = 1).

Scheme 2

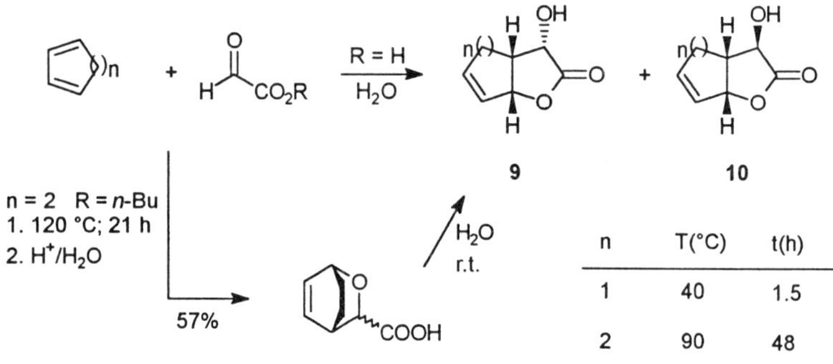

11 (*exo/endo* = 1:9)

n	T(°C)	t(h)	**9/10**	Yield (%)
1	40	1.5	73/27	83
2	90	48	60/40	85

Similarly the cycloaddition of cyclohexadiene gives lactones **9** and **10** (n = 2, Scheme 2). In this case the origin of these lactones from the Diels-Alder adduct **11** has been proven because the acid **11**, prepared by Diels-Alder reaction in a sealed tube by using butyl glyoxylate and subsequent hydrolysis, spontaneously rearranges to **9** and **10** (n = 2) in water at room temperature.

It should be noted that, in the absence of solvent the rearrangement occurs at 150 °C. These results have been used to synthesize naturally occurring compounds.[13]

The enantioselective synthesis of 3-deoxy-D-manno-2-octulosonic acid (KDO), an essential constituent of the outer membrane lipopolysaccharide of all gram-negative bacteria was carried out [13a] (Scheme 3) taking advantage of the heterocyclization of diene **12** with sodium glyoxylate **13** and butyl glyoxylate **14** in water and neat conditions respectively. The stereospecific dihydroxylation of newly created double bound is the second key step of the sequence that leds to KDO.

Scheme 3

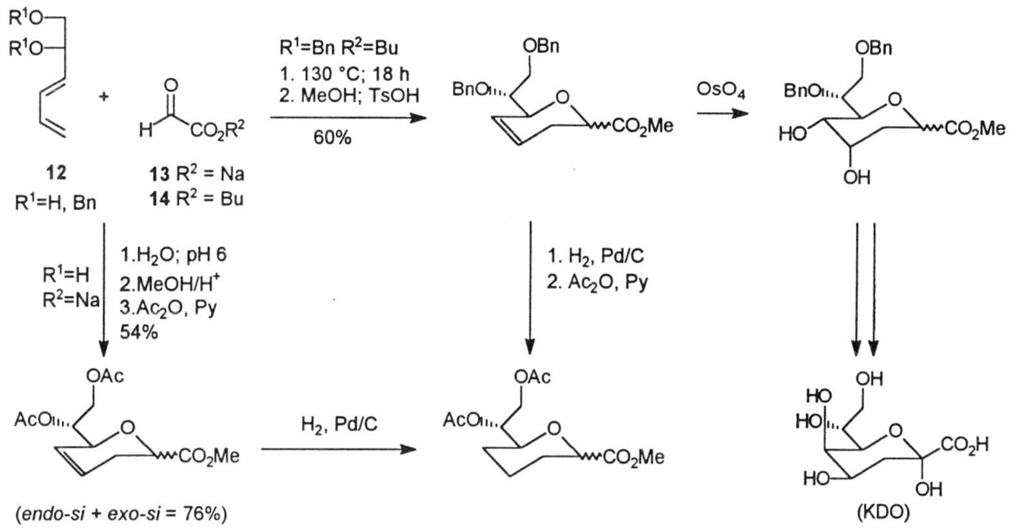

Cycloaddition of (E)-penta-2,4-dienol **15** with sodium glyoxylate **13** (Scheme 4) in water provided the starting material for a concise synthesis [13b] of ketodeoxyheptulosonic acid derivatives **16** which are compounds of relevant biological importance.

Scheme 4

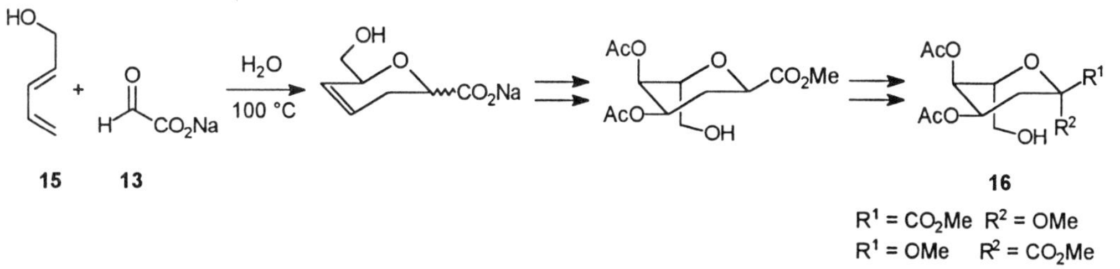

The main bicyclic lactone **9** (n = 1) coming from the cycloaddition of cyclopentadiene with glyoxylic acid in water, was used as starting material for a multi-step synthesis of (±)-sesbanimide A and B which are potent natural antitumoral compounds.[14]

It has recently been found that lanthanide complexes [15] have many applications in organic synthesis as water-tolerant Lewis acids [2e,16]. The chiral yttrium bis-trifluoromethanesulfonylamide (yttrium bis-trifylamide complex, Y-BTA) prepared by reaction of yttrium III triflate and (+)-1,2-diphenylethylene-bis-trifylamide, is an efficient catalyst for the asymmetric hetero Diels-Alder reaction of butyl glyoxylate **18** with Danishefsky diene **17**. By using sodium hydride as a base and water as additive (11 equiv. with respect to the chiral complex) to the toluene solution of Y-BTA, the enantioselectivity and the yield of the reaction are increased considerably (Scheme 5).

Scheme 5

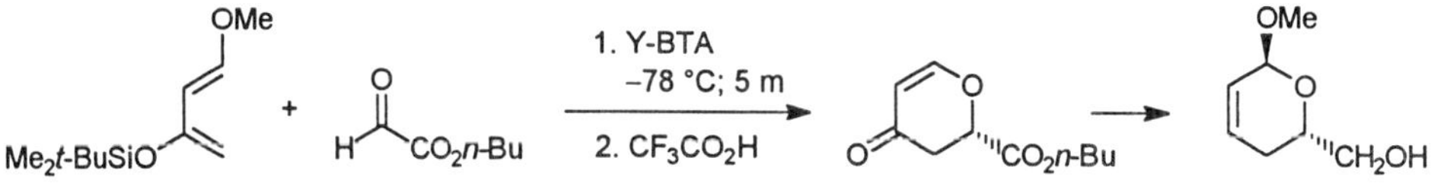

17 18 19 20

Solvent	Yield (%)	*ee* (%)
CH_2Cl_2	54	43
PhMe	74	54
PhMe-H_2O	88	66

The dihydropyrone ester **19** is easily converted to hydroxymethyl-dihydropyran **20** which is a useful intermediate for the synthesis of monosaccharides and other natural products, such as mevinolin and compactin.[17]

2.3. Imino Function

The imino function is a very useful dienophile for constructing key compounds for the synthesis of alkaloids by Diels-Alder reaction. Simple imines however are poor dienophiles and must be activated by protonation or by attaching an electronwithdrawing group to the nitrogen atom.

Grieco and Larsen [18] have found that benzyliminium ion **21**, generated from an aqueous solution of benzylamine hydrochloride and 37% aqueous formaldehyde, reacts with cyclopentadiene under heterogeneous conditions to give the bicyclic amine **22** in quantitative yield.

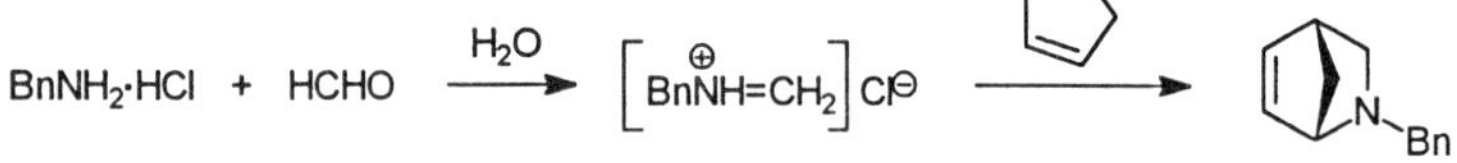

21 22

Less reactive dienes require longer reaction times and higher temperatures (Table 3). Methylamine hydrochloride and ammonium chloride have also been successfully employed.

Table 3. Aza Diels-Alder Reactions of Benzyliminium Ion **21** with Various Dienes

Diene	T(°C)	t(h)	Yield of adduct (%)
Cyclopentadiene	25	3	100
Cyclohexadiene	55	42	35
2,3-Me_2-1,3-butadiene	35	48	64
Isoprene	35	70	59
1,4- Me_2-1,3-butadiene	55	96	62
1,3- Me_2-1,3-butadiene	25	23	45

By using acetaldehyde, the reaction of benzyliminium ion **23** with cyclopentadiene affords a 47% isolated yield of a mixture of *exo* and *endo* adducts **24** and **25**.

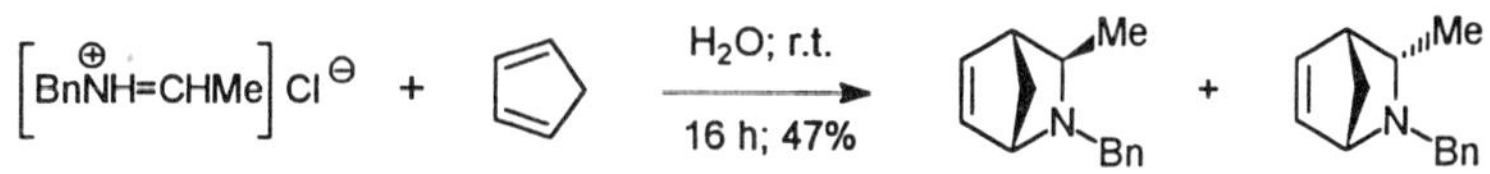

23 **24** **25**

The intramolecular iminium ion Diels-Alder reaction has also been investigated [18] examining both the cycloaddition *via* dienyl amines **26** and the cycloaddition *via* dienyl aldehyde **27** and **28** (Scheme 6). The procedure [19] has been used to synthesize the racemic dihydrocannivonine **29** (Scheme 6).

Scheme 6

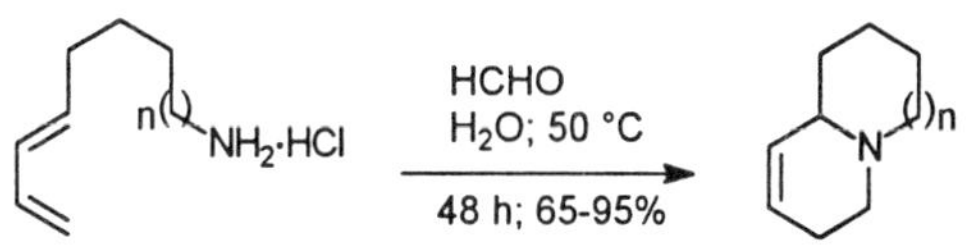

26 (n = 0,1)

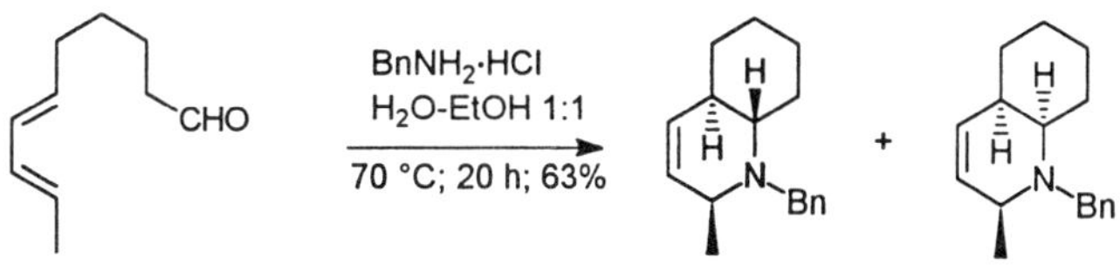

27 (2.5 : 1)

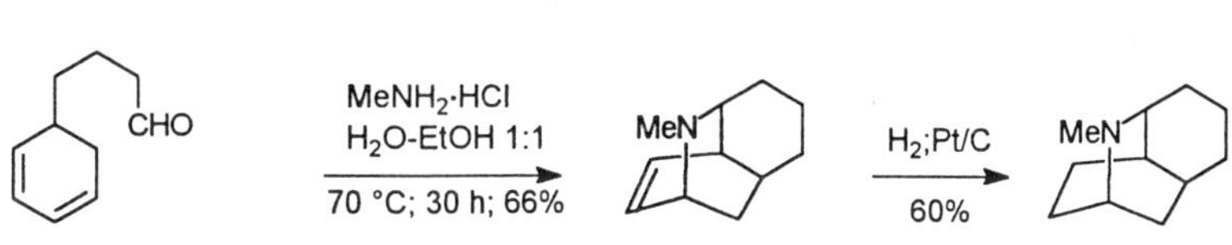

28 **29**

Iminium ions **30** generated from phenylglyoxal (R = Ph) or pyruvic aldehyde (R = Me) with benzylamine (R^1 = Bn) or methylamine (R^1 = Me) hydrochlorides and pyruvyl iminium itself **30** (R = Me, R^1 = H), generated from ammonium chloride and pyruvic aldehyde, are very reactive dienophile and react with cyclopentadiene in water at room temperature giving bicycloamines **31** and **32** with good *exo/endo* diastereoselectivity and good yield (Table 4).[20]

If glyoxylic acid is used, the formation of iminium salt **30** (R = $O^{\ominus}$, R^1 = Me) does not require the amine hydrochloride, but the free amine.

The Aza Diels-Alder cycloadditions with iminiun ions in aqueous medium are already reversible at room temperature and therefore the *exo/endo* ratio observed may not represent the kinetic but rather the thermodynamic diasteroselectivities.[21-23]

Table 4. Aza Diels-Alder Reactions of Glyoxyl Iminium Ions with Cyclopentadiene

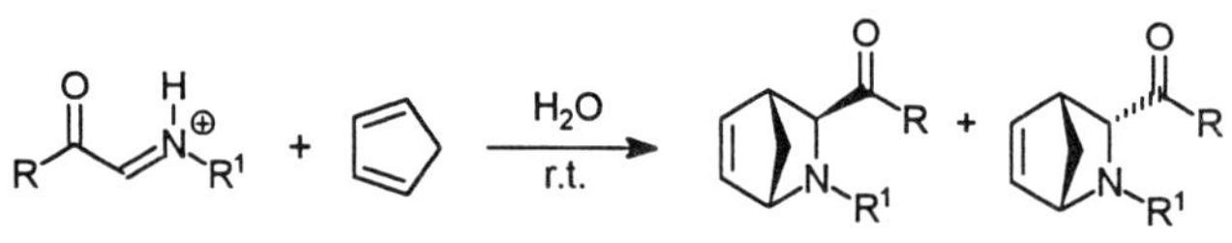

R	R^1	t(h)	**31/32**	Yield (%)
Ph	Me	22	4.2/1	82
Me	H	22	1/2	84
Me	Me	20	3.6/1	67
Me	Bn	18	10/1	65
OH	Me	22	1.9/1	86

The low reactivity of iminium ions generated from benzylamine hydrochloride and alkyl and aryl aldehydes and the low yield of the heterocycloaddition in aqueous mediun are increased considerably by catalyzing the reaction with lanthanide III trifluoromethanesulfonates [$Ln(OTf)_3$, triflates].[24] Some examples are reported in Table 5.

$Pr(OTf)_3$ and $Yb(OTf)_3$ are the most effective catalysts and, comparing the catalyzed with uncatalyzed reactions, the $Ln(OTf)_3$ do not change the *exo/endo* ratio of the products.

The interest in asymmetric Diels-Alder reaction has prompted to explore the chiral induction by using optically active iminium ion.

An aqueous solution of (-)-α-methyl benzylamine hydrochloride and 37% aqueous formaldehyde reacts with cyclopentadiene to give a 4:1 mixture of separable diastereoisomers in 86% yield after 20 h at 0 °C.[18]

Iminium ions **33** derived from (R)- and (S)-aminoacid methyl ester hydrochlorides react with cyclic and open-chain dienes in the presence of formaldehyde in aqueous THF (Scheme 7) to form heterocyclic amines in good yield.[21] The cycloadditions in water only [22] are faster (0.5-2 h) and occur with higher yields. The reactions can even be carried out on a 30 g scale and the diastereoisomers are easily separated by chromatography.[21] Removing the amino acid side chain, the adducts are converted into enantiomerically pure azabicyclo derivatives and piperidines (Scheme 7).

Table 5. Aqueous Aza Diels-Alder Reactions of Alkyl and Arylbenzyliminium Ions with Cyclopentadiene Catalyzed by $Ln(OTf)_3$

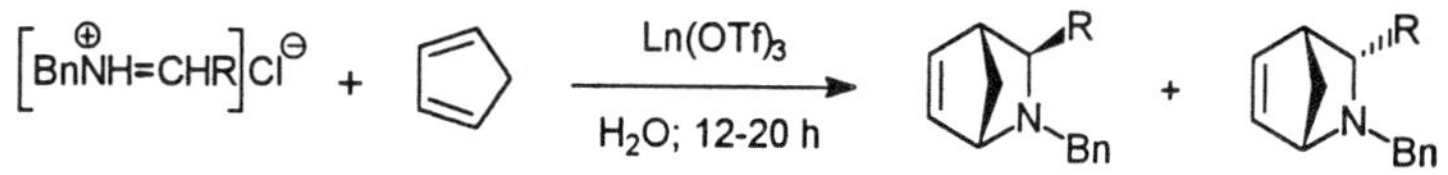

R	Ln	exo/endo	Yield (%)[a]	
$Me(CH_2)_4$	Pr	2.9/1	68	7
$MeCH_2$	La	2.5/1	64	4
Ph CH_2	Yb	4/1	72	3
Ph	Yb	>99/1	7	0

[a] The second column reports the yield for the reaction executed without lanthanide.

Scheme 7

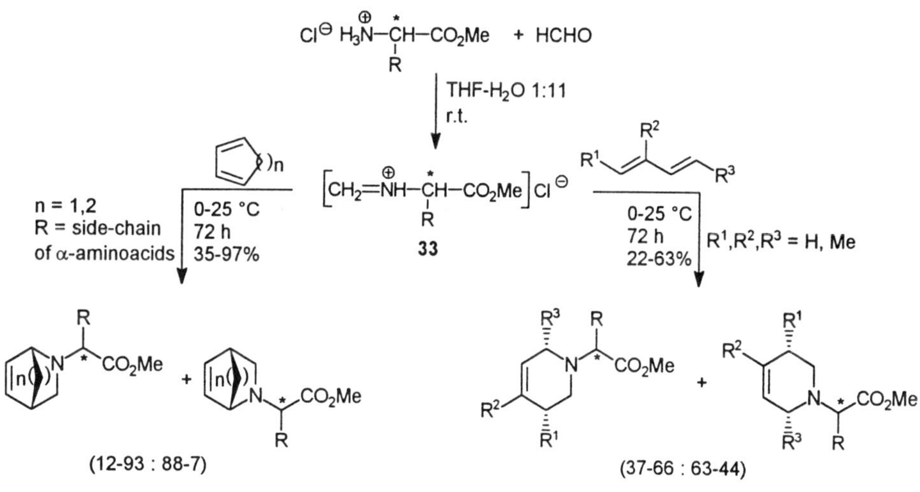

Similarly aqueous solutions of hydrochloride salts of small peptide methyl esters (dipeptide, tripeptide) expose to 37% aqueous formaldehyde and cyclopentadiene give, after 1-2h at room temperature, the expected 2-azanorbornene derivative as a mixture of diasteroisomers.[22] The adducts give an easy retro Diels-Alder reaction under non aqueous conditions, allowing a simple method for the N-methylation of peptides.[22]

Scheme 8 illustrates the reaction of preparation of N-L-leucine-L-phenylalanine-2-azanorbornenes and the heterocycloreversion to N-methyl-L-leucine-L-phenylalanine.

Scheme 8

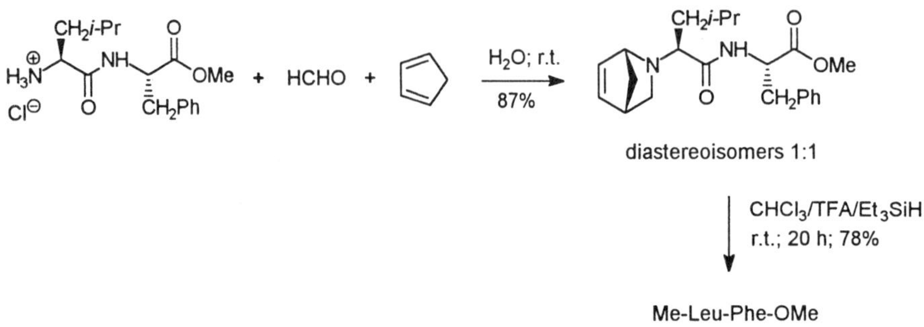

Cycloadditions in aqueous medium between the iminium ions generated from methyl glyoxylate and (S)-valine and (S)-isoleucine methyl ester hydrochlorides and cyclopentadiene fail, while cycloadducts are obtained [25] with iminiun ions **35** formed *in situ* with methyl and ethyl glyoxylate and (S)-(+)-alanine methyl ester hydrochloride **34** [R* = (S)-$MeCHCO_2Me$] and (R)-(+)-α-methylbenzylamine hydrochloride **34** [R* = (R)-PhCHMe] (Scheme 9).[25] The approach of the diene with *exo* orientation with respect to the glyoxylic acid ester group is always favoured.

Scheme 9

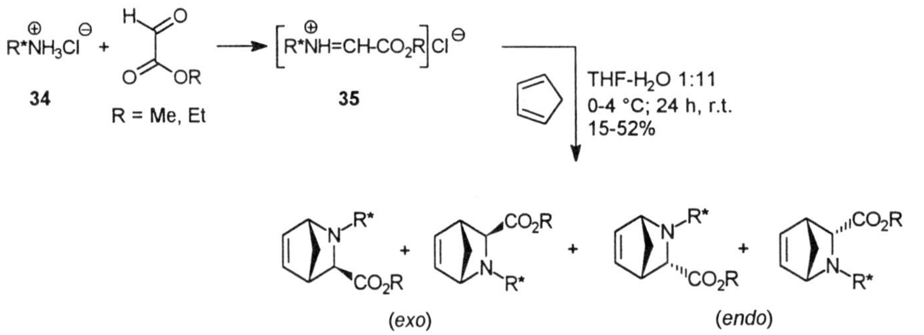

R*	R	exo/endo	exo *de* (%)
(R)-PhCHMe	Et	6/1	80
(S)-$MeCHCO_2Me$	Me	7/1	66

Bailey and co-workers [26-28] found that the reaction yield of iminium ion, prepared *in situ* from benzylamine hydrochloride and ethyl glyoxylate, with cyclopentadiene is strongly dependent on the solvent (Scheme 10). DMF gives the best results, but in water the yield of the reaction is higher than in THF and

EtOH. With in mind to improve the yield of the reaction, attempts were made to remove the water generated during the preparation *in situ* of the iminium ion but no reaction occurred under anhydrous conditions. In contrast, by isolating the imine and performing the reaction in the presence of TFA (1 equiv.) and with a catalytic amount of water, the adduct with 2,3-dimethylbutadiene was isolated in 92% yield (Scheme 10).[26,27] The presence of water is therefore essential for a successful reaction but in excess it reduces the reaction yield presumably because of the hydrolysis of the imine species.

Scheme 10

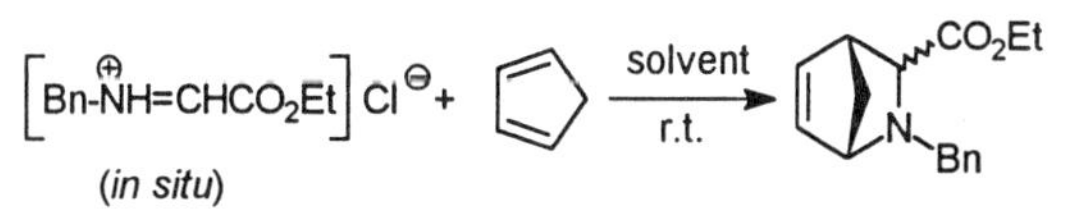

solvent	THF	EtOH	H_2O	DMF
Yield (%)	0	17	52	89

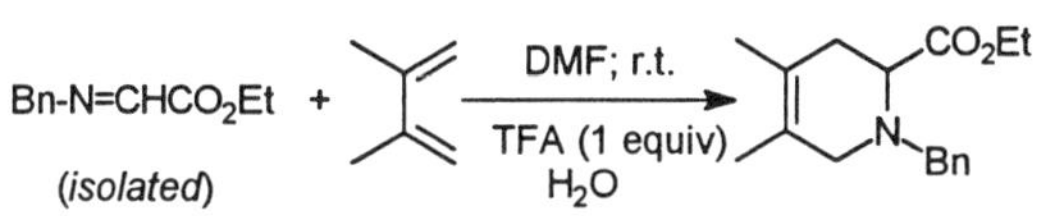

H_2O (mol%)	1	10	100	1000
Yield (%)	94	94	29	9

The procedure allows regiospecific and highly diasteroselective cycloadditions to be achieved and it has been used to the synthesize chiral bicyclo- and monocyclo-heterocylic amines (Tables 6 and 7) from ethyl glyoxylate and (R)-1-α-methylbenzylamine.[27,28]

The cycloaddition with isoprene allows a short general route to optically pure substituted pipecolic acid derivatives to be achieved.[27,28]

The role of water is not easy to explain. The water probably increases the π-character of the iminium ion (the TFA has also been shown to be essential) by forming a cyclic intermediate *via* two hydrogen bonds.[27]

Table 6. Aza Diels-Alder Reactions between $R^*N{=}CHCO_2Et$, [R^* = (R)-PhCHMe] and Cyclic Dienes at Room Temperature Catalyzed by Water

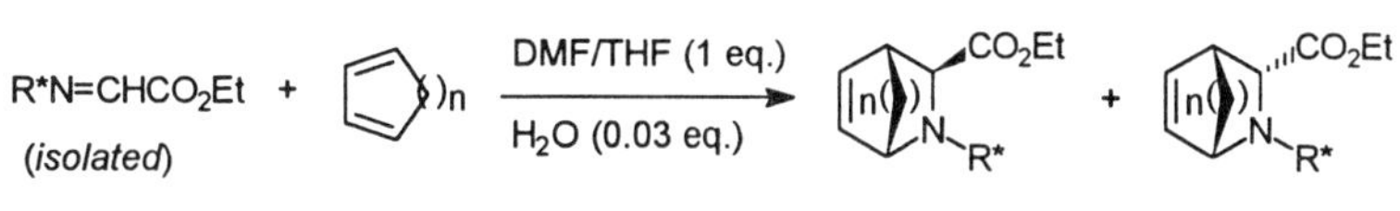

n	t(h)	exo/endo	exo *de* (%)	endo *de* (%)	Yield (%)
1	24	93/7	89	100	82
2	60	92/8	84	100	31

Table 7. Aza Diels-Alder Reactions between R*N=$CHCO_2Et$, [R* = (R)-PhCHMe], and Methyl Butadienes at Room Temperature Catalyzed by Water

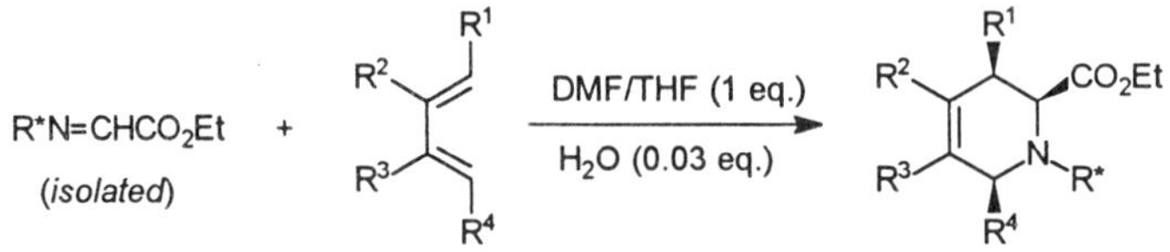

R^1	R^2	R^3	R^4	exo/endo	*de* (%)	Yield (%)
H	H	H	Me	0/100	38	35
H	H	Me	H		70	44
H	Me	Me	H		72	69
Me	H	H	Me	0/100	24	55

3. Heterodienes

Many heterodienes are known [29] but very few have been used in Diels-Alder reaction carried out in aqueous medium.

Engherts and co-workers [30] investigated the kinetics of Diels-Alder cycloadditions of di-(α-pyridyl)-1,2,4,5-tetrazine **36** (a 2,3-diazadiene) with *para*-substituted styrenes **37** in aqueous media and organic solvents. The cycloaddition is followed by a rapid retro Diels-Alder reaction which affords the final diazine **38** after hydrogen-shift (Scheme 11).

Scheme 11

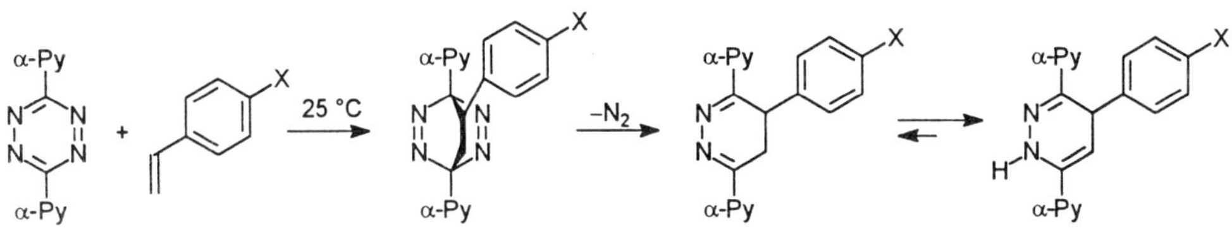

36 **37** **38**

α-Py = α-pyridyl X = Me, H, Cl, NO_2, OMe

The reaction is controlled by LUMO-diene HOMO-dienophile interaction (inverse electron demand Diels-Alder reaction) and therefore is accelerated by electron-donating substituents in the dienophile.

Table 8 reports the relative rates k_2(solvent)/k_2(MeCN) in various solvents at 25 °C. The reaction rate in aqueous medium (H_2O/*t*-BuOH, mole fraction of water = 0.95) increases dramatically and a lower pH accelerates the reaction even more.

Table 8. Relative Rates of the Diels-Alder Reactions of **36** with **37** in Various Solvents at 25 °C

Solvent	k_2(solvent)/k_2(MeCN)				
	X = MeO	X = Me	X = H	X = Cl	X = NO_2
H_2O/*t*-BuOH	127	130	120	87.5	33.4
CF_3CH_2OH	7.66	8.74	6.49	4.30	1
$MeCH_2CH_2OH$	5.37	4.99	4.68	3.81	2.10
PhMe	1.64	1.71	1.97	2.23	2.02
MeCN	1	1	1	1	1
THF	0.88	0.96	1.02	1.15	1.29

The rate constants are roughly correlated with solvent properties such as the polarity and the acceptor number. Correlations have been found with Hammett σ-values, (different σ-parameters are required for the various solvents in order to obtain linear correlations) and with the electronic demand η-values. These observations suggest that the water induces an increased charge separation in the transition state stabilizing it through hydrogen-bond interactions thereby accelerating the reaction considerably.

An example of synthetic applications comes from our laboratories.[31] 3-Nitrocoumarin **39** (Scheme 12)

Scheme 12

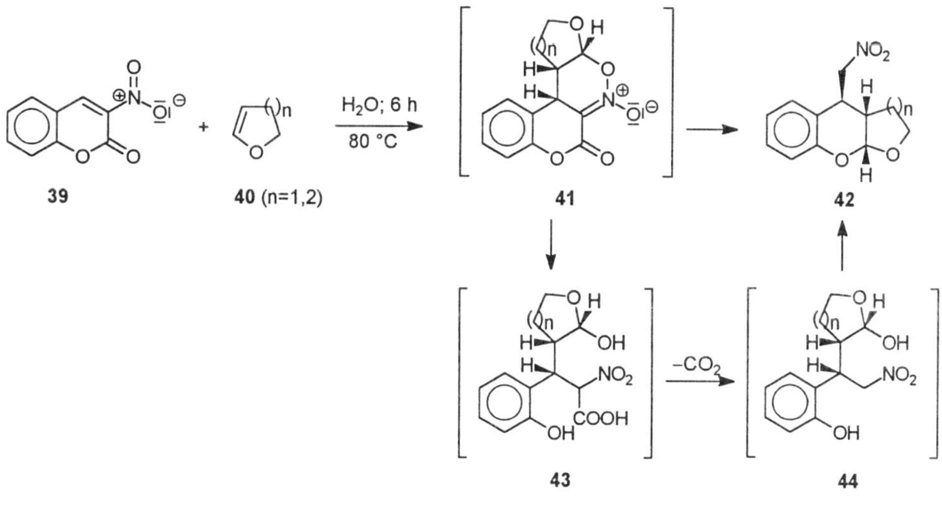

behaves like a diene (oxa-aza-diene) with vinyl ethers **40** such as 2,3-dihydrofuran and 3,4-dihydro-2H-pyran. The reaction occurs in water at 80 °C but the reaction product was not the expected nitrone **41** (coming from an *endo* regiospecific cycloaddition) but rather the nitroacetal **42** that was isolated in a 40-70% over-all yield. The reaction does not occur in benzene and in methanol the adduct **41** is isolated in low yield. The cycloadduct **41** in water gives the acetal **42**.

The origin of **42** can be explained by considering that the acetal and ester functionalities of Diels-Alder adducts **41** hydrolize to give the nitro-carboxy-hemiacetal **43**. Subsequent decarboxylation affords the nitrohemiacetal **44** which undergoes intramolecular acetalization to give the tricyclic nitroacetal **42**.

4. Retro Hetero Diels-Alder

To date the only available literature data of retro hetero Diels-Alder in aqueous medium concern 2-azanorbornenes.

The cycloreversion of these compounds is usually carried out,[32] in the absence of solvent, at 400-600 °C and 10^{-5} tor. In water the reaction occurs at 20-50 °C.[33]

The retro aza Diels-Alder reaction of 2-azanorbornenes in aqueous medium works well [33] on the hydrochloride salts in the presence of N-methylmaleimide which acts either as a trapping agent of cyclopentadiene or as catalyst presumably through the formation of ammonium salt such as **45**. Some examples are reported in Table 9.

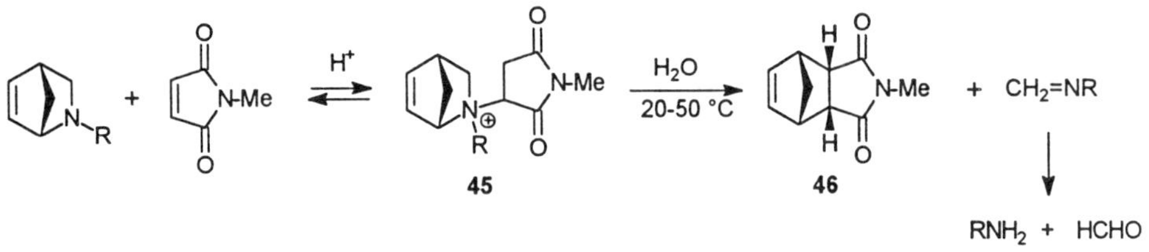

45 **46** **47**

Table 9. Acid Catalyzed Retro Aza Diels-Alder Reactions of N-Alkyl-2-Azanorbornenes.

R	T(°C)	t(h)	Yield (%)	
			46	**47**
H	50	5	80	
Bn	50	2	91	81
Me	50	2	69	
CH_2CO_2Me	50	5	89	
(L)-*i*-$PrCH_2CHCO_2Me$	23	4		95

The cycloreversion of N-benzyl-2-azanorbornenes does not proceed in PhH, THF and MeCN at 50 °C. Traces of N-methylmaleimide-adduct **46** are detected at higher temperatures. Interestingly the retro Diels-Alder of 2-azanorbornene prepared from L-leucine methyl ester affords the parent amino acid as methyl esters in 95% yield without racemization.

Subsequently, two methods [23] have been developed that do not require the use of N- methylmaleimide for affecting the cycloreversion of N-alkyl-2-azanorbornenes.

One method employs copper (II) as catalyst at 70 °C in alcohol-water. An example is the cycloreverserion of 2-azanorbornene **48**.

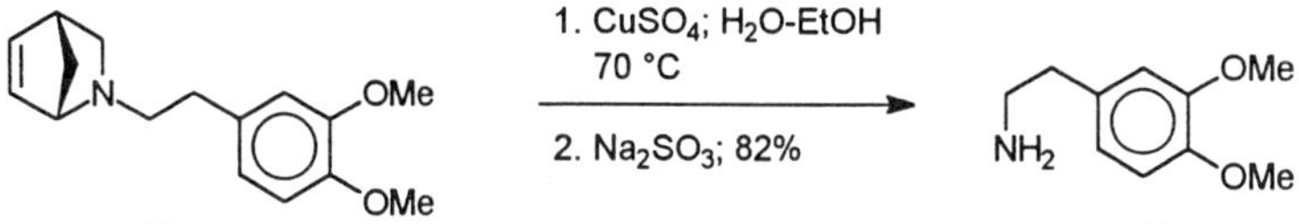

48 **49**

The procedure allows the unmasked amine **49** to be obtained directly, but when it is applied to 2-azanorbornenes derived from dipeptide methyl esters, a considerable hydrolysis of the ester group occurs. This problem is overcome by using [23] sulfonic acid based ion exchange resins. 2-Azanorbornene of L-phenylalanyl-L-leucine methyl ester **50** in H_2O-MeOH at 40 °C with the ion exchange resin Bio-Rad AG50W-X2 200-400 mesh, gives L-phenylalanyl-L-leucine methyl ester **51** in 80% yield without racemization.

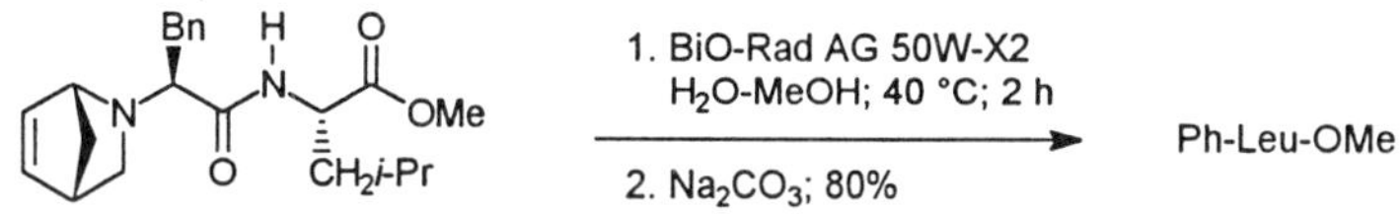

50 **51**

The facile and efficient Diels-Alder reaction in water between an imminium salt and cyclopentadiene, which allows to incorporate an amino function into an azanorbornene unit, coupled with the equally facile retro aza Diels-Alder in water, suggests that the azanorbornene moiety may be used as protecting group for primary amincs in aqueous medium.

5. The Role of the Water

To date there are no definite answers but only hypotheses to explain the unusual acceleration and selectivities of Diels-Alder and other organic reactions when carried out in aqueous medium, frequently under heterogeneous conditions.

In the realm of solvents, water is unique. The molecular structure of water is well-known but the fine inner structure of liquid water is still the subject of current research.[34] Liquid water consists of bound ordered regular lattice regions and regions where the water molecules are hydrogen-bonded in a random array. Monomeric units of water are interspersed with random holes, lattice vacancies and cages. There are chains and small polymers as well as bound, free, and trapped water molecules.

Water has an exceptionally small molecular volume, is highly polar (see ε, μ and E parameters), has a very high value of cohesive pressure (also called cohesive energy density) and a low value of internal pressure.[34a]

In his pioneer work of 1980, Breslow suggested [1] that the "water effect" must be attributed to the hydrophobic interaction, the well-known entropy-driven aggregation of apolar molecules or apolar groups of large molecules in water, which minimizes their exposure to the water. This hypothesis is based on kinetic data, activation parameters and experiments carried out in the presence of "salting-out" and "salting-in" solutes and cyclodextrins.

Reaction rates and reaction selectivities correlate with the solvophobicitiy parameter Sp supporting the role of the hydrophobic interaction but the correlations are not observed in a broad series of solvents.[35]

The importance of the high cohesive pressure of water has also been suggested [36] which could act as an external pressure and therefore greatly favour processes having a negative activation volume. The acceleration in water could be due to steric compression of the transition state in a cavity of the water structure. This is supported by the fact that most bimolecular reactions of apolar reactants accelerated in water are characterized by a negative volume of activation.

Solvent polarity has also been considered [37] and it certainly has a marked influence on the reaction rate however the poor correlations between the solvent polarity parameters and the constant reaction rates, indicate that polarity alone cannot explain the acceleration in water.

A micellar catalytic effect with amphiphilic molecules has also been hypothesized.[38]

Engbert and co-workers [29,39] observe that the hydrophobic packing of reactants in aqueous medium cannot be held responsible for the "water effect" because intramolecular Diels-Alder cycloaddition and Claisen rearrangement, in which the reactants are already "packed", are also considerably faster in water than in organic solvent. They suggest that the hydrogen bond between the water and the activated complex (a stabilizing interaction) and the reduction of hydrophobic surface area of the reactants during the activation process (enforced hydrophobic interactions), are the driving force of the reaction and explain the high diastereoselectivities observed. Theoretical studies confirm this hypothesis.[40]

Water, in any case, is a unique and extraordinary reaction medium and perhaps it is not by chance that nature has chosen it for many biochemical processes.

Acknowledgements. The Ministero della Ricerca Scientifica e Tecnologica (MURTS) and Consiglio Nazionale delle Ricerche (CNR-Rome) are thanked for financial support.

References

1. Rideout, D. C.; Breslow, R. *J. Am. Chem. Soc.* **1980**, *102*, 7816.
2. For pertinent reviews see:
   (a) Fringuelli, F.; Pizzo, F. *Seminars in Organic Synthesis*, Polo Ed. Ch. Milano **1992**, 289; (b) Li, C. J. *Chem. Rev.* **1993**, *93*, 2023; (c) Lubineau, A.; Augé, I.; Queneau, Y. *Synthesis* **1994**, 741; (d) Chan, T. H.; Li, C. J.; Lee, M.C.; Wei, Z. Y. *Can. J. Chem.* **1994**, *72*, 1181; (e) Kibayashi, S. *Synlett* **1994**, 689.
3. Mc Carrick, M. A.; Wu, Y. D.; Houk, K. N. *J. Org. Chem.* **1993**, *58*, 3330.
4. Kibayashi C.; Aoyagi S. *Synlett* **1995**, 873.
5. Naruse, M.; Aoyagi, S.; Kibayashi C. *Tetrahedron Lett.* **1994**, *35*, 595.
6. Naruse, M.; Aoyagi, S.; Kibayashi C. *J. Org. Chem.* **1994**, *59*, 1358.
7. Naruse, M.; Aoyagi, S.; Kibayashi C. *Tetrahedron Lett.* **1994**, *35*, 9213.
8. Naruse, M.; Aoyagi, S.; Kibayashi C. *J. Chem. Soc., Perkin Trans. 1* **1996**, 1113 and 2077.
9. Breslow, R. *Acc. Chem. Res.* **1991**, *24*, 159.
10. (a) Carruthers, W. *Cycloadditions Reactions in Organic Synthesis*, Pergamon Press, Oxford **1990**; (b) Kametami, T.; Hibino, S. *Advances in Heterocyclic Chemistry* **1987**, *42*, 245.
11. Lubineau, A.; Augé, J.; Lubin, N. *Tetrahedron Lett.* **1991**, *32*, 7529.
12. Lubineau, A.; Augé, J.; Grand, E.; Lubin, N. *Tetrahedron* **1994**, *50*, 10265.
13. (a) Lubineau, A.; Augé, J.; Lubin, N. *Tetrahedron* **1993**, *49*, 4639; (b) Lubineau, A.; Queneau, Y. *J. Carbohydr. Chem.* **1995**, *14*, 1295.
14. Grieco, P. A.; Henry, K. J.; Nunes, J. J.; Matt, Jr. J. E. *J. Chem. Soc., Chem. Commun* **1992**, 368.

15. Molander, G. A. *Chem. Rev.* **1992**, *92*, 29.
16. Mikami, K.; Kotera, O.; Motoyama, Y.; Sakaguchi, H. *Synlett* **1995**, 975.
17. Mikami, K.; Motoyama, Y.; Terada, M. *J. Am. Chem. Soc.* **1994**, *116*, 2812.
18. Larsen, S. D.; Grieco, P. A. *J. Am. Chem. Soc.* **1985**, *107*, 1768.
19. Grieco, P. A.; Larsen, S. D. *J. Org. Chem.* **1986**, *51*, 3553.
20. Grieco, P. A.; Larsen, S. D.; Fobare, W. F. *Tetrahedron Lett.* **1986**, *27*, 1975.
21. (a) Waldman, H. *Angew. Chem. Int. Ed. Engl.* **1988**, *27*, 274; (b) Waldman, H. *Liebigs Ann. Chem.* **1989**, 231.
22. Grieco P. A.; Bahsas, A. *J. Org. Chem.* **1987**, *52*, 5746.
23. Grieco P. A.; Clarck, J. *J. Org. Chem.* **1990**, *55*, 2271.
24 Yu, L.; Chen, D.; Wang, G. *Tetrahedron Lett.* **1996**, *37*, 2169.
25. Waldam, H.; Braun, M. *Liebigs Ann. Chem.* **1991**, 1045.
26. Bailey, P. D.; Wilson, R. D.; Brown, G. R. *Tetrahedron Lett.* **1989**, *48*, 6781.
27. Bailey, P. D.; Brown, G. R.; Korber, F.; Reed, A.; Wilson, R. D. *Tetrahedron: Asymm.* **1991**, *2*, 1263.
28. Bailey, P. D.; Wilson, R. D.; Brown G. R. *J. Chem. Soc., Perkin Trans. 1* **1991**, 1337.
29. Fringuelli, F.; Taticchi, A. *Dienes in Diels-Alder Reaction*, J. Wiley and Sons, N.Y. **1990**.
30. Vijnen, J. W.; Zavarise, S.M; Engberts, J. B. F. N. *J. Org. Chem.* **1996**, *61*, 2001.
31. (a) Fringuelli, F.; Piermatti, O.; Pizzo, F. *Proceedings Euchem Conference "Cycloadditions and Related Reactions: Theory and Practice"*, Vulcano **1995**, 74; (b) Fringuelli, F. *Proceedings 5th IBN SINA International Conference on Pure and Applied Heterocyclic Chemistry*, Il Cairo **1995**, 207; (c) Fringuelli, F.; Piermatti, O.; Pizzo, F. *Proceedings RSCI-SCI Joint Meeting on Heterocyclic Chemistry*, Numana **1996**, O-2.
32. Braillon, B.; Lasne, M. C.; Rippol, J. L. *Nouv. J. Chim.* **1982**, *6*, 121.
33. Grieco, P. A.; Parken D. T.; Fobare, W. F.; Ruckle, R. *J. Am. Chem. Soc.* **1987**, *109*, 5859.
34. (a) Reichardt, C. *Solvents and Solvent Effects in Organic Chemistry*; VHC, Cambridge **1990**; (b) Benson, S. W.; Seibert, E. D. *J. Am. Chem. Soc.* **1992**, *114*, 4269.
35. (a) Schneider, H. J.; Sangwan, N. K. *J. Chem. Soc., Chem. Commun.* **1986**, 1787; (b) Scheneider, H. J.; Sangwan, N. K. *Angew. Chem. Int. Ed. Engl.* **1987**, *26*, 896; (c) Sangwan, N. K.; Schneider, H. J. *J. Chem. Soc., Perkin Trans. 2* **1989**, 1223.
36. (a) Lubineau, A. *J. Org. Chem.* **1986**, *51*, 2142; (b) Lubineau, A.; Meyer, E. *Tetrahedron* **1988**, *44*, 6065.
37. (a) Breslow, R. *Acc. Chem. Res.* **1991**, *24*, 159; (b) Samii, A.-Z.; de Savignac, A.; Rico, I.; Lattes, A. *Tetrahedron* **1985**, *41*, 3683.
38. Grieco, P. A.; Garner, P.; Yoshida, K.; Huffman, J. C. *Tetrahedron Lett.* **1983**, *24*, 3807.
39. Blokijl, W.; Engberts, J. B. F. N. *Angew. Chem. Int. Ed. Engl.* **1993**, *32*, 1545.
40. (a) Blake, J. F.; Jorgensen, W. L. *J. Am. Chem. Soc.* **1991**, *113*, 7430; (b) Blake, J. F.; Lim, D.; Jorgensen, W. L. *J. Org. Chem.* **1994**, *59*, 803.

REACTIVITY AND TAUTOMERISM OF 2-THIAZOLEAMINE DERIVATIVES

Luciano Forlani

Dipartimento di Chimica Organica "A. Mangini", Università di Bologna, Viale Risorgimento 4, I-40136 Bologna, Italia.

Abstract. *Some spectroscopic properties and reactivities of 2-thiazoleamine derivatives toward electrophilic and nucleophilic reagents are reviewed and discussed. 2-Thiazoleamine may exists in at least two tautomeric forms as illustrated by the following equilibrium*

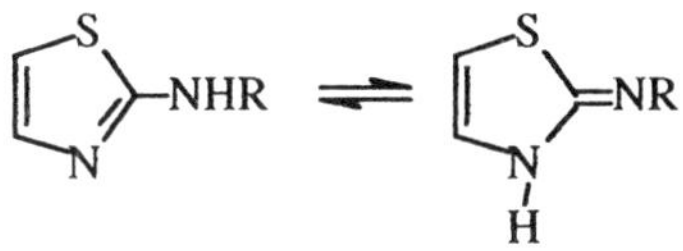

Spectroscopic investigations on 2-thiazoleamine derivatives reveal that the amino aromatic form is the more populated form, with the exception of derivatives with strong electronwithdrawing groups bonded to the amino group. pKa Values of 2-thiazoleamine derivatives are in agreement with this conclusion and show the strong influence of the electronic effect of the substituent on the basicity of the "aza" nitrogen. In principle, the electronic effects of substituents bonded to a small-sized cycle (five-membered) are stronger than those on a six-membered cycle. The reactivity toward electrophilic reagents (E^+) reveals that 2-thiazoleamine (and its derivatives) is a tridentate nucleophile.

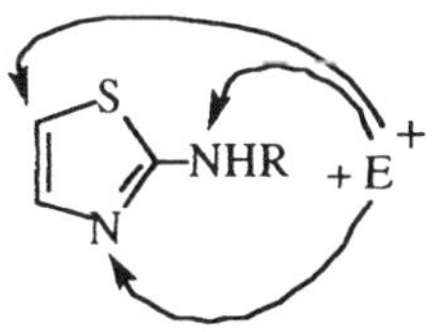

The surprising reactivity towards nucleophilic reagents of 5-halogeno-2-thiazoleamines is due to a simple nucleophilic substitution reaction which may be followed by the addition reaction of a solvent molecule to the thiazole ring. This reaction discloses some interesting peculiarities of 2-thiazoleamine derivatives, in particular the importance of the interactions between solvent or nucleophiles with the aza nitrogen.

1. Introduction

Physical and chemical properties of derivatives of 1,3-thiazole[1] (thiazole) are of interest because the penta-atomic heterocycle shows peculiarities which are unusual in homocyclic series.The presence of sulphur in a penta-atomic cycle produces an electron rich aromatic system (thiophene-like). On the contrary, the aza-nitrogen presence serves as an electronwithdrawing group, and the resulting p system when both heteroatoms (N and S) are present is a balance of electron-rich and electron-poor aromatic systems. Consequently, the properties of thiazole derivatives[2] differ from the properties of benzene derivatives and also from properties of pyridine derivatives.

The presence of an sp^2 "aza" nitrogen, which is a basic centre, is the most relevant peculiarity of the thiazole moiety. In fact, it is responsible for interactions between the thiazole derivatives with the medium (or the reagents) when protic medium is used or, in general, when electrophiles are present in the reaction mixtures. For instance, hydrogen bonding interactions (as well as the other interactions with electron acceptors) produce an electronic situation different from that of the free "aza" nitrogen. The equilibrium reported in Scheme 1 may be a salification equilibrium or it may be an interactions hydrogen bonding-like, depending on the proton releasing nature of the partner (which may be the simple solvent).

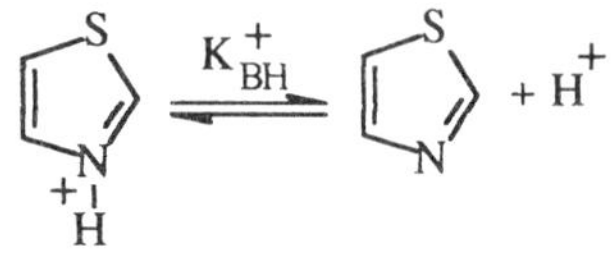

Scheme 1

Clearly, the introduction of a positive charge on the nitrogen atom in position 3 changes the electronic charge distribution of the ring atoms.

Among the three thiazoleamine isomers, 5-thiazoleamine and 4-thiazoleamine have been investigated to a lesser degree than 2-thiazoleamine,[2] probably because of some difficulties in their preparation and in their stability. 2-Thiazoleamine however, has been extensively studied while its properties do not even parallel those of aniline or 2-pyridineamine.

2. Basicity

Owing the importance of the endocyclic sp^2 nitrogen presence, the basicity of thiazole and its derivatives has been carefully considered. Substituent electronic effects on the position of the equilibrium shown in Scheme 1 have been investigated[3,4] in water at 25°C for monosubstituted thiazoles **1** and **2**.

1 **2**

X = H, Me, Et, Bu^t, NO_2, SO_2Me, SO_2Ph, SOPh, Cl, Br, OMe, NHAc, Ph, NH_2, NMe_2; Y = Me, NO_2, SO_2Me, Cl, Br, OMe.

The pKa value is very sensitive to changes of substituent: 2-thiazoleamine is about 10^{10} more basic than 2-nitrothiazole. For 2-X-substituted thiazoles, **1**, the position of the equilibrium in Scheme 1 (as expressed by pKa values) is linearly correlated with σ_m values in a Hammett plot[5] (ρ = -9.9).

The use of σ_m values reveals that the inductive effect is more important than the mesomeric electronic effect. Comparison with the pKa values of 5-substituted thiazoles **2** excludes the steric requirements for the protonation of 2-substituted thiazoles; 5-Y-thiazoles **2** are also well correlated with σ_m values in a Hammett plot; (ρ = -6.0). This is expected since position 5 of the thiazole ring is comparable (with respect to the aza nitrogen) to a meta position in the benzene series.

The use of σ_m values for 2-X-thiazoles **1** is in agreement with that of 2-substituted pyridine and 2-substituted imidazole systems.

In these systems the "aza" nitrogen (which is the proton acceptor centre) and the substituent X are bonded to the same carbon atom.

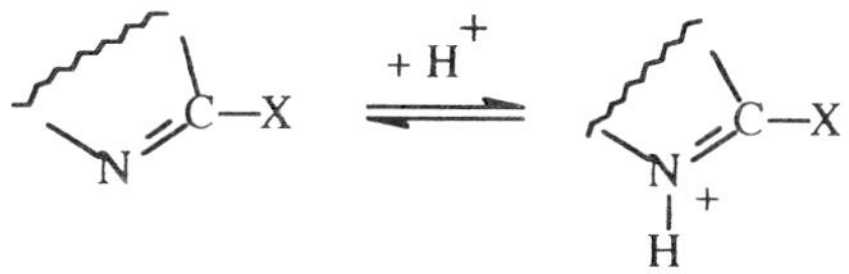

Scheme 2

Consequently, the superposition of the mesomeric and inductive electronic effects leads to a balance in which the inductive effect is stronger than the mesomeric one.

It is noteworthy that the 2-substituted heterocycles aza-containing cannot be considered as ortho-like derivatives in estimating the "ortho" electronic effects[6] in the benzene series. For instance, 2-substituted benzoic acids have the reaction centre far from the carbon bearing the substituent (see Scheme 3): the geometrical assessment of 2-substituted benzoic acids strongly differs from that of 2-substituted thiazoles (or other aza-containing heterocycles).

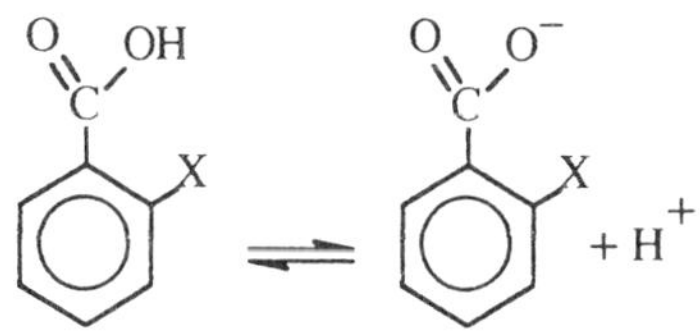

Scheme 3

In principle, the sensitivity of five-membered heterocycles to the electronic effects of substituents is higher than that observed for six-membered aromatic heterocycles (pyridines) or homocycles. ρ Values for pyridine system (-11.9)[6] and for 2-substituted benzimidazoles (-11.1)[7] are both close to the ρ value observed for 2-substituted thiazoles. The proximity of the substituent with respect to the reaction centre, as emphasized by Scheme 2, is the main factor which explains this aspect of the substituents' electronic effect on the pKa values of the considered systems, which have similar geometries.

3. pKa Measurements and tautomeric properties

In principle, there are two possible protonation centres in 2-thiazoleamine derivatives: the endocyclic "aza" nitrogen and the exocyclic nitrogen of the amino group.

If the protonation centre is the nitrogen of the amino group, bonded to position 2 of the thiazole ring, the electronic effect of the substituents bonded to position 5 is more aptly expressed by σ_p values[8,9] or by other σ_p values which involve extra-conjugations.

The plot of the pKa values of 5-X-2-thiazoleamines (X = H, Me, OMe, Ph, SPh, Cl, Br, CO_2Et, SO_2Ph, NO_2) against σ_{para} values does not give an acceptable correlation. The pKa of 5-substituted-2-

thiazoleamines show a linear Hammett plot against σ_{meta} values.[10] This fact is expected if the protonation centre is the endocyclic nitrogen, as illustrated in Scheme 4.

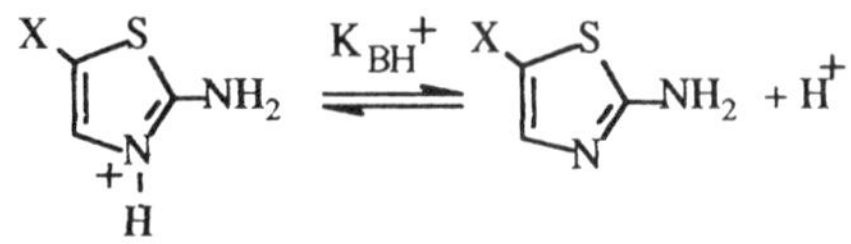

Scheme 4

ρ Values calculated from pKa values of 5-X-2-thiazoleamines and for 5-X-2-N,N-dimethylthiazoleamines are -6.0 and -6.6, respectively.[10] These values are very close to the value calculated from pKa values of 5-substituted thiazoles (by using σ_m values) of -6.0.[4]

There is however,a significant deviation in the case of the nitro group due to an electronic interaction (or an internal charge transfer) between the amino and the nitro group, as depicted by **3**.

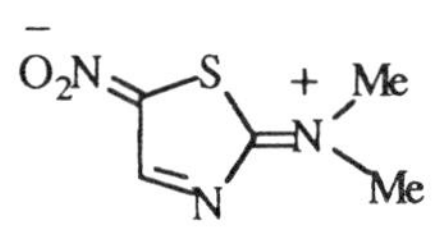

3

In fact, a restricted rotation of the N,N-dimethylamino group has been observed by dynamic nuclear magnetic resonance.[11] The free energies of activation (ΔG* in kcal mol^{-1}) for the rotation of the N,N-dimethylamino group of 2-N,N-dimethylthiazoleamine and of 5-nitro-2-N,N-dimethylthiazoleamine have been calculated at 7.7 and 12.4 respectively.

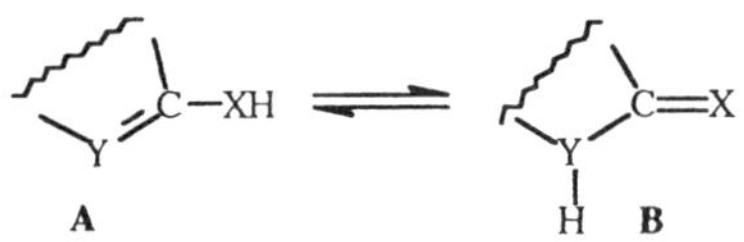

X = O, S, NR

Y = CH, N

Scheme 5

The tautomeric equilibria in heterocyclic and homocyclic series are of great importance in explaining the physical and chemical properties of the considered systems. Potential hydroxy and mercapto groups are usually reported to exist in the form **B** of the Scheme 5, while the amino (aromatic) form **A** is usually considered the more populated form of amino derivatives.

In principle, 2-thiazoleamine derivatives can exist in mainly 2 tautomeric forms, as illustrated by the equilibrium in Scheme 6.

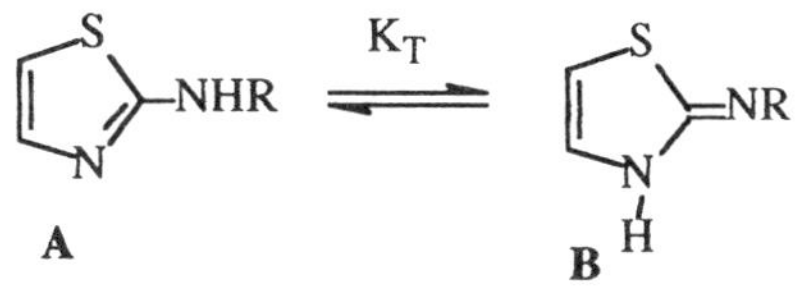

Scheme 6

Several spectroscopic investigations[12] indicate that the most populated form of 2-thiazoleamine is the amino aromatic form **A**. The presence of the imino form (**B** in Scheme 6) was predicted to be detected by usual spectroscopic methods when strong electronwithdrawing groups are present in the thiazole moiety to facilitate the loss of H^+ from the amino group.[2]

The pKa values of the fixed amino parent of **A** are very different[13] from the pKa values of the fixed imino parent of **B** (where the hydrogen atom is replaced by an alkyl group).

The plot in Figure 1 shows a comparison between pKa values of 5-X-2-thiazoleamines and of the corresponding 5-X-2-N,N-dimethylthiazoleamines.[10] The slope of the straight line of this plot is very close to unity (slope = 1.07±0.05). This fact clearly indicates that in 2-thiazoleamines the amino aromatic form **A** predominates over the imino form **B**, even when strong electronwithdrawing groups are bonded to the position 5 of the thiazole ring. In all cases the protonation centre is the endocyclic nitrogen.

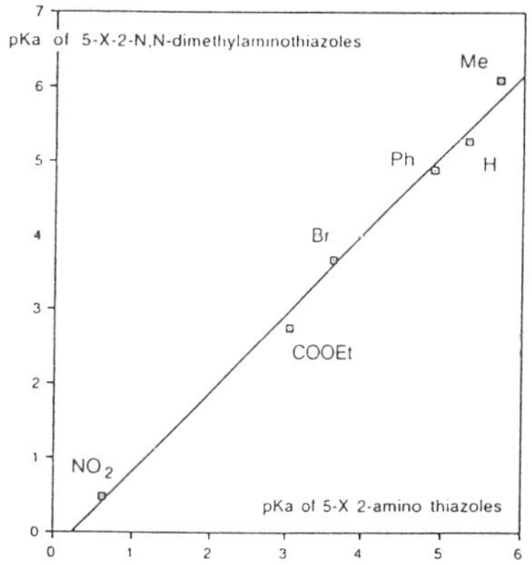

Figure 1. pKa Values of 5-X-2-N,N-dimethyl-thiazoleamines *versus* pKa values of 5-X-2-thiazoleamines.

pK_{BH^+} Values[14] of 2-thiazoleamines and of their methiodide and of some fixed imines (in water at 25°C) also indicate that the amino aromatic form predominates in the 2-thiazoleamines considered, while 2-*p*-tosylthiazoleamine **4** exists in the imino form **4B**, as indicated by X-ray analysis[15] (see Scheme 7).

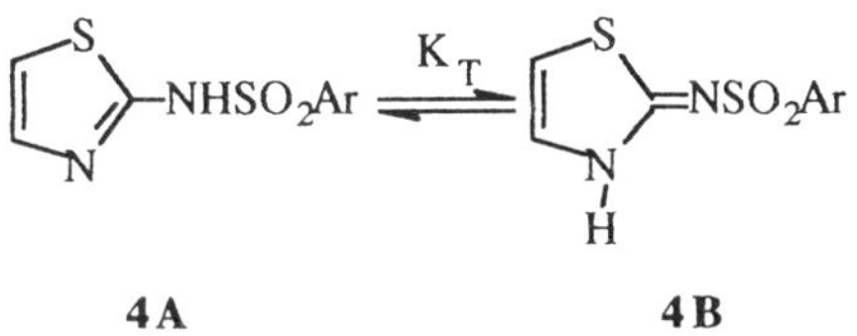

Scheme 7

For a number of derivatives, the pK_{BH^+} values of the amino and the imino fixed forms, allow K_T values to be calculated by means of Equation 1.[13] K_T Values are provided in Table 1.

Equation 1 $$K_T = K_a^{amino}/K_a^{imino} = [amino]/[imino]$$

From the data in Table 1, two major considerations are worthy of attention: the first is the moderate anellation effect (the K_T value of the 2-aminobenzothiazole is slightly lower than that of 2-thiazoleamine); the second one refers to the observation that the introduction of a phenyl group on the amino group reduces the K_T values by about 200 times. In other words, in 2-N-phenylthiazoleamine, the two tautomers come close to having comparable stability. 2-Anilinopyridine is indicated to show very similar K_T values to that of 2-aminopyridine[13,16]: the five-membered ring appears to be more sensitive to phenyl substitution than the six-membered heterocycle. Consequently, thiazole derivatives are more suitable as models in investigating the tautomerism of the amino group in the heterocyclic series than pyridine derivatives.

Table 1

K_T values {= [**A**]/[**B**] of Scheme 6} of some 2-thiazoleamine derivatives, in water at 25 °C, calculated by Equation 1, from pK_{BH^+} values

substrate	K_T
2-thiazoleamine	2.1×10^4
5-methyl-2-thiazoleamine	1.7×10^4
2-N-benzylthiazoleamine	6.8×10^3
2-N-phenylthiazoleamine	9.3×10
2-aminobenzothiazole	6.9×10^3

On the other hand, the introduction of substituents on the thiazole ring of 2-N-phenylthiazoleamine ($pK_{BH^+} = 4.33$), does not produce effects on the position of the tautomeric equilibrium: pK_{BH^+} values of 5-nitro-2-anilinothiazole and 5-methyl-2-anilinothiazole are -0.099 and 4.57, respectively, and are analogous to the corresponding values of 5-substituted 2-thiazoleamines. In particular, the introduction of a nitro group on position 5 of the thiazole ring does not change the fact that the more populated form is also the amino aromatic form for 5-nitro-2-phenylthiazoleamine.

The electronwithdrawing groups on the exocyclic nitrogen shift the position of the tautomeric equilibrium toward the imino tautomer. In the case of 2-phenylthiazoleamine as well, the presence of electronwithdrawing groups on the phenyl ring of the 2-phenylthiazoleamine effect interesting modifications on the position of the tautomeric equilibrium.

Similar conclusions[17] have been obtained from ^1H NMR spectra of 5-substituted 2-thiazoleamines in esa-deuterodimethylsulphoxide. The amino aromatic form is the more populated form even when electronwithdrawing groups are bonded to the position 5 of the thiazole ring.

The electronic effect of substituents on the chemical shift of the signals related to the protons of the amino group is related to the σ values in the Hammett plot ($\rho = 1.33 \pm 0.1$). The same behaviour (and a very similar ρ value = 1.36) have been reported in the case of the chemical shift of the amino groups of substituted anilines in dimethylsulphoxide.[18] The regularity of the chemical shifts of the amino group agrees well with the conclusion that the amino aromatic form is the morepopulated form for all the considered substituted 2-thiazoleamines. However, the comparison of ^1H NMR spectral data (in DMSO-d_6) of fixed amino and imino parents of sulphathiazole derivatives **4**, confirms that the imino form **4B** predominates.

4. Tautomerism of 2-N-(polynitrophenyl)-thiazoleamines

The crystallographic properties[19] of 2-N-(2,4,6-trinitrophenyl)-4-methylthiazoleamine **5**, 2-N-(2,4,6-trinitrophenyl)-pyridineamine **6** and 2-N-(2,4,6-trinitrophenyl)-pyrimidineamine **7** reveal some interesting aspects regarding the structures of these compounds. These derivatives can also exist either in the imino non-aromatic form **B** or in the amino aromatic form **A** (see Scheme 8).

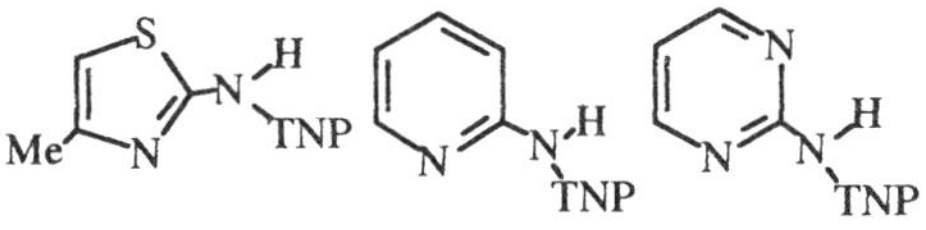

5 6 **7**

TNP = 2,4,6-trinitrophenyl

In the solid state, the amino aromatic form **A** clearly predominates in all the three considered derivatives.

In principle, the considered derivatives (as well as 2-thiazoleamine and related heterocycles) are able to form both cyclic, such as reported in **8**, or non-cyclic dimeric (or polymeric) aggregates.

Dimeric forms of 2-amino azoles[20] (as well as dimeric forms of related hydroxy compounds[21,22]) may be in the pathways of tautomeric transformations.

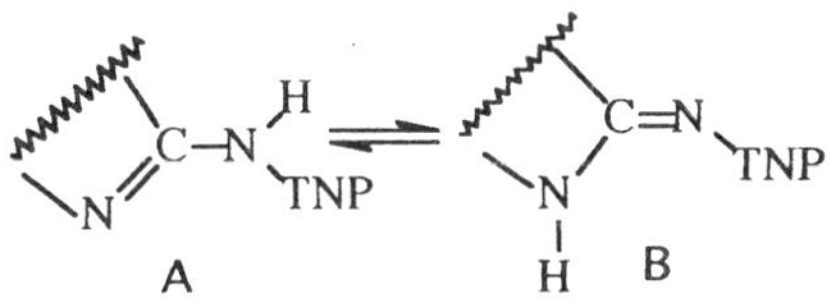

TNP = 2,4,6-trinitrophenyl

Scheme 8

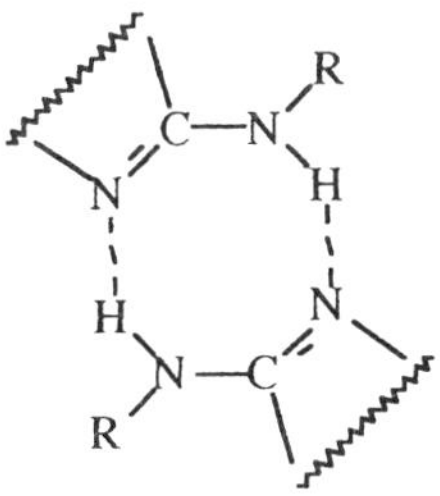

8

In the considered structures (**5**, **6**, **7**) the amino group (N1, see **9**) interacts by hydrogen bond with the nitro group (N2), while the "aza" nitrogen (N5) interacts with the nitrogen atom of the nitro group (N4).[19]

It is interesting that in these examples the geometry of the nitro groups and of the amino group bonded to the heterocyclic moiety does not agree with the classical conjugation theory (between the nitro and amino groups bonded to the same aromatic ring), in explaining the spectroscopic properties, such as uv/vis spectroscopic data, of nitro anilines.[21,24] In particular, the geometry of the amino group changes when the hydrogen atom is replaced by an alkyl group (usually a methyl group). For example, the spectroscopic properties of derivatives **10**, **11**, **12** differ from those of **5**, **6**, **7** respectively.

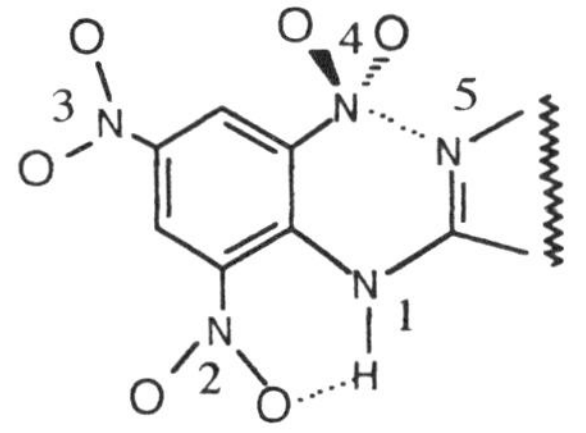

9

This fact can be explained by an internal charge transfer which allows for electron donor-acceptor interactions (between amino and nitro groups) even when these groups are not co-planar.

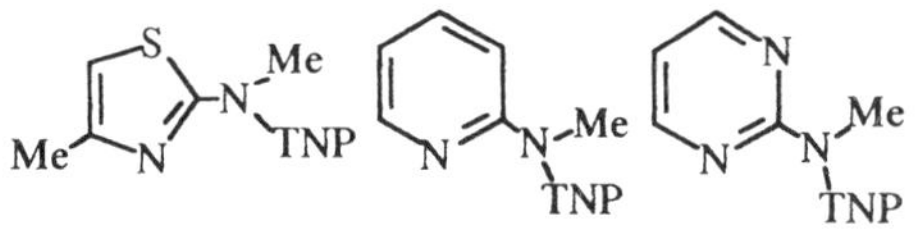

10 **11** **12**

TNP = 2,4,6-trinitrophenyl

A method frequently used in investigating the relative population of tautomeric species[25] involves the comparison of the spectroscopic properties of the potential tautomers with the properties of fixed parents obtained by substitution of a hydrogen atom with a methyl group.

This method starts from the assumption that the physical and spectroscopic properties of tautomers are very similar to those of the fixed parents. This assumption (which often is reasonably corrected) may rise to some misinterpretations. In the literature, there are reports of the importance of the replacement of the hydrogen atom with a methyl group as in the case of 2,4,6-trinitrodiphenyl amines[23] and in 2-N-(2,4,6-trinitrophenyl) heteroaryl amines.[19]

Recently,[20] a simple method for the quantitative evaluation of the position of tautomeric equilibrium (K_T values of equilibrium in Scheme 6) without the direct use of spectroscopic properties of fixed parents was proposed. This method grows out of the observation that the two tautomers differ in "polarity", and the population of the two isomeric species may depend on the "polarity" of the medium. Changes in medium polarity produce variations in the position of tautomeric equilibrium. In fact, when **A** is the more populated form in a poorly polar solvent and if the position of the tautomeric equilibrium is not too large to favour **A**, the addition of polar substances to a solution of **A** may shift the equilibrium to reach a situation which favours the presence of **B**.

Comparison with fixed imino and amino forms clearly indicates that compounds **5**, **6**, **7** are in amino aromatic form **A** in apolar solvents (CCl_4, toluene), while in polar solvents (DMSO) the imino form **B** predominates for compound **5**.

Addition of dimethylsulphoxide to solutions of **5** in toluene produces a red shift of λ_{max} in the uv/vis spectrum from λ_{max} = 389 nm (ε = 9.4 x 10^3) (in toluene) to λ_{max} = 447 nm (ε = 4.47 x10^4) (in DMSO). The fixed amino form **10** gives a λ_{max} = 421 nm (ε = 4.7 x 10^3) in toluene and λ_{max} = 421 nm in DMSO (ε = 3.9 x10^3). The fixed imino form **13** gives a λ_{max} = 419 nm (ε = 8.3 x 10^3) in toluene and λ_{max} = 427 nm (ε = 9.3 x 10^3) in DMSO.

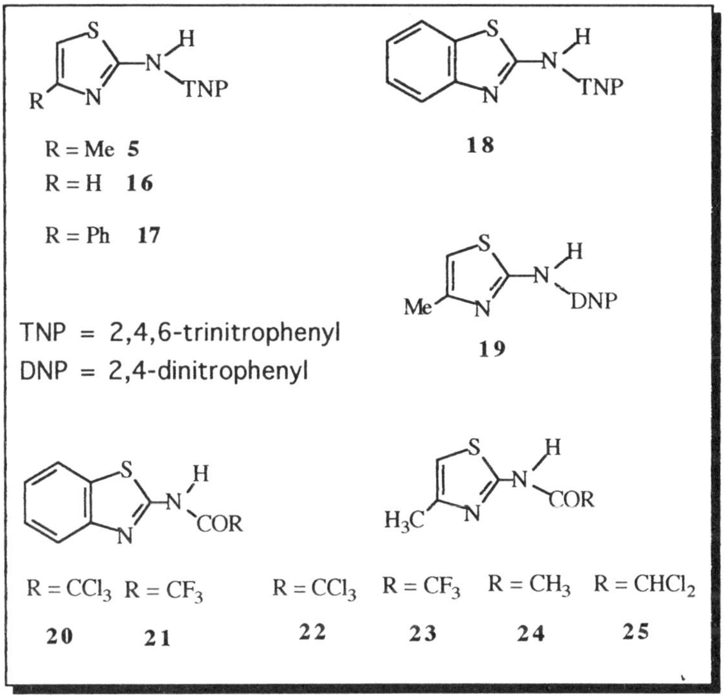

Chart 1

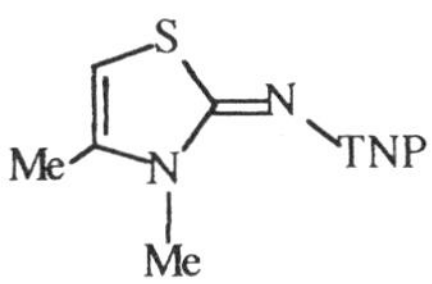

TNP = 2,4,6-trinitrophenyl

13

Addition of tetrabutylammonium bromide (TBAB) to solutions of 2-N-(2,4,6-trinitrophenyl)-4-methylthiazoleamine **5**, in toluene produces a red shift of the maximum of the uv/vis spectrum, similar to that of DMSO, as depicted in Figure 2.

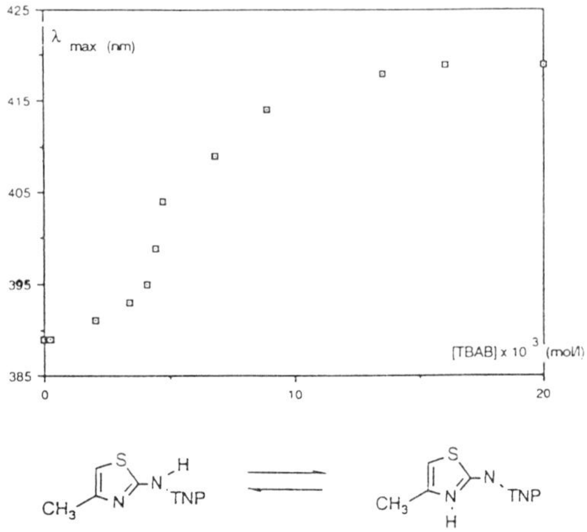

Figure 2. Effect of the addition of tetrabutylammonium bromide (TBAB) on the uv/vis spectrum of a solution of 4-methyl--(2,4,6-trinitrophenyl)thiazoleamine (**5**) in toluene at 25 °C [**5**] = 1.0 x 10^{-4} (mol l^{-1}).

5. Tautomerism of 2-acetamidothiazoles

Some 2-acetamido thiazole (and benzothiazole) derivatives show similar behaviours to those reported above for 2-N-(trinitrophenyl)-thiazoleamine.

In 1969,[26,27] the isolation of both tautomers in the equilibrium in Scheme 9, was reported when R = -CHClCH$_3$. On the basis of spectral data, structure **14A** was assigned to the compound with melting point 114-115°C and structure **14B** to the compound which melts at 124-125 °C.

Recently,[28] the structure of a compound with m.p. 124-125 °C was investigated by X-ray diffraction and structure **14A** was subsequently assigned.

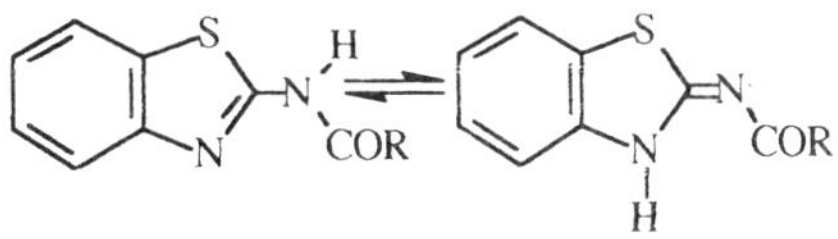

14A **14B**

Scheme 9

The structure assignment to the compound with m.p.114-115 °C was more complicated since crystals for X-ray diffraction analysis were not available. By considering that both isomers show indistinguishable spectroscopic properties in solutions, but that they differ in IR spectra (in solid KBr) and in their m.p.s, either the compound with m.p. 114-115 °C is the actual tautomer of **14A** (*i.e.* its structure is **14B**) or it is a polymorphic form of **14A**. On the basis of the spectroscopic properties of these compounds and of related derivatives, the hypothesis of a polymorphic species is more reasonable than that a of tautomeric species.[28] In fact, the electronwithdrawing power of the considered R is too low to produce tautomeric displacement (see below). In addition, a third compound (m.p. 132-133 °C) was obtained by slow crystallization of **14A** from light petroleum. This compound is an isomer of **14**, a fixed imino form **15**.

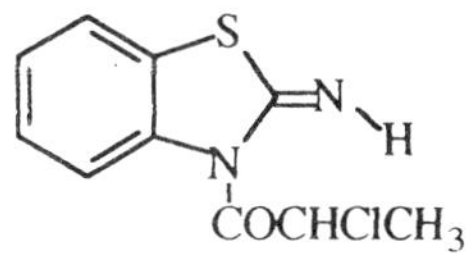

15

Compound **15** is most likely derived from a slow migration of the acyl group of **14** from the exocyclic to the endocyclic nitrogen.

For the sake of clearity, the formulae of the investigated potential tautomers are reported in the Chart 1.

For 2-acetamidothiazole and benzothiazole derivatives, the comparison between spectroscopic properties of potential tautomers and fixed parents indicates that the apolar solvent favours the amido aromatic form, while the polar solvent favours the imido form.[29] An example of the effect of the addition of DMSO to a solution of **20** in CCl_4 is provided in Figure 3. Spectrum 2 (see Figure 3) is very similar to the spectrum of the fixed imido parent **26**.

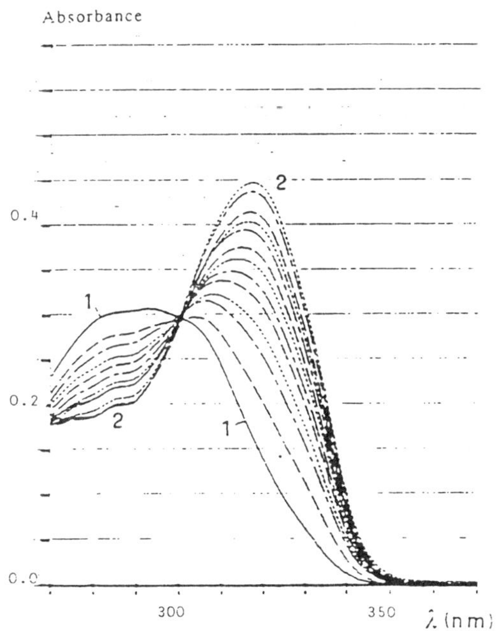

Figure 3. Effect of the addition of DMSO on the uv/vis spectrum of a solution of 2-trichloroacetamidobenzothiazole (**20**) in CCl_4 at 25 °C [**20**] = 2.7 x 10^{-5} (mol l^{-1}); [DMSO] = from 0 (spectrum 1) to 1.2 x 10^{-2} (mol l^{-1}) (spectrum 2).

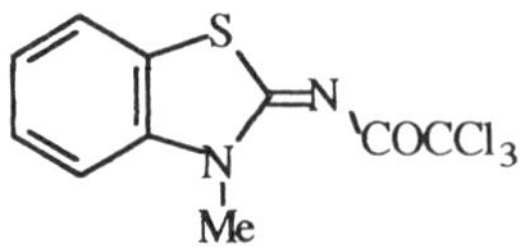

26

The λ_{max} changes of solution of **21** in CCl_4 operate by addition of DMSO, as shown in Figure 4. The plot of Figure 4 indicates further that at low [DMSO] values the amino form predominates, and at high [DMSO] value the imido form is the more populated form (λ_{max} value of the fixed imido form is 315 and 319 nm in CCl_4 and in DMSO, respectively) with a very similar value to that recorded for the potential tautomer in CCl_4 and in the presence of DMSO (λ_{max} = 314 nm, see Figure 4). The [DMSO] values which fall in the range 0.2-0.4 mol l^{-1} favour the contemporaneous presence of both tautomers.

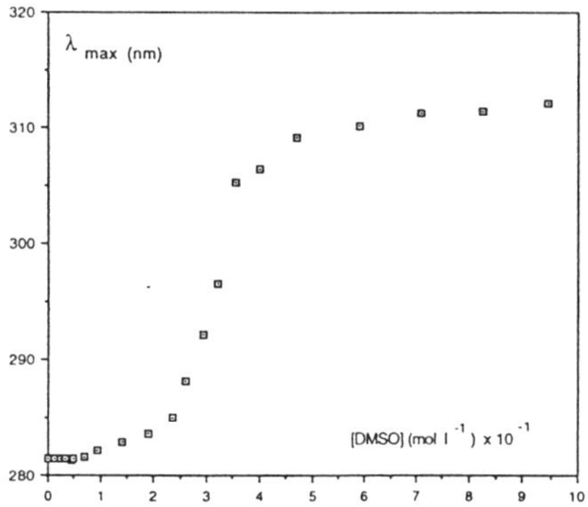

Figure 4. Effect of the addition of DMSO on the λ_{max} value of uv/vis spectrum of a solution of 2-trifluoroacetamidobenzothiazole (**21**) in CCl_4 at 25 °C [**21**] = 5.0 x 10^{-5} (mol l^{-1}).

The evaluation of the position of the tautomeric equilibrium K_T expressed by the [**A**]/[**B**] ratio, is possible (*without the use of spectroscopic data for fixed parents*) by using the spectroscopic data in the apolar solvent (toluene) (without the addition of polar substances) for the amino aromatic form and spectroscopic data for the solution with excess salts (or of DMSO). The slope of the plots of the [**A**]/[**B**] ratio values against the concentration value of the added polar substances may be considered an evaluation of the sensitivity of the tautomeric equilibrium to changes of medium polarity.

These slope values for some derivatives of 2-thiazoleamine and of 2-aminobenzothiazole (see Chart 1) are summarized in Table 2. When the substituted acetyl group is less electronwithdrawing than the trichloro and trifluoro methyl groups, the addition of polar solvent does not show evidence for the presence of the imino form. When R = $CHCl_2$ (**25**), addition of DMSO produces slightly significant variations of the uv/vis spectrum; when R= CH_2Cl or CH_3, the addition of DMSO does not produce changes of the spectrum. This fact confirms the attribution of the structure **14**: the uv/vis spectrum of **14** is unaffected by the addition of DMSO (in CCl_4).

In the same way, when the electronwithdrawing group is the 2,4,6-trinitrophenyl group, the imino form may be populated as the amino form, while when the 2,4-dinitrophenyl group is bonded to the exocyclic nitrogen, the amino form **A** is the more populated form.[20,30,31]

Table 2

Slopes (S) of plots of [B]/[A] values against the [X] values {X = DMSO or TBAB} in toluene at 25°C for some 2-thiazoleamine and benzothiazoleamine derivatives.[18, 23]

Compound	[**X**]	**S**
5	DMSO	0.41
16	DMSO	0.28
17	DMSO	0.15
18	DMSO	0.58
5	TBAB	3.2
16	TBAB	4.0
17	TBAB	1.8
18	TBAB	45.5
20	DMSO	4.5
21	DMSO	13.2
22	DMSO	2.2
23	DMSO	3.2

If the concentration of added DMSO (or TBAB) is 10^{-1} mol l^{-1}, the values (**S**) reported in Table 2 are the apparent constant values (K_T) of the tautomeric equilibrium in the considered medium. All the trinitrophenyl derivatives considered in Table 2 are more sensitive to TBAB addition than to the addition of DMSO. Probably, the salt addition is more able to enhance the medium's polarity than the addition of DMSO. furthermore, it is possible that some specific interactions operate on the tautomeric equilibrium, and further quantitative investigations are necessary to explain the mechanism of these proton shifts. For six-membered heterocycles, 2-N-(2,4,6-trinitrophenyl)-pyridineamine **6** and 2-N-(2,4,6-trinitrophenyl)-pyrimidineamine **7**, the amino aromatic forms strongly predominate in both solvents and no evidence for the presence of imino forms was obtained.

These conclusions are in agreement with those of pKa measurements[13] which indicate that 2-thiazoleamine is more prone to exist in the imino (non-aromatic) form than 2-aminopyridine. The fact that the resonance energy of six-membered heterocyclic derivatives is higher than that of five-membered rings most likely explains this difference in behaviour: contrary to six-membered derivatives, the five-membered derivative **5** tends show evidence of the presence of the imino tautomer (**B**) (see Scheme 8) since the loss of stabilisation from amino to imino is less relevant.

A particular modification of 2-thiazoleamine properties arises when another amino group is bonded in position 4 of the thiazole ring such as in 2,4-diamino thiazole **27**.[32]

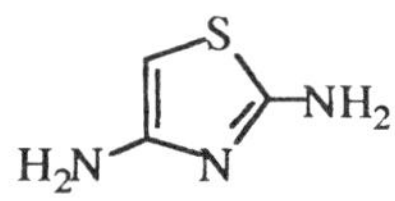

27

While analysis of ^{1}H NMR spectroscopic data indicates that the more populated form of 5-(4)-substituted 2-thiazoleamines is the amino aromatic form, in the presence of a second amine group bonded to the position 4 of the 2-thiazoleamine, an imino form is prevalent.[33]

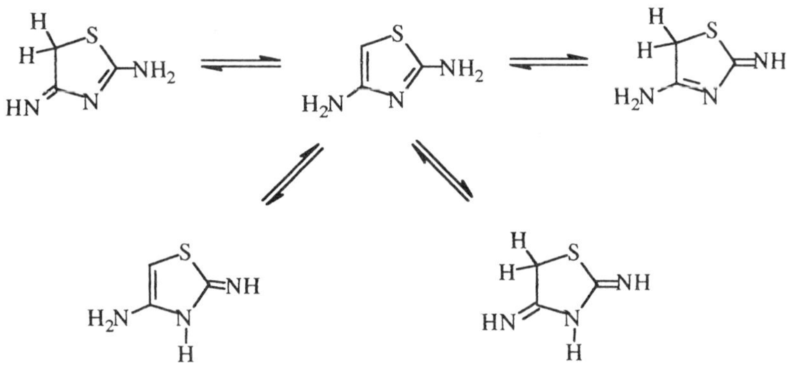

Scheme 10

In principle, several tautomeric forms of **27** are possible as depicted in Scheme 10. Comparison of the ^{1}H NMR spectroscopic data of hydrohalide salt derivatives of 4-amino-2-(N,N-dimethylamino)thiazole **28** reveals that the imino form predominates, as a consequence of the presence of the amino group bonded in position 4, while the amino group bonded in position 2 scarcely participates in the tautomerism. The more probable structure of the hydrochloride of **27**[33] is **29**.

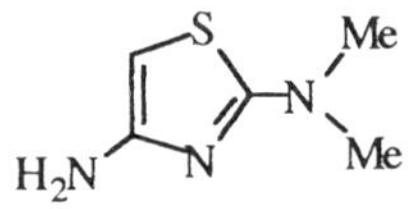

28

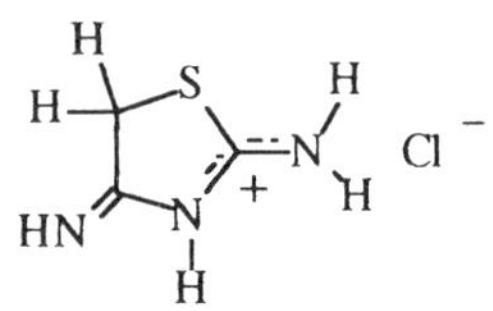

29

6. Reactivity of amino-thiazole derivatives towards nucleophilic reagents

The reactivity of 2-halogenothiazoles and of related heterocycles, toward nucleophiles[9] has been extensively investigated and explained by the usual electronwithdrawing effect of the aza nitrogen in activating the attack of nucleophiles. Some examples are provided in Scheme 11, in which L represents the leaving group.

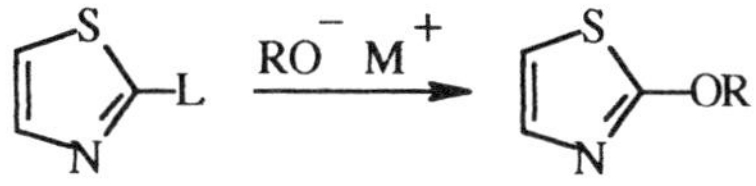

L = F, Cl, Br, I; M = Li, Na, K.R= alkyl

Scheme 11

5-Halothiazoles react with anionic nucleophiles[34] (such as sodium alkoxides, see Scheme 12) despite of position 5 cannot be directly activated by the electronwithdrawing power of the aza nitrogen.

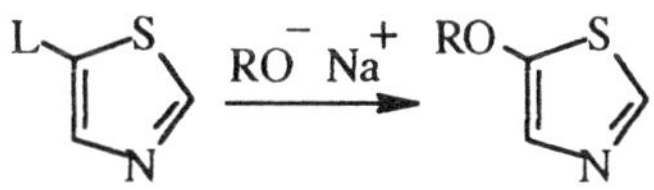

L = Cl, Br, R = alkyl

Scheme 12

Position 5 of the thiazole ring (as well as positions 3 and 5 of the pyridine ring) is a "meta-like" position with respect to the nitrogen atom. Nevertheless, the k_5/k_2 ratio (k in $s^{-1}mol^{-1}$ l) of 5-chlorothiazole and 2-chlorothiazole for the nucleophile $CH_3O^- Na^+$ in CH_3OH is 2.3; for bromothiazoles in CH_3OH the k_5/k_2 ratios (k in $s^{-1}mol^{-1}$ l) are 2.2, 0.16 and 0.05 for $CH_3O^- Na^+$, $CH_3S^- Na^+$ and $PhS^- Na^+$, respectively.[35]

A possible explanation for this unexpected reactivity may be the formation of an hetaryne intermediate (**30** of Scheme 13): in this case the reaction is via elimination addition pathway.

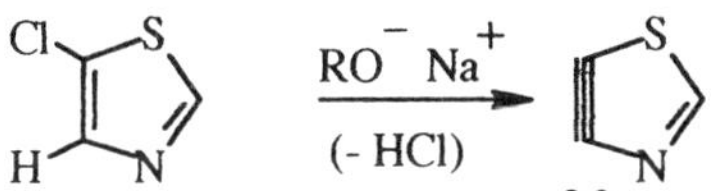

30

Scheme 13

On the contrary, the hetaryne pathway is highly improbable because the reactivity of 4-phenyl-5-chlorothiazole (which, obviously, cannot eliminate HCl) is very similar to that of 5-chlorothiazole.

A reasonable explanation of the reactivity toward nucleophilic reagents of 5-halogenothiazoles involves the inductive electronwithdrawing power of both heteroatoms. On the other hand, other reactions carried out in both positions 5 and 2, such as the oxidation of phenylsulphynylthiazoles with perbenzoic acid,[34,36] base-catalyzed H/D exchange,[37,38] decarboxylation reactions,[39] pKa values of carboxylic acids,[40] indicate that the electron density at C5 is close to the electron density at C2. Furthermore, the medium effect may be relevant in activating the reactions (with nucleophiles) of position 5. In fact, the presence of hydrogen bonding interactions affects the charge separation of the p bond C5-C4, as shown in Scheme 14

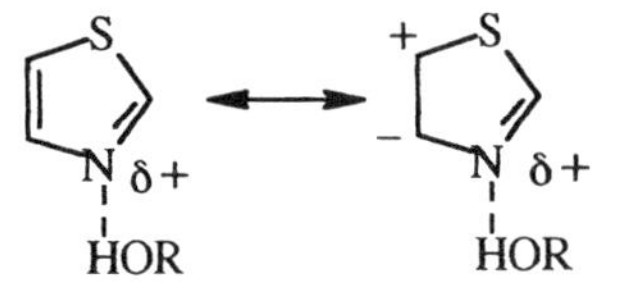

Scheme 14

The importance of interactions between the aza nitrogen of halogenothiazoles and protic media (or protic reactives) is well emphasized by the reactions of halogenothiazoles and undissociated benzenethiols. 2-Halogenothiazoles are known to react not only with sodium benzenethiolate,[9] but also with un-ionized benzenethiols[41] (without addition of bases), as depicted in Scheme 15.

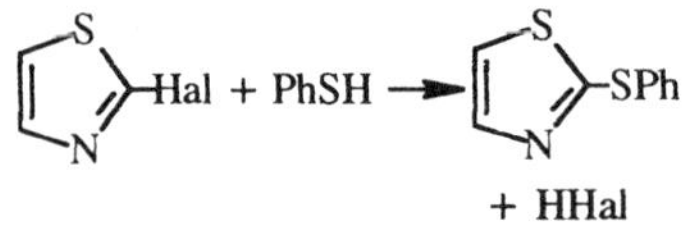

Hal = Cl, Br.

Scheme 15

The reaction of Scheme 15 is possible because of the presence of an equilibrium (acid-base like) which precedes the substitution process, leading to two activated species: the protonated thiazole derivative and the benzenethiolate ion. 5-Bromothiazole reacts with sodium benzenethiolate as well as with benzenethiol (Scheme 16).[42]

The reaction with benzenethiol does not occur with 4-bromothiazole which, on the contrary, is reactive toward sodium benzenethiolate.[42]

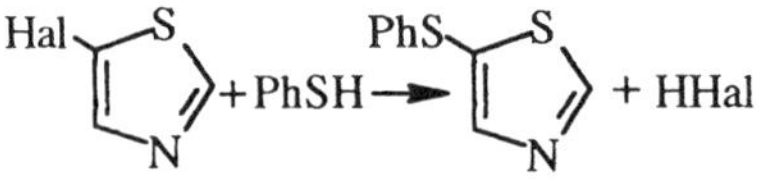

Hal = Cl, Br.

Scheme 16

The reactions in Scheme 16 (and the reactions of 5-bromothiazole and sodium benzenethiolate) do not show evidence for the presence of a mechanism involving radical species. The first step of the reactions in Scheme 16 is the pre-equilibrium reported in Scheme 17.

The equilibrium in Scheme 17 may afford more or less intimate ion pairs and is in agreement with the electronic effects of substituents on the benzenethiol and on the thiazolic substrate.[41,42]

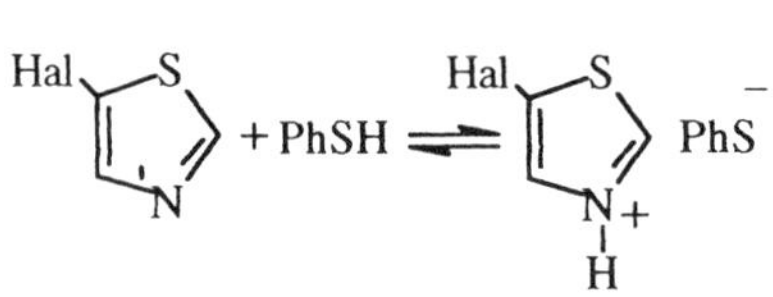

Scheme 17

Structure **32** in Scheme 18 illustrates the separation of charges of the protonated thiazoles regarding the bond between C4 and C5. Mesomeric form **32** may explain the reactivity of 5-halogenothiazoles and of the unreactivity of the 4-halogenothiazoles toward the benzenethiol.

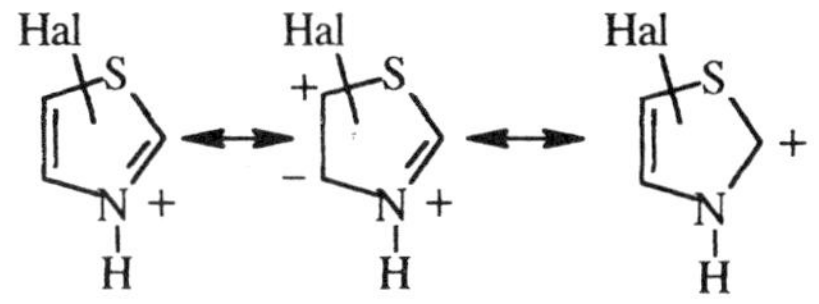

31 **32** **33**

Scheme 18

Obviously, the reactivity of position 2 of the thiazole ring toward the undissociated benzenethiol is enhanced by the protonation of aza nitrogen as illustrated by the mesomeric form **33**.

This picture of the reactivities of the three positions of the thiazole ring, explains the unusual activation of position 5 which has been indicated to be negatively charged by orbital calculation data which, obviously, cannot take into account the solvents or the particular interactions between the substrate and the reagents.

In principle, the reactivity of the 5-halogeno-2-thiazoleamines toward nucleophiles was completely unexpected[44] because of the electron-releasing effect of the amino group (by a mesomeric mechanism). In fact, position 5 of 2-thiazoleamine has usually been indicated[2] to be activated toward electrophilic reagents. The reactions of 5-halogeno-2-thiazoleamines afford, in almost quantitative yields, the de-halogenation substitution products[44,45] in position 5, as depicted in Scheme 19.

The same reaction was previously observed for 5-bromo-2-acetylthiazoleamine and sodium benzenethiolate[46] (R = $COCH_3$ in Scheme 19).

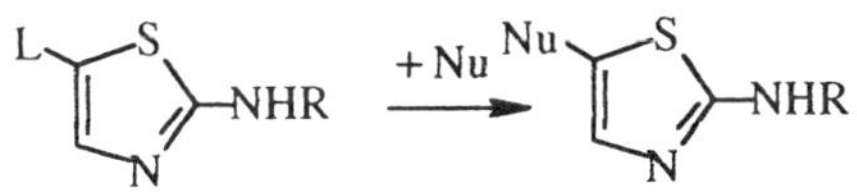

L = Cl, Br. Nu = nucleophiles (sodium methoxide, sodium benzenethiolate, piperidine); R = H, benzyl, *sec*.butyl, phenyl.

Scheme 19

The presence of the amino group dramatically enhances the reactivity of position 5 of the thiazole ring. For instance, the reaction toward sodium methoxide has[44] the $k_{5\text{-bromo-2-thiazoleamine}}/k_{5\text{-bromothiazole}}$ ratio = 10^4.

Since 2-N,N-diethylamino-5-bromothiazole is unreactive toward nucleophilic reagents, this reactivity cannot be ascribed to some particular electronic effects of the amino group, but rather to the presence of a hydrogen atom bonded to the amino group. Consequently this reactivity may be explained by the presence of tautomeric equilibria such as those reported in Scheme 20 which lead to the reactive species **34** and **36**.

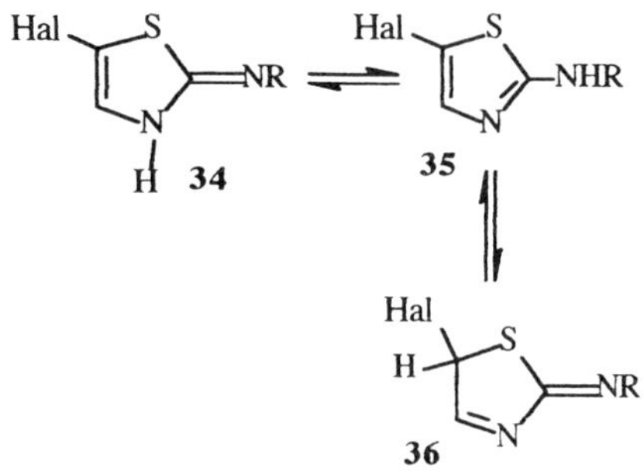

Scheme 20

The base catalyzed H/D exchange[38a] (in CD_3OD/DMSO-d6, 1:1 v/v) of the thiazole occurs via an anionic intermediate as well as in the thiazolium salts,[38b] in both positions 2 (see Scheme 21) and 5 (see Scheme 22).

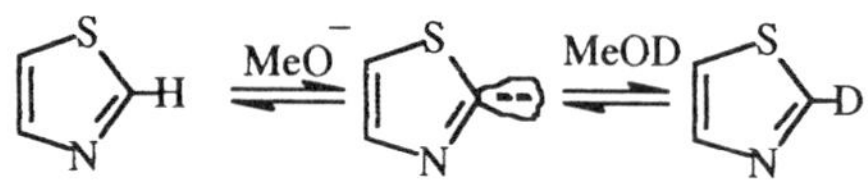

Scheme 21

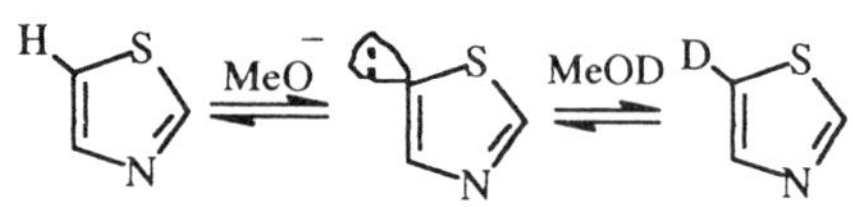

Scheme 22

The equilibria in Scheme 22 are supported by the H/D exchange rates of 2-substituted thiazoles: the ρ value (= 5.7) for the electronic effect of substituents bonded in position 2, agrees well with the formation of a negative charge, as illustrated in the Scheme 22.

2-Thiazoleamines may echange by a tautomeric form **37**, as indicated in Scheme 23.

On the contrary, the rate of H/D exchange in the MeOH/MeO^- system, indicates the improbability of the equilibria in Scheme 23, at least for considered substituents.

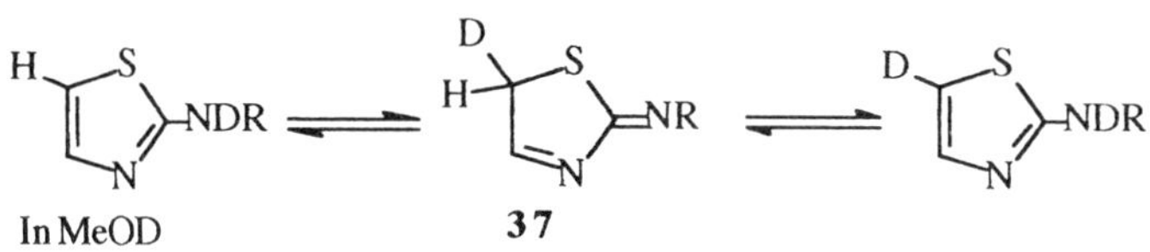

R= alkyl, phenyl

Scheme 23

In agreement with Scheme 22, when 2-aminothiazolium salts are considered, the H/D exchange (base catalyzed) in position 5 is much faster than for neutral bases. H/D exchange of **38** is 3 x 10^3 times faster than the rate of exchange of 2-N,N-dimethylthiazoleamine.

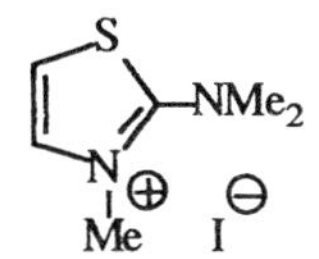

38

Consequently, the reactivity of 5-halogeno-2-thiazoleamines is more probably related to the presence of the tautomeric species **34** (see Scheme 20) than to the presence of the species **36**.

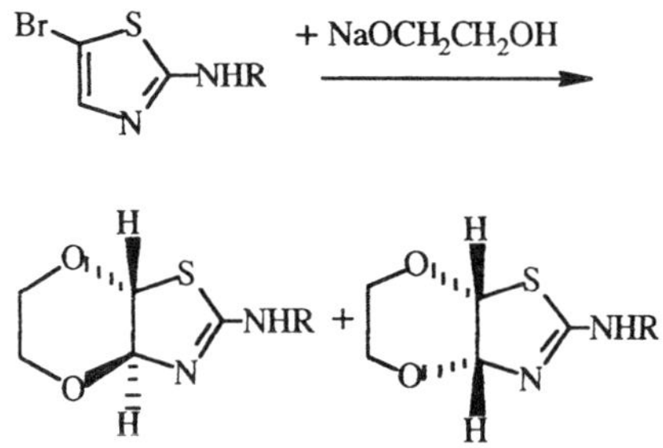

39 **40**

R = PhCHMe, $PhCH_2$, Ph, m.ClC_6H_4, p.$MeOC_6H_4$.

Scheme 24

When the reactions of 5-bromo-2-thiazoleamines are carried out with sodium 2-hydroxy-ethoxide, partially saturated compounds[47] (2-amino-3a,5,6,7a-tetrahydro-[1,3]thiazolo-[1,4]dioxins) **39** and **40** are obtained, according to Scheme 24.

The *trans* isomer **39** is obtained in higher yields than the *cis* isomer **40**.

The formation of partially saturated thiazole derivatives by reactions of nucleophiles with 5-halogeno-2-thiazoleamine is confirmed by the reactions between 5-bromo-2-thiazoleamines and sodium methoxide.[45] When sodium methoxide is used in an excess with respect to the thiazole derivative, the simple substitution products are obtained, but when equimolecular ratios are used, compounds **41** and **42** are obtained (see Scheme 25). Both **41** and **42** are unstable in the presence of bases.

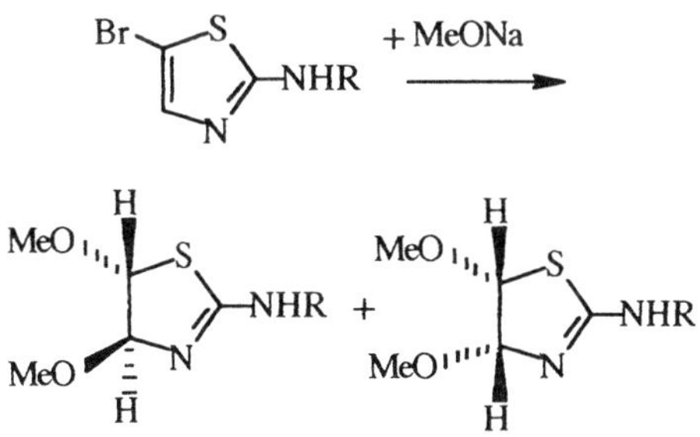

41 **42**

R = $PhCH_2$, MeCHEt , Ph, *m*.ClC_6H_4

Scheme 25

This reaction is indicated to be a nucleophilic substitution of the halogen atom on a tautomeric form of 5-halogeno-2-thiazoleamine followed by a rapid addition of the solvent to the substitution product (5-methoxy-2-thiazoleamine **43**), probably by base catalysis.

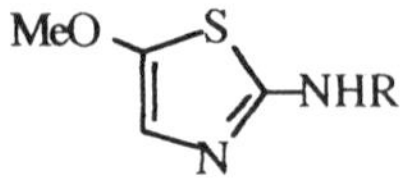

43

2-Amino-5-bromopyridine and 2-amino-5-bromopyrimidine in the same experimental conditions showed no reaction and the starting materials were recovered unchanged after long reaction times.

7. Reactivity toward electrophilic reagents

7.1. Introduction

The reactions of 2-thiazoleamine derivatives with electrophilic reagents display some surprising and interesting behaviours.

Apparently, electrophiles react toward 2-thiazoleamine derivatives at position 5 (see Scheme 26) as in common electrophilic aromatic substitution reactions activated by the strong electrondonating power of the amino group bonded in position 2. Usually, this position is considered to be directly conjugated with position 5.[2]

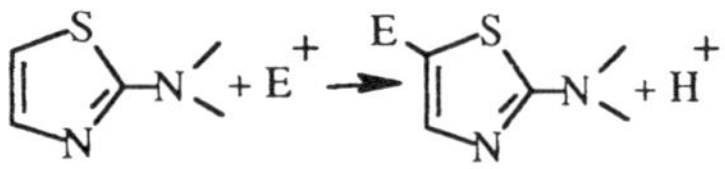

Scheme 26

Actually, 2-thiazoleamines acts as a tridentate nucleophile toward electrophilic reagents, as illustrated in Chart 2 (where PicCl is 2,4,6-trinitrophenyl-chloride).

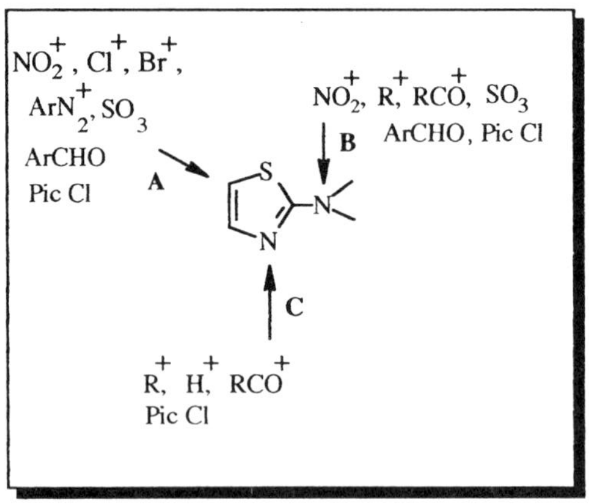

Chart 2

The explanation of the regioselectivities of some electrophiles is not completely clear and is therefore a challenging subject of investigation. Most reactions occurring at the exocyclic nitrogen (reactions B) are reversible reactions, while reactions at C5 of the thiazole ring are irreversible.

Protonation of 2-thiazoleamines (reactions C) occurs at the endocyclic nitrogen (see Section 3) which is, probably, the more basic and hard centre. Carbon electrophiles may react with both nitrogen atoms of 2-thiazoleamines, depending on the steric requirements of the attack.

The nitration reaction (with H_2SO_4/HNO_3 mixtures at low temperature) of 2-thiazoleamines affords nitramine **44** derivatives[48] which easily rearrange to 5-nitro-2-thiazoleamines[49], as reported in Scheme 27.

7.2. Reactions with halogens

The halogenation reactions of 2-thiazoleamine derivatives represent an apparent electrophilic aromatic substitution while it is an electrophilic addition to the C4-C5 double bond. This reaction was usually considered a simple attack of the electrophilic reagent on the heterocyclic "aromatic" substrate activated by the amino group (see Scheme 28).

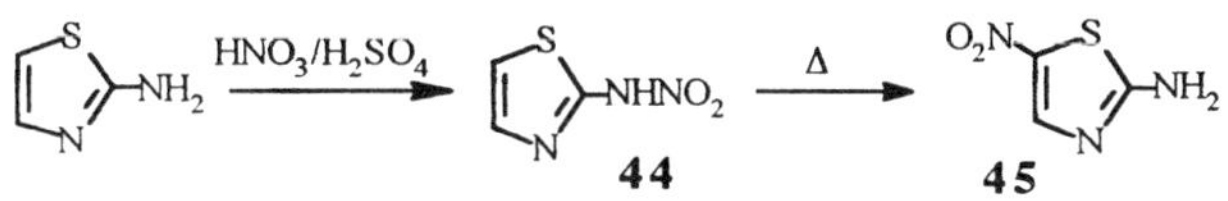

Scheme 27

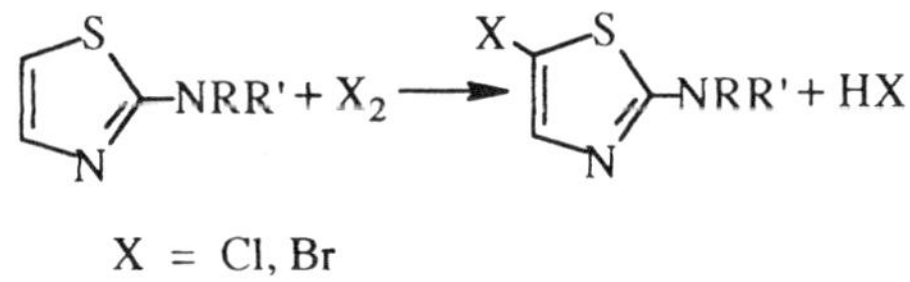

Scheme 28

When the bromination of 2-thiazoleamine derivatives is carried out in nucleophilic solvents (ROH) and at low temperature (less than 0° C), the partially saturated derivatives **46** in Scheme 29 were isolated in high yields.[50] The common halogenated 2-thiazoleamine is obtained by simply heating **46**, as indicated in the Scheme 29. The elimination reaction is the subsequent step.

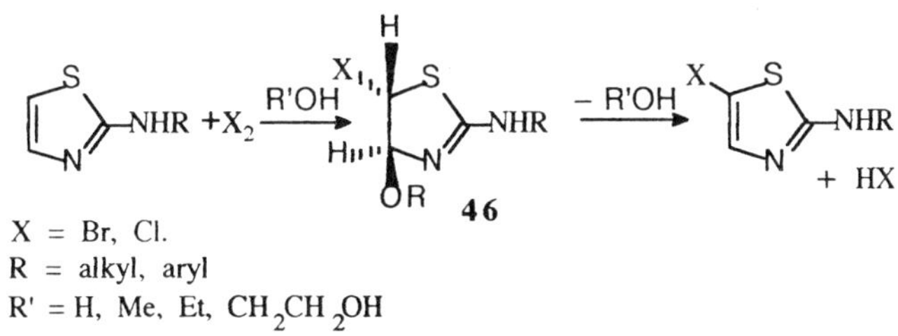

Scheme 29

The partially saturated derivative **46** is obtained only in the trans form, probably because **47** is the intermediate halonium ion toward the product of the addition reaction to C=C double bond.

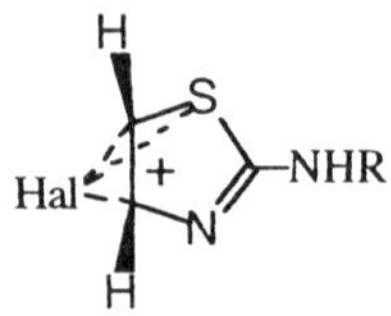

Hal = Cl, Br

47

This behaviour of 2-thiazoleamine derivatives parallels the behaviour of simple ethylene derivatives. The presence of sulfur probably stabilizes the halonium ion.

7.3. Reactions with the sp^2 carbon of aldehydes

The reactions of carbonyl compounds with ammonia derivatives afford the imino derivative **48**, but the C=N double bond may react faster than the C=O group toward amines to form 1,1-diamino derivatives **49**. The complete reaction of aromatic aldehydes with amines is shown in Scheme 30.

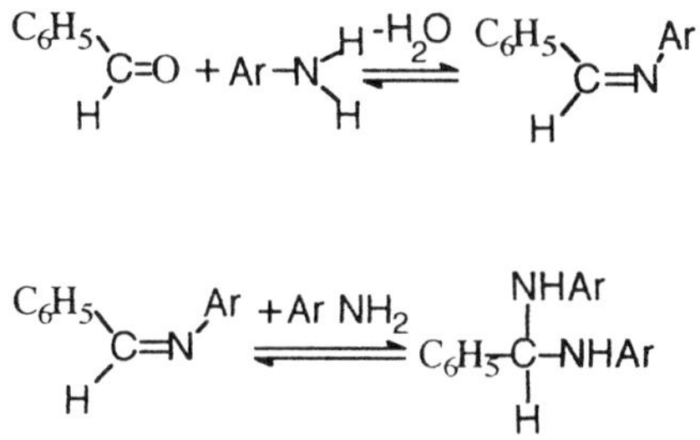

48 **49**

Ar = p.nitrophenyl, 2-thiazolyl, 2-benzothiazolyl, 2-pyridyl

Scheme 30

Generally, the C=N double bond is less reactive toward neutral nucleophiles than the C=O bond and the reactions in Scheme 30 afford the Schiff bases **48**. However when Ar (in Scheme 30) is an electronwithdrawing system, such as 4-nitrophenyl, 2-pyridyl, 2-thiazolyl and 2-benzothiazolyl, 1,1-diamino derivatives **49** are obtained in high yields.[51]

The electronwithdrawing power of the heterocyclic moiety of imines **49** most likely activates the C=N bond, as shown in the mesomeric system **50**.

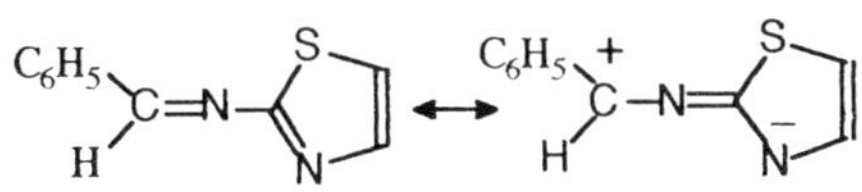

50

Consequently, the carbon atom of C=N bond is more prone to nucleophilic attack than is the C=O bond (without electrophilic catalysis). This explanation is in agreement with the findings that a secondary amine (piperidine) easily forms, with some aromatic aldehydes, the 1,1-diamino derivative **51**.[52]

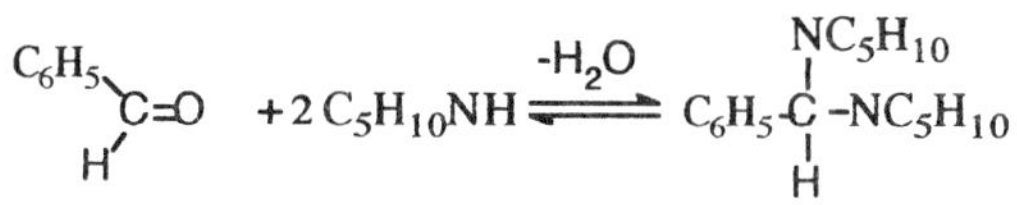

$C_5H_{10}NH$ = Piperidine **51**

Scheme 31

The formation of **51** probably arises from the immonium cation **52** which is generally accepted as a precursor in enamine formation. Primary amines give Schiff bases without further complications.

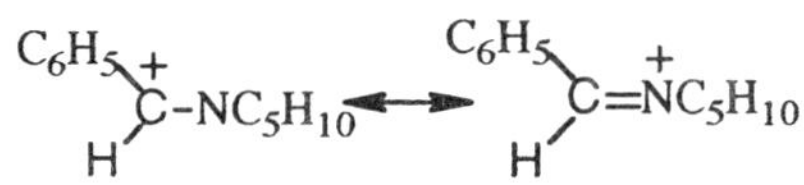

52

Aldheydes may react with the amino group of 2-thiazoleamine to yield imines or aminals by reversible reactions, as above reported, but aromatic aldehydes can also react toward the position 5 of the thiazole ring of 2-thiazoleamines.

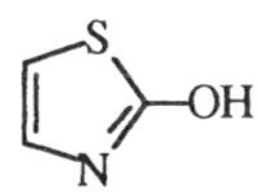

53

The condensation reaction between aromatic aldehydes and 2-thiazoleamines (or 2-hydroxythiazole **53**) derivatives[52] affords stable bis-adducts **54**, as depicted in Scheme 32. The nucleophilic center is carbon 5 of the thiazole ring.

The attack on position 5 is acid catalyzed. Probably the electrophilic reagent is the protonated aldehyde, and the first step of the reaction is the formation of the carbinol **55** which may repeat the attack on another thiazole ring by the carbocation arising from the carbinol **55** (see Scheme 33). This reaction is an irreversible process.

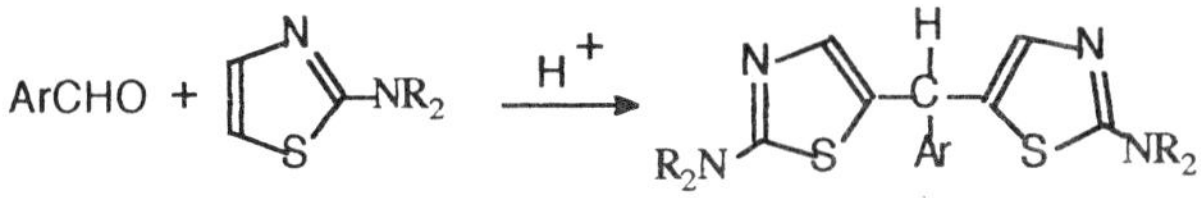

54

ArCHO = benzaldehyde, 4-methylbenzaldehyde, 4-methoxybenzaldehyde

R = H, Me

Scheme 32

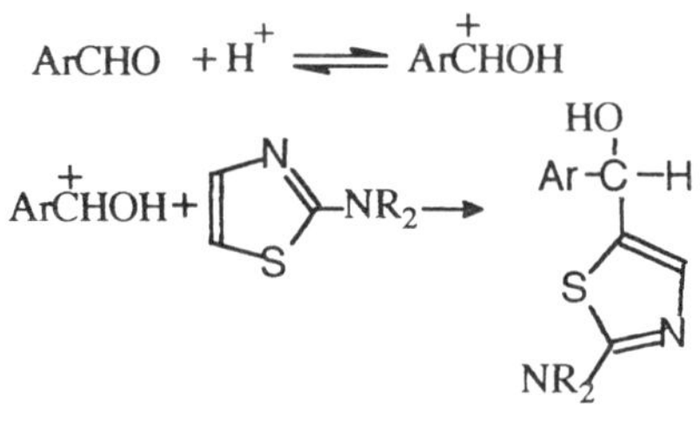

55

Scheme 33

7.4. Reactions with the sp² carbon of aromatic substrates (S_NAr reactions)

2-Thiazoleamine derivatives easily react (as nucleophiles) with activated aromatic substrates, such as 2,4-dinitro-halobenzenes and 2,4,6-trinitro-halobenzenes.

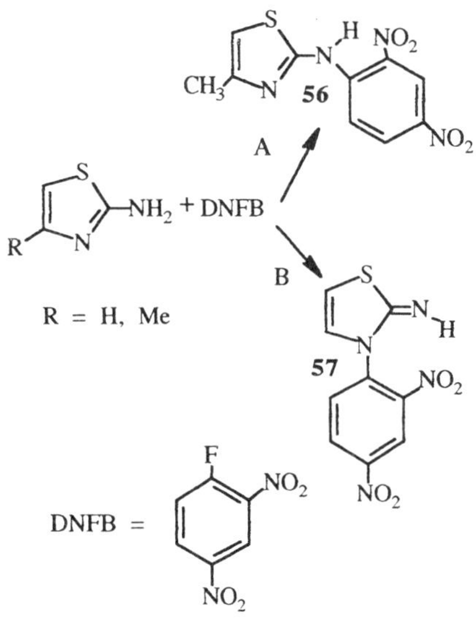

Scheme 34

The reactions of 2-4-dinitrofluorobenzene[30] (DNFB), in dimethylsulphoxide, occur at both nitrogen atoms of 2-thiazoleamine (see Scheme 34).

Compounds **56** and **57** are obtained under the experimental conditions $[TZ]_o >> [DNFB]_o$, *i.e.* when the initial concentration of the thiazoleamine is higher than the initial concentration of the dinitrofluorobenzene. Not only hen the DNFB is initially in large excess with respect to the 2-thiazoleamine $\{[TZ]_o << [DNFB]_o\}$, but also with the ratio $[TZ]_o:[DNFB]_o = 1/2$, the major product (90% of yields) recovered from the reaction mixtures is **58** which arises from the double attack of the DNFB on the thiazoleamine (see Scheme 35).

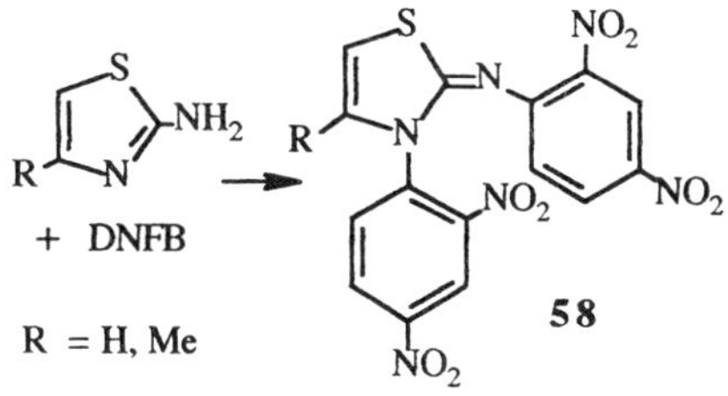

Scheme 35

The structure of **58** has been ascertained by X-ray diffraction analysis.[30]

The reactions in Scheme 34 occur at the endocyclic nitrogen or at the exocyclic nitrogen depending on the steric requirements of the thiazole ring: 4-methyl-2-thiazoleamine reacts at the exocyclic nitrogen (reaction A of Scheme 34), while 2-thiazoleamine reacts at the endocyclic nitrogen (reaction B of Scheme 34). Steric requirements of the endocyclic nitrogen are probably the major factors which explain the behaviours reported in Chart 1.

2,4,6-Trinitro-chlorobenzene reacts[54] with 2-thiazoleamine derivatives in a similar way to that of 2,4-dinitro-fluorobenzene (see Scheme 34). In this case the regioselectivity of the reaction also arises from steric hindrance of the nitrogen atoms of 2-thiazoleamine derivatives. When both nitrogen atoms are sterically hindered, the electrophile (the picryl chloride) attacks the carbon 5 of the thiazole ring,[54] as reported by Scheme 36.

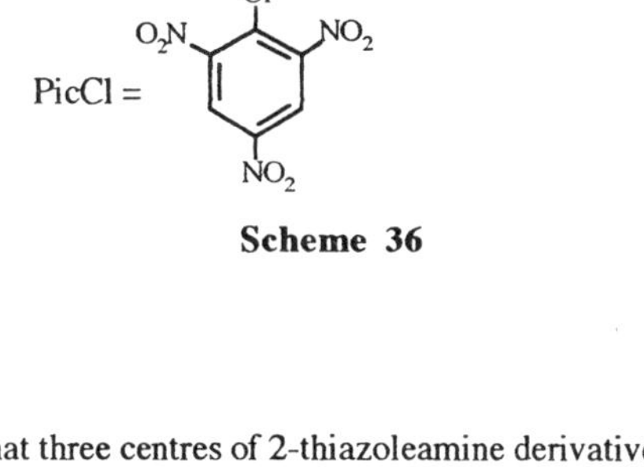

Scheme 36

This reaction confirms that three centres of 2-thiazoleamine derivatives are active in reactions toward electrophilic reagents, as illustrated in Chart 1 for a number of electrophilic reagents.

2-Pyridineamine derivatives, under the same experimental conditions of 2-thiazoleamine derivatives, afford only the product of condensation of the exocyclic nitrogen, as illustrated in Scheme 37

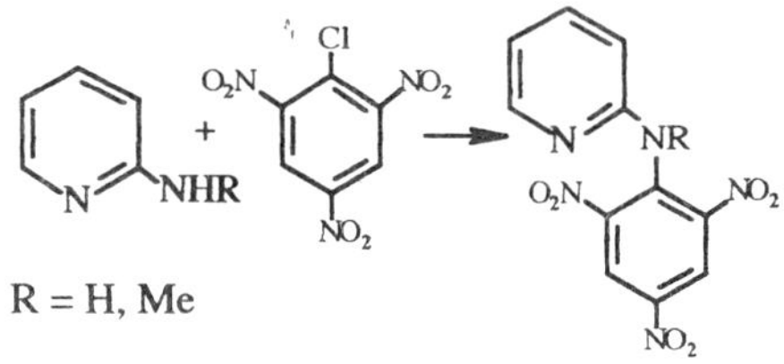

Scheme 37

Kinetic measurements[55] of the reactivities relative to the endocyclic and the exocyclic nitrogen indicate that the aza nitrogen is more reactive than the amino (exocyclic) nitrogen in 2-thiazoleamine. The k_B/k_A ratio = 32 is reported,[55] and imino forms are more reactive than the amino form. The reactions in Scheme 38 are 1,500 and 54 times faster, in dimethylsulphoxide[55] and in benzene,[56] respectively, than the reactions A in Scheme 34: the imino nitrogen is therefore a better nucleophile than the amino nitrogen. Consequently, the formation of **58** arises from the imino form **56** because the first attack is on the aza nitrogen yielding the imino form with the exocyclic nitrogen sp^2 hybridized.

In benzene, the reactions in Schemes 34 and 38 follow simple second order kinetics in both reactants. When the reactions of DNFB with amines are carried out in benzene, the common kinetic behaviour is characterized by autocatalytic phenomena: k_{obs} ($s^{-1}mol^{-1}l$) values increase upon increasing the initial value of the concentration of the nucleophile. The intermediates in the reactions are represented by **59** in the reaction in Scheme 34 and by **60** for reactions of Scheme 38.

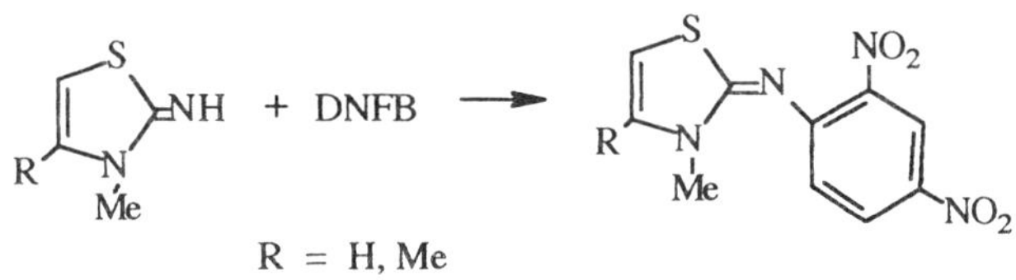

Scheme 38

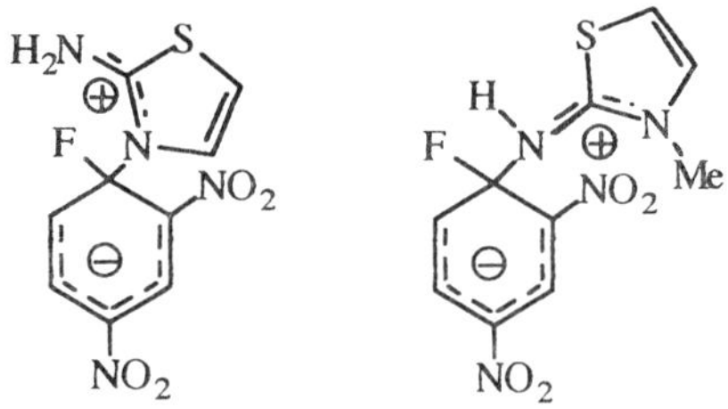

59 **60**

In both cases, the auto-catalytic behaviour may be expected (on the basis of the comparison with of the more common reactions of simple aliphatic and aromatic amines) from the zwitterionic intermediate toward the reaction products.[58]

On the other hand, both reactions are catalyzed by 1,4-diaza[2,2,2]bicyclooctane **61**, 2-hydroxypyridine **62** and δ-valerolactame **63**, which are catalysts in S_NAr reactions in non-polar solvents.[57,58]

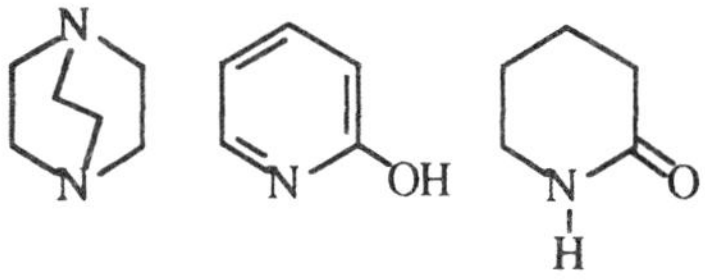

61 **62** **63**

The lack of auto-catalysis in the reactions between DNFB and thiazoleamine derivatives carried out in benzene is unexpected because the intermediate **59** and **60** are very similar to the zwitterionic intermediates in S_NAr reactions, and the pKa and nucleophilicities of the thiazole derivatives considered as nucleophiles are very similar to those of the amines displaying strong auto-catalytic behaviours. We explain the lack of self catalysis by considering that the mixture of DNFB and the thiazole derivatives does not show the presence of the molecular complexes which are observed in the reaction mixtures of DNFB and other more common aliphatic and aromatic amines.

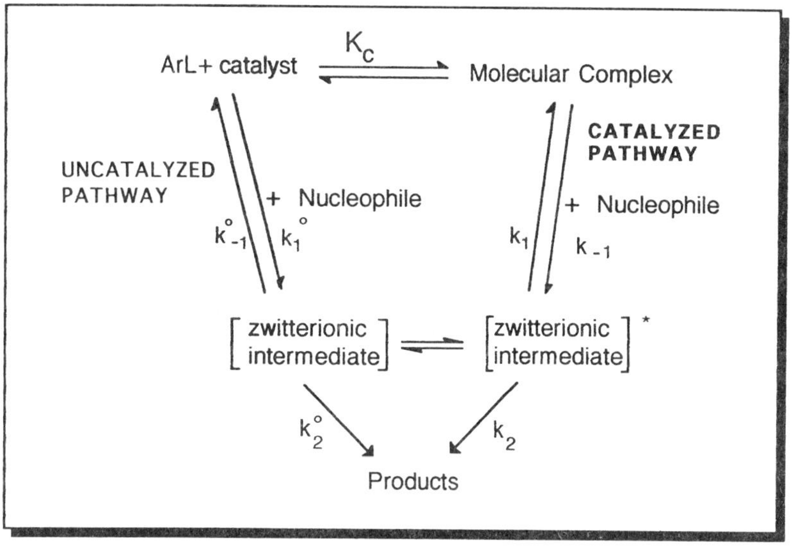

Chart 3

This fact is further support for the explanation that the catalytic kinetic behaviour in S_NAr reactions in apolar solvents cannot be attributed to the HF abstraction from the zwitterionic intermediates. However, it is reasonable to explain this behaviour with the presence of interactions between the nucleophile and the substrate in an equilibrium preceding the substitution attack. The reaction pathways are illustrated in Chart 3.

In fact, when the catalyst is prone to complex with the substrate, (as in the cases of the DABCO, 2-hydroxypyridine and of other reacting or non-reacting catalysts, k_{obs} increases when the initial concentration of the catalyst increases. The interactions between the catalysts (which are polar substances) and the substrate enhance the rate of the substitution reaction by favouring the charge separation on the reaction pathway toward the zwitterionic intermediate.[58]

The geometries of the considered zwitterionic intermediates **59** and **60** appear to be considerably different, but this fact is unimportant with respect to the absence (or the presence) of the catalyzed pathway. On the other hand, substantial evidence has been reported on the presence of interactions between the catalyst (or the nucleophile) and the nitro derivatives yielding a situation favourable (with respect to the polarity of the medium surrounding the reaction centre) for the attack of the nucleophile when the solvent is not likely to favour the charge separation as the reaction is going on toward the transition state preceding the zwitterionic intermediate.[59]

In a recent review, the assumption[60] about catalytic behaviours of catalysts like **61**, **62** and **63** involves the association between the hydroxypyridine and the nucleophile by formation of hydrogen-bonding interactions by considering **62** and **63** as proton acceptors and the amine as a proton donor. Contrary to this assumption, pKa values of the considered pairs (amines/amides) strongly support *the inverse* hydrogen-bonding interaction as a part of the explanation of the kinetic behaviour: the amine is the proton acceptor and the amides (or the 2-hydroxypyridine) is a proton donor.[59]

In conclusion, the reaction in Chart 3 is a reasonable pathway to explain the kinetic behaviour in the case of the reactions of thiazoleamines (as nucleophiles) which are not able to undergo auto-catalysis because they are not able to associate the substrate.

8. Conclusion

The main statements inferable from the reported reactions of thiazole derivatives may be summarized as follows:

i) In principle, geometrical assessment and charge distribution put the thiazole ring in a favourable situation for the observation of some chemical behaviours which cannot be observed by investigations on more classical aromatic systems (benzene, pyridine, ...).

ii) 2-Thiazoleamines may exist in the imino forms depending on two main parameters: a) the presence of strong electronwithdrawing effects on the amino group; b) the polarity of the medium.
2-Pyridineamine derivatives, in the same experimental conditions show the presence of the only amino, aromatic form.

iii) The position of the tautomeric equilibrium may be detected (by small changes in the medium's polarity) without the use of spectroscopic properties of fixed imino/amino parents: this possibility allows to avoid some assumption (and errors) arising from the use of physical parameters of fixed parents.

iv) The reactions of 5-halogeno-2-thiazoleamine derivatives toward nucleophilic reagents are unusual and can be explained by the tautomerism of the thiazoleamine. In some cases partially saturated compounds are obtained.

v) The reactions of 2-thiazoleamine derivatives toward electrophilic reagents indicate that 2-thiazoleamine presents three nucleophilc centres: the exocyclic nitrogen, the endocyclic nitrogen and the carbon 5 of the thiazole ring.

vi) The explanation of observed regioselectivities is based on the reversibility of the considered reactions, the polarizability and the basicity of reaction centres, the steric requirements of the reactions and the steric hindrance of the involved centres.

vii) Halogenation reactions of 2-thiazoleamine derivatives occurs by an addition/elimination pathway in which the addition step is similar to the addition to the double bond of alkenes. The reaction does not occur through the traditional mechanism of aromatic electrophilc attack at C5 of the thiazole ring.

References

1. Metzger, J. V., Ed.; "*Thiazole and its Derivatives. Part one.*"; John Wiley & Sons: New York, 1979.
2. Metzger, J. V., Ed.; "*Thiazole and its Derivatives. Part two.*"; John Wiley & Sons: New York, 1979.
3. Phan-Tan-Luu, R.; Surzur, J. M.; Metzger, J. V.; Aune, J. P.; Dupuy, C. *Bull. Soc. Chim. France* **1967**, 3274.
4. Forlani, L.; Breviglieri, G.; De Maria, P. *J. Chem. Soc. Perkin Trans. 2* **1979**, 163.
5. Chapman, N. B.; Shorter, J. "*Advances in Linear Free Energy Relationships*"; Plenum Press, London: 1972.
6. Charton, M. *J. Amer. Chem. Soc.* **1964** *86*, 2033.
7. Charton, M. *J. Org. Chem.* **1965**, *30*, 3346.
8. Bartoli, G.; Sciacovelli, O.; Bosco, M.; Forlani, L.; Todesco, P. E. *J. Org. Chem.* **1975**, *40* , 1275.
9. Bosco, M.; Forlani, L.; Liturri, V.; Riccio, P.; Todesco, P.E. *J. Chem. Soc. (B)* **1971**, 1373.
10. Forlani, L.; De Maria, P.; Fini, A. *J. Chem. Soc. Perkin Trans. 2* **1980**, 1156.
11. Forlani, L.; Lunazzi, L.; Medici, A. *Tetrahedron Lett.* **1977**, *18*, 4525.
12. Elguero, J.; Marzin, C.; Katritzky, A. R.; Linda, P. "*The Tautomerism of Heterocycles*"; Academic Press, London: 1976.
13. Angyal, S. J.; Angyal, C. L. *J. Chem. Soc.* **1952**, 1461
14. Forlani, L.; De Maria, P. *J. Chem. Soc. Perkin Trans. 2* **1982**, 535.
15. Kruger, G. J.; Gofner, G. *Acta Crystallogr.* **1972**, *B28*, 272.
16. Takahashi, T.; Otsuka, S.; Masuda, H.; Hirata, M.; Ito, Y.; Hamada, Y. *Bull. Chem. Soc. Jpn.* **1976**, *49*, 2770.
17. Forlani, L., *Gazz. Chim. Ital.* **1981**, *111*, 159.
18. Tabes, H.; Searles, S. *J. Amer. Chem. Soc.* **1959**, *81*, 2600.
    Holmes, J.; Kivelson, D.; Drinkard, W. *J. Amer. Chem. Soc.* **1962**,*84*, 4677.
    Traynham, J. G.; Knesel, G.A. *J. Org. Chem.* **1966**, *31*, 3355.
    Tribble, M. T.; Traynham, J. G. in "*Advances in Linear Free Energy Relationships*"; Chapman, N. B.; Shorter, J., Eds.; Plenum Press, London: 1972.
19. Forlani, L.; Battaglia, L.; Bonamartini Corradi, A.; Sgarabotto, P. *J. Cryst. and Spectr. Chem.* **1992**, *22*, 705.
20. Forlani, L. *J. Heterocycl. Chem.* **1992**, *29*, 1461.
21. Kracov, M. H.; Lee, C. M.; Mautner, H. G. *J. Amer. Chem. Soc.* **1965**, *87*, 892.
22. Davidovics, G.; Chouteau, J. *Spectrochim. Acta*, **1966**, *22*, 703. Metzger, J., Bonzom, A.; *Spectrochim. Acta* **1966**, *22*, 719
23. Forlani, L.; Battaglia, L.; Bonamartini Corradi, A.; Pelosi, G. *J. Cryst. and Spectr. Chem.* **1990**, *20*, 499.
24. Wheland, G. W. "*Resonance Theory in Organic Chemistry*"; J. Wiley and Sons, New York: 1955.
    Yokoyama, T.; Taft , R. W.; Kamlet, M. J. *Spectr. Acta* **1984**, *40A*, 669.

Kamlet, M. J.; Abboud, J. L.; Taft, R. W. *J. Amer. Chem. Soc.* **1977,** *99*, 6027.
Hiberty, C.; Ohanessian, G. *J. Amer. Chem. Soc.* **1984,***106*, 6963.
Krygowski, T. M.; Maurin, J. *J. Chem. Soc. Perkin Trans. 2* **1989**, 695.

25. Katritzky, A. R.; Lagowski, J. M. "*Advances in Heterocyclic Chemistry*": Academic Press, London: 1963, vol. 1.
26. Costakis, E.; Canone, P.; Tsata, G. *Canad. J. Chem.* **1969**, *47*, 4483.
27. Elguero, J.; Marzin, C.; Katritzky, A. R.; Linda, P. *"The Tautomerism of Heterocycles"*; Academic Press, London: 1976, p. 429.
28. Annese, M.; Bonamartini Corradi, A.; Forlani, L.; Rizzoli, C.; Sgarabotto, P. *J. Chem. Soc. Perkin Trans. 2* **1994**, 615
29. Birkinshaw, T. N.; Harkin, S. A.; Kaye, P. T.; Meakins, G. D.; Smith, A. *J. Chem. Soc. Perkin Trans. 2* **1982**, 939.
30. Forlani, L; De Maria, P.; Foresti, E.; Pradella, G. *J. Org. Chem.* **1981**, *46*, 3178
31. Forlani, L.; Guastadisegni, G.; Raffellini, L.; Todesco, P.E.; Foresti, E. *Gazz. Chim. Ital.* **1990**, *120*, 493.
32. Ceder, O.; Beljer, B. *Tetrahedron* **1975**, *31*, 963.
33. Forlani, L.; Magagni, M.; Todesco, P. E. *Gazz. Chim. Ital.* **1979**, *109*, 377.
34. Bosco, M.; Forlani, L.; Todesco, P. E.; Troisi, L. *J. Chem. Soc. Perkin Trans. 2,* **1976**, 398.
35. Forlani, L.; Todesco, P. E. "*Thiazole and its Derivatives. Part one.*" Metzger, J. V., Ed.; John Wiley & Sons: New York, 1979, Chapt. 5.
36. Bosco, M.; Forlani, L.; Sapone, D.; Todesco, P. E. *Boll. Sci. Fac. Chim. Ind. Bologna* **1969**, *27*, 86.
37. Coburn, R. A.; Landesberg, J. M.; Kemp, D. S.; Olofson, R. A. *Tetrahedron* **1970**, *26* 685.

38a. Forlani, L.; Magagni, M.; Todesco, P. E. *J. Chem. Soc. Perkin Trans. 2,* **1979**, 1145.
38b. Washabaugh, M. W.; Jenks, W. P. *J. Amer. Chem. Soc.* **1989**, *111*, 674.

39. Haake, P.; Bausher, L. P.; McNeal, J. P. *J. Amer. Chem. Soc.* **1971**, *93*, 7045.
40. Otsuji, Y.; Kimura, T.; Sugimoto, Y.; Imoto, E.; Okawara, T. *Nippon Kagaku Zhassi* **1959**, *80*, 1021.
41. Bosco, M.; Liturri, V.; Troisi, L.; Forlani, L.; Todesco, P. E. *J. Chem. Soc. Perkin Trans. 2* **1974**, 508.
42. Forlani, L. *Gazz. Chim. Ital.* **1984,** *114*, 279.
43. Forlani, L.; Todesco, P. E.; Troisi, L. *J. Chem. Soc. Perkin Trans. 2* **1978**, 1016.
44. Forlani, L.; Medici, A.; Todesco P. E. *Tetrahedron Lett.* **1976**, *17*, 201.
45. Forlani, L.; Medici, A. *Gazz. Chim. Ital.* **1983,** *113*, 807.
46. Mahajanshetti, C. S.; Basanagoudar, L. D. *Can. J. Chem.* **1967**, *45*, 1808.
47. Forlani, L.; Medici, A.; Ricci, M.; Todesco, P. E. *Synthesis* **1977,** *113*, 321.
48. Viron, S.; Taurins, A. *Can. J. Chem.* **1953**, *31*, 885.
49. Dickey, J.; Towne, E.; Wight, G. *J. Org. Chem.* **1955**, *20*, 499.
50. Forlani, L.; Medici, A. *J. Chem. Soc. Perkin Trans. 1* **1978**, 1169.
51. Forlani, L.; Sintoni M.; Todesco, P. E. *Gazz. Chim. Ital.* **1986**, *116*, 229.
52. Forlani, L.; Marianucci, E.; Todesco, P. E. *J. Chem. Res. (S)* **1984**, 126.
53. Forlani, L. *J. Chem. Res. (S)* **1990**, 70.

54. Forlani, L.; Guastadisegni, G.; Raffellini, L.; Todesco, P. E.; Foresti, E. *Gazz. Chim. Ital.* **1990**, *120*, 493.

55. Forlani, L.; Sintoni, M. *J. Chem. Res. (S)* **1986**, 110.

56. Forlani, L.; Sintoni, M. *J. Chem. Soc. Perkin Trans.2* **1988**, 1959.

57. Bernasconi, C. F."*Mechanism and Reactivity in Aromatic Nucleophilic Substitution Reactions*'"; M.T.P. Int. Rev.; Org. Chem., Ser. One, 1973, *3*, 33.

58. Forlani, L.; Marianucci, E.; Todesco, P. E. *Gazz. Chim. Ital.* **1992**,*122*, 349.

59. Forlani, L. in "*The Chemistry of Amino, Nitroso, Nitro and Related Groups*"; Patai, S., Ed.; J. Wiley and Sons, London: 1996, Chapt. 10.

60. Nudelman, S. N. in "*The Chemistry of Amino, Nitroso, Nitro and Related Groups*"; Patai, S., Ed.; J. Wiley and Sons, London: 1996, Chapt. 26, p. 1253.

Combinatorial Models and Polarity Control Rules in Heterocyclic Design

Eugene V. Babaev

Chemistry Department, Moscow State University, Moscow, 119899, Russia

Abstract. *The review is devoted to applications of combinatorial models for classification and design of heterocyclizations, recyclizations, and ring opening reactions of heterocycles. These models are based on the combination of concepts of reaction graphs and various polarity control rules. The existing models of reaction graphs and polar classifications of reagents are critically overviewed. The model of polar cyclization graphs is suggested for design of novel routes to the consonant heterocycles. The models of local and global polarity control in the heterocyclic ring opening reactions and their applications are discussed. The model of the ring bonds redistribution graphs for the recyclization reactions and its applications to the computer-assisted search of novel ring transformations are reviewed. Elaborated computer programs and examples of experimental confirmation of some predictions are discussed.*

1. Introduction

The ring with heteroatoms is the central subject of heterocyclic chemistry. Since the cycle is the fundamental topological concept, the processes of cycle formation and destruction are among the central topics in the chemistry of heterocycles. The reactions with such essential topological changes are the well-known processes of heterocyclizations (*appearance* of a cycle), ring opening reactions (*disappearance* of a cycle), and recyclizations (*transformation* of one cycle to another). Many generations of chemists have contributed to understanding peculiarities of these three fundamental types of reactions, but even nowadays the discovery of new examples of such reactions is occasional. Due to the lack of exact mathematical models the chemistry of heterocycles is rather an art than a strict science. Of course, there are many attempts to rationalize the logic of heterocyclic chemistry by applying various theoretical models and implement them as computer programs (see e.g., the comprehensive review[1] on this topic). In spite of these efforts, the spacious factual material on heterocyclic rings (trans)formation and cleavage still requires better classification.

In our early works we suggested few combinatorial models helpful for classification and design of reactions in heterocyclic chemistry. These models, based on the combination of two important concepts of the *reaction graphs* and of the *polarity control rules,* have been applied to heterocyclizations, recyclizations and ring opening reactions of heterocycles. The goal of this paper is to review our methodology and to prove its usefulness in solving some selected problems of heterocyclic chemistry. The plan of the review is the following. In Section 2 we shall briefly review the concept of reaction graphs. Then in Section 3 we shall critically review the concept of polarity control in the heterocyclization reactions and suggest the model of polar cyclization graphs as a useful tool in classification and design of novel cyclizations. In Section 4 the problem of local and global polarity control in the ring opening reactions will be discussed. In the last Section 5 the model of the ring bonds redistribution graphs for the recyclization reactions will be reviewed, and our attempts of the computer-assisted search of novel ring transformations will be discussed.

2. Reaction Graphs as the Diagrams of Bonds Redistribution

2.1. What is the reaction graph?

During the last two decades methods of graph theory became useful instruments for the analysis of various problems in theoretical and experimental chemistry[2,3]. A traditional application of graphs in chemistry is the description of molecular structures. Indeed, a graph, as a set of vertices and edges, corresponds to the classical notion of a structural formula as a sequence of atoms and bonds. An unusual important application of graph theory is the use of a graph as a representation of a chemical reaction, namely, for the description of bond redistribution in the course of an organic reaction. The first mathematical and graph-theoretical approaches to the description of structural changes during a chemical reaction were developed in the 70's by Balaban[4], Ugi and Dugundji[5,6], Hendrickson[7], and Trach and Zefirov[8,9]. Later publications in this direction are represented by the research of Fujita[10,11] and others, see reviews[1,12,13].

How can the graph be used as an image of chemical reaction? The central idea of virtually all the approaches (see examples in Scheme 1) consists in the establishment of a correlation between the atoms of the

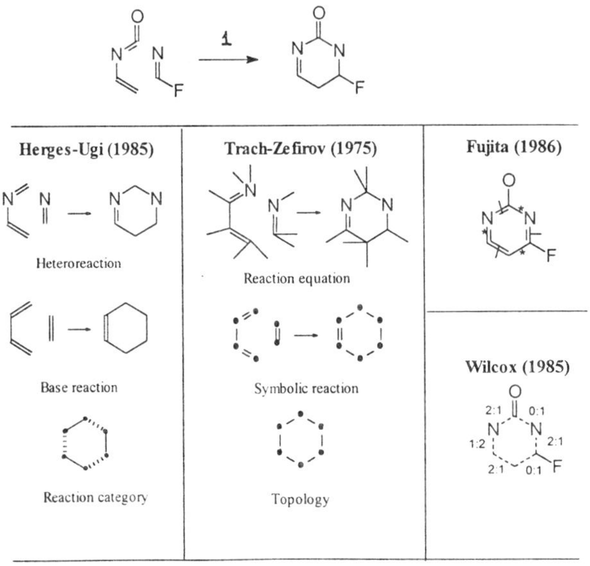

Scheme 1. Examples for description of the same reaction by various reaction graphs[1].

the reactants and the products by means of their "superimposition" upon one another. After this confluence (or superposition), one can unambiguously describe the *changes in the bonds* during a chemical reaction, expressing this redistribution of bonds by a graphical diagram. Such diagrams have different names (for example, symbolic equations, reaction categories, and even virtual transition states), and we shall use below the term *reaction graphs*. As an example of how the diagrams of the same reaction look in the languages of the various approaches, one can consider a conceivable heterocyclization 1 and the corresponding reaction graphs in Scheme 1. (For various aspects of representing the reactions see reviews[12,14].)

2.2. The use of reaction graphs for classification and design of new reactions

The reaction graphs (like those reported in Scheme 1) substantially simplify the usual chemical equation which is changed to the graphical aspect ("skeleton") of the reaction. Such skeleton, as a rule, contains atoms that change the oxidation state (or any environment) and bonds that change their multiplicity during the reaction. Thus chemical reactions can be reduced to a finite number of bond redistribution diagrams. A comparison of such diagrams for various reactions makes it possible, first and foremost, to reveal the similarity or the degree of novelty of the reactions themselves[5,8].

Another way to use bond redistribution diagrams is the prediction of new reactions. For this purpose one is required to build on real substituents, chains, rings, etc. to these diagrams. In a number of cases this sort of predictions was heuristic. The indicated principle was the cornerstone of the computer programs SYMBEQ[15,16] and IGOR[17] capable to generate the reaction graphs. The first example of computer-assisted discovery of new reaction was the novel synthesis of the furan ring (equation 2 in Scheme 2) predicted by the SYMBEQ program[15]. Another example of new reaction, that has been predicted by means of the IGOR program and discovered experimentally, is the novel diene synthesis as the result of unknown heterocyclic ring opening reaction (equation 3 in Scheme 2)[18].

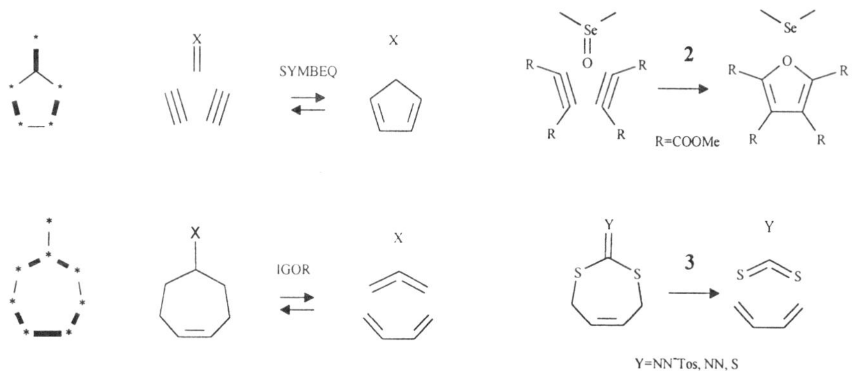

Scheme 2. Examples of prediction of novel reactions by use of reaction graphs methodology. Left: the reaction graphs; middle: "skeletal" equations restored from reacttion graph; right: actually found examples of new reactions[15,17].

2.3. Advantages and disadvantages of the reaction graph models

The concepts of reaction graphs have introduced into the chemistry a novel idea of a *function* (or operator) that is acting on the set of molecular structures. Evidently, this function can be equally applied to either starting ensemble of reactants (predicting the products) or to the final set of products (predicting the reactants). Hence, having been first claimed, the concept of reaction graphs looked as a new promising tool for resolving key problems of organic synthesis planning. This forecast, however, turned out to be too optimistic, and the reaction graphs still have not become the natural language of chemistry.

Although the idea of reaction graphs was developed to rigorous mathematical level[15], it is of rather limited use in the chemistry of heterocycles. Thus, in many important reactions, e.g., usual cyclocondensations and recyclizations, it is either difficult to built simple reaction graphs or to apply them fruitfully for classification and design of novel reactions. The first problem is that the multicomponent reactions (that are widespread in heterocyclic synthesis) involve many leaving groups, and considering of all such groups may dramatically complicate the structure of the resulting bond redistribution diagram. The second serious problem lies in the tautomerism of heterocycles (and some their acyclic precursors), which makes difficult to assign certain reaction graphs for such processes. Indeed, taking into account various tautomeric pairs results in many "waste" reaction graphs, while the important information about the bonds from the heterocyclic rings is often lost.

It is therefore not surprising that the concept of reaction graphs may find (and has really found) most useful application in the specific area of classification and design of the concerted processes. In such reactions the bonds are redistributed along a cyclic framework (actual or imaginary), and the location of all formed/broken bonds along the monocyclic reaction graph is the definite procedure. For more complex reactions the (potentially fruitful) idea of reaction graphs requires some natural modifications.

3. The Graphs of Heterocyclization Reactions

3.1. Classical disconnection schemes of heterocycles as the specific type of reaction graphs

Although heterocyclic chemists have no habit to use the reaction graphs for classification and design of heterocyclization reactions, they do intuitively use quite similar principle of *mapping* the atoms of a heterocyclic product into the atoms of starting reactants. Indeed, the standard graphic method used to demonstrate the diversity of possible cyclizations leading to the given heterocycle is the disconnection of the skeleton of the heterocycle into fragments. On paper, this is done by simply removing (or marking by a dashed line) the skeletal bonds that arise during the course of the reaction. By now it is difficult to establish the authorship of this construction, which is intuitively obvious: as far back as the 1950s, the use of such "diagrams" or "schemes of synthesis" was extremely popular[19]. Today it is simply impossible to review the methods of synthesis of a specific heterocycle without such disconnection schemes (see for example[20]); and in the classic handbook[21] these diagrams are methodically enumerated and used for a comprehensive classification of the synthesis of diverse heteroaromatic systems.

Drawing of a disconnection scheme means direct and reverse mapping of matching atoms of the acyclic reactants and cyclic product. Therefore these schemes resemble the reaction graphs, although they are in some sense "bad reaction graphs." Indeed, such diagrams bear incomplete information on the total redistribution of bonds and changes in atoms' environment and display only those cyclic bonds that are formed. (Thus, the leaving groups of reactants eliminated in the course of reaction usually are not presented in the disconnection schemes.) Nevertheless, let us call them the *cyclization graphs*.

All of the advantages of describing heterocyclic syntheses by such graphs are obvious at first glance. If there are no permutations of atoms via cyclization it is easy to establish the structural kinship between different syntheses of a given heterocycle, or even the syntheses of different heterocycles. Hence, analogously to the reaction graphs, the oversimplified cyclization graphs are helpful in classification of reactions. The combinatorial character of such diagrams is also obvious, and it is easy to enumerate either manually (or by a programmable microcalculator) all the conceivable schemes for the synthesis of simple heterocycles. Theoretically deduced cyclization graphs can be also compared with real syntheses, and for the still unknown schemes an attempt can be made to devise likely reagents that might be suitable for use in a new type of synthesis. A discussion of such attempts (particularly for synthesis of thiazoles and indazoles) can be found in the reviews[1,12].

Meanwhile, an obvious disadvantage of the cyclization graphs is that they do not offer any information whatever on the chemistry involved in the formation of the skeletal bond. In fact, a single diagram may describe processes that are basically different. For example, in the synthesis of quinolines[22] exactly the same skeletal bond C3-C4 is formed in the Camps reaction 4 and in the Madelung reaction 5 (Scheme 3).

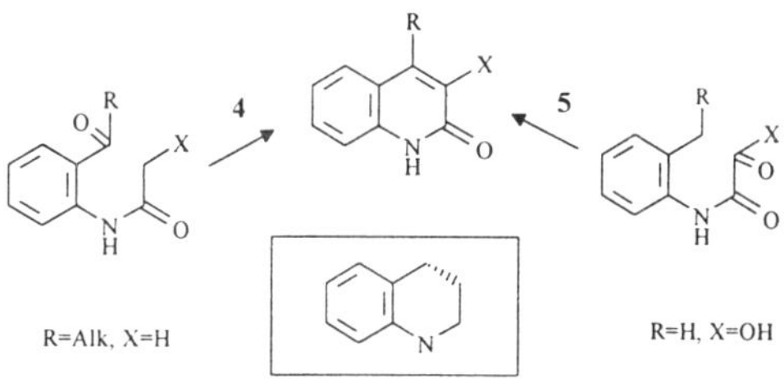

Scheme 3. Synthesis of quinolines by the Camps reaction 4 and the Madelung reaction 5. Both reactions are described by the same disconnection scheme (cyclization graph) shown in rectangle

It would appear that the synthetic schemes should be expressed by a common diagram. However, it is obvious to any organic chemist that in these intramolecular condensations the carbonyl and methylene components (i.e., the electrophilic and nucleophilic centers) have simply changed places relative to the heteroatom (or the benzene ring). As a result, the heuristic and classifying function of cyclization graphs is seriously limited, and it is advisable to modify them by supplementing with information on the polar nature of the reaction centers.

3.2. How to account for the nature of reaction centers in the reagents?

3.2.1. Classifications of reagents according to the polarity of terminal atoms. In order to avoid paradoxical identification of different reactions (as in the example above) by the same cyclization graph it is necessary to distinguish the electro- and nucleophilic nature of the reaction centers in reactants. This problem has attracted special attention in the reviews on heterocyclic synthesis. Thus, Potts' classification[23] of reagents involved in heterocyclizations included the concept of bielectrophiles (in particular, 1,1-, 1,2-, and all the way up to a 1,5-bielectrophile) and also the concept of binucieophiles (in particular. 1,2-, 1,3-, and 1,4-binucleophiles). In addition, Potts mentioned in passing the importance of a third category of reagents that are simultaneously electro- and nucleophiles, but focused primary attention on the design of bielectrophiles and binucleophiles by varying the degree of unsaturation (hybridization) of the reaction center. This type of classification in regard to the importance of the electronic nature of the reagents' reaction centers in describing heterocyclization was developed further, for example, in the handbook of Katritzky[24] and paper of Jorgensen[25].

In order to provide more heuristic utilization of this concept in the design of heterocyclic structures or reactions, let us generalize the Potts' classification in two directions. First, let us add to the existing classification of 1,*k*-bielectrophiles and 1,*k*-binucleophiles the no less rigorous concept of a 1,*k*-"ambiphile," that is a *k*-atomic chain with the opposite polarities of terminal atoms. (Natural particular case should be a 1,1-ambiphile with simultaneous combination of electro- and nucleophilicity in one and the same atom, like in carbenes, or halogen derivatives of CH-acids.) Second, we should consider the case concept of a 1,1-binucleophile, which for unknown reason was missing in early classification tables [23-25]. Typical 1,1-binucleophiles (like water, hydrogen sulfide, amines, and aliphatic nitro compounds) are often used in heterocyclizations with 1,4- or 1,5-bielectrophiles, forming five- or six-membered hetarenes or a benzene ring.

As a result of this slight modification[26], the original Potts' classification of reagents for heterocyclic synthesis acquires the previously lacking harmony and completeness. For simplicity, we will name a reagent with *k* atoms and a given *E* or *N* nature of the terminal reaction centers as a reagent of a particular *EN* type (*E*-electrophilic, *N*-nucleophilic). Therefore, any reagent can be assigned to either 1,*k*-*EE*, or 1,*k*-*NN*, or 1,*k*-*EN* type. We should emphasize the main feature of reagent classification on the basis of *EN* type: this classification reflects the behavior of the particular reagent *in a given reaction only*. For example, hydroxylamine is of a 1,2-*NN*-type in syntheses of isoxazoles, but it will naturally be of a 1,1-*NN* type in syntheses of N-oxides of pyridines. Acetone, depending on the reaction, may act as reagent of a 1,1-*EE*, 1,2-*EN*, or 1,3-*NN* type. The flexibility of this sort of classification consists mainly of the following: the potential ambiguity (multiplicity) of reactivity modes of a given reagent is compensated by the absolutism of definition of its *EN* type if the reaction mechanism is known.

3.2.2. Classifications of reagents according to the polarity distribution along the chain.

Additionally to the *EN*-classification (based on the polarity of the terminal atoms in reactants) there is another general classification of reagents based on the arrangement of all polar centers along the reagent's chain. Such arrangement may be either consonant (with alternating sequence of donor and acceptor centers along the chain) or dissonant (with violation of such an alternation). The initial terminology first suggested by Evans in 1971 was further used in the works related to the computer-assisted organic synthesis[27-33]. Initial idea of such a classification was claimed in 1920 by Lapworth in his principle of alternating polarities induced by a heteroatom along the chain[34]. (Lapworth classified reagents in similar sense using the terms "homogeneous" and "heterogeneous".) An analogous classification of reagents was discussed by Seebach in his reactivity umpolung concept[35]. He used the terms of "normal" structures (i.e., with an alternating sequence of donor-acceptor centers in the chain) and structures, containing an "umpolung" (inversion of polarity of some donor or acceptor carbon center in relation to the heteroatom at the end of the chain). The same classification principle (in terms of "conjoint and disjoint" structures) has been also discussed by Ho[36,37]. More references can be found in the author's review[38].

3.3. Statistical analysis of polar types of reagents used in pyridine ring synthesis

The existence of rational polar classifications of open chain reagents for heterocyclic synthesis leads to an intriguing problem. Suppose, we have a certain heterocycle and select a certain disconnection scheme of its synthesis. The question arises: can we arbitrarily choose the polarity of the open chain reagents (say, varying their *EN*-type or the consonant and dissonant properties)? Theoretically, there are no restrictions for polar structure of reagents, but in practice there are.

After examining the rich and extremely diverse material on the syntheses of six-membered heterocycles, McKillop and Boulton[39] concluded that in this field of synthesis the concept of *E*- and *N*- centers (readily identifiable in the original reagents) plays an important role: in many cases, electrophilic and nucleophilic centers *alternate around the ring*. Unfortunately, these authors never formulated clearly the exact content of the "alternation rule." Instead, without any further reference to the alternation phenomena, the cited review covered many examples of disconnection schemes with the participation of an electrophilic heteroatom, making unclear the initial thesis.

It is necessary to clarify this statement. Consider example of possible ring synthesis of the pyridine, the simplest representable member of the heteroaromatic azines. Let us analyze the following problem: how are reagents distributed by their *EN*-type in the ever reported syntheses of the pyridine ring? Are there any *polarity control* rules for the pyridine ring synthesis that can be definitively verified by careful statistical analysis of empirical facts?

Resolving of such a problem has required extremely precise criteria for the selection of the literature sources that are used, the choice of actual structures that are classed as "pyridines," and the methods of pyridine ring construction. Details and rigid restrictions we finally used are published[26,40]. As the result we have set up a unique computer database on those syntheses that are one- and two-component heterolytic cyclizations of maximally unsaturated pyridines, including those containing annelated alicycles[40]. Following the selection principles we classed as heterolytic cyclizations only those examples of cycloaddition and electrocyclization in which the influence of the substituents clearly indicated the *EN* type of the reagent. (The recyclizations, frequently used in design of pyridines, have been ignored.)

Every reagent used for the synthesis was classified according to its *EN*-type in given reaction. For the total 638 reactions the distribution of reagents to polar types was the following[26]:

Polar type of reagent(s)	Number of examples
(1,1-*NN*)+(1,5-*EE*)	130
(1,3-*NN*)+(1,3-*EE*)	289
(1,4-*EN*)+(1,2-*EN*)	81
(1,5-*NN*)+(1,1-*EE*)	8
(1,6-*EN*)	130

As it is evident, there are no examples of reagents that serve as 1,2- or 1,4-bielectro- or binucleophiles, and as 1,1- or 1,3-ambiphiles. Therefore, there is a real polarity control rule that connects oddness of the reactants chain and the polarities of terminal atoms. It is easy to understand that the nature of the terminal reaction centers of the reactants that are used for the synthesis of pyridines *can be established by truncating the hypothetical chain with alternating charges*. For the best of our knowledge there are no exceptions from *this* alternation rule.

The next generalization arises from the analysis of permissible position of the nitrogen heteroatom in the skeleton of "allowed" reagents. From total 638 reagents with nitrogen atom (as the precursor of the pyridine ring heteroatom) in 634 cases the N atom was either nucleophilic center or was located in consonant position to the polar (donor or acceptor) terminal atoms of reagent. In only 4 reported cases this "heteroalternation rule" (but not the initial alternation rule!) was violated.

An attempt to generalize the rule to the multicomponent syntheses of pyridines and more hydrogenated structures has resulted in the similar trend[26]. In some multicomponent synthesis of saturated structures the "forbidden" dissonant open chains (like alpha-haloketones of 1,2-*EE*-type or alpha-aminocarbonyl compounds of ambiphilic 1,3-*EN* type) were used. However, just for these cases the pyridine skeleton appeared as a by-product of the competitive formation of odd membered rings. Moreover, the heteroalternation rule was strictly valid for one- and two-component synthesis of the quinoline[26], see Scheme 4. Perhaps the only exception is the above mentioned reaction 5 (Scheme 3), and it is remarkable that the main direction of this "forbidden" reaction is the formation of an odd-membered ring. (This is actually the Madelung synthesis of indoles, not of quinolines!) The rule is also applicable (although with more exceptions) for synthesis of the isoquinoline, and the exceptions are explainable[26]. Qualitative analysis of the polarity of reagents traditionally used in the synthesis of other azines (and their benzoderivatives) gives rise to very simple conclusion: the heteroalternation rule is most strictly valid for those six-membered rings where the heteroatoms are located in the alternating sequence around the ring.

The provided analysis defines more accurately the content of the alternation rule. However, it can be further generalized. Following Evans and others[27,30,31], let us expand the consonant or dissonant classification from open chains to the cyclic structures. The consonant/dissonant property of the cycle depends on the existence of odd and even chains between the heteroatoms and oddness of the ring. Odd-membered cycles (e.g. azoles) are always dissonant. Even-membered cycles (e.g. azines) may be either consonant or dissonant, depending on mutual arrangement of heteroatoms around the ring as well as polar exo-substituents. Thus, pyridine, pyrimidine, pyridone-2, and resorcinol are consonant, while pyrazine, pyridazine, 3-hydroxypyridine, and pyrocatechole are dissonant.) From this viewpoint, the observed literature trends for heterocyclizations clearly confirm the following thesis[41]: the predominant precursors of the consonant heterocycles are the acyclic consonant structures. This rule was confirmed even for the synthesis of carbocyclic benzene ring[42] by polar cyclocondensations. In turn, by combination of only consonant chains it is impossible to obtain the dissonant heterocycle, say 3-hydroxypyridine or pyrrole, and the dissonant precursors are required.

3.4. Polar cyclization graphs

Let us emphasize that the discussed rules accentuate combinatorial properties of the initial and final *structures*. Then the question is how to introduce the concept of electro- and nucleophilic centers into classical disconnection schemes, that, as we mentioned, are simple reaction graphs (cyclization graphs). We do not have the right to place E- and N-labels on the *mapping* until we have at our disposal counter-information on the nature of the reaction centers in the final product of cyclization. For consonant cyclic structures the arrangement of polarities around the ring is well known: it is simply alternation of electro- and nucleophilic centers (cf. known reactivity of positions 2, 3, and 4 of the pyridine nucleus). Hence, in the course of heterocyclizations from consonant chains to consonant cycles the initial "philicity" of centers (electro- or nucleophilicity) is conserved. Therefore, it is possible to define (only for the consonant heterocycles) the *polar cyclization graphs* as the new sort of reaction graphs by placing *E* and *N* labels in the corresponding cyclization graphs. Examples of such diagrams are presented in Scheme 4 for the quinoline ring, where the *EN* type of reaction centers is reflected by hollow and black dots.

Of course, the black and white (or *N* and *E*) labels can be arbitrarily located on any cyclization graph of any (consonant or dissonant) heterocycle, thus helping to enumerate possible combinations of reactants' polarities. Such diagrams, however, have nothing common to the idea of reaction graphs, and they are not heuristic in estimating the optimal polarity of the precursors. Only in the case of consonant heterocycles the polar cyclization graphs are the desired tool to predict optimal polarities of precursors for the unknown synthesis schemes: the required polarity is manifested by the diagram itself.

3.5. Heuristic utilization of the rules in novel quinoline synthesis

Early we elaborated the computer program "Heterocycland"[1a,b] capable for generating and ranking polar cyclization graphs for the given consonant target. Results of generation of such diagrams for one- and two-component synthesis of the quinoline ring are presented in Scheme 4. Comparison of theoretically possible and experimentally discovered synthetic pathways allows one to conclude, which reaction types are poor or better investigated. The numbers of known reactions near every diagram in Scheme 4 are taken from reviews[21,22,26].

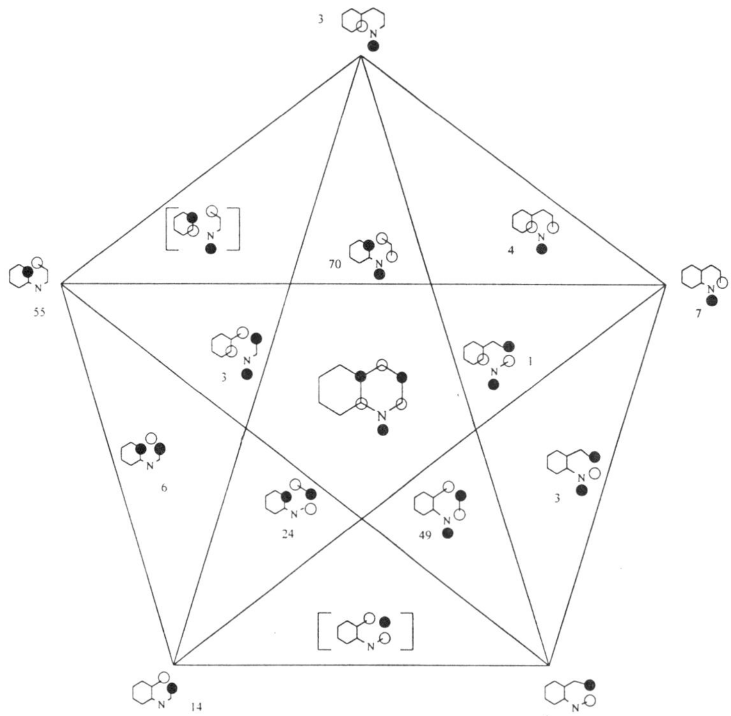

Scheme 4. Actual number of examples and polarity of terminal atoms in reagents (*N* - black and *E* - white dots) used for one- and two-component cyclizations leading to the quinolines illustrated by the polar cyclization graphs. Unknown schemes are in square brackets.

It is rather elusive to consider that all possible combinations of consonant precursors have been exhausted in the synthesis of consonant quinolines. Thus, there are still no examples for two cyclization graphs, namely for the CCCNC + C combination of reagents (such cyclization graph is also still unknown for pyridine) and for the combination CCCN + benzene fragment.

The concept of polar cyclization graphs immediately suggests that for the first case the mono-carbon fragment should be of 1,1-*NN* type, and the other fragment must be of 1,5-*EE* type[43]. One possible combination of reagents was suggested to be the 1,1-binucleophilic nitromethane and 1,5-bielectrophilic amidine of anthranilic acid, see reaction 6 in Scheme 5. This predicted reaction -- the novel synthesis of the quinoline nucleus -- was indeed confirmed experimentally in 1993 by the author[43]. In 1994 another research group independently rediscovered the same cyclization graph with the same (1,1-*NN* + 1,5-*EE*) arrangement of polarities of reactants[44] as is shown by reaction 7 in Scheme 5.

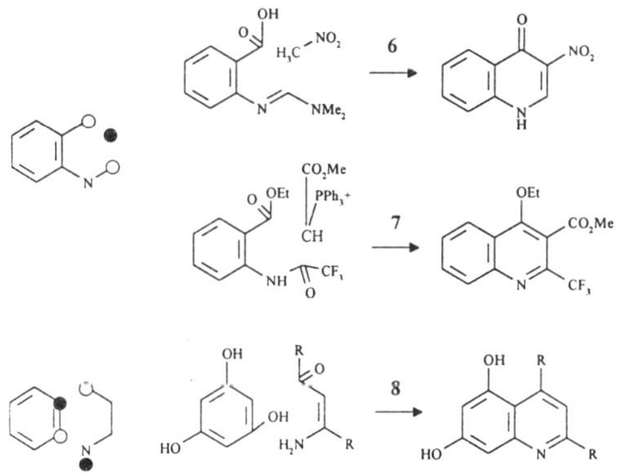

Scheme 5. The uknown polar cyclization graphs for the quinoline synthesis (left), the recently discovered novel strategies of qunoline ring synthesis (reactions 6,7), and the predicted combination of consonant reactants for the still unknown reaction 8.

The "last" still unknown two-component cyclocondensation leading to the quinoline ring (more precisely, to the pyridine fragment of the nucleus) is the combination "CCCN fragment + benzene fragment". The heteroalternation rule predicts the 1,4-ambiphilic type for the CCCN fragment. (Such reactivity type is well known and is typical, say for enamine of 1,3-dicarbonyl compounds and enamides.) The benzene fragment, obviously, should serve as 1,2-ambiphile. Although it is somewhat difficult to imagine a benzene with opposite reactivity of atoms in ortho-position, such reactivity type is realized in the chemistry of phloroglucinole. (The last compound readily reacts with electrophiles and is also able to undergo substitution of a hydroxyl group with amines.) The final hypothetical reaction 8 is presented in Scheme 5, and the author offers the confirmation of this prediction to any interested heterocyclic chemist.

4. Combinatorial Models and Polarity Control for Ring Opening Reactions

One actual branch of organic synthesis is the use of heterocyclic structures as masked precursors of open chain structures with necessary arrangement of polar groups along the chain. The topic is well reviewed (see, e.g.[45-48]), and there are two still unanswered general questions associated with the ring opening reactions. The first one (related to synthesis planning) is: which heterocycles may serve as precursors of the given open chain fragment. The second question (related to the reaction planning) is: how to predict the direction of bonds cleavage for a given heterocycle with few heteroatoms and/or fused rings. An ideal solution of both problems would be the idea of reaction graphs with mapping of the starting cycle to the final open chain(s). However, except the prediction of the ring opening reaction 3 (see Section 2), there are no such approaches in the literature.

It is rather easy to suggest how the reaction graphs for ring opening reactions may look like. One can simply label the bond(s) to be broken in the initial cycle, say, by dashed line(s), and this will be just the oversimplified example of mapping of the atoms in the initial cycle into the atoms of final open chain. Defined in such a manner the ring opening graphs are simply reversal of the idea of the cyclization graphs discussed in the previous chapter. Such graphs can result useful in two respects: first, to enumerate all possible ring opening reactions (say, for a complicated heterocycles with many heteroatoms) and second, to compare similarity of such reactions for different heterocycles. However, a care should be taken in attempts to displace polarity labels on such reaction graphs.

4.1. Correspondence of polarities in heterocyclic precursor and open chain structures

As we proved above in the Section 3, the global arrangement of polarities in the acyclic precursors and final cycle appeared important for better understanding of heterocyclizations. Hence, we may put reversed question: what is the correspondence in polarities of atoms in the initial cycle and the resulting open chain in the ring opening reactions? Let us avoid discussion of photochemical, oxidative and reductive processes of ring cleavage and consider only thermal polar reactions (usually under action of polar species) of heteroaromatic (may be fused) azoles and azines.

Developing the early idea of Stirling[47b], let us subdivide the polar ring opening processes into two groups[49]. Actually, the cleavage of a ring bond is accompanied by the formation of a new bond. This can be either a new bond between initial cycle and external (exo) group or a new bond between (endo) atoms of initial cycle. The first group will be called *exo-ring opening* (equation 9a) and the second group -- *endo-ring opening* reactions (equation 10a in Scheme 6). The exo-processes resemble usual nucleophilic addition/substitution reactions, while the endo-processes resemble the elimination reactions.

Polarity preservation in the exo-ring opening reactions. As a rule, the exo-ring opening reactions (cleavage of cyclic Y-Z) bond occur as the result of nucleophilic attack on the carbon atom (Y is carbon and Z is heteroatom on Scheme 6). Of course, the carbon atom is the electropositive center of the polar bond C-Z. If the external nucleophilic center W is more electronegative than the carbon atom, then the polar nature of the cyclic atoms C and Z should conserve in the formed open chain. Hence, in exo-ring opening reactions the appropriate *dissonant or consonant type of an open chain should be the same as was in the initial cycle* [41]. Thus, hydrolysis of consonant pyrillium salt results in the consonant glutaconic dialdehyde, while dissonant furane under hydrolysis result in the dissonant open chain, see reactions 9b, 9c in Scheme 6.

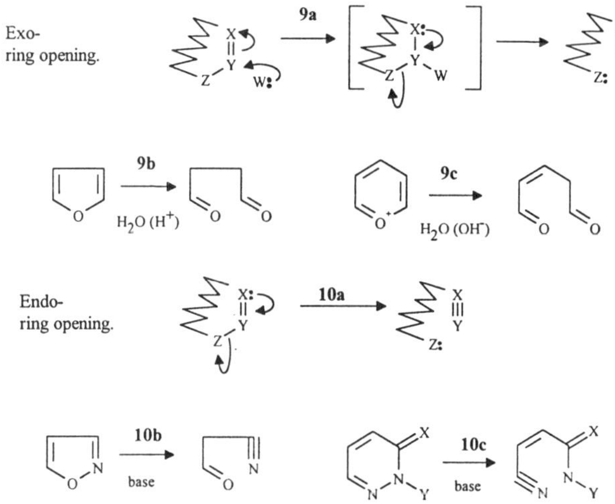

Scheme 6. Examples of exo- and endo-ring opening and polarity changes via reaction with corresponding changes or conservation of the initial consonant and dissonant nature of a cycle.

Polarity changes in the endo-ring opening reactions. The endo-ring opening reactions often require strong bases and frequently lead to the cleavage of bonds between heteroatoms. Consider azine and azole with Y-Z bond (Y, Z - electronegative heteroatoms like N and O), for instance pyridazine and isoxazole. Both initial heterocycles are evidently dissonant. It is easy to imagine a ring opening with the cleavage of the cyclic bond Y-Z, as in examples 10b, 10c in Scheme 6, so that the atoms Y and Z of the open chain are the nucleophilic centers. Ring opening of the *dissonant* oxazole results in the *consonant* alpha-ketonitrile, while the *dissonant* pyridazine is converted into the *dissonant* gamma-aminonitrile.

In the two above examples the initial polar nature of the cycle is either conserved or changed *via* ring opening reaction. The only way to elaborate rational combinatorial model for this phenomenon is to consider the dissonance to be not only qualitative but quantitative property. Let us call a "polar conflict" an even chain between a pair of heteroatoms. The neighborhood of heteroatoms (even-membered “zero size chain”) is the particular case of the polar conflict. There is one polar conflict in isoxazole (indeed, only the bond NO is dissonant, while the NCCCO fragment is consonant), and there are two such conflicts (NN and NCCCCN even fragments) in pyridazine. Hence in the case of oxazole the polar conflict disappeared via ring opening giving rise to the consonant product. However, for pyridazine the ring opening causes disappearance of only one conflict from two, and the final product is still dissonant. This "arithmetic" of conflicts and their changes is under our extensive current investigation (see[41,42]), and the just discussed examples illustrate particular importance of this idea for heterocyclic chemistry.

Difference in polarity changes via exo- and endo-ring opening reactions is essential. It is possible to put polarity labels on the centers of a ring opening graph (as the mapping) only for the case of exo-ring opening reaction.

4.2. Selectivity of azole ring opening reactions: the "polar triads" rule

A heteroaromatic structure (monocyclic or fused) with few heteroatoms may undergo cleavage of different bonds, and the type of cleavage may be either the endo- or exo-ring opening. Thus, two C-O bonds in oxazoles may undergo cleavage according to exo- or endo-type, and all four possibilities have been found, see Scheme 7. Although the topic is well reviewed for separate heterocycles, we failed to find simple, easily formalized rules, that may permit one to predict (or even estimate) the direction of bond cleavage for given heterocyclic structure. Early we have collected and analyzed available empirical data on the selectivity of ring opening for azoles (for azines the data are less representable). As we proved early, the fruitful approach to explain (and predict) selectivity of ring opening in azoles is to consider *local* environment of an atom in the cycle, i.e. to analyze the *polarity in triads* of cyclic atoms. (Although the exo-cyclic groups are ignored, one can reasonably estimate their influence on the adjacent atom of a cyclic triad.) Ranking the triads permits us to order the bonds in given azole structure according to the highest probability of their cleavage. The following algorithm can be suggested[49]:

As the first approximation, the bond to be broken was considered as the single bond. We also considered that any ring opening needs at least three ring atoms (numbered 1,2, and 3) responsible for this reaction. The single bond to be broken is the 1-2 bond (bond 2-3 may be either single or double). For the case of azoles (with two double and three single bonds) there are either 3 or 6 (depend on symmetry) possibilities to find such 1-2-3 triads. Every atom of such a triad is labeled as a donor (d) or an acceptor (a) center. Thus, any heteroatom (N, S, O - either of pyrrole or pyridine type) is (d) center, while carbon atom is usually (a) center. Of course, an exo-substituent may emphasize either (a) or (d) nature of the adjacent endocyclic carbon atom.

The possibility of ring opening according to exo- and endo- type depends on the *polar nature of all atoms in triad 1-2-3*, namely on the arrangement of (d) and (a) labels along a vector triad. Empirical data confirm the following qualitative rules[49]. Most probable consequence of atoms for endo-opening to be (d)(d)(d) (i.e. neighborhood of three heteroatoms), while the less probable to be the combination (a)(a)(a), (i.e. fragment with three carbons). Most probable consequence of atoms for exo-type of opening is another one, namely

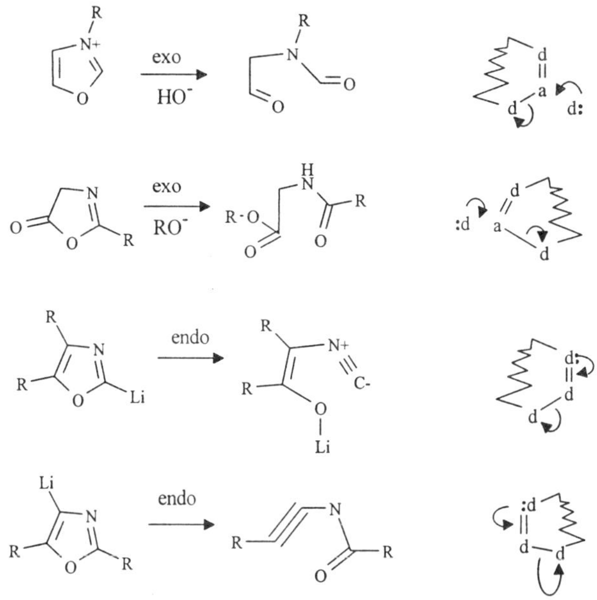

Scheme 7. Different selectivity for endo- and exo-ring opening reactions observed for the oxazole ring qualitatively treated by different arrangements of polarities in triads (d - donor, a - acceptor center).

(d)(a)(d), while the less probable is the combination (a)(d)(a). Other possible combinations have intermediate priority, and for every such combination the numerical value of probability (one set of numbers for endo- and another for exo-case) has been assigned.

The discussed principles illustrate an example of the *local polarity control rule*. The suggested algorithm permits one (by selecting all triads and comparing their rating) to provide immediate conclusion on the most probable selectivity of ring opening. These simple rules have been implemented into the computer program FROG[49]. Input of a heterocyclic structure results in two rating lists of most probable bond cleavage: for exo- and for endo-ring opening. As an example, the highest rating suggested by the program for the selectivity of the oxazole ring opening coincides with the experimentally observed directions of exo- and endo-reactions (see Scheme 7). Hence, the program really selects (by rating criteria) just those ring openings that are found experimentally and may be also useful to predict previously unknown directions of ring cleavage.

4.3. Selectivity of ring opening reactions of bridgehead azolo-azines: the alternation effect in the non-alternant system

Due to our long year experience in the area of ring opening and transformations of bridgehead azolopyridines (particularly, indolizines)[50,51] we have been interested in the theoretical analysis of selectivity of their ring opening reactions. Cleavage of both 5- and 6-membered rings is possible for this class. The general trends of the exo- and endo-ring opening reactions of bridgehead azoloazines and the influence of substituents on the selectivity of these reactions have been recently reviewed[52]. As it was proved, the exo-type of ring opening is the most representative for azaindolizines, and especially with the cleavage of a CN bond adjacent to bridgehead nitrogen atom, i.e. either C5-N bond of the azine fragment or C3-N bond of the azole ring. (The numbering we use corresponds to the conventional numbering of atoms in the prototype indolizine ring.) These two directions of rings cleavage are presented in general form in Scheme 8.

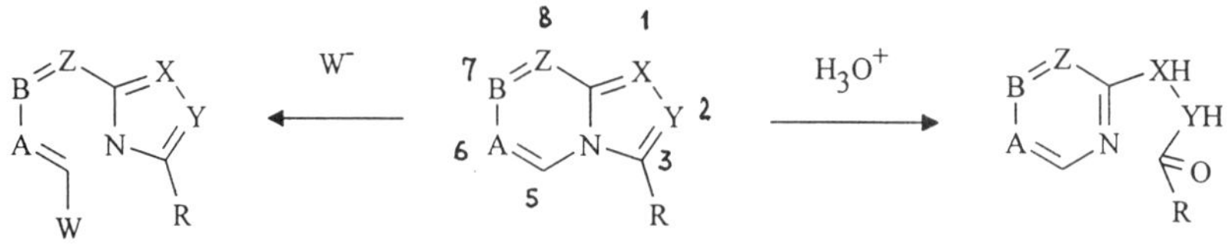

Scheme 8. Schematic representation of possibility of C5-N (left) and or C3-N (right) bonds cleavage in the series of bridgehead azoloazines. The ring opening products may be either isolated structures or intermediates in further rearrangements. (Numbering of atoms as in the parent indolizine nucleus).

Although both endo- and exo-types of ring cleavage are observed for this family, let us concentrate over the exo-processes. The observed qualitative trends of substituent influence on the selectivity of exo-ring opening are the following[52]:

(i) The C5-N bond cleavage is promoted by the donating nature (azasubstitution, the presence of an external acceptor) of positions 6, 8, 1, and 3 in the bicycle and by the accepting nature (the presence of external donating and oxo substituents at positions 5, 7, and 2). The chain atoms closest to the opening assembly (primarily 5 and 6 and then 7 and 8 and so forth) have the greatest effect. Quaternisation at positions 1 and 3 also promotes this bond cleavage.

(ii) Cleavage of the C3-N bond is observed more rarely. It occurs in the structures with donating groups at positions 2, 5, and 7 and the accepting nature of atoms 3, 1, 8, and 6. As in the previous case the polarity of the centers closest to the opening assembly (i.e. of atoms in the five-membered ring) is most significant.

Some observed trends can be easily explained with the help of polar triads rule, i.e. in terms of more or less preferable (d)(a)(d) fragments in the bicycle. However, it was somewhat difficult to treat the influence of polar substituents located far along from the bonds to be broken. A simple approach helpful to resolve this problem [52,53] is based on the idea of alternating effect of substituent along the parent indolizine framework. One can separate in the parent indolizine bicycle three types of chains A, B, and C (as shown on the Scheme 9). These chains are obtained by formal removal of two (from three) CN-bonds. One may assume that the nitrogen atom adjacent to the carbon tetraene fragment induces an alternation of the donor (d) and acceptor (a) centers along the chain. The arrangement of (d) and (a) centers in the chains A and C match each other, and this clockwise alternation will be called the direct alternation chain. The arrangement of polarities in the chain B is opposite to the cases A and C, and this counter-clockwise alternation will be called the reversed alternation chain.

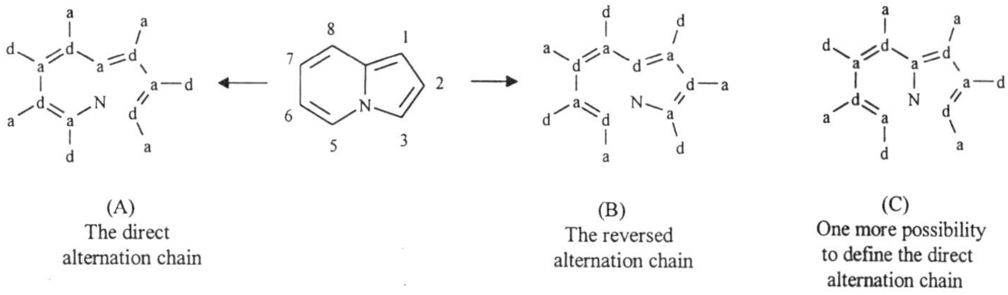

Scheme 9. Qualitative model of substituent influence along the parent indolizine framework. The direct and reversed alternation chains help to understand the influence of donor and acceptor groups on the polarity of atoms C5 and C3, responsible for the ring cleavage in the azaindolizines.

One may assume that any polar substituent attached to the indolizine bicycle (heteroatom or exo-group) according to its own polarity should intensify one or another alternation chain. Thus, the direct alternation chain is intensified by heteroatoms and exo-acceptor groups in positions 6, 8, 1, 3 and exo-donor groups in positions 5, 7, 2. In contrast, the reverse alternation chain is picked out by heteroatoms and exo-acceptor groups in positions 5, 7, 2 and exo-donor groups in positions 6, 8, 1, 3. In turn, intensifying of one or another chain means increase of the positive charge on either C5 or C3 atom and, therefore, assist either C5-N or C3-N bond cleavage.

One may conclude that experimentally observed trends (i) and (ii) of substituent influence on the ability of C5-N and C3-N bonds cleavage can be quite clearly explained in the terms of intensifying of either direct or reversed alternation chains. Namely, the substituents that emphasize the direct alternation chain promote C5-N bond cleavage, whereas the substituents that increase the reverse alternation chain assist C3-N bond cleavage.

The semiempirical SINDO1 calculations[53] for the families of isomeric aza-, nitro-, and methoxy-substituted indolizines and azaindolizinium cations support this simple model. The changes of charges of skeletal atoms caused by the insertion of substituent displayed pronounced alternation in accordance with the simple model of strengthening of either direct or reverse alternation chain. The final confirmation of this simple rule follows from the analysis of the energies for pairs of 3- and 5-OH anionic sigma-complexes calculated by SINDO1 for isomeric azaindolizines. The data of energies for isomers display the same principle: the azaindolizines with a nitrogen atom at positions 6, 8, 3, 1 (increase of direct alternation chain) prefer to form the C5-adducts, whereas those aza-substituted at positions 2, 5, 7 (increase of reverse alternation chain) favor to form the C3-adducts.

4.4. Experimental study of unknown ring cleavage reactions

The models suggested in Sections 4.2 and 4.3 may serve as examples of *local* and *global polarity control rules* useful for explaining and predicting the selectivity of ring opening reactions. It was attractive to use these rules in the experimental search of unusual selectivity of ring opening reactions for a certain heterocycle. Our choice was the bridgehead aromatic system of oxazolo[3,2-a]pyridinium cation (Scheme 10). Both fused rings have known capacity to undergo exo-ring opening, however the only reported reactions for this bicycle were the exo-cleavage of C9-O bond for cation and of C2-O bond for mesoionic 2-oxoderivatives[52]. The local triad rule confirmed the observed direction of cleavage and suggested possibility of exo-cleavage for any of three C-N bonds, although with lower rating. The common sense excluded the possibility of C9-N cleavage (due to the loss of aromaticity of both fused rings). The global alternation rule (strengthening of the direct alternation chain in bicycle by oxygen heteroatom) favored to the C5-N cleavage and excludes the possibility of C3-N cleavage.

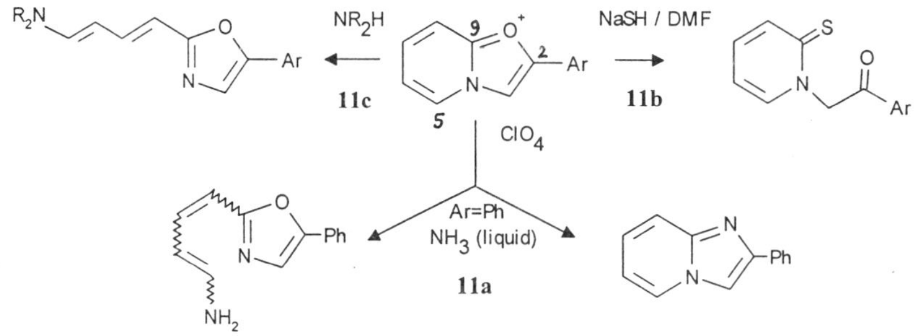

Scheme 10. Ambident reactivity of the oxazolo[3,2-a]pyridinium salts in the ring opening reactions.

The efforts, therefore, were concentrated over the discovery of the exo-opening of fused pyridine ring at the C5-N bond. The quantum chemical calculations have predicted[54] that for O- and S-nucleophiles the energy of adducts favored exclusively the C9-O ring opening, whereas the C5-N cleavage could be awaited for nitrogen nucleophiles. Indeed, the use of liquid ammonia as reactant resulted in the mixture of products of opening of both 5- and 6-membered rings (reaction 11a in Scheme 10). On the other hand, the action of hydrosulfide anion caused cleavage of C9-O bond (reaction 11b) Finally, the use of secondary amines gave high yields of aminobutadienes as the result of the desired selective C5-N bond cleavage (reaction 11c in Scheme 10).

5. Recyclization Graphs - New Language for Description and Classification of Heterocyclic Ring Transformations

5.1. Early attempts to classify recyclizations

An important class of organic reactions that play major part in the entire heterocyclic chemistry are the ring transformations (or recyclizations) of heterocycles. These elegant reactions (frequently discovered by a lucky chance) often lead to heterocycles with unusual disposition of heteroatoms and substituents, or products not available by any other synthetic methods. Well-known examples of recyclizations are the "named" reactions[55] of Yur'ev, Zinke-Koenig, Hafner type, or the Dimroth, Cornforth, Boulton-Katritzky, Kost-Sagitullin rearrangements. Nowadays the ability to undergo ring transformations is found for the overwhelming majority of known heterocycles, particularly for heteroaromatic cycles. The mechanisms of transformations of this sort are diverse, and it is often difficult to propose the structural scheme of bonds redistribution, by considering only the initial and final heterorings. Despite the enormous amount of factual material in this area and the profusion of reviews on this topic (see e.g.,[55]), one must conclude that there is no general rational classification of heterocyclic ring transformations in accordance with a structural principle. As a result, it is often difficult, for example, to establish the real degree of novelty of a recyclization or rearrangement declared as being "novel". Evidently, it is necessary first to define somehow the degree (or measure) of structural similarity of the existed reactions.

In most of reviews devoted to recyclizations the general question on the structural classification of such reactions has not arisen; only unique papers touched upon this problem. For instance, in the classical monograph[55a] of Van der Plas the heterocyclic recyclizations were not classified, but conveniently ordered, according to the size of initial and final cycles, and the nature and number of heteroatoms. Meanwhile, in other reviews the idea of structural classification has appeared in different forms. Schwaika suggested[56] the most important structural feature for comparison of recyclizations to be the size of fragment of initial ring incorporated into the final heterocycle, and similar principle was applied to classification of various pyrillium ring transformations[57]. In the review of L'abbé[58] convenient classification of azoles rearrangements was proposed, taking into account another criterion - the size of the side chain of initial heterocycle, that is incorporated into the final cycle. (Following this idea, the size of side chain for the Dimroth, Cornforth, and Boulton-Katritzky rearrangements should be considered as 1, 2, and 3, respectively.) In the other reviews only some general reaction schemes of ring transformations for particular heterocycles (or some families of analogous heterocycles differed by heteroatoms or annelated rings) were discussed.

We would like to mention, that even in the most developed of the approaches cited above chemists utilized the principle of structural similarity of the molecules (or their different substructures), but not the similarity of reactions. As we mentioned in the Section 2, the reaction graphs are just the appropriate objects for comparison the similarity of reactions. In our early series of papers[59-62] we suggested novel type of reaction graphs for description of heterocyclic rearrangements and ring transformations. These graphs -- the recyclization graphs (or graphs of cyclic bonds redistribution) -- have been utilized to classify recyclizations in the hierarchic system, to compare structural similarity of recyclizations, and to predict interesting examples of recyclization of previously unknown structural types.

5.2. Definition of recyclization graphs

Let us briefly recall our definition of recyclization graphs. We call recyclization (or ring transformation) any transformation of a heterocycle, that include the steps of ring opening and ring closure in any sequence. We shall consider mainly monocyclic systems, and among condensed heterocycles - only those containing one joint bond between the annelated rings (as it is in indole or acridine). Let us call ring transformation to be *simple ring transformation* (or SRT) if: (a) there is no formation of any transient cycles via recyclization, except the final one; (b) there is no permutation of atoms or groups via ring transformation; (c) the initial cycle is transformed into no more than one cycle of the final structure. Most of known recyclizations are SRTs, particularly, the large family of ANRORC-reactions; boundaries of the term are discussed early[59,60]. Below we shall be limited only by SRTs.

The main idea of our approach is rather simple. Since the mechanism of a recyclization is established, the skeleton of the initial cycle can be easily found among the atoms and bonds of final products and vice versa, the skeleton of the final cycle can be found among the atoms and bonds of the starting reagents. Paying all the attention only to these cyclic skeletal substructures (and ignoring all other details of molecular structure), we can get significant simplification of the chemical equation. The reaction mechanism establishes the correspondence between matching atoms and bonds of the initial and final cycles. Therefore, just these cyclic (initial and final) substructures can be chosen for superposition to define the graph of ring transformation reaction.

Recyclization graph as superposition of molecular graphs. Let us give more strict definitions following Ref. 60. Let us define for any SRT (with the mechanism known beforehand) two types of graphs - the molecular graphs (that are not coincide with initial and final structures and are determined only by the given type of transformation) and the recyclization graph (for any reaction that is SRT). Let us define the molecular graph M_S of starting reagents as the set of vertices and edges, that correspond only to those V atoms and R_S skeletal bonds that either exist in the initial cycle or are incorporated in the final cycle. By analogous manner, the molecular graph of final products M_f (with V vertices and R_f edges) contains the same V atoms and only those skeletal bonds, that either exist in the final cycle or were presented in the initial cycle. Let us keep the symbols of heteroatoms as the labels of vertices in the graphs M_S and M_f and omit all other symbols (like hydrogen atoms, multiple bonds, and any substituents including annelated rings). Although the same V vertices are chosen to construct both graphs M_S and M_f, the number and/or distribution of edges in these graphs is obviously different.

Let us enumerate the vertices of M_S -graph; then the mechanism of SRT immediately permits us to enumerate by corresponding numbers the matching vertices of the M_f -graph. Let us make mental superposition of the structures of graphs M_S and M_f according to the matching vertices (and edges) with identical numbers. The pairs of vertices with the same number should be identified into one new vertex, as well as the pairs of edges - into one new edge. Let us define such superposition as the new recyclization graph with V vertices. Let us designate the edges of recyclization graph (presented also in the graphs M_S and/or in M_f) by the following manner:

(a) by the *dashed* line (edge) if the edge is presented only in one (but not another) M-graphs;
(b) by the *bold* line if the edge belongs to both (initial and final) cycles;
(c) by the *usual (solid)* line if the edge belongs to only one cycle of M-graphs.

Examples. In order to illustrate the idea of recyclization graphs in pictorial form let us use examples of SRTs discovered by the author's research group[63-65] and presented in Scheme 11.

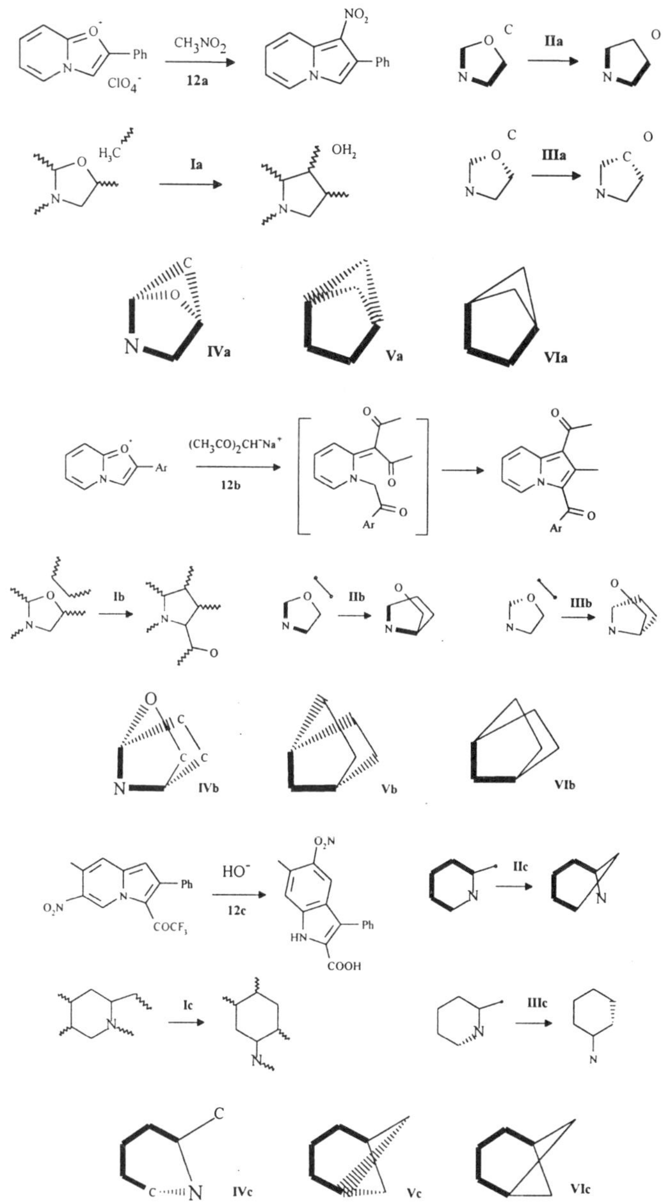

Scheme 11. Assignment of recyclization graphs to reactions 12a-12c. Ia-c: Choice of skeletal atoms that appeared in both (initial and final) cycles. IIa-c: M-graphs with bold fragment common to both cycles. IIIa-c: The cyclization and ring opening graphs. IVa-c: G_2-graph as the result of superposition of M-graphs. Va-c: G_1-graphs. VIa-c: G_0-graphs. -graphs.

For every reaction 12a-12c the pairs of M-graphs II and the resulting recyclization graph IV are shown. In the first reaction[63] the 1,1-binucleophilic carbon atom substitutes oxygen atom in the oxazole ring, and both M_S and M_F graphs are disconnected. (The result is the appearance of four dashed edges in the recyclization graph IVa.) For the second reaction[64] 12b formally the same conversion of one heterocycle to another occurs. However in this case the incorporated 1,2-ambiphilic fragment consists of two atoms, and initially disconnected M_S graph is converged to the connected M_F graph with quite different recyclization graph. Of course, in both reactions the fused pyridine fragment is considered as an "inessential substituent" in the 5-membered ring and, therefore, is not shown in the recyclization graph. In the third example[65] of "haloformic recyclization" 12c the transformation of the indolizine ring to indole (actually, of the pyridine fragment to benzene) is followed by the haloformic reaction. However, the resulting recyclization graph IVc does not reflect any changes in the "inessential" substituent, since only atoms of initial and final six-membered rings are used to construct the diagram.

From the formal viewpoint, the recyclization graph is nothing but a labeled graph. The labels are on the vertices (symbols of the heteroatoms) and on the edges (three above mentioned types or "colors" of edges - dashed, bold, and solid). Including of so many labels into the same recyclization graph permits one to follow the changes in any cyclic skeletal bond of heterocycle via SRT at the same -- united -- diagram. As it follows from the above definition, the structure of recyclization graph does not depend on the order of superposition of M-graphs; hence, the recyclization graphs of direct and reversed reactions coincide. With evidence, skeletal pairs of M-graphs can be immediately restored from a recyclization graph (by drawing two diagrams and reversed procedure of restoring/removal of corresponding cyclic bonds from dashed lines in each of them).

Recyclization graph as superposition of the ring opening graph and cyclzation graph. Let us compare recyclization graph with the discussed above ring formation graphs (disconnection scheme) and ring opening graphs. The disconnection scheme (the superposition of cyclic and acyclic molecular graphs) is evidently a labeled monocycle with two sorts of labeled edges (formed and retained via cyclization). The ring opening graph is the exact reversal of the ring formation graph, and is again the labeled monocycle. In contrast, the recyclization graph is the superposition of a cycle to cycle, so it is by definition a bicycle (since the fragment of one cycle is identified with fragment of another cycle).

Actually, the mechanism of recyclization is the combination of ring opening and ring closure (generally, in any sequense). Each step, as we discussed above, can be described by a graph of reaction, namely by ring opening graph and by cyclization graph, where only the cyclic bonds (formed or broken) are presented. Therefore, the total redistribution of cyclic bonds -- recyclization graph -- is the *superposition of these two reaction graphs*. This is evident from Scheme 11, where the pair of reacton graphs IIIa-c is shown for every reaction 12a-c. Such an alternative definition of recyclization graphs is extremely important. First, we may "extract" not only skeletal M-graphs, but also elementary steps from the final superposition diagram. Second, the polarity rules, separately defined for the cyclization graphs and ring opening graphs, can be jointly utilized for understanding the polarity control via recyclizations.

The idea of superposition of two reaction graphs into a new one allows us to simplify the procedure of obtaining the recyclization graph. Let us draw the skeleton of the initial cycle only with that side chain (or a disconnected external fragment) that will participate in the final heteroring. Let us draw the ring opening graph, indicating by dashed line(s) those bond(s) that are broken in the initial cycle. Let us draw at the same diagram the cyclization graph by simply adding new bonds (also labeling them by dashed lines) that are the novel bonds of the forming cycle. To emphasize the fragment common to both cycles let us also label it, say by bold line. The resulting diagram is the recyclization graph.

5.3. Recyclization graphs and the structural similarity of reactions

The bicyclic feature of any recyclization graph is essential. It is not a problem to compare labeled monocycles (for heterocyclizations or ring opening processes), thus establishing kinship or difference of the corresponding reactions. It is, however, somewhat difficult to compare labeled bicycles in order to establish classification of recyclizations.

Recyclization kinds. Analogously to cycizations, the recyclizations can be multicomponent, and this is the first criteria for classification. M-Graphs are connected graphs only in the case of rearrangements. In all other cases (i.e. for reactions where an external reagent is included into the final cycle and/or a fragment is eliminated from the initial cycle) at least one of M-graphs should be disconnected and consists of more than one component. The number of components in M-graphs is the first natural criteria to classify SRTs into kinds. Designating the initial and final heterocycles by symbols A and B, and the external reagents (or eliminated fragments) -- by symbols X,Y..., one can immediately find three main kinds of recyclizations, presented by the following schemes:

A => B (i.e. rearrangements);

A + X => B or A => B + X (reactions with elimination *or* incorporation of ring fragments);

A + X => B + Y (reactions with elimination *and* incorporation of ring fragments).

Recyclizations of the same kind evidently have the same number of dashed edges in the structure of their recyclization graphs. Examples 12a-12c are of different kinds, and the last reaction has the same kind as any rearrangement.

Recyclization families, classes and types.

Comparison of various recyclization graphs for transformations of azoles[59,60] and azines[61,62] led to conclusion, that recyclization graphs of SRTs differ from one another either by the topological structure of graphs, or by the number and/or distribution of labeled vertices and dashed edges. These characteristics have been used as the most important ones to establish the general structural similarity of SRTs. The following three different types of recyclization graphs have been introduced[60]:

G_2-graph is the initially defined recyclization graph with labels on vertices and edges.

G_1-graph is G_2-graph where the symbols of heteroatoms are omitted.

G_0-graph is G_1-graph where the dashed labels of edges are omitted.

Examples of G_0-, G_1- and G_2-graphs (diagrams IV, V, and VI) are shown for recyclizations on Scheme 11.

The identity of G_2 -graphs corresponds to the closest structural similarity of SRTs; let us say in this case that the SRTs belong to the same *family*. The family is determined uniquely by the skeletons of initial and final heterocycles and the position of heteroatoms (i.e. by the pair of M-graphs). Theefore, any variation in the degree of unsaturation, tautomerism, presence of any substituents (including condensed rings) should keep the reaction at the same family of SRTs. Another type of structural similarity of SRTs is the identity (more strictly, isomorphism) of their G_1-graphs; let us say SRTs with the same G_1-graphs belong to the same *class*. In classes the heterocycles are different, but the structural diagram of their recyclizations (without heteroatoms) is the same. At least, let us say reactions belong to the same *type* if their G_0-graphs coincide. Recyclization type is the crudest similarity level. Evidently, reactions of the sâme type have the same bold-labeled fragment in the structure of their G_0-graphs, i.e. the fragment presented both in the initial and final cyclic structures. The idea of classification into types is important if one compares SRTs of different kinds, say inter- and intramolecular recyclizations. Many illustrations of similarity and dissimilarity of famous recyclizations are collected in[59-61]. Particularly, the Dimroth-, Cornforth-, and Boulton-Katritzky-type rearrangements of azoles (see early classification[58]) belong to different types, while the furoxane rearrangement differs from the Dimroth-type transformation in the class, not in the type (see Schemes 12, 13).

The hierarchical classification of recyclization graphs. The graphs G_0, G_1, and G_2 are combinatorial objects: all they are bicyclic graphs, that are differently labeled. The manner of labeling can be treated in terms of a hierarchy[60] (see Scheme 12). Indeed, at the top there are G_0-graphs. Next level is formed by G_1-graphs, that are labeled G_0-graphs, and the labels are dashed edges that can be differently placed on the solid edges of G_0-graphs. On the lowest level there are G_2-graphs, that are labeled G_1-graphs with the set of heteroatoms as the labels. From the last level the corresponding pair of M-graphs can be immediately restored, and familiar equation of recyclization can be obtained (of course, only with skeletal atoms of cycles and heteroatom labels). The symmetry of a graph on higher level rigidly determines the possibility of non-equivalent labeling at the lower level.

The suggested idea of the hierarchical classification seems very fruitful: all theoretically possible skeletal equations for SRTs can be once and forever enumerated and classified by arranging their recyclization graphs into structural types, kinds, classes, and families. The classification, in turn, can be used as an organizational principle to create database on known recyclizations, to establish degree of structural similarity between SRTs and real novelty of discovered examples, and to assist search of really new recyclizations in respect to the chosen hierarchical level. This ideal project has indeed been realized in practice[1,60] with the help of our computer program GREH (Graphs of REcyclizations of Heterocycles). The program is easily available from WWW (see Ref. 1b).

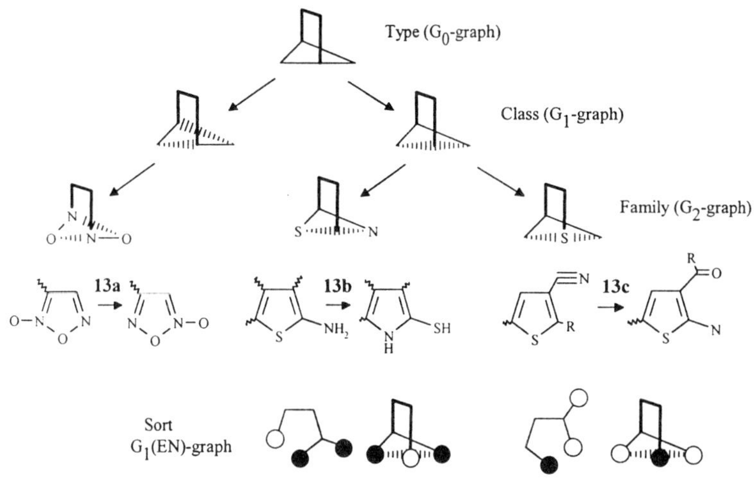

Scheme 12. Illustration of the hierarchic classification of recyclizations (see text). For reactions 13b and 13c the polarity of centers in open chains and the corresponding G_1(EN)-graphs are shown (labels: black for *N*-centers, white for *E*-centers).

The enumeration and codes for recyclization graphs. The concept of recyclization graphs was further developed in few directions. Particularly, two combinatorial problems has been resolved, namely:

(i) How to assign a specific code to every recyclization graph, so that the chemist can immediately call any recyclization without referencing to the diagram.

(ii) How much ring transformations are possible for every branch of the classification tree

Solution of the first problem lied in the assignment of a unique code for every type and class. The unique code for G_0-graph is the vector XYZ (X and Y are the size of initial and final heterocycles and Z - number of vertices in the bold-labeled bridge of G_0-graph). The codes for recyclization types of reactions 12a-12c are 554, 553, and 665 respectively. The codes for classes or G_1-graphs require precise definition of arrangement of the dashed edges in the ancestoring graph G_0. For rearrangements there are only two dashed edges. The position of dashed edges in the smaller cycle of the G_1-graph is indicated in the following manner: the edge closest to the bold-labeled bridge should be designated by letter a, the next one - by letter b etc. The solid edges in the greater cycle should be numbered analogously by the same symbols a,b,c..., starting from the edge adjacent to the a-edge of the smaller cycle. The resulted code for rearrangements looks like XYZ-$i_X j_Y$, (i and j - relative position of dashed labels in cycles X and Y) and the above example of recyclization 12c has the code 665-aa. For recyclization graphs of various kinds the notation XYZ-($i_X j_X$)($i_Y j_Y$) was suggested[62]. The letters in the brackets (defined in the manner similar to rearrangements) reflect the relative disposition of broken/formed bonds in the initial and final cycle. For SRTs 12a, 12b the codes of classes are 554-(ab)(ab) and 553-(a)(ac).

The problem of enumeration of recyclizations was resolved in different ways. Combinatorial formulas for enumeration of recyclization classes, and families have been suggested; the formulas were obtained by application of the Cauchy-Frobenius-Burnside lemma to this specific sort of graphs. One particular result of such enumeration is that for combinations of three heteroatoms (C,N,S) the number of theoretically possible Dimroth rearrangements for azoles is 486. In addition to the enumeration formulas, the full list of all possible codes for recyclizations of various kinds for 5-, 6-, and 7-membered rings was generated with the help of computer program for those SRTs with no more than 4 bonds changed via reaction. Therefore, this was the desired full structural classification of any (known and still unknown) SRTs. Meanwhile, the most important result was the program GREH, that permits chemist to visualize the branches of the classification tree.

How the program GREH works. The program GREH was designed[60] in order to generate, visualize, and investigate the SRT hierarchic tree. Each new branch of the tree is generated when the user attempts to display it. All the branches are stored in separate files on the hard disk for the subsequent viewing and search. For the user's convenience, recyclization graphs are shown only at the top level (G_0-graphs); all the deeper levels present information as differently labeled pairs of molecular graphs. Thus, starting from any chosen G_0 graph the user can go down the levels with the unique possibility to scan all pairs of molecular graphs on the given level and to search the necessary (hetero)cycle and its side chain for reactants and products. At the bottom level the degree of saturation is varied, and it is possible to edit the structures and add references, i.e. to create the database entries. After such an input of a recyclization the information is saved in such a manner that on every higher level it is clear how much data exists in the lower level(s). This methodology permits one to immediately conclude on the real degree of novelty of any SRT in respect to its type, family, and class.

Degenerated rearrangements. The model of recyclization graphs has appeared to be extremely useful for the description and enumeration of the specific type of SRTs -- the degenerated and quasi-degenerated heterocyclic rearrangements[62]. One example of such reaction is the well-known (quasi)degenerated Dimroth-type rearrangement reported for many heterocycles, particularly for 1-alkyl-2-iminopyridine derivatives, imidazo[1,2-a]pyridines etc.[19,20,52,55]. Specific of recyclization graphs G_1 and G_2 for this case of SRTs is that the initial skeletal structures (M-graphs) are isomorphic. Therefore, superposition of two isomorphic M-graphs results in the diagram that with necessity should have the certain elements of symmetry. As we proved early, *only three types of symmetry groups* (more precisely, the automorphism groups of a graph) are permitted for these recyclization graphs. These symmetry rules have been used to enumerate all theoretically possible examples of degenerated rearrangements for 6- and 5-membered rings with one heteroatom[62]. Analogous atlas for the still unknown degenerated rearrangements of quinoline was discussed recently[66].

Recyclization sorts. In many cases the SRTs are polar on nature. The opening of the initial ring occurs as the heterolytic exo-process of bond cleavage (see Section 3), and the formation of novel ring is the usual heterocyclization between appropriate *E*- and *N*-reaction centers. One can consider the arrangement of polarity labels in the initial cycle, open chain intermediate (real or hypothetical), and final cycle. If in all cases the polaritiy is the matching property, then the appropriate labels may be included into the structure of recyclization graphs.

Therefore, it appeared reasonable to introduce one more type of reaction graphs -- $G_1(EN)$-graphs[59-61], that are the G_1-graphs with *E* and *N* labels on the vertices. Examples of $G_1(EN)$-graphs are presented in Scheme 12. These $G_1(EN)$-graphs can play the same role as the above discussed polar cyclization diagrams. Of course, these graphs are combinatorial objects, and they can be arranged as an independent level of the hierarchical tree.

5.4. The model and reality

Actual arrangement of SRTs into types and classes. In order to understand actual distribution of SRTs into types, classes, and families, and with the goal to precise the criteria of "novelty" of recyclizations we performed analysis of the literature data. We have limited ourselves by the simplest cases of SRTs - the heterocyclic rearrangements of azoles to azoles, azines to azines, and interconversion reactions between azoles and azines, according to the strict limitation "one cyclic bond is broken, one cyclic bond is formed". Following the logic of the hierarchic classification it was extremely interesting to understand what types and classes of SRTs are still unknown.

For rearrangements of azoles the comprehensive analysis was published[59], and it resulted in conclusion that all possible types (G_0-graphs) have already been discovered. For interconversions of azines and azoles there are few recyclization types that have never been realized[61]. These still unknown types have the codes 563, 663, and 662 (see definition above), and their diagrams are presented in Scheme 12.

Analysis of rearrangements' distribution into classes (G_1-graphs) resulted in rather surprising picture. For overwhelming majority of all rearrangements *the unique predominating class has the code* XYZ-aa. By other words, in the pattern of bonds redistribution for rearrangements the formed and broken bonds prefer to be adjacent[59-61]. It is noteworthy, that the letter code aa for G_1-graphs corresponds to some unique feature of recyclization graphs: only for this code the total number of reaction centers (between which the skeletal ring bonds are redistributed) is equal to three. (It is clear, that for any other permitted combination of letters in the code the total number of the centers should increase up to four.) Hence, the high frequency of aa-classes appearance among the heteroaromatic rearrangements seems to follow from peculiar magic rule of "minimal number of reaction centers". The violation of this rule may cause the concurrent processes and formation of by-products, for example due to intermolecular condensations, oligomerizations etc. Although some classes with the code other than XYZ-aa are known, they appeared rather rarely[59]. To the best of our knowledge, among the classes with the aa-code there are three still unknown classes presented in Scheme 13.

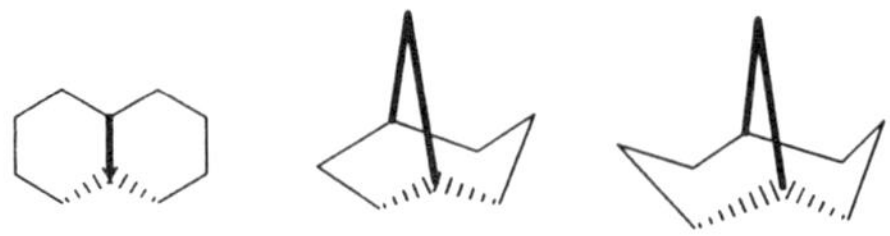

Scheme 13. Unknown types of SRTs with aa-code for rearrangements of 5- and 6-membered rings.

Prediction of unknown sorts of rearrangements. As we mentioned above, it is difficult to suggest a new rearrangement for azoles at the top of hierarchical tree: all the types and all aa-classes are already discovered. Then the question is: how the azole rearrangements are distributed in sorts?

As it follows from the combinatorial considerations, any G_1-graph with the code XYZ-aa can give rise only to the pair of $G_1(EN)$ graphs: the possible structures should contain either two *N* and one *E* label or vice versa. (Indeed, either *E*-center is rearranged between two *N*-centers, or vice versa, the *N*-center is redistributed between two *E*-centers.) Let us add these indices (either *NNE* or *EEN*) to the codes of G_1-graphs in order to establish the codes of recyclization sorts. The rearrangements of the first *NNE*-sort turn out to be more widespread in azoles' chemistry, and they are also better studied. In particular, the well-known examples of the Dimroth (see Scheme 12), Cornforth, and Boulton-Katritzky rearrangements have the codes 552-aa-*NNE*, 553-aa-*NNE*, and 554-aa-*NNE*, see diagrams in Scheme 14.

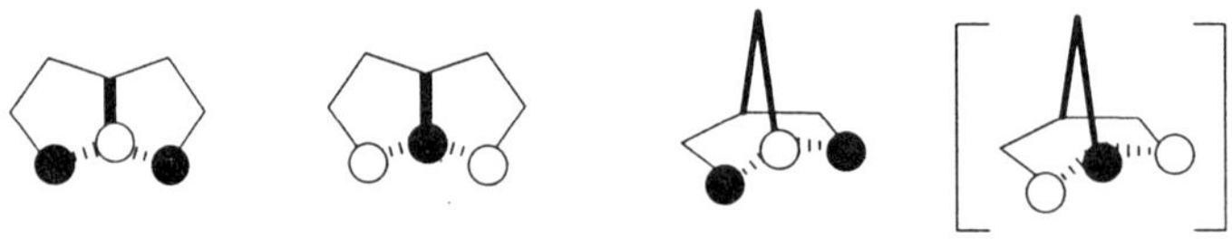

552-aa-*NNE* **552-aa-*EEN*** **552-aa-*NNE*** **552-aa-*EEN***

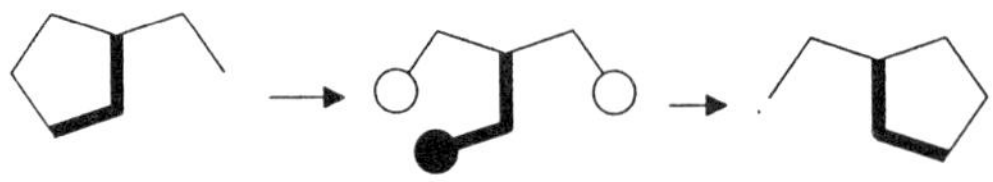

Z = NR, O, S; X, Y = CR, N; A, B = CCR, NN^+

Scheme 14. Examples of some known and unknown recyclization sorts. For unknown sort (shown in square brackets) the skeletal structures and arrangement of polar centers in the initial and final cycles (and intermediate open chain) can be restored from the recyclization graph, and the concrete reaction 14 of unknown sort can be predicted[60].

Two from these three named reactions have "opposite" sort, and examples of 552-aa-*EEN* and 554-aa-*EEN* rearrangements can be found in literature. However, the sort, opposite to the Cornforth rearrangement, with the code 553-aa-*EEN* is still unknown. It is easy to extract from the code the total structural equation and the appropriate polarities required for the design of recyclization (see Scheme 14) and hence, to predict the novel elegant rearrangement sort illustrated by reaction 14. The choice of various concrete initial heterocycles able to undergo the rearrangement 14 was discussed early[60].

The sorts of still unknown rearrangements "azole-azine", "azine-azole", and "azine-azine" (for which we failed to find at least one reference in the literature), are the following [61]: 563-aa-*NNE*, 563-aa-*EEN*, 562-aa-*EEN*, 653-aa-*NNE*, 663-aa-*NNE*, 663-aa-*EEN*, 662-aa-*EEN*, 662-aa-*NNE*. These codes are just the predictions of novel recyclization sorts recommended for experimental studies.

5.5. Computer-assisted predictions and experimental discovery of unknown recyclization families

Our interest in the experimental chemistry of the bridgehead azoloazines stimulated our efforts in predicting novel examples of recyclizations in this class with the help of program GREH and to verify the predictions experimentally. The bridgehead azoloazines (particularly the heteroaromatic cations) seemed to be attractive models for study new recyclizations, due to their ability to open either 5- or 6-membered ring (see ref. (52) and Section 4.3) and variety of modes for ring closure.

The methodology we used has been recently reviewed[67]. After careful analysis of the literature data the structures selected to study recyclizations were the derivatives of 1-heteroindolizines (neutral, mesoionic, and cationic). The GREH program has suggested for every cycle of the bicyclic 1-heteroindolizines the appropriate location of certain functions around the ring that are necessary for closure of new ring.

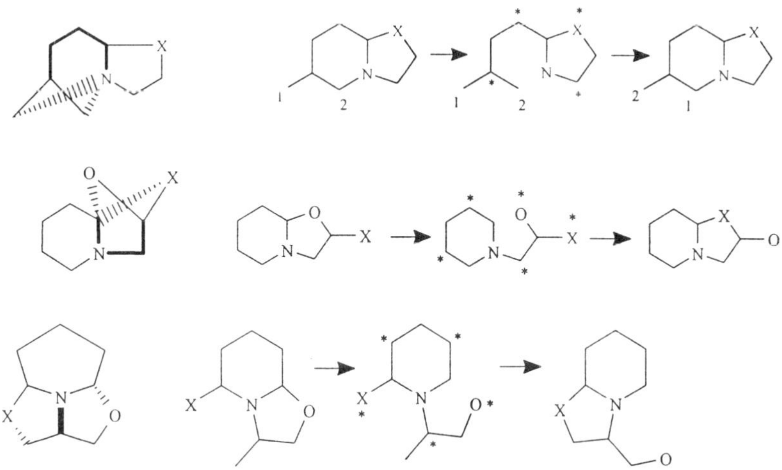

Scheme 15. Selected predictions of novel recyclization families for 1-heteroindolizines by the GREH program. From the recyclization graph (left, shown with annelated ring) the skeletal equation is restored. The design of polarity for exo- and endo-cyclic atoms achieved by application of the principle of the direct alternation chain. Labels in the open chain structures indicate the nucleophilic centers.

In order to avoid "combinatorial explosion", the strict limitations were used, particularly, invariance of cycle size, minimal number of broken/formed bonds, and the aa-classes of recyclization graphs. The last requirement was the cleavage of only the most probable bonds in 1-X-indolizines (X - heteroatom), i.e. of C9-X, C2-X, and C5-N bonds (see Section 4.3). After applying of such restrictions the predicted number of recyclizations decreased from few hundreds to few tens. Selected sixteen examples were published as the forecast[67], and few of them are presented in Scheme 15.

Design of substituent polarity for the concrete structures, capable to undergo the predicted transformations, was achieved by applying the principle of the *direct* polarity chain to 1-heteroindolizines (see Section 4.3) as it is illustrated in Scheme 15 this Synthesis of some structures with the desired polarities of substituents around the ring was achieved, particularly of 5-methyloxazolo[3,2-a]pyridinium salts[68]. The obtained structure undergoes elegant recyclization to 5-aminoindolizines (reaction 15 in Scheme 16) with complete agreement of the initial forecast[69,70]. It should be mentioned that neither for monocyclic nor for fused oxazoles such structural class has never been observed. We also failed to find any related analog with exactly the same sort. Another example of confirmation the forecast[67] was the recyclization of the mesoionic structure to the cationic oxazolopyridine[71], reaction 16 in Scheme 16. The feature of this reaction is that the final cationic structure, in turn, is able to undergo further recyclizations (see e.g. Scheme 11), and the principle of such "recyclization tandems" is rather rare case of heterocyclic chemistry.

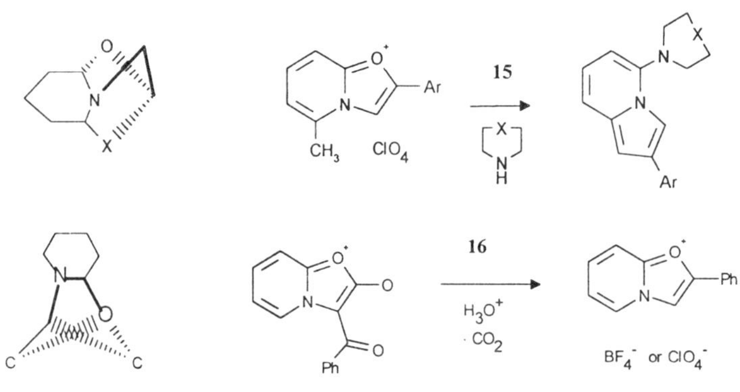

Scheme 16. Examples of novel recyclization families for fused oxazoles predicted early and discovered recently. On the left: initial recyclization graph used for verification.

6. Conclusion

As we tried to prove, many qualitative observations in heterocyclic chemistry can be more strictly reformulated with the help of the reaction graphs. These graphs are not something external, that comes from outside the topic of the chemistry of heterocycles: these models are inside the discipline. The well reviewed idea of polarities arrangement (along the cycles and chains) should be supplemented by simple considerations on the conservation and change of polarities via heterolytic reactions. Adding combinatorics to this idea results in helpful tool for education and rational classifications of reactions, particularly of heterocyclic formation, cleavage and transformation reactions. As we proved, the discussed models, implemented as the computer programs, may be really useful in rational *reactions planning* -- search and discovery of unknown reactions.

Acknowledgements

Research leading to the results discussed in this review was generously funded by International Science Foundation (in 1994-95), Volkswagen - Stiftung (in 1996-97), GRACENAS foundation (St. Petersburg, Grant Grant No.95-0-9.4-222), and Russian Foundation of Basic Research (Grant No.96-03-32953). The author wish to thank his co-authors for contribution.

References

1. (a) Lushnikov, D. E.; Babaev, E. V. *Chem. Heteroc. Compounds (Engl. Transl.)* **1993**, *10*, 1111. Original: *Khim. Geterotsikl. Soedin.* (in Russian), 1299 (*Chem. Abstr.* **120**: 216194k). (b) Computer programs mentioned in the review (programs GREH and Heterocycland for IBM PC; DOS or MS Windows 3.x) are now free available from WWW (http://org.chem.msu.su/~babaev).
2. Trinajctic, N. *Chemical Graph Theory.* CRC Press: Boca Raton, FL, 1991; second edition.
3. *Chemical Graph Theory: Introduction and Fundamentals;* Bonchev, D.; Rouvray D., Eds.; Abacus Press/Gordon and Beach: New York, 1991.
4. Balaban, A.T. *Rev. Chim.* **1967**, *12*, 875.
5. Dugundji, J.; Ugi I. *Topics Curr. Chem.* **1973**, *39*, 19.
6. Ugi, I.; Gillespie P. *Angew. Chem. (Int. Ed. Engl.)* **1971**, *10*, 914.
7. Hendrickson, J. B. *Angew. Chem. (Int. Ed. Engl.)* **1974**, *13*, 47.
8. Zefirov, N. S.; Trach S. S. *Zh.Org.Khim.* **1975**, *11*, 1785 (in Russian).
9. Zefirov, N. S. *Acc. Chem. Res.* **1987**, *20*, 237.
10. Fujita, S. *J. Chem. Inf. Comput. Sci.* **1986**, *26*, 205.
11. Fujita, S. *J. Chem. Soc. Perkin Trans. II.* **1988**, 597.
12. Barone, R.; Chanon, M. In *Computer Aids to Chemistry.* Vernin, G.; Chanon M., Eds.; Chichester: E. Horwood, 1986.
13. Wilcox, C. S.; Levinson, R. A. In *Artificial Intelligence Applications in Chemistry;* Pierce, T.H.; Hohne, B.A., Eds.; Washington: Amer. Chem. Soc. 1986, p.209.
14. Hendrickson, J. B. *Rec. Trav. Chim.* **1992**, *111*, 323.
15. Zefirov, N. S.; Tratch, S. S. *Anal. Chim. Acta.* **1990**, *235*, 115.
16. Zefirov, N. S.; Baskin, I. I.; Palyulin, V. A. *J. Chem. Inf. Comput. Sci.* **1994**, *34*, 994.
17. Bauer, J. *Tetrahedron Comput. Methodol.* **1989**, *2*, 269.
18. Herges, R.; Hoock C. *Science.* **1992**, *255*, 711.

19. *Heterocyclic Compounds;* Elderfield, R. C., Ed.; Wiley: NY, 1950; Vols. 1-7.

20. *Comprehensive Heterocyclic Chemistry;* Katritzky, A. R.; Rees, C. W., Eds.; Pergamon Press: Oxford, 1984; Vols. 1-8.

21. *Houben-Weil Methoden der organischen Chemie. Bd. E 7a (Hetarene II);* Kreher, R. P., Ed.; Stuttgart, N.Y.: Thieme Verl., 1991; Teil 1.

22. Jones, G. In *Quinolines;* Jones, G. Ed.; L.:Wiley, 1977; Part 1, p. 93

23. Potts, K. In *Comprehensive Heterocyclic Chemistry;* Katritzky, A. R.; Rees, C. W., Eds.; Pergamon Press: Oxford, 1984; Vol. 5, p. 111.

24. Katritzky, A. R. *Handbook of Heterocyclic Chemistry;* Oxford: Pergamon, 1985; p. 382.

25. Bures, M.G.; Jorgensen, W.L. *J. Org. Chem.* **1988**, *53,* 2504.

26. Babaev, E.V. *Chem. Heteroc. Compounds (Engl. Transl.)* **1993**, *7*, 796. Original: *Khim. Geterotsikl. Soedin.* (in Russian), 937. (*Chem. Abstr.* **120**: 77095p)

27. Evans, D. A. *UCLA Physical Organic Chemistry Seminar*, May 6, 1971. (Manuscript 242A: Evans, D. A. An Organizational Format for the Classification of Functional Groups).

28. Evans, D. A.; Andrews G. C. *Acc. Chem. Res.* **1974**, *7*, 147.

29. Ugi I., Gillespie, P. *Angew. Chem.* **1971**, *23*, 985.

30. Doenges, R.; Groebel, B.-T.; Nickelsen, H; Sander J. *J. Chem. Inf. Comput. Sci.* **1985**, *4*, 425.

31. Serratosa, F. *Buttleti Societat Catalana Cienc. Fis., Quim., Mat.* **1977**, *1(2)*, 75 (In Catalan). (*Chem. Abstr.* **89** 196341s.)

32. Wagener, M.; Gasteiger, J. In *Software Developments in Chemistry*; Gasteiger, J., Ed.; Springer-Verlag: Berlin, 1990; Vol. 4, p. 265.

33. Serratosa, F.; Xicart, J. *Organic Chemistry in Action. The Design of Organic Synthesis*. 2nd Edition, Elsevier: Amsterdam, 1996.

34. Lapworth, A. *Manchester Memoirs Lit. Phil. Soc.* **1920**, *64(3)*, 1.

35. Seebach, D. *Angew. Chem. (Int. Ed. Engl.)* **1979**, *4*, 239.

36. Ho, T.-L. *Reviews of Chem. Intermediates.* **1988**, *9*, 117.

37. Ho, T.-L. *Polarity Control for Synthesis*. Wiley: Chichester, 1991.

38. Babaev, E. V. *Mendeleev Chemistry Journal (Engl.Tansl.)*, **1994**, *38(6)*, 78. Original: *Ross. Khim. Zh. [Zhurnal Ross. Khim. Ob-va im. Mendeleeva]* (in Russian), 54 (*Chem. Abstr.* **122**: 186630w).

39. McKillop, A.; Boulton, A. J. In *Comprehensive Heterocyclic Chemistry;* Katritzky, A. R.; Rees, C. W., Eds.; Pergamon Press: Oxford, 1984; Vol. 2, p. 67.

40. Babaev, E. V.; Cytovskij, S. V. *Moscow University Chem. Bull. (Engl. Transl.)* **1994**, *49(1)*, 82. Original: *Vestnik Mosk. Univ. Ser.2: Khim.* (in Russian), *35(1)*, 85 (*Chem. Abstr.* **120**: 2298437d).

41. (a) Babaev, E. V. *AIP Conference Proceedings. (ECCC 1. Computational Chemistry. Nancy, France May 1994.)*. AIP Press: Woodbury, NY; **1995**, *330*, 746. (b) Babaev, E. V., *7-th International*

Conference on Mathematical Chemistry, Girona (Spain), May 26-31, 1997 (Plenary lecture); Proceedings **1997**, 17.

42. Efimov, A. V.; Babaev, E. V. *ibid.* 751.
43. (a) Babaev, E. V. *Chem. Heteroc. Compounds (Engl. Transl.)* **1993**, *7*, 818. Original: *Khim. Geterotsikl. Soedin.* (in Russian), 962 (*Chem. Abstr.* **120**: 106734d).
44. Latham, E. J.; Murphy, S. M.; Stanforth, S. F. *Tetrahedron Lett.* **1994**, *35*, 3395.
45. Meyers, A. I. *Heterocycles in Organic Synthesis*; N.Y.: Wiley Interscience, 1974.
46. Kozikowski, A. P. In *Comprehensive Heterocyclic Chemistry;* Katritzky, A. R.; Rees, C. W., Eds.; Pergamon Press: Oxford, 1984; Vol. 1, p. 413.
47. (a) Lipshutz, B. H. *Chem. Rev.* **1986**, *86*, 795. (b) Stirling, Ch. J. M. *ibid.* **1978**, *78*, 517.
48. Gilchrist, T. L. *Adv. Heterocycl. Chem.* **1987**, *41*, 41.
49. (a) Maiboroda, D. A.; Scherbukhin, V. V.; Babaev, E. V. *AIP Conference Proceedings. (ECCC 1. Computational Chemistry. Nancy, France May 1994.).* AIP Press: Woodbury, NY; **1995**, *330*, 757. (b) Simple computer program FROG (for DOS, IBM PC) described in this paper is available upon request from the author.
50. Babaev, E. V.; Bobrovskii S. I.; Bundel Yu. G. *Chem. Heteroc. Compounds (Engl. Transl.)* **1988**, *11*, 1307. Original: *Khim. Geterotsikl. Soedin.* (in Russian), 1570 (*Chem. Abstr.* **111**: 97031g).
51. (a) Babaev, E. V. *Bull. Soc. Chim. Belg.* **1992**, *9*, 823. (b) Babaev, E. V.; Torocheshnikov V. N.; Bobrovskii S. I. *Chem. Heteroc. Compounds (Engl. Transl.)* **1995**, *9*, 1079. Original: *Khim. Geterotsikl. Soedin.* (in Russian), 1235 (*Chem. Abstr.* **124**: 342267g).
52. Maiboroda, D. A., Babaev E. V. *Chem. Heteroc. Compounds (Engl. Transl.)* **1995**, *11*, 1251. Original: *Khim.Geterotsikl.Soedin.* (in Russian), 1445 (*Chem. Abstr.* **125**: 33500r).
53. (a) Maiboroda, D. A.; Babaev, E. V.; Jug K. *J. Org. Chem.* **1997** (in press). (b) For SINDO method see Nanda, D. N.; Jug, K. *Theor. Chim. Acta* **1980**, *57*, 95.
54. Babaev, E.V.; Pasichnichenko K. A.; Maiboroda, D. A. *Khim. Geterotsikl. Soedin.* (in Russian), **1997**, *3*, 397.
55. (a) van der Plas, H. C. *Ring Transformation of Heterocycles;* Academic Press: London, N.Y., 1973; Vols. 1, 2. (b) Other comprehensive reviews on the "named" reactions are Refs. 17, 24, 58 - 60.
56. Schvaika, O. P.; Artyomov, V. N. Usp. Khimii (in Russian) **1972**, *41*, 1788.
57. *Pyrillium Salts. Syntheses, Reactions and Physical properties*. Balaban, A. T., Ed.; Academic Press: N.Y., 1982.
58. L'abbé, G. *J. Heterocycl. Chem.* **1984**, *21*, 627.
59. Babaev, E. V.; Zefirov, N. S. *Bull. Soc. Chim. Belg.* **1992**, *101*, 67.
60. Babaev, E. V.; Lushnikov D. E.; Zefirov N. S. *J. Amer. Chem. Soc.* **1993**, *115(6)*, 2416.

61. Babaev, E. V.; Zefirov, N. S. *Chem. Heteroc. Compounds (Engl. Transl.)* **1992**, *6*, 671. Original: *Khim. Geterotsikl. Soedin.* (in Russian), 808 (*Chem. Abstr.* **118**: 101252u).

62. Babaev, E. V.; Zefirov, N. S. *J. Math. Chem.* **1992,** *11*, 65.

63. Babaev, E. V.; Bozhenko, S. V.; Maiboroda, D. A. *Russ. Chem. Bull. (Engl.Transl.)*, **1995,** *11*, 2203. Original: *Izv. Akad. Nauk, Ser. Khim.* (in Russian), 2298. (*Chem. Abstr.* **124**: 316926b).

64. Babaev, E. V.; Bozhenko, S. V. *Khim. Geterotsikl. Soedin.* (in Russian), **1997**, *1*, 141.

65. Terenin, V. I.; Babaev, E. V.; Yurovskaya, M. A.; Bundel, Yu. G. *Chem. Heteroc. Compounds (Engl. Transl.)*, **1992**, *6*, 658. Original: *Khim. Geterotsikl. Soedin.* (in Russian), 792 (*Chem. Abstr.* **118**: 168899d).

66. Babaev, E. V.; Jernigan, K. In *4-th Internatonal Conference on Chemical Structures. June 2-6, 1996*; Noordwijkerhout. Netherlands, 1996;. Abstr. of papers, 31.

67. Babaev, E. V.; Zefirov, N. S. *Khim. Geterotsikl. Soedin.* (in Russian), **1996**, *11-12*, 1564.

68. Babaev, E. V.; Efimov, A. V.; Maiboroda, D. A. *Chem. Heteroc. Compounds (Engl. Transl.)*, **1995**, *8*, 962. Original: *Khim. Geterotsikl. Soedin.* (in Russian), 1104 (*Chem. Abstr.* **124**: 232301f).

69. Babaev, E. V.; Bozhenko, S. V.; Efimov, A. V. In *12-th Symposium on Chemistry of Heterocyclic Compounds and 6-th Blue Danube Symposium on Heterocyclic Chemistry*. Brno, Czech Republik, Sept. 1-4, 1996; Conference Proceedings, P21.

70. Babaev, E. V.; Efimov, A. V.; Maiboroda, D. A.; Jug, K. *Liebigs Ann.* **1997** (in press).

71. Babaev, E. V.; Orlova, I. A. *Khim. Geterotsikl. Soedin.* (in Russian), **1997**, *4* (in press).

FIVE- AND SIX-MEMBERED HETEROCYCLES FROM NUCLEOPHILIC ADDITION TO NITRILE OXIDES

Giovanni Grassi and Francesco Risitano

Istituto di Chimica dei Composti eterociclici, Università, Vill. S. Agata, I- 98166 Messina, Italy

Abstract. *Nitrile oxides and suitably functionalized nucleophiles lead to a variety of five- and six-membered heterocycles. In general, the reaction involves nucleophilic addition at the nitrile oxide carbon and subsequent cyclization at the other reaction site of the original nucleophile. This useful synthetic methodology allows the construction of heterocycles in which the nitrile oxide participates with its entire CNO skeleton or only one C-unit. Limitations in the applicability of these reactions can arise with nitrile oxides which easily tend towards autotransformation.*

1. Introduction

The study of nitrile oxide addition reactions has almost exclusively concerned dipolar 1,3-cycloaddition to unsaturated systems like olefins, acetylenes, carbonylic compounds, Schiff bases, nitriles, etc.[1,2] These are now well-established as attractive methods for the highly regio and stereocontrolled synthesis of N-O pentatomic heterocycles as well as useful routes for the synthesis of complex products through chemical modification and ring cleavage of primarily formed heterocycles.[3,4] Even the intramolecular version of these cycloadditions with olefins (INOC), which provokes the simultaneous double annulation of two rings, is enjoying rapid expansion as a result of the fact that the selective opening of the isoxazoline moiety of the fused heterocycle produces a new cyclic molecule with functionalized groups that are favourably positioned for the synthesis of complex natural molecules.[5,6]

Nucleophilic addition represents the other important class of reactions typical of nitrile oxides so far neglected probably because of the limited synthetic interest of many of these additions.[1,7] For instance, neutral aprotic nucleophiles like pyridine or trimethylamine in polar solvents determine the simple dimerization of aromatic nitrile oxides to 3,6-diaryl-1,4,2,5-dioxazines or to 3,5-diaryl-1,2,4-oxadiazole 4-oxides respectively (Scheme 1).[8]

However, by using suitable nucleophiles excellent results have been obtained in the construction of five- and six-membered heterocyclic ring systems containing one or more heteroatoms. It seemed appropriate, therefore, to focus on those reactions which offer an interesting synthetic methodology for heterocyclic systems.

The present article reviews these advantageous nucleophilic additions to nitrile oxides which evidence the extraordinary capacity of this dipole to synthesize heterocycles also by other routes than conventional concerted 1,3 cycloaddition reactions.

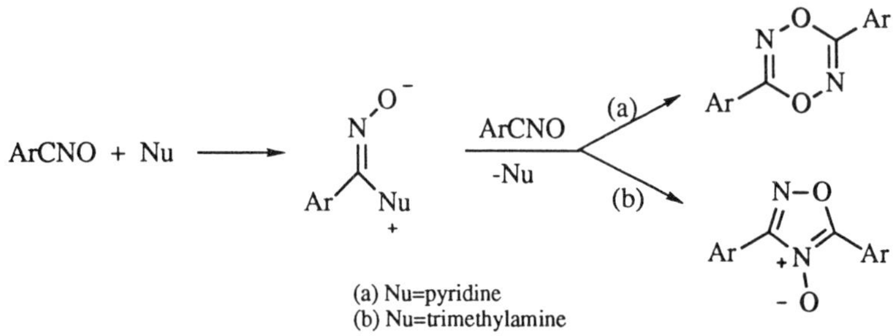

Scheme 1

2. General aspects of addition-cyclization reactions

The addition reaction of nucleophilic agents to nitrile oxides occurs exclusively at the carbon atom of the dipole and generally proceeds with the stereospecific formation of the Z-adduct even when this kinetic product is thermodynamically less stable than its *E*-isomer (Scheme 2).[9,10]

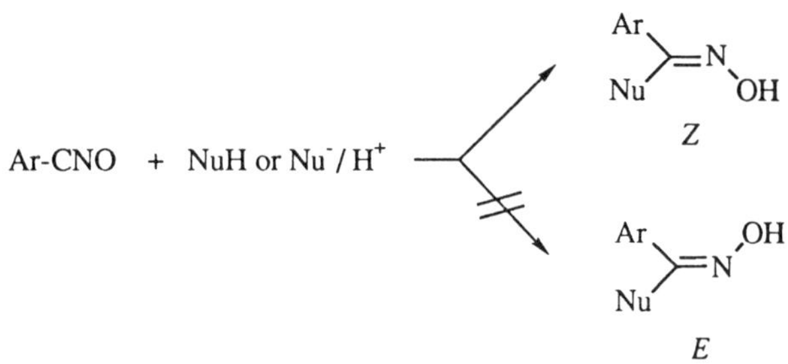

Scheme 2

NuH type protic nucleophiles lead exclusively to Z-amidoximes, whose formation in cases where they are not isolated, is confirmed by chemical evidence.

The relative addition mechanism has been the subject of some discussion. The most recent results are in the main consistent with a concerted process which occurs via a single transition structure connecting the reactants and the oxime adduct.[11] The subsequent fate of the open-chain Z-adduct is dependent on the reaction conditions and, above all, on the nitrile oxide and the nucleophile used.

When the nucleophilic agent **1** bears a further reaction site Y (Scheme 3), since the hydroxyl group and the nucleophile moiety are disposed in a cis relationship across the C=N bond, the relative acyclic Z-adduct can undergo easy subsequent intramolecular reaction.

This affords versatile access to heterocyclic systems of varying size depending on the distance between the reaction sites of the nucleophile molecule and the number of atoms with which the nitrile oxide participates in the construction of the ring.

In particular heterocycles can be obtained via the pathways illustrated in Scheme 3, which lead to molecules in which the contribution of the atoms of the nitrile oxide fundamental skeleton (CNO) is total or partial.

In the first case (Y=E), the presence of an electrophilic center on the nucleophilic substrate allows the formation of five- and six-membered heterocycles depending on the relative position of the two reaction sites; the oxime nucleophilic moiety, given its ambident nature, can react both via the oxygen (path i) and the nitrogen (path ii).

In the second case (Y=X'), in the presence of a further nucleophilic center on the original nucleophile, the nitrile oxide participates in the construction of the heterocycle only with the substituted C-unit via an apparent complex reaction sequence.

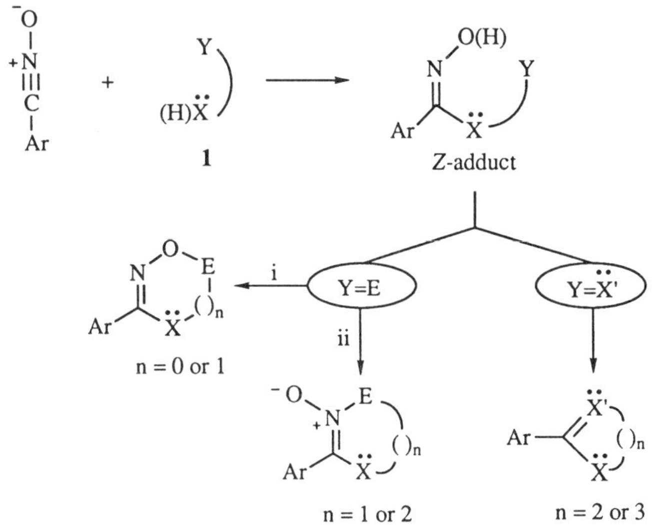

Scheme 3

Obviously, several variations on this general scheme can be effected. For example, by using nitrile oxides which have in turn been functionalized, it is possible to obtain unexpected heterocyclic syntheses.

All these addition-cyclization reactions prove synthetically useful when the competitive dimerization and/or rearrangement reactions of the nitrile oxide used are sufficiently limited.

3. Addition-cyclization of nucleophiles with electrophilic or nucleophilic center

Addition to nitrile oxides of nucleophiles with an electrophilic center can determine cyclization with retention of the N-O group of the starting nitrile oxide and relative location of the oxygen atom in endo or exocyclic position.

Anionic nucleophiles like thiocyanate **2** (X=S)[12] and cyanate **3** (X=O)[13] with arylnitrile oxides lead to the formation of 3-aryl-5-imino-1,4,2-oxathiazolines **4** and 3-aryl-1,2,4-oxadiazol-5(4H)-ones **5** respectively (Scheme 4).

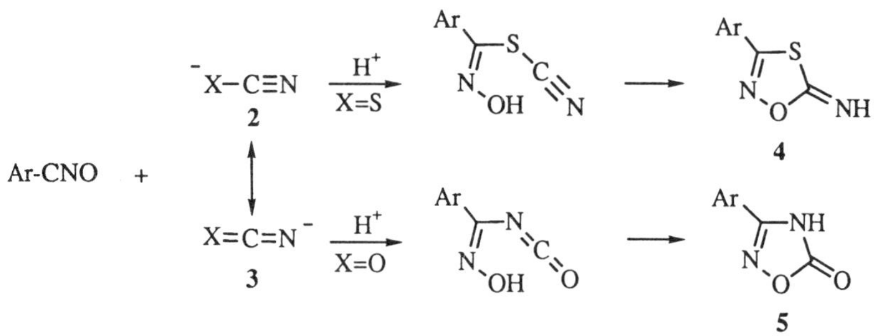

Scheme 4

α-Amino acid esters **6** react in the expected manner with nitrile oxides and the resulting Z-amidoxime esters cyclize exclusively in an allowed 6-Exo-Trig process to the corresponding 1,2,4-oxadiazin-6-ones **7** (Scheme 5).[14]

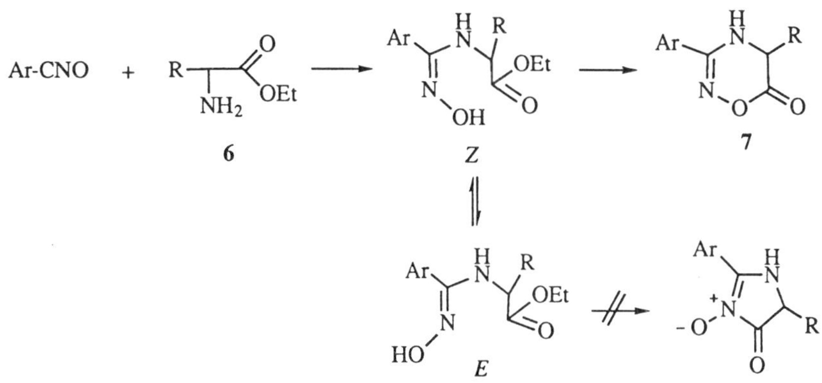

Scheme 5

The action of anion **2** confirms a similar previous result.[15] What can be observed is the selective attack of the more nucleophilic sulfur atom at the C-atom of the nitrile oxide followed by cyclization and the complete absence of isomeric cycloadducts to the C≡N bond.

In contrast, no support was given to the reaction mechanism with isocyanates **3**, so it cannot be excluded that oxadiazolone **5** might originate from a 1,3- cycloaddition of nitrile oxide to the C=N bond of **3**. Indeed similar cumulated C=N bond systems of aromatic carbodiimides have been reported to react as dipolarophiles with nitrile oxides.[16]

In the same way, the unusual formation of 1,2,4-oxadiazoles **10** from benzamide **8** and nitrile oxides[17] can be interpreted differently (Schema 6): either as a concerted 1,3-cycloaddition to the C=N bond of enolic form **8**(B), in agreement with similar reactions with alkylimidates,[18] or as 5-Exo-Trig process of the open-chain Z-oxime **9**. The latter brings to mind the five-membered ring intermediate supposedly formed in the rearrangement of the primarily formed acyl hydroximic acid from carboxylic acid and nitrile oxides to acyl hydroxamate.[19]

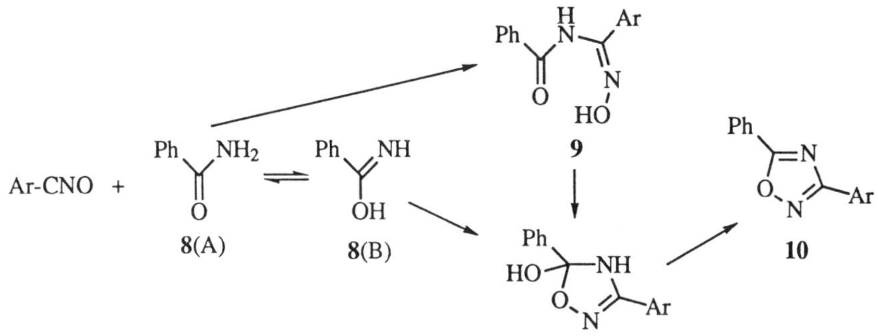

Scheme 6

However, the addition of α-amino esters **6** proceeds with the formation of only one of the two possible cyclization products: the six-membered one requiring the *Z*-configuration of the intermediate oxime (Scheme 5).

Similar behaviour can be observed for the reaction of arylnitrile oxides and benzimidazole **11** (Scheme 7).[20]

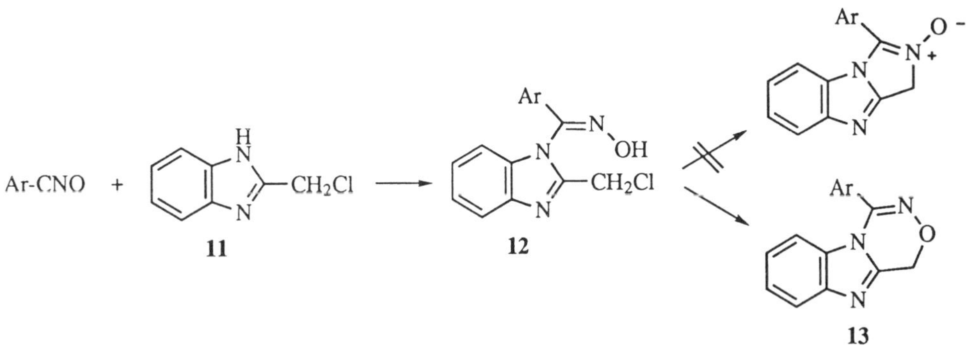

Scheme 7

Indeed, the initial amidoxime **12** cyclizes exclusively through the oxygen atom to the tricyclic derivative **13** via a 6-Exo-Tet process.

The reaction with N-methylhydrazones of aldehydes and ketones **14** (Scheme 8) is more complex.[21,22,23] Addition to the nitrile oxide produces the kinetically favoured *Z*-adduct **15** which can undergo either isomerization to the thermodynamically stable *E*-adduct **16** or tautomerization to the 5,6-dihydro-4H-1,2,4,5-oxatriazine **17** or irreversible cyclization to the 1-methyl-1,2,4-triazoles **18**. The distribution of these products depends on the reaction procedure and the substituents.

No product of 1,3-cycloaddition of the nitrile oxide to the C=N bond of the N-Me derivatives **14** has ever been observed; while with simple hydrazones or phenylhydrazones, the 1,3-cycloadducts become the preferred products. Thus, in the reaction of hydrazonic systems (C=N-N) in apolar solvents, the attack site of the dipole seems determined by the nucleophilicity of the terminal hydrazonic N-atom.[24]

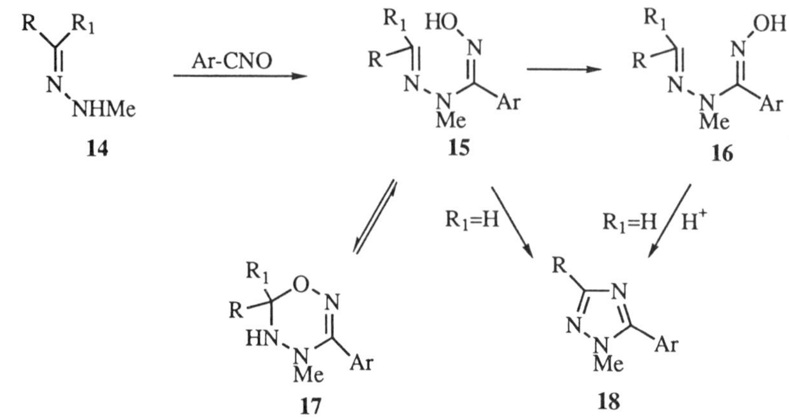

Scheme 8

In particular, using methylhydrazones of aromatic aldehydes **14** (R_1=H),[21] triazoles **22** and *E*-oximes **21** are the only products when reaction conditions, for example on contact with silica gel, favour the tautomerization of the *Z*-oximes **19** to the nitroso intermediate **20** (Scheme 9).

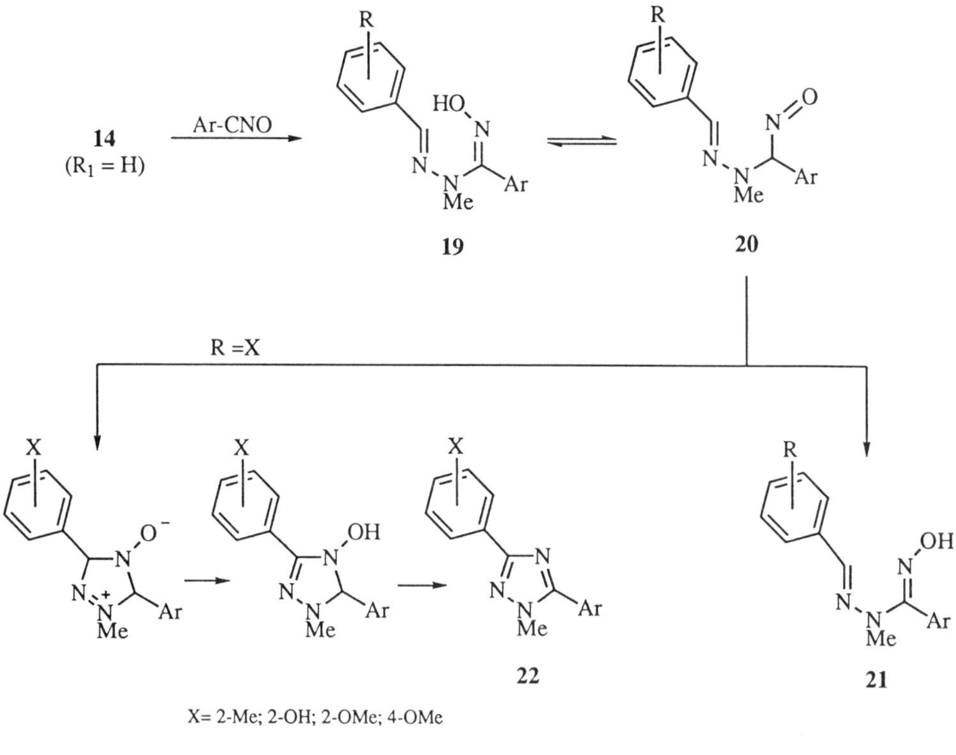

Scheme 9

The latter either produces *E*-isomer **21** or, if it possesses the sufficiently nucleophilic azomethyne carbon atom, cyclizes to triazole **22** by means of a facile 5-Exo-Trig process. In all other cases acidic treatment is necessary to obtain triazoles **22**.

C-nucleophiles, like β-keto sulfoxide **23a** and ylidene malonitrile **23b** anions and like sulfoxonium allylides **23c**,[25,26,27] with nitrile oxides permit the synthesis of 4-sulfinyl isoxazoles **25**, pyridine N-oxides **26** and 6H-1,2-oxazines **27** respectively (Scheme 10).

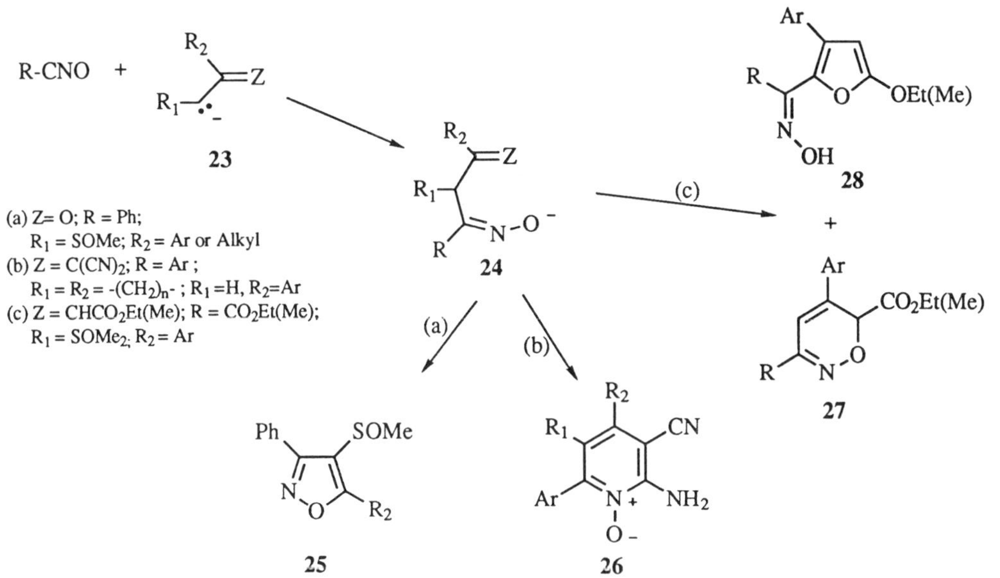

Scheme 10

The formation of isoxazoles **25** could also be interpreted as a concerted 1,3-cycloaddition to the C=C double bond of the sulfoxide in enolic form, in agreement with results reported for the synthesis of similar systems.[28] Nevertheless, the two step pathway preferred here is supported by the reaction of the nitrile oxide with β-ketoesters and malonic derivatives which, in aqueous alkali, leads to isoxazolones via the isolated open-chain intermediate.[29] Similar acetylenic oximes, produced by acetylides and nitrile oxides,[30] are easily obtained since subsequent cyclization to isoxazoles requires a more difficult 5-Endo-Dig process.

Formation of aromatic cyanopyridine **26** constitutes an unusual example of cyclization via the nitrogen of the oxime functional group. This route is favoured because the alternative mode of cyclizing with the oxygen via a 5-Exo-Trig or a 6-Endo-Trig process would have given rise to non-aromatic heterocyclic systems.

Reaction with sulfoxonium allylides **23c** in the main leads to furanylglyoxylate oxime **28c**, but under acid conditions these derivatives can be transformed, expanding quantitatively to the 1,2-oxazinic derivative **27c** (Schema 11).

This result is undoubtedly linked to the nature of the nitrile oxide used (R = COOEt or COOMe) since a similar addition product obtained as a minor component from furan and PhCNO neither cyclizes to a fused heterocycle nor expands to oxazine.[31]

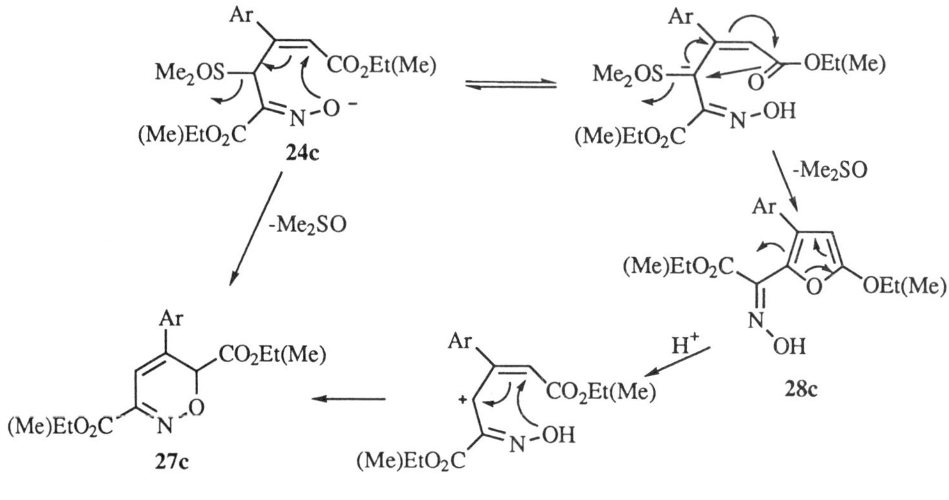

Scheme 11

Addition to nitrile oxides of nucleophiles with suitably positioned electrophilic centers also allows the synthesis of benzo-fused heterocycles via two conventional approaches based on an *ortho*-disubstituted benzene or on a monosubstituted benzene with a free *ortho* position.

N-Aryl S,S-dimethylsulphimides **29**, possessing a highly nucleophilic N-atom which bears a good leaving group, were used successfully in reactions with nitrile oxides for the synthesis of the benzo-fused heterocycles **30, 31** and **32** (Scheme 12).[32,33,34]

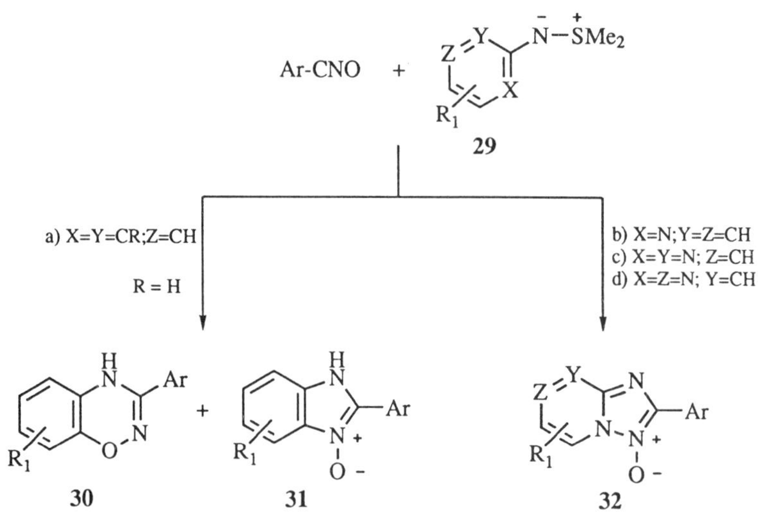

Scheme 12

With 2,6-unsubstituted arylnitrile oxides and sulphimides **29a** only benzo-1,2,4-oxadiazines **30** are obtained via the formation of the nitroso-imine intermediate **33** (Scheme 13).[35] The alternative mode of cyclizing with an allowed 6-Endo-Trig process can be rejected on the grounds that the intermediacy of the nitroso-imine **33** has been conclusively demonstrated. Indeed its 1:1 adduct with the nucleophilic diene thebaine was isolated releasing the thebaine and the expected cycloaddition product when heated.[32]

The resulting benzo-1,2,4-oxadiazines **30** exist in two tautomeric forms (A) and (B): an X-ray study carried out on 7-chloro-3-*p*tolylbenzo-1,2,4-oxadiazine **30** (R_1=7-Cl;Ar=*p*tolyl) showed that in solid state these heterocycles are present only as 4H (A).[36]

The above reaction was also observed for *ortho*-disubstituted N-aryl S,S-dimethylsulphimides **29a** (R≠H) (Scheme 13); with R=Me in particular the fixed benzo-fused product **34** was isolated but it seems not to be very stable because of the loss of aromaticity of the fused benzene ring.[37]

With a good leaving group, for example R=Br, the analogous system initially formed evolves spontaneously into aromatic benzoxazole **35**. In this case, the removal of the bromine and the rearrangement into benzoxazole, via the breaking of the weak N-O bond, are favoured by recovery of aromaticity in the fused ring.[38]

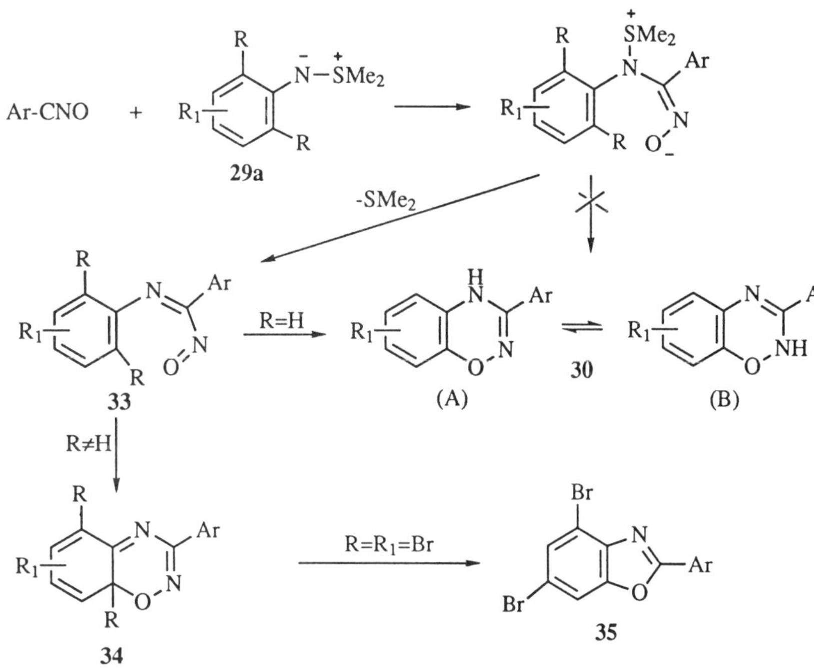

Scheme 13

In subsequent studies with sterically stabilized nitrile oxides, benzimidazoles 3-oxides **31** as well as the expected benzo-oxadiazines **30** were isolated. The formation of these new products was rationalized on grounds of the presence of the nitroso-imine intermediate in its *s-trans* form too.[33]

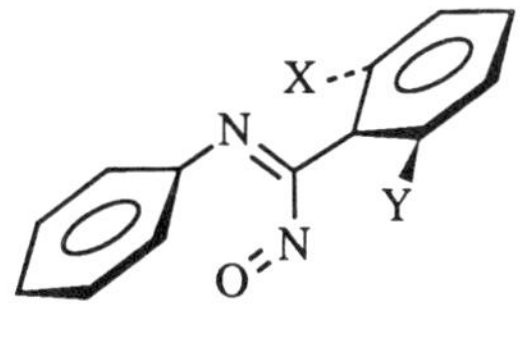

s-cis form

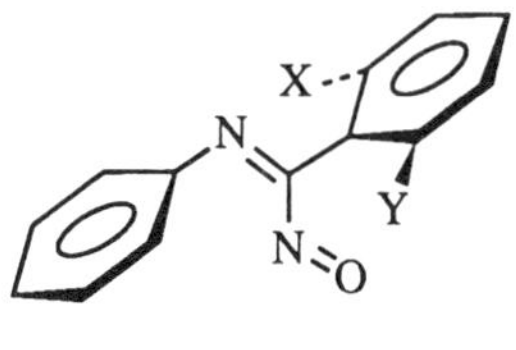

s-trans form

Five-membered cyclization with formation of the benzo-fused triazole N-oxide **32** (Scheme 12), is the only cyclization observed when the N-aryl of the sulphimide bears a favourably incorporated nucleophilic nitrogen atom, cases b,c,d (X=N).[34] In the same way, S,S-dimethyl-N-(N-arylbenzimidoyl)sulphimides **36** react with nitrile oxides to cyclize by a nitroso-imine intermediate **37** and give 1,2,4-triazole 2-oxides **38** (Scheme 14).[34]

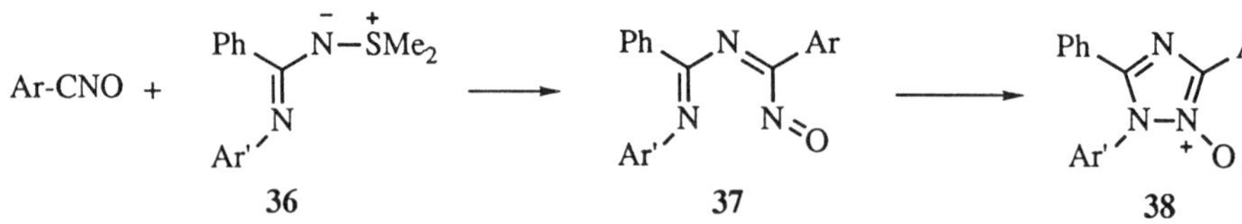

Scheme 14

A similar reaction was observed between nitrosoarenes **39** and nitrile oxides where the resulting 1:1 adduct **40**, an α-nitrosonitrone, can cyclize depending on the reaction conditions and on the nature of the substituents, forming either five- and six-membered benzo-fused systems[39,40] or fused heterocycles[41] with bridgehead nitrogen (Scheme 15).

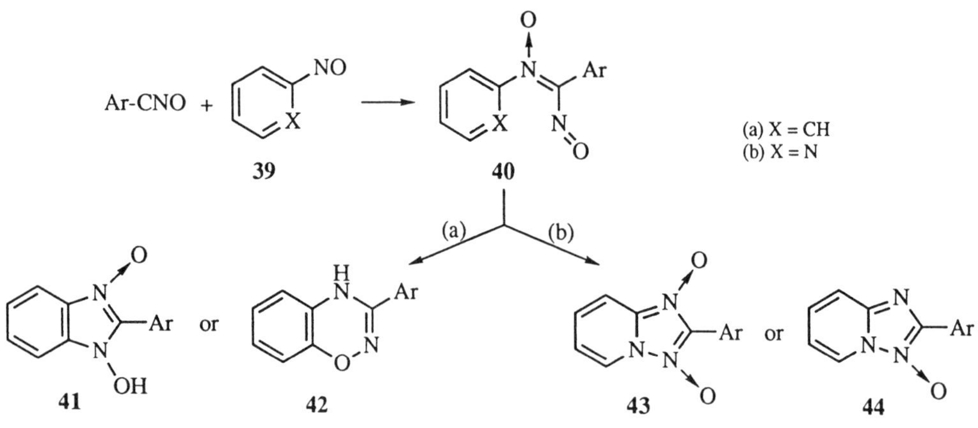

Scheme 15

1-Hydroxybenzimidazole 3-oxide **41** or 3-aryl-4H-1,2,4-benzoxadiazines **42** are the products isolated from nitrosobenzene **39a**. The expected six-membered-ring products of electrocyclic closure of the nitrosonitrones, the 4-hydroxy-1,2,4-benzoxadiazines, were not found but were assumed to be intermediates in the formation of deoxygenated benzoxadiazine.

The extension of this reaction seems limited to 2,6-unsubstituted nitrosoarenes. Indeed, the presence of *ortho* blocking groups inhibits cyclization. 2-Nitrosopyridines **39b** obviously give rise exclusively to five-membered ring closure, with formation of 1,2,4-triazolo[1,5-a]pyridine 1,3-di-N-oxides **43** or 3-oxides **44** (Scheme 15). The specific 1-deoxygenation stage was not clarified.

A nitrosonitrone intermediate **46** was invoked to justify the expansion of benzofurazan N-oxides **45** to benzo-as-triazine tri-N-oxides **47** in the reaction with nitrile oxides (Scheme 16).[42]

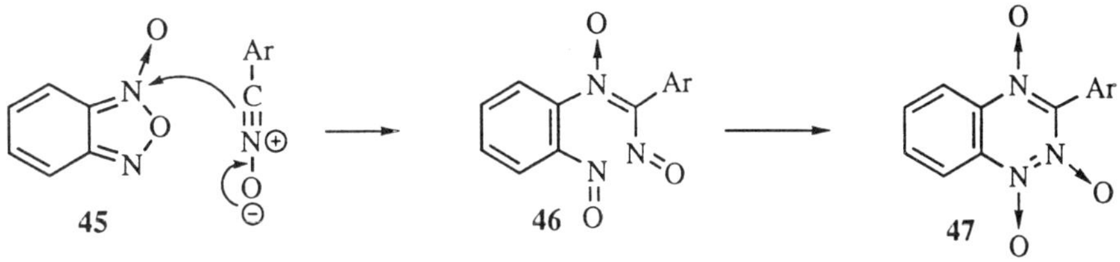

Scheme 16

In such a case, the intermediate **46** would be produced following an unusual attack of the nitrile oxide as a C-nucleophile. Since it cannot be excluded that the same intermediate **46** might originate from the ring-chain tautomer *o*-dinitrosobenzene **48** (Scheme 17), it could reasonably be maintained that the nitrosonitrones **40** might also derive from a similar C-nucleophile nitrile oxide attack.[43]

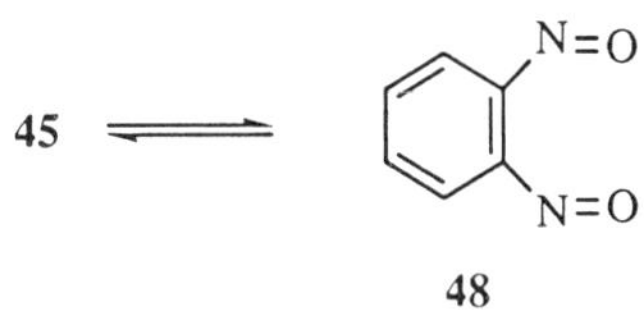

Scheme 17

This is supported by the fact that 2-nitrosopyridines **39b** are subject to attack exclusively at the nitroso N-atom, unlike the 2-substituted pyridines, which give expected nucleophilic addition with the pyridine N-atom at the nitrile oxide C-atom.

N-substituted benzamidoximes **49**, which, like the sulphimides observed, contain an amidine moiety in their molecule, display more complex behaviour in reaction with nitrile oxides depending on how the primarily formed non-isolated O-benzoyloximic intermediate **50** evolves (Scheme 18). At room temperature, in ether or chloroform, the O-benzoylderivate **51** is obtained; whereas in toluene at 100°C it produces 1,2,4-triazole 4-oxide **52** and benzotriazine **53**.[44,45]

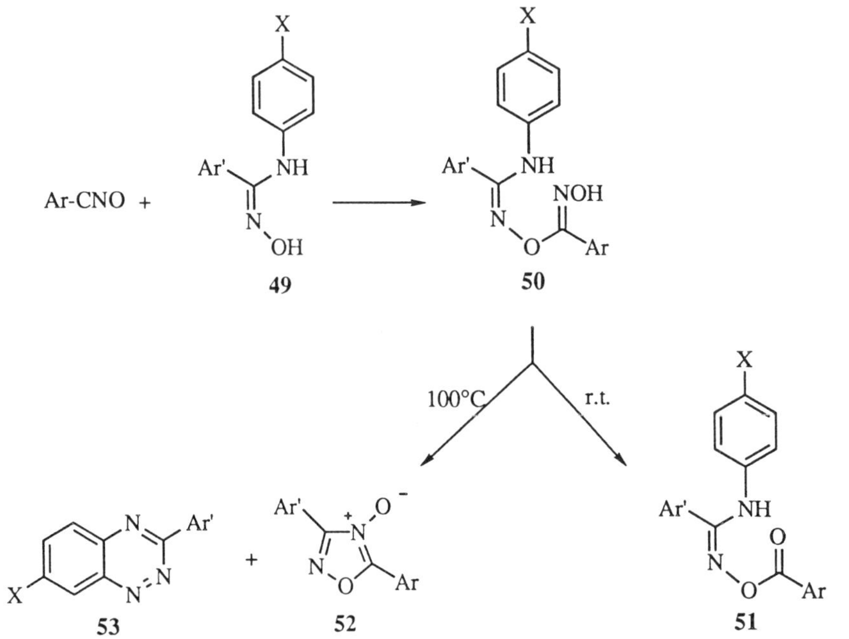

Scheme 18

Benzotriazines **53** are the unexpected products in these reactions. These are formed mainly when the N-aryl benzamidoximes **49** bear an electron donor group in *para*. In such a case (Scheme 19) conditions favourable to the loosening of the weak N-O bond of the O-benzoyloxime intermediate **54** would be created, provoking annelation at **53** by means of an unusual [3,5] sigmatropic rearrangement in competition with 5-Exo-Trig cyclization at **52** and hydrolytic scission at **51**.[45]

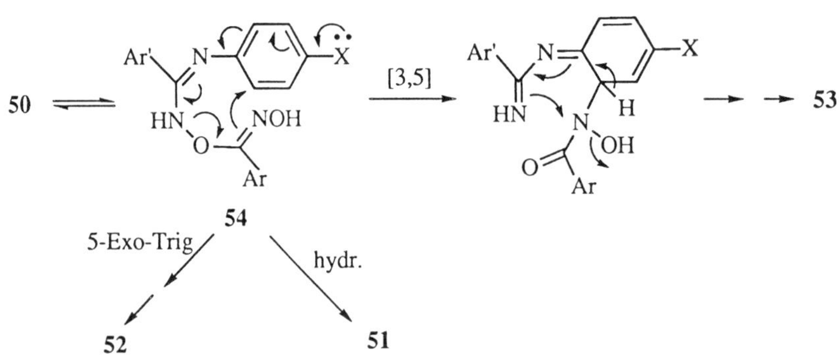

Scheme 19

The influence of the substituents on the nitrile oxide aryl has been proved by making benzamidoxime **49** react with mesitonitrile oxide **55** (Scheme 20).

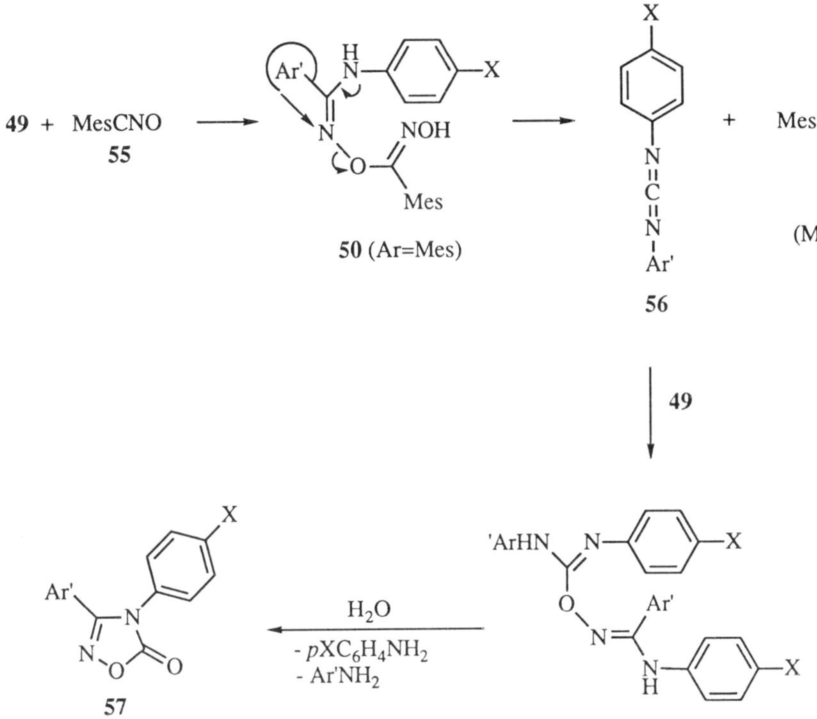

Scheme 20

None of the previously identified products were observed with this reactant, only 1,2,4-oxadiazol-5-one **57**: steric hindrance on the intermediate **50** (Ar=Mesitil) seems to determine a Beckmann-type rearrangement and the formation of carbodiimide **56**, which in the presence of the original amidoxime cyclizes at **57**.

O-benzoylderivatives **58**, which can also be prepared separately from benzamidoximes and benzoyl chloride in high yields,[46] evolve by subsequent reaction with nitrile oxides into the above-mentioned 1,2,4-triazole 2-oxides **60** (Scheme 21).[47] The latter route is more profitable and in more general use and occurs via the formation of the common intermediate nitroso-imine **59**.

Similar 1,2,4-triazole 2-oxides can be obtained by autotransformations of the nitrile oxide via a mechanism which involves a total of four molecules of the dipolar reactant alone and proceeds in final cyclization as shown in Scheme 21.[48]

These derivatives are often isolated as by-products together with variable amounts of other known products of dimerization or rearrangement of the 1,3-dipole in reactions between nucleophiles or even dipolarophiles that are not particularly reactive and not very stable nitrile oxides like benzonitrile oxide.

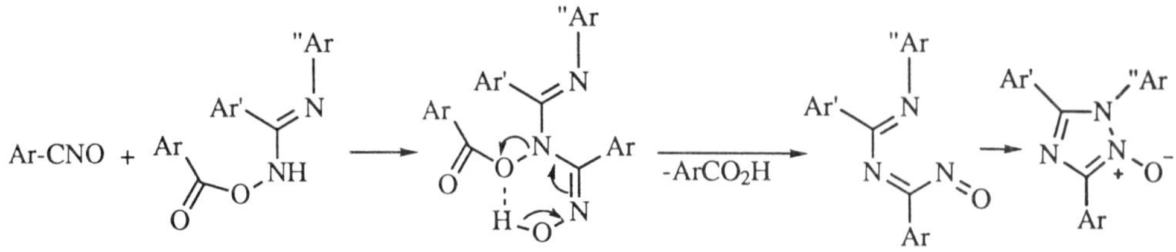

58 59 60

Scheme 21

Considering the fact that benzamidoximes **49** can be prepared from the appropriate arylamine and nitrile oxide,[49] the previous reactions could be carried out using the amines themselves as starting substrates with an excess of the 1,3-dipole. Unfortunately this procedure is not very efficient because of the low yields of the desired final products.

By using aniline with *ortho* positioned nucleophilic groups **61**, the reaction occurs differently. The formation of the expected amidoxime **62** is followed by the easy ring closure to benzoazoles **63** with loss of hydroxilamine (Scheme 22).[50,51]

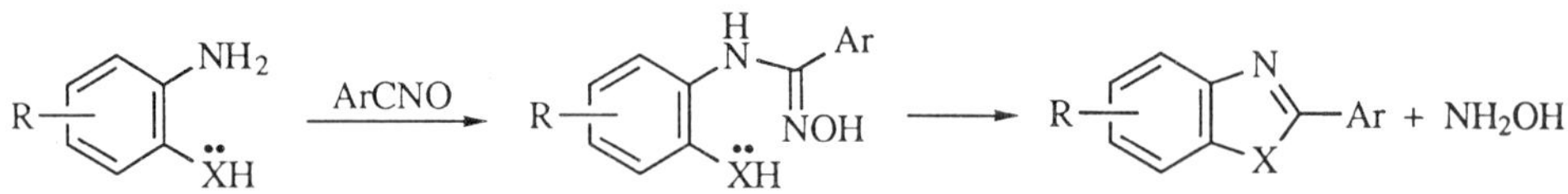

61 62 63

X = NH, O, S

Scheme 22

The reaction mechanism was further confirmed by the fact that the supposed 1:1 adduct intermediate **62** (X=NH) was isolated and quantitatively transformed simply by heating into the corresponding benzimidazole **63**.[52]

This result shows how, with substrates which present another suitably positioned nucleophilic site, the nitrile oxide contributes to the construction of the heterocycle furnishing only a C-unit.

In this way, tautomerizable isoxazolones **64**, where the nucleophilic centers are contiguous and part of the same cycle, undergo either expansion to an oxazinonic system or rearrangement into an imidazolic system as a result of the subsequent action of the nitrile oxide on the Z-amidoxime **65**.[53]

Thus, depending on the position of the isoxazolonic carbonyl and of the substituents, 4-oxazinones **66** were obtained from 3-isoxazolones **64** (x=0; y=1) and 6-oxazinones **67** and/or imidazoles **68** and **69** from 5-isoxazolones **64** (x=1; y=0) (Scheme 23).[54,55,56,57]

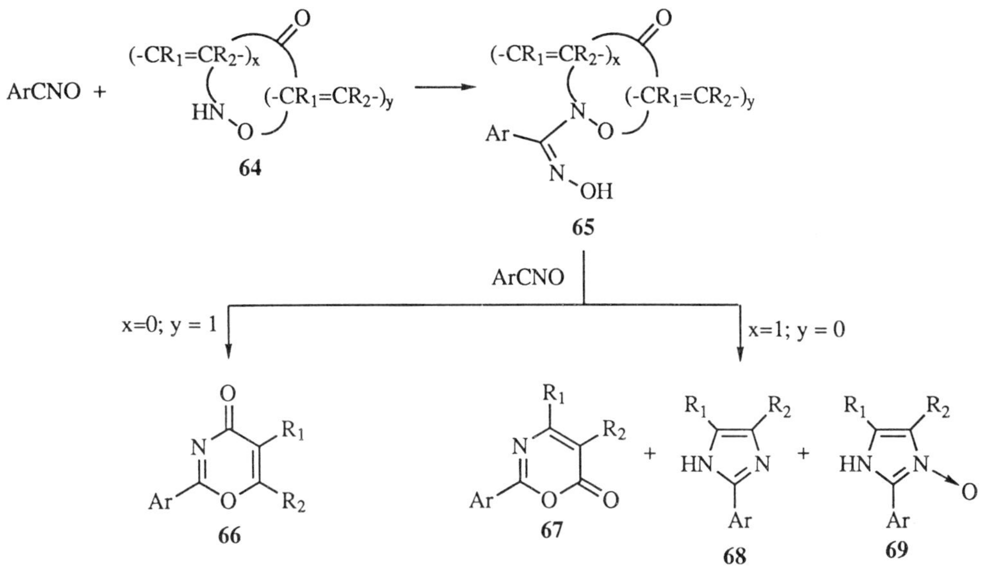

Scheme 23

The reaction is observed for the isoxazolones **64** which can exist in NH form; the relative 4-ones **70**, for which this tautomeric form is not possible, do not react (Scheme 24).[55]

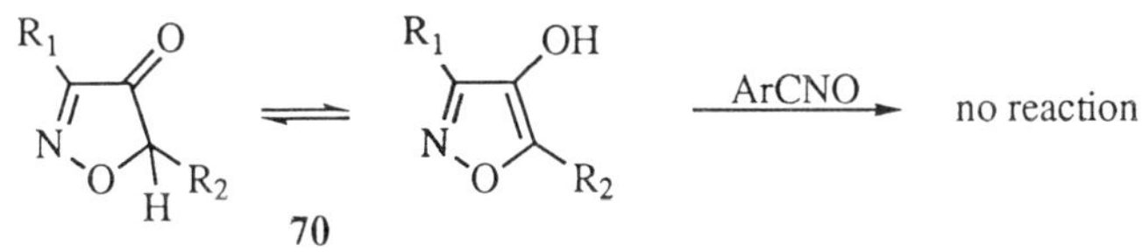

Scheme 24

Thus, interesting behaviour was observed in an arylmethyleneisoxazol-5-one; this type of substrate with nitrile oxides usually produces spirocompounds derived from 1,3- or 1,1-cycloaddition to the exocyclic C=C bond[56] depending on the steric hindrance on the relative methylene.[57]

3-methyl-4-methylphenylmethyleneisoxazol-5-one **71**, however, reacts with nitrile oxide exclusively in N-H form (B) giving isoxazolinyloxazin-6-one **72** (Scheme 25).[54]

In confirmation of this, a similar reaction carried out on the equally hindered 4-diphenylmethylene-3-methyl-isoxazol-5-one, which cannot exist in the above mentioned N-H form, showed a site-specific attack of the nitrile oxide to the endo C=N bond without production either of the expected spirocompounds or of expansion or rearrangement derivatives.[60]

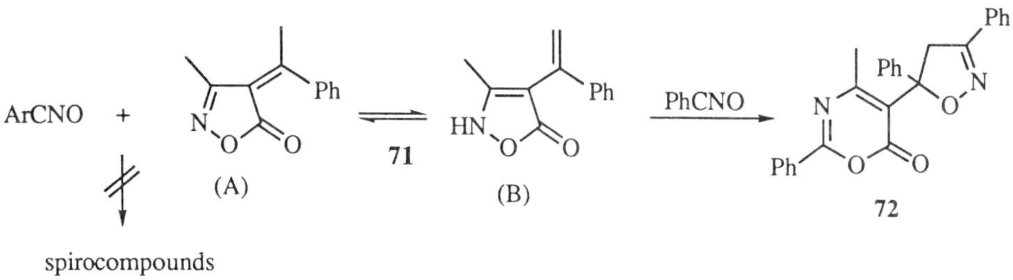

Scheme 25

The behaviour of non-oxygenated N-unsubstituted pentatomic heterocycles towards nitrile oxide was also studied.

1,2,4-triazoles **73** (X=N)[61] and 2-substituted imidazoles **73** (X=CR2;R≠H)[62] give rise to isomeric mixtures of N-aroyloximes **74** and **75** (Scheme 26) that cannot be further transformed.

2-Unsubstituted imidazoles **73** (X=CR2; R=H) produce the expected oxime **75** which, in this case, can rearrange and transform itself quantitatively into imidazolone **77** in the presence of an excess of the dipolar reactant.[63]

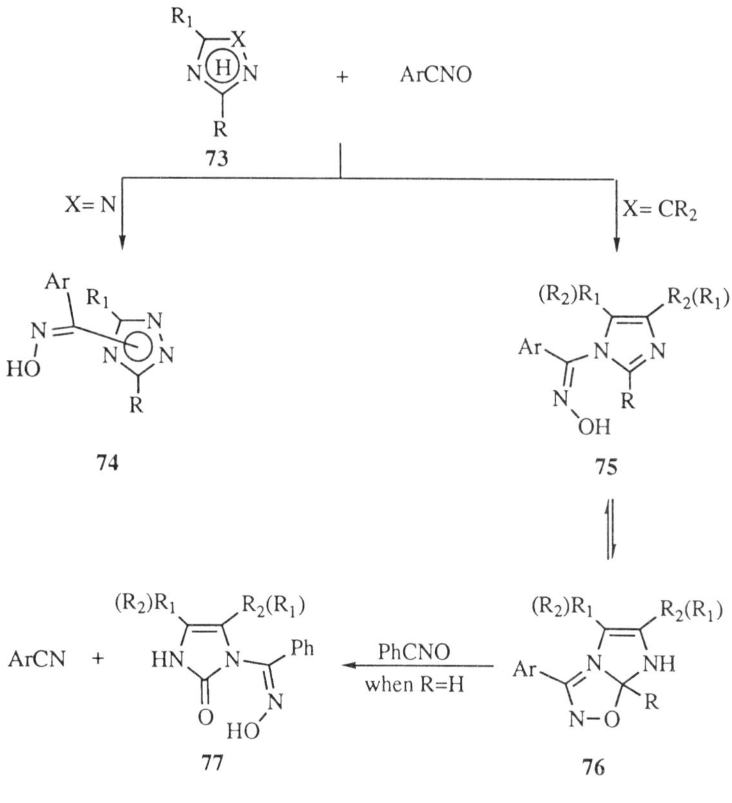

Scheme 26

The imidazolone derives from the selective degradation of the oxadiazolinic cycloadduct **76**,[64] present in solution in tautomeric equilibrium with the *Z*-oxime form **75**, which is the only form observed in solid state.

4. Addition-cyclization reactions with functionalized nitrile oxides

The possibility of synthesising heterocycles by the use of functionalized nitrile oxides was not explored in any depth.

Nitrile oxides derived from 2-substituted 2-methylpropanehydroxamoyl chloride **78** lead to five- **79** or six-membered rings **80** as a result of reaction with thioureas and ethylenediamines respectively (Scheme 27).[65]

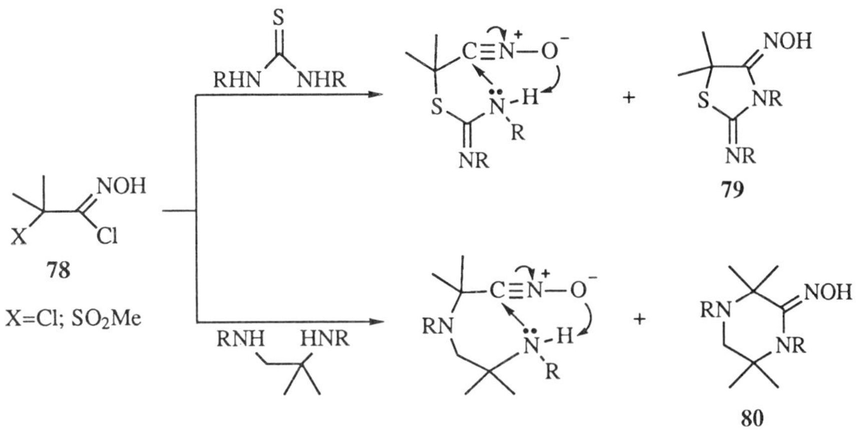

Scheme 27

α-Amino azines and azoles **81** react with aroylnitrile oxides **82** to give nitrosoderivatives of imidazo-fused heterocycles **84** (Scheme 28).[66]

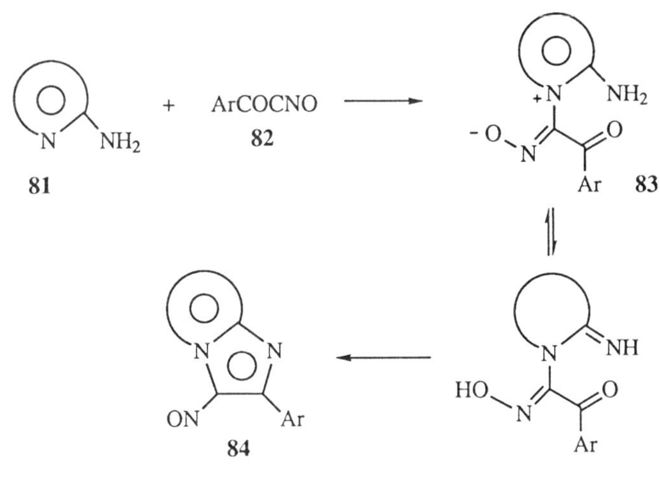

Scheme 28

An analogous zwitterionic 1:1 adduct **83**, obtained with α-unsubstituted pyridine does not cyclize and slowly reverts to the addends, whereas the more reactive isoquinoline is able to give a moderately stable cycloadduct.[67]

Again with pyridine but using benzonitrile oxide, the relative fragile zwitterion can evolve either leading to the unusual dimerization of the nitrile oxide to 1,4,2,5-dioxadiazine **87** or cyclizing.[68,69] Contrary to this hypothesis, the intermediacy of the zwitterionic adduct **86** in the cycloaddition of nitrile oxide to diazines and relative benzoderivatives **85** was rejected (Scheme 29).[70]

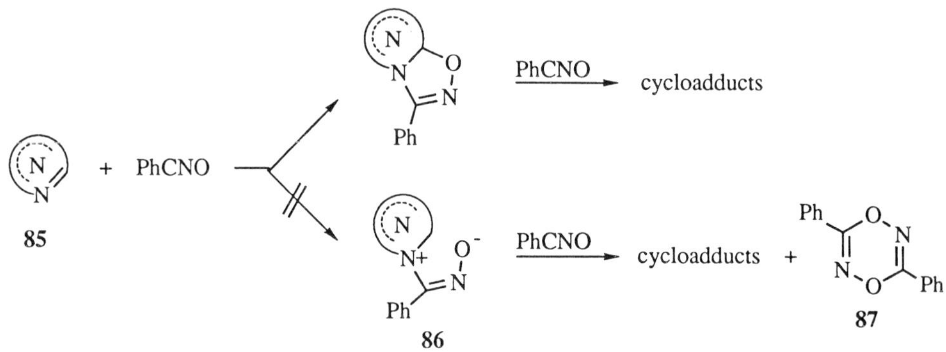

Scheme 29

5. Concluding remarks

The present concise review has shown how nucleophilic additions to nitrile oxide can lead to an interesting variety of potentially useful heterocyclic systems. The flexibility of the methodology and its applicability in synthetic projects are linked to the identification of suitably functionalized nucleophile agents. Modifications to this general scheme can be introduced by appropriately functionalizing the dipolar reactant too. Some of the addition-cyclization reactions described here might seem of limited synthetic interest as they do not afford satisfactory yields. In most cases, they can be successfully optimized by using more stable nitrile oxides for which the competitive dimerization and/or rearrangement reactions are very limited or even inexistent.

References

1. Grundmann, C. and Grünanger, P. *The Nitrile Oxides*, F. Springer Verlag New York, 1971.
2. Caramella, P. and Grünanger, P. In *1,3-Dipolar Cycloaddition Chemistry* ; Padwa A.; John Wiley and Sons New York; 1984; Chapt.3 pp. 291-392.
3. Curran, D.P. In *Advances in Cycloaddition*; JAI Press Greenwich Connecticut; 1988; pp.129-189,.
4. Kanemasa, K. and Tsuge, O. *Heterocycles* **1990**, *30*, 719.
5. Kozikowski, A.P. *Acc. Chem. Res* **1984**, *17*, 410.
6. Hassner A. In *Heterocycles in Bio-organic Chemistry*; Bergman, J.;Van der Plas, H.C.; and Simonyi, M.; The Royal Society of Chemistry; 1991; pp. 130-143.

7. Torssell, K.B.G. In *Nitrile Oxides, Nitrones, and Nitronates in Organic Synthesis*; VCH New York; 1988; p.69.
8. De Sarlo, F. and Guarna, A. *J.C.S. Perkin II*, **1976**, 626; *J.C.S. Perkin I*, **1976**,1826.
9. Hegarty, A.F. *Acc. Chem. Res.* **1980**, *13*, 448.
10. Leroy, G.; Nguyen, M.T.; Sana, M.; Dignam, K.J.; and Hegarty, A.F. *J. Am. Chem. Soc.* **1979**, 1988.
11. Nguyen, M.T.; Malone, S.; Hegarty, A.F. ; and Williams I.I. *J. Org. Chem.* **1991**, *56*, 3683.
12. Hussein, A. Q.; El-Abedelah, M.M.; and Sabri, W.S. *J. Heterocyclic Chem* **1983**, *20*, 301.
13. Hussein, A. Q. J. *Chem. Eng. Data* **1987**, *32*, 127.
14. Hussein, A. Q.; El-Abedelah, M.M.; and Sabri, W.S. *J. Heterocyclic Chem.* **1984**, *21*, 455.
15. Armand, J. *Bull. Soc. Chim. Fr.* **1966**, 882.
16. Grundmann, C. and Richter, R. *Tetrahedron Letters* **1968**, 963.
17. Nagahara, K; Tabagi, K.; and T. Ueda, T. *Chem. Pharm. Bull.* **1975**, *23*, 3178.
18. See ref.(1) pp.126-127 and references cited therein.
19. See ref.(1) p.156 and references cited therein
20. Rao, B. and Ahmed, K. *Indian J. Chem.* **1977**, *Sect. B, 15,* 509.
21. Risitano, F.; Grassi, G.; and Foti, F. J. *Chem. Research (S)* **1981**, 65.
22. Hussein, A. Q.; El-Abedelah, M.M.; Hodali, H.A.; Kamal M.R.; and Aduf, M.M. *Heterocycles* **1987**, *26*, 2199.
23. Risitano, F.; Grassi, G.; Foti, F.; Bruno, G.; and Nicolò, F. *Tetrahedron* **1993**, *49*, 10427.
24. Grassi, G.; and Risitano, F. unpublished results.
25. Bravo, P. and Ponti P. P. *J. Heterocycl. Chem.* **1973**, *10*, 669.
26. Gewald, K. and Hain, U. *Z. Chem.* **1986**, *12*, 434.
27. Nakada, Y.; Hata, T.; Tamura, C.; Iwaoka, T.; Kondo, M.; and Ide, J. *Tetrahedron Letters* **1981**, *22*, 473.
28. Quilico, A. and Speroni, G. *Gazz. Chim. Ital.* **1946**, *76*, 148.
29. Stagno d'Alcontres, G.; Lo Vecchio, G.; and La Monica, G. *Gazz. Chim. Ital.* **1961**, *91*, 1005.
30. Morocchi, S.; Ricca, A.; Zanarotti, A; Bianchi, G.; Gandolfi, R.; and Grünanger, P. *Tetrahedron Letters* **1969**, 3329.
31. Caramella, P.; Cellerino, G.; Corsico Coda, A.; Gamba Invernizzi , A.; Grünanger, P.; Houk, K.N.; and Marinone Albini, F. *J. Org. Chem.* **1976**, *41*, 3349.
32. Gilchrist, T.L.; Harris, C.J.; King, F.D.; Peek, M.E.; and Rees, C.W. *J.C.S. Perkin I* **1976**, 2161.
33. Shiraishi, S.; Shigemoto, T.; and Ogawa, S. *Bull. Chem. Soc. Jpn.* **1978**, *51*, 563.
34. Gilchrist, T.L.; Harris, C.J.; Hawkins, D.G.; Moody, C.J.; and Rees, C.W. *J.C.S. Perkin I* **1976**, 2166.
35. Gilchrist, T.L.; Harris, C.J.; and Rees, C.W. *J.C.S. Chem. Comm.* **1974**, 485.
36. Forbes Cameron, A. and Freer, A.A. *Acta Crystallogr.* **1976**, *Sect.B, 32,* 1995.
37. Gilchrist, T.L.; Peek, M.E.; and Rees, C.W. *J.C.S. Chem. Comm.* **1975**, 914.
38. Shigemoto, T.; Imamura, K.; Hayakawa, T.; Ogawa, S.; Matsumoto, K.; and Shiraishi, S. *Chem. Lett.* **1981**, *7*, 843.
39. Minisci, F.; Galli, R.; and Quilico, A. *Tetrahedron Letters* **1963**, 785.
40. Gilchrist, T.L.; Gordon, P.F.; and Rees, C.W. *J. Chem. Research (S)* **1988**, 148.
41. Gallos, J.K.; Lioumi, M.I.; and Lekka, A.N. *J. Heterocycl. Chem.* **1993**, *30*, 287.
42. Argyropoulos, N.G.; Gallos, J.K. *J.C.S. Perkin I* **1990**, 3277.
43. Abushanab, E. and Alteri Jr., N.D. *J. Org. Chem.* **1975**, *40*, 157.

44. Risitano, F.; Foti, F.; and Grassi, G. *Atti Acc. Peloritana Pericolanti* **1981**, Cl.I, *58*, 165.
45. Risitano, F.; Grassi, G.; Foti, F.; and Filocamo, F. *Tetrahedron* **1997**, *53*, 1089.
46. Foti, F.; Grassi, G.; Risitano, F.; and Caruso, F. *Atti Acc. Peloritana Pericolanti* **1981**,Cl.I, *58*, 152.
47. Foti, F.; Grassi, G.; Risitano, F.; and Caruso, F. *J. Chem. Research (S)* **1982**, 302.
48. Grassi, G.; Risitano, F.; and Foti, F. *J. Chem. Research (S)* **1981**, 56.
49. See ref.(1) pp. 161-169 and references cited therein.
50. Sasaki,T.; Yoshioka, T.; and Suzuki, Y. *Bull. Chem. Soc. Jpn.* **1969**, *42*, 3335.
51. Abdelhamid, A. O.; Parkanyi, C.; Khaledur Rashid, S. M.; and Lloyd, W.D. *J. Heterocycl. Chem.* **1988**, *25*, 403.
52. Risitano, F.; Grassi, G.; Foti, F.; and Caruso, F. *J. Chem. Research (S)* **1983**, 52.
53. Lo Vecchio, G.; Foti, F.; Grassi, G.; and Risitano, F. *Tetrahedron Letters* **1977**, *24*, 2119.
54. Risitano, F.; Grassi, G.; Foti, F.; Caruso, F.; and Lo Vecchio, G. *J.C.S. Perkin I* **1979**, 1522.
55. Grassi, G.; Foti, F.; and Risitano, F. *J. Chem. Research (S)* **1983**, 102.
56. Grassi, G.; Foti, F.; Risitano, F.; and Caruso, F. *J. Chem. Research (S)* **1983**, 172.
57. Lo Vecchio, G.; Caruso, F.; Risitano, F.; Foti, F.; and Grassi, G. *J. Chem. Research (S)* **1983**, 76.
58. Lo Vecchio, G.; Grassi, G.; Risitano, F.; and F.Foti. *Tetrahedron Letters* **1973**, 3777.
59. Lo Vecchio, G.; Risitano, F.; Grassi, G.; Foti, F.; and Caruso, F. *Atti Acc. Peloritana Pericolanti* **1981**, Cl.I, *54*, 297.
60. Foti, F.; Grassi, G.; and Risitano, F. unpublished results.
61. Bruno, G.; Foti, F.; Grassi, G.; Caruso, F.; and Risitano, F. *Acta Crystallogr.* **1989**, Sect.C, *45*, 653.
62. Bruno, G.; Nicolò, F.; Foti, F.; Grassi, G.; and Risitano, F. *Acta Crystallogr.* **1994**, Sect.C, *50*, 428.
63. Grassi, G.; Foti, F.; Risitano, F; Bruno, G.; Nicolò, F.; and De Munno, G. *J.C.S. Perkin I* **1993**, 2615.
64. Srimannarayana, G.; Srivastava, R.M.; and Clapp, L.B. *J. Heterocycl. Chem.* **1970**, *7*, 151.
65. Bushey, D.F. and Hoover, F.C. *J. Org. Chem.* **1980**, *45*, 4198.
66. Parkanyi, C.; Abdelhamid, A. O.; Cheng, J.C.S.; Shawali, A.S. *J. Heterocycl. Chem.* **1984**, *21*, 1029.
67. Caramella, P.; Bandiera, T.; Marinone Albini, F.; Gamba, A.; Corsaro, A.; and Perrini, G. *Tetrahedron* **1988**, *44*, 4917.
68. Corsaro, A.; Perrini, G; Caramella, P.; Marinone Albini, F.; and Bandiera, T. *Tetrahedron Letters* **1986**, *27*, 1517.
69. Marinone Albini, F.; De Franco, R.; Bandiera, T.; Grunanger, P.; and Caramella, P. *Gazz. Chim. Ital.* **1990**, *120*, 1.
70. Grassi, G.; Risitano, F.; and Foti, F. *Tetrahedron* **1995**, *51*, 11855.

THE CHEMISTRY OF TRINEMS

Chiara Ghiron and Tino Rossi
Glaxo Wellcome Medicines Research Centre, Via A. Fleming 4, 37100, Verona, Italy.

Abstract *An overview of the chemistry involved in the synthesis of the trinems, a new class of potent antibacterial agents, is presented in the following chapter.*

1. Introduction

When penicillin was discovered, it earned the title of a miracle drug. It gave control over infections that were expected to kill patients and appeared to be the long-sought "magic bullet". But penicillin became more and more frequently ineffective in the treatment of infections and this fact - already observed by Fleming at the time of penicillin's discovery - was often caused by the rapid evolutionary selection of resistant strains.[1] The emergence of bacterial infections caused by strains that are resistant to several different antibacterial agents makes this research field a very challenging one. In order to provide effective yet safe drugs, research must continuously follow the 'moves' of bacteria and try and keep one step ahead. In this field it is necessary to be fast, focused and effective, and yet it is very likely that a new drug, especially when used inappropriately and without control (as it has happened with penicillin for example) will exert such a selective pressure on bacteria that within a few generations new strains will arise, that are resistant even to that drug.

Penicillin was the first example of the ß-lactam class of antibiotics. Since then, several other classes have been found,[2] either as natural products or by chemical synthesis. Some examples include: cephalosporins, carbacephems, oxacephems, monobactams, penems, carbapenems, and more recently the trinems[3] (formerly referred to as tribactams).

Trinems are a class of totally synthetic antibacterials[4] having a very good spectrum of activity against both Gram-positive and Gram-negative aerobic and anaerobic bacterial strains, and stability to hydrolytic enzymes such as bacterial ß-lactamases and human dehydropeptidases.

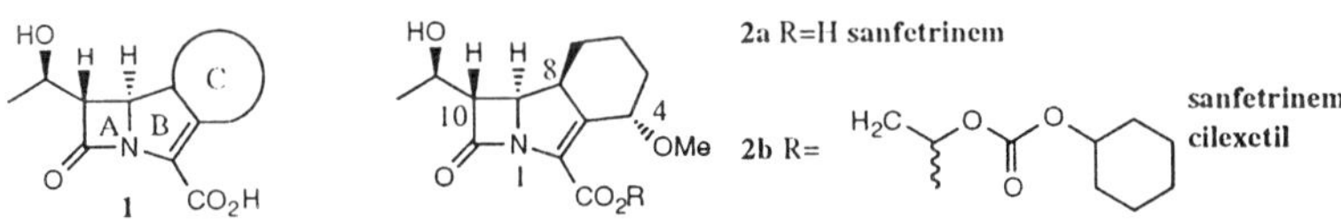

Figure 1

The basic structure of trinems **1** (Figure 1) consists of a tricyclic ß-lactam nucleus bearing a (*R*)-hydroxyethyl side-chain on ring A, a carboxylate group on ring B and a third five, six or seven membered carbocyclic or heterocyclic ring C. The presence of additional substituents around ring C contributes to the stereochemical complexity of this family, displaying at least four stereocentres. It has been observed in other classes of ß-lactams, particularly in the carbapenems,[5] that very little stereochemical variation is allowed without having a detrimental effect on the biological activity. A similar situation has been found to exist for the trinems.

The best known example from the trinem class is sanfetrinem **2a**, which possesses an extremely good biological profile and is currently undergoing Phase II clinical trials as its orally available, metabolically labile cilexetil ester **2b**.

Sanfetrinem bears a methoxy group at the 4-position (trinem numbering, see Figure 1) with (4*S*, 8*S*) absolute stereochemistry around the third ring. Extensive structure-activity relationship studies have proven this stereochemical configuration to be the best in terms of biological profile.

Since the discovery of sanfetrinem, research has progressed both in Glaxo Wellcome and other laboratories with the aim of discovering other potent molecules as well as providing more viable routes to sanfetrinem itself.

The scope of this chapter is to provide a comprehensive review of the work that has been done in the trinem and related systems area. Particular emphasis on the chemical problems is given, while for biological activity data direct consultation of the original literature is recommended.

2. General chemical methodologies

As mentioned in the introduction, the stereochemical arrangement around the ß-lactam ring (C9 and C10) as well as the configuration of the hydroxyethyl side-chain are very strict requirements for biological activity. For this reason, most of the syntheses start from *tert*-butyldimethylsilyl-protected 4-acetoxyazetidinone **3** shown in Scheme 1.

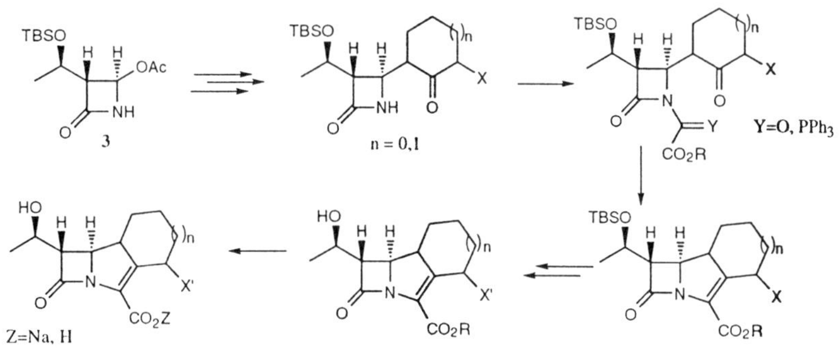

Scheme 1

Introduction of ring C is often accomplished by acid- or base- catalysed condensation of the appropriate cycloalkanone derivative and **3**, followed by the introduction of a side-chain precursor at position 4. Due to the complexity of the trinem system, most of the research effort has been focused on the synthesis of advanced intermediates amenable to provide a large number of final compounds. In order to do so, modification of the 4-position has been accomplished at late stages of the synthesis, where possible after the closure of the tricyclic system. This ring closure is accomplished either *via* intramolecular Wittig reaction or *via* oxo-oxo reductive cyclisation (*vide infra*). Further modification of the side-chain and a series of deprotection steps follow to provide the target compounds.

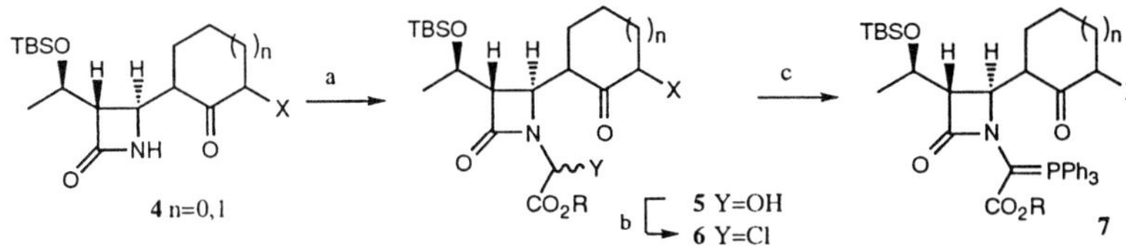

a) RO_2CCHO, toluene, reflux; b) $SOCl_2$, 2,6-lutidine, THF, -40°C; c) PPh_3, 2,6-lutidine, THF

Scheme 2

Scheme 2 outlines the phosphorane based intramolecular Wittig cyclisation to the trinem nucleus. This strategy was first developed by Woodward in his syntheses of penems.[6] Azetidinone **4** is condensed with the appropriate glyoxylate to give a diastereomeric pair of aminals **5** which is then transformed into the unstable chloro derivatives **6** by treatment with thionyl chloride and 2,6-lutidine. Displacement with triphenylphosphine follows, and the presence of 2,6-lutidine in the reaction medium allows for the direct isolation of phosphorane intermediate **7**. This intermediate is usually quite stable at room temperature and can be further modified or directly cyclised to the trinemic system.

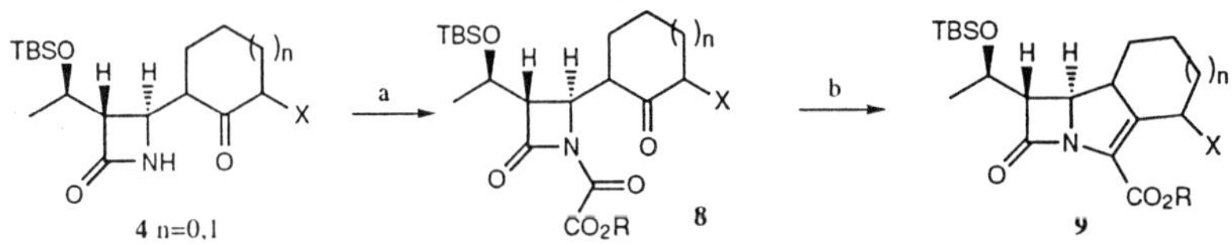

a) $RO_2COCOCl$, Et_3N or pyridine, toluene; b) $(EtO)_3P$ or $MeP(OEt)_2$, xylene, Δ

Scheme 3

Alternatively, the azetidinone **4** is treated with the appropriate alkyl oxalylchloride to afford the oxalimide derivative **8** which then undergoes a triethyl phosphite-mediated oxo-oxo cyclisation (Scheme 3).[7] It is assumed that the first step in the reaction sequence involves the formation of a carbene intermediate. This reacts further with excess triethyl phosphite to yield a phosphonate which then undergoes *in situ* cyclisation to **9**.

Generally after further modification of the side-chain, the removal of the silyl and ester protecting groups must be accomplished (Scheme 4).

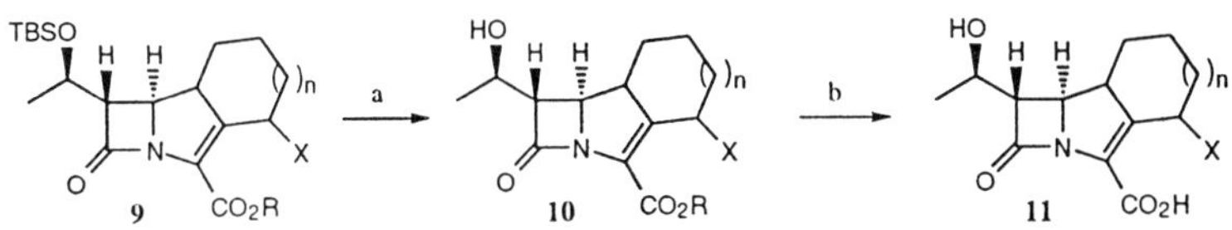

R = allyl, benzyl, *p*-^{t}butylbenzyl

a) Bu_4NF, AcOH, THF; b) see Table 1

Scheme 4

Upon deprotection of the secondary hydroxyl group the molecules generally become less stable, and particular precautions must therefore be taken during this step. The strong basicity of the fluoride ion can lead to ring opening reactions of the strained trinem system, with very poor yields in the desired product. It was found during a thienamycin synthesis that buffering the reaction medium with acetic acid results in better yields of the tetrabutylammonium fluoride mediated desilylation.[8] This procedure, which was successfully applied to the trinems, has been further modified by running the reaction at slightly higher temperature (40°C), with yields often greater than 80%.[9] The dramatic difference in rate - from more than 72 hours to less than two - often avoids degradative pathways taking place. Other modifications avoids the use of the extremely expensive tetrabutylammonium fluoride by replacing it either with tetrabutylammonium bromide and potassium fluoride[10] or triethylamine trihydrogen fluoride,[11] without affecting the yield of the reaction.

In the synthesis of sanfetrinem, the silyl deprotection was accomplished before the cyclisation step on the phosphorane intermediate 7.[12] Desilylation was carried out by treatment of the corresponding stable phosphonium salt with aqueous HCl in AcOEt, and the phosphorane was then regenerated by treatment with base ($NaHCO_3$).

The choice of the ester moiety is very much dependent upon the scale of the synthesis.

Final Product	Ester	Removal
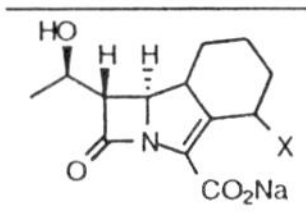	benzyl, *p*-^{t}butylbenzyl	H_2/Pd(C), H_2O/^{i}PrOH, $NaHCO_3$ (1eq)
	allyl	Pd(PPh_3)$_4$, PPh_3, Na^+ 2-ethylhexanoate, THF or CH_2Cl_2
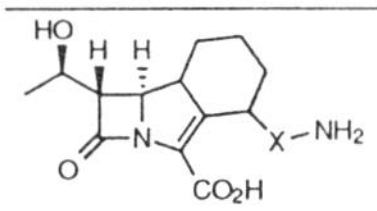	benzyl, *p*-^{t}butylbenzyl (a)	H_2/Pd(C), H_2O/^{i}PrOH, ($NaHCO_3$ (1eq))
	allyl (b)	Pd(PPh_3)$_4$, PPh_3,dimedone, THF or CH_2Cl_2
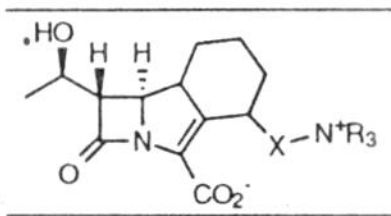	allyl	Pd(PPh_3)$_4$, PPh_3, Bu_3SnH, DMF

(a) usually the amino group is protected as the benzyl carbamate, and is removed under the same conditions;(b) usually the amino group is protected as the allyl carbamate, and is removed under the same conditions

Table 1

In laboratory scale preparations, it proved more convenient to use an allyl ester (see Table 1) which can be removed *via* a palladium (0) catalysed trans-allylation.[13] As the allyl acceptor, sodium or potassium 2-ethylhexanoate is usually employed when the trinem sodium (or potassium) salt is the desired product,[14] whereas dimedone (2,2-dimethylcyclohexane-1,3-dione) is used in the case of an amino acid formation.[15] In the case of the formation of an inner salt, tributyltin hydride is generally the reagent of choice. On the process research scale, both for cost reasons and for the necessity of dealing with highly crystalline intermediates, it is found more convenient to use a benzyl-type ester. In this latter case, the final step is a catalytic hydrogenation.

It has been mentioned already how extensive structure-activity relationship studies have shown the best biological profile to be possessed by the (8*S*) isomers in the six-membered ring C series. This stereocentre is generated at a very early stage of the synthesis, namely in the condensation of a cycloalkanone derivative **12** on 4-acetoxyazetidinone **3** (Scheme 5).

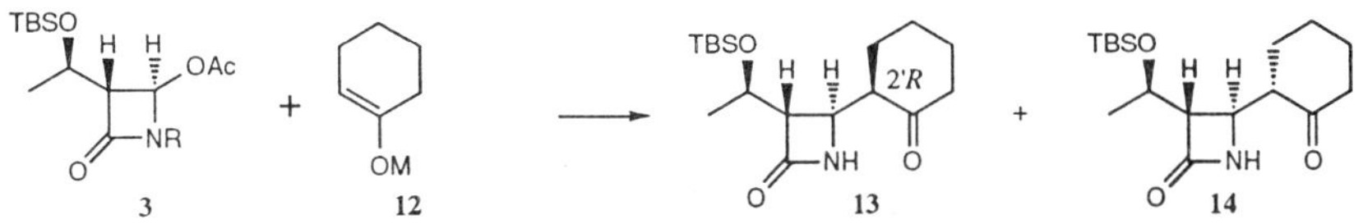

Scheme 5

While the *trans* stereochemistry around the ß-lactam ring is usually dictated by the steric bulk of the protected hydroxyethyl side-chain, the stereochemical control of the aldol condensation leading to **13** towards the formation of the (2'*R*) isomer depends on the conditions chosen (see Table 2).

R	M (base)	Catalyst	Solvent	Ratio 13:14	Yield% 13+14
H	Li (LiHMDS)	-	THF	1:1	75
H	Me_3Si	$ZnCl_2$	CH_2Cl_2	3:7	80
Me_3Si	Me_3Si	Me_3SiOTf	MeCN	7:3	70

Table 2

Under basic conditions, the azetinone intermediate formed by action of the lithium base (or excess of the lithium enolate) on the 4-acetoxyazetidinone reacts very rapidly at -78°C with very poor or no selectivity with the metal enolate to afford a *ca.* 1:1 mixture of **13** and **14**.[16] Using cyclohexanone silyl enol ether, under Lewis acid catalysed conditions, a slower reaction occurs at room temperature.[17] The choice of the Lewis acid as well as the protection of the azetidinone nitrogen is crucial to the stereoselectivity of the reaction. It has been postulated that coordination of the zinc (see **A**, Figure 2) to the imine nitrogen and the enolate oxygen would lead to a closed transition state which favours the formation of **14** with *syn* selectivity. When the azetidinone nitrogen is protected as its trimethylsilyl (or *tert*-butyldimethylsilyl) derivative and reacted under Lewis acid catalysed conditions (such as trimethylsilyl triflate), a reversal of stereoselectivity in favour of **13** is observed. This has been assumed to be due to the prevalence of an open transition state **B** leading to *anti* selectivity, the enolate geometry being fixed by the ring.

A similar trend is seen in the condensation of other substrates (see Table 3),[18] where the use of the silyl enol ether derivative in the acid catalysed aldol-type condensation leads to some selectivity, while the rapid enolate reaction results in a mixture of all possible stereoisomers (as well as regio-, in the case of 3-oxacyclohexanone).

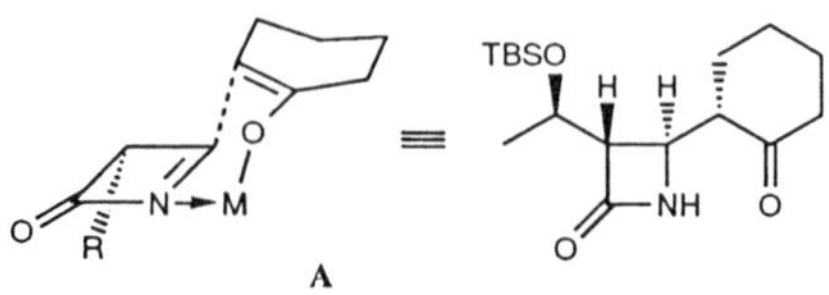

A

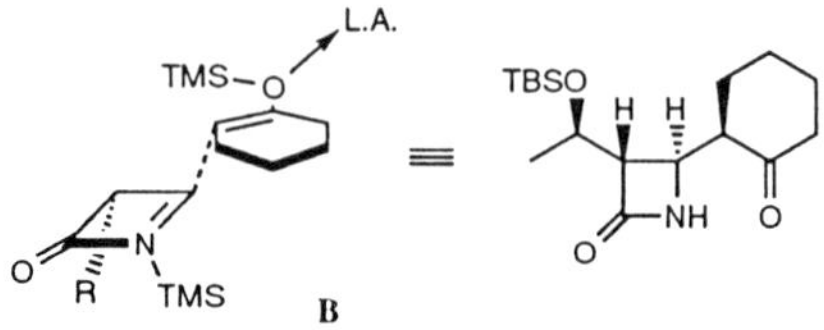

B

Figure 2

Cycloalkanone	M (base)	Catalyst	Solvent	Ratio (2'*R*):(2'*S*)	Yield % (2'*R*):(2'*S*)
cyclopentanone	Me_3Si	Me_3SiOTf^a	CH_2Cl_2	2:1	75
2-thiacyclopentanone	Li (LiHMDS)	-	THF	1:1	91
3-oxacyclohexanone	Li (LiHMDS)	-	THF	b	65
	Me_3Si	Me_3SiOTf^a	CH_2Cl_2	7:3	62

a) reaction with *N*-TMS protected **3**; b) mixture of all regio- and stereoisomers

Table 3

In a study of the stereoselective synthesis of some tetracyclic derivatives,[19] scientists at Hoechst studied the direct condensation of some tetralone derivatives with **3** (Scheme 6).[20]

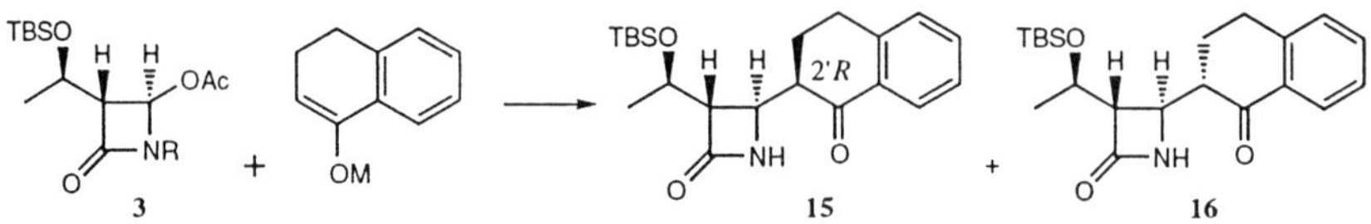

Scheme 6

It was found that the desired (2'*R*) isomer could be obtained in good yield when the titanium enolate was used. This species was generated by transmetallation of the lithium enolate with chlorotitanium tri*iso*propoxide. The use of the lithium enolate, Mukayama's tin enolate[21] or the Lewis acid catalysed condensation of the silyl enol ether generated *in situ* afforded preponderantly the undesired α isomer **16** (see Table 4).

In the case of the Hoechst tetracyclic derivatives,[22] a milder oxalimide cyclisation procedure was adopted, where the more reactive diethyl methylphosphonite was used instead of trimethyl or triethyl phosphite. This milder cyclisation reagent was found to give higher yields in the final product of the sequence, a highly strained system.

R	M (base)	Catalyst	Solvent	Ratio 15:16	Yield % 15+16
H	Li (LiHMDS)	-	THF	34:66	20
H	Ti (LDA + $ClTi(O^iPr)_3$)	-	THF	78:22	56
H	Sn(II)(Et_3N+$Sn(OTf)_2$)	-	CH_2Cl_2	3:97	56
Me_3Si	Me_3Si	Me_3SiOTf	CH_2Cl_2	1:1	79
Me_3Si	Me_3Si	$ZnCl_2$	CH_2Cl_2	5:95	86

Table 4

3. 4-Substituted trinems

The presence and proximity of a substituent to the unsaturated, strained bicyclic ß-lactam system was found to impart favourable biological properties to the basic trinem nucleus in terms of stability and biological profile.

For this reason, the effect of a series of carbon and heteroatom-containing side-chains at position 4 was studied.

3.1. Intermediates to 4-substituted trinems

In order to obtain a large number of derivatives and to fully evaluate the antibacterial potential of the class, it was necessary to develop robust routes to versatile intermediates. The original synthesis of epoxide **19** (Scheme 7) started from cyclohexanone derivative **13**.

The olefin moiety was introduced through a Shapiro reaction and stereoselective epoxidation was accomplished with *meta*-chloroperoxybenzoic acid. Interestingly it was possible to isolate the diastereomeric epoxide **20** through protection of the azetidinone nitrogen as the *tert*-butyldimethylsilyl derivative **18**.[23] This inversion of stereoselectivity has been ascribed to hydrogen bonding of the peroxyacid to the free azetidinone NH with stereoselective delivery of the electrophilic oxygen.[24] More recently, magnesium monoperoxyphthalate as been used as the oxidising agent with improved stereoselectivity.[25]

Although this route has been applied with success to the synthesis of several derivatives, the moderate stereoselectivity in the aldol-type reaction leading to **13**, as well as the number of steps leading to **19**, led to the development of a direct olefination reaction.

Early studies demonstrated that 3-cyclohexenylzinc bromide, generated *in situ* from commercially available 3-cyclohexenylbromide and zinc,[26] could react with **3** to give a 1:1 mixture of **17** and **22** in more than 90% overall yield.[27] In spite of the high efficiency of the reaction, the isolation of **17** from the reaction mixture could not be efficiently achieved by all the conventional purification techniques.

Better results were obtained *via* an intramolecular Sakurai-type reaction between **3** and 3-cyclohexenyldimethylchlorosilane **A** (Scheme 8).[28]

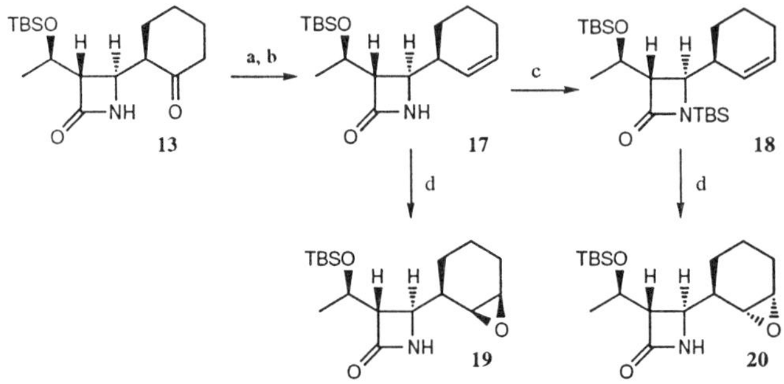

a) $TsNHNH_2$, AcOH; b) LDA (3eq), THF, -78°C; c) TBSCl, Et_3N, DMF; d) *m*CPBA, CH_2Cl_2, 0°C.

Scheme 7

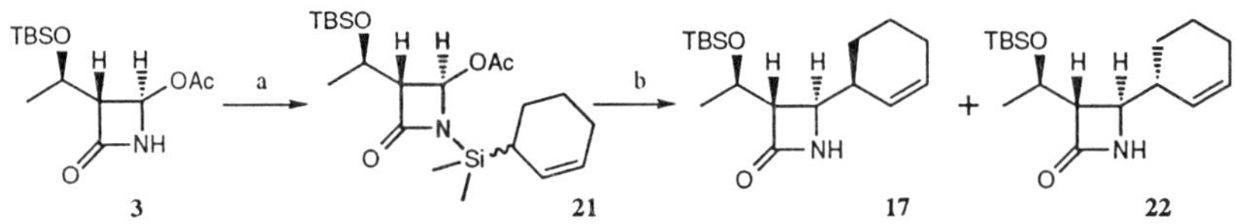

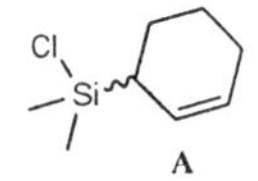

A

a) A, Et_3N, THF; b) TMSOTf, MeCN, 0°C, *then* Et_3N, MeOH

Scheme 8

Under these conditions a more favourable 4:1 of **17** and **22** could be isolated in 50% overall yield from **3**.
A detailed analysis of the most likely reaction mechanism (Figure 3) showed that the availability of enantiomerically pure *(R)*-**A** could render this reaction amenable to further scale-up studies, but 1,3-cyclohexadiene and dimethylchlorosilane failed to yield **A** in any measurable enantiomeric excess under a number of asymmetric hydrosilylation conditions.

Further studies led to the discovery of a highly stereoselective amidoalkylation of **3** by means of *B*-cyclohex-2-enyl-dialkylboranes **A** (Scheme 9).[29]

Initial studies with racemic *B*-cyclohex-2-enyl-9-borabicyclo [3.3.1] nonane showed that the reaction is highly stereoselective yielding a 1:1 mixture of two *syn* products **17** and **23** in >90% isolated yields (see Table 5).

There is no precedent in the literature of a C-C bond formation at the *si* face of acetoxyazetidinone **3** yielding 3,4-*cis* products. Apparently the reaction occurs under reagent control and no detectable influence of the silyloxyethyl side-chain is observed. In order to explain the results, a "closed" transition state (Figure 4) has been proposed for this reaction.

The reacting species, formed under the influence of a mild Lewis acid,[30] reacts with *syn* selectivity at C-4 and C-1' through a favoured "chair-like" transition state, with *trans* compound **17** originating from the borane having an (*R*) absolute configuration at the stereogenic centre of the cyclohexenyl moiety through transition state **A** (Figure 4). The *cis* isomer **23** would be generated from the borane reagent having an (*S*) absolute configuration *via* transition state **D**.

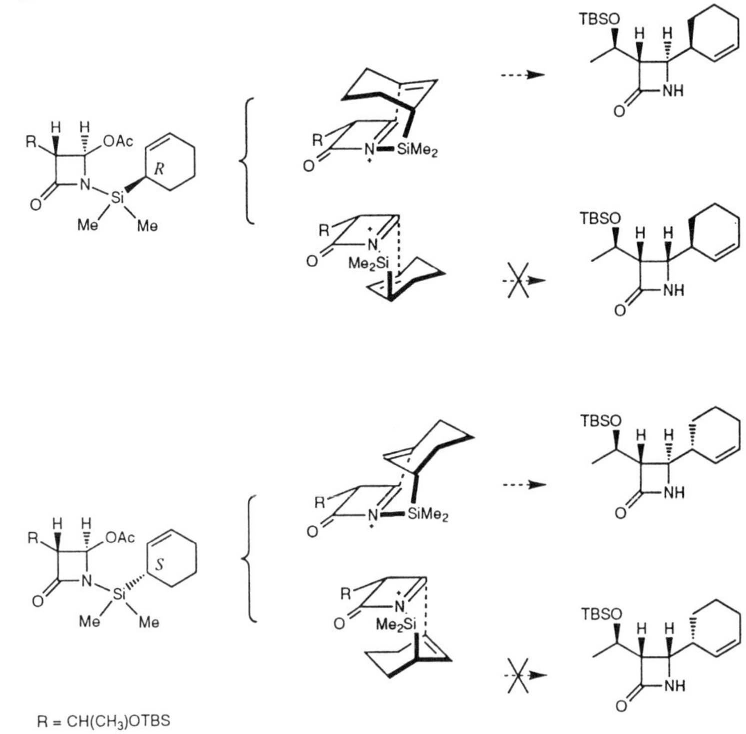

Figure 3

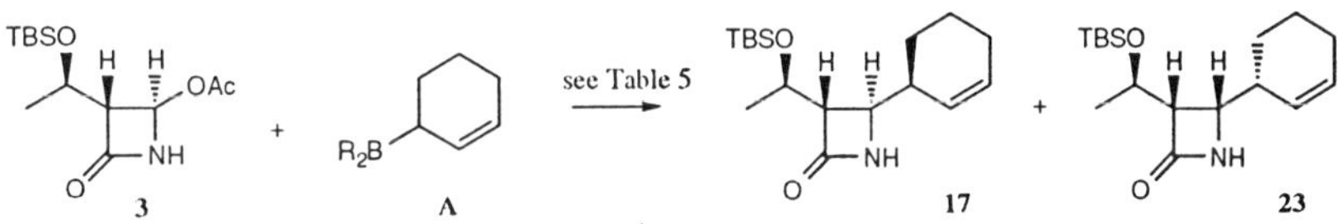

Scheme 9

The formation of by-product **23** was avoided through the use of the appropriate homochiral *B*-(cyclohex-2-enyl)-diisopinocampheylborane, obtainable with high enantiomeric excess through well established methodology.[31] A 19:1 ratio of **17** and **23** in 95% overall yields could be obtained.

The use of the optically pure borane was later rendered unnecessary through the observation that the two enantiomers of hindered dialkyl boranes should react at different reaction rates with **3** so that, in principle, **17** could be produced with high stereoselectivity with the use of the appropriate racemic 1-cyclohex-2-enyl dialkylborane.

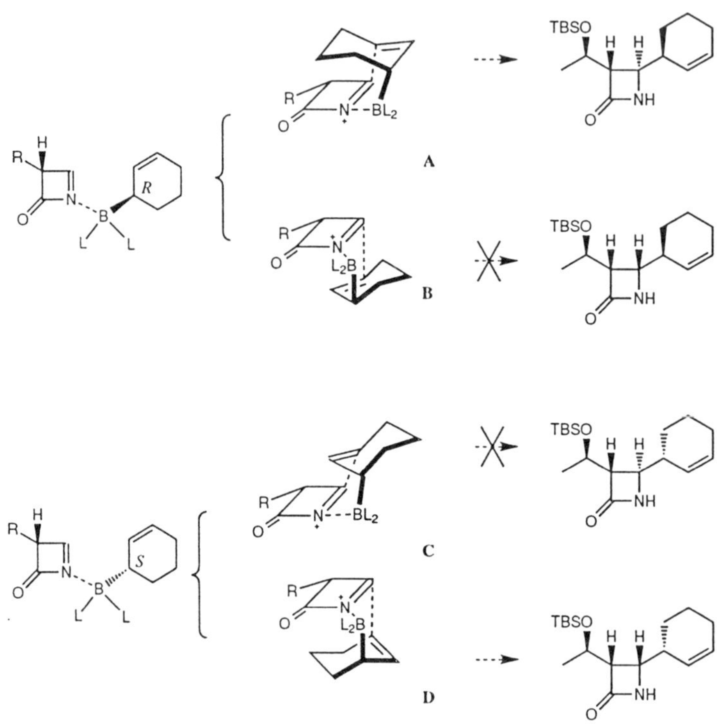

R = $CH(CH_3)OTBS$

Figure 4

Final optimisation studies led to the selection of the borane obtained from inexpensive 1-methylcyclohexene, that under standard reaction conditions gave **17** in >90% isolated yield (based on **3**) and a 19:1 ratio with **23** (see Table 5, Entry 4).

Borane A	Lewis acid	Yield %	Ratio 17:23
L_2B= 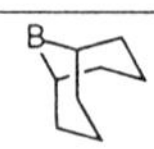	Et_2Zn	95	49:51
L= 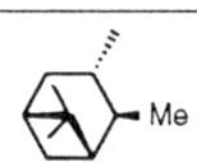	Et_2Zn	95	95:5
L= 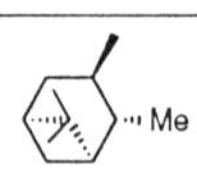	Et_2Zn	85	25:75
L= (±) 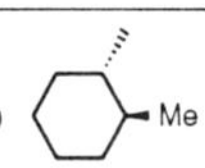	Et_2Zn	96	95:5

Table 5

3.2. Epoxide opening reactions

The opening reaction of epoxide **19** with various nucleophiles has been one of the more exploited routes to 4-substituted trinems (Scheme 10).[32]

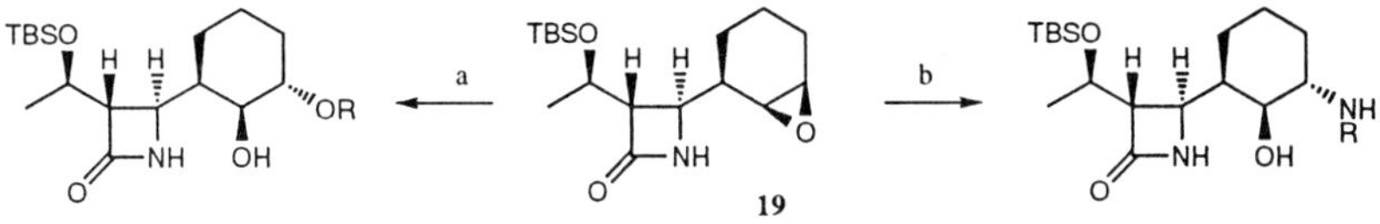

a) ROH, catalyst; b) RNH_2, catalyst, see Table 6

Scheme 10

A list of amine and alcohol nucleophiles is reported in Table 6.[33]

Nucleophile	Solvent	Catalyst	Temperature	Yield%
MeOH	-	*p*TSA	reflux	80
$HOCH_2CH_2OH$	$HOCH_2CH_2OH$/ CH_2Cl_2 20/1	*p*TSA	room temp.	50
$HOCH_2CH_2F$	-	-	room temp.	21
$HOCH_2CH_2CN$	-	CAN	room temp.	20
NaN_3	$EtOH/H_2O$	NH_4Cl	reflux	73
$MeNH_2$	$EtOH/H_2O$	NH_4Cl	reflux	quantitative
Me_2NH	$EtOH/H_2O$	NH_4Cl	reflux	55
${}^{cy}C_5H_9NH_2$	$EtOH/H_2O$	NH_4Cl	reflux	45

Table 6

As can be seen from the results reported in Table 6, acidic catalysis is required and various acids can be used effectively. Very good results have been obtained with amines when using $LiClO_4$ in MeCN.[34] The steric bulk of the nucleophile as well as the presence of electron-withdrawing groups have a strong effect on the yields of the reaction.

Cerium ammonium nitrate has been also used as a reagent for the direct formation of nitrate esters, useful intermediates to 4-alkoxy trinems,[35] from epoxide **19**.

Of particular interest is the reaction with ethylene glycol (Scheme 11), as the primary hydroxyl group could be selectively silylated and then selectively deprotected, after the construction of the trinem skeleton of **26**. Oxidation and modification of the side-chain *via* Wittig olefination reaction afforded a series of potent compounds.[36]

Selective deprotection of **26** has been also exploited in the synthesis of a series of thienamycin and imipenem analogues[37] as shown in Scheme 12;[38] in this case, formation of the azido derivative *via* Mitsunobu-type reaction and full deprotection to the sodium salt was followed by Lindlar hydrogenation of the azide to give the amino acid **30**. Reaction in buffered aqueous system with various imido esters led to compounds **31a,b**.

Another versatile intermediate was identified in epoxyphosphate **33** (Scheme 13), synthesised from the cyclohexanone derivative **13**.[39]

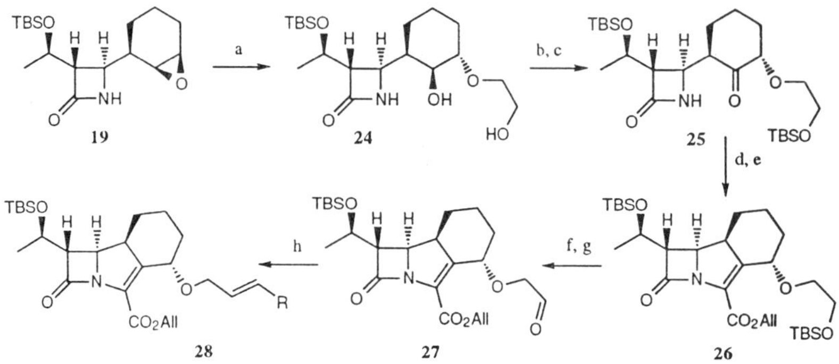

a) $HOCH_2CH_2OH$, CH_2Cl_2, TsOH, room temp.; b) TBSCl, imidazole, DMF; c) Swern oxidation; d) $ClCOCO_2All$, Et_3N, CH_2Cl_2; e) $P(OEt)_3$, *o*-xylene, reflux; f) Bu_4NF, AcOH, THF; g) Swern oxidation; h) Wittig olefination

Scheme 11

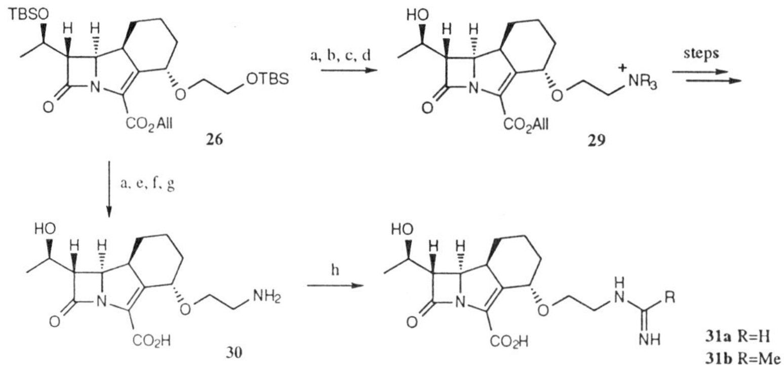

a) Bu_4NF, AcOH, THF; b) TsCl, Et_3N, CH_2Cl_2; c) KI, acetone; d) R_3N, MeCN; e) PPh_3, CBr_4, NaN_3, DMF; f) $Pd(Ph_3)_4$, PPh_3, K^+ 2-ethyl hexanoate, THF; g) $H_2/Pd(CaCO_3)$, phosphate buffer pH=7; h) Ethyl formimidate·HCl or ethyl acetimidate·HCl, phosphate buffer pH=8.5

Scheme 12

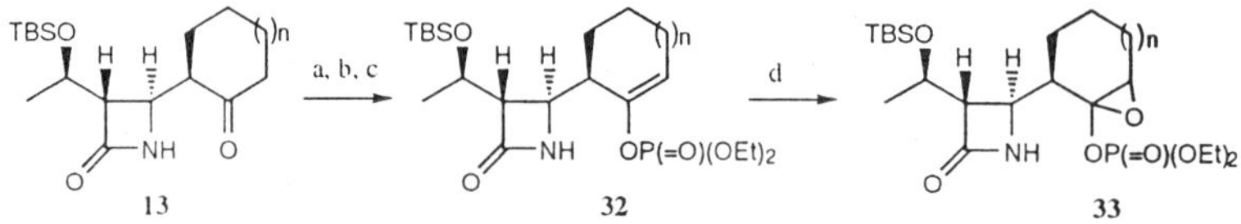

a) TBSCl, Et_3N, DMF; b) LHMDS, THF, -78°C, *then* $ClP(=O)(OEt)_2$; c) KF, MeOH; d) *m*CPBA, CH_2Cl_2, 0°C

Scheme 13

Formation of the enolphosphate is accomplished on *tert*-butyldimethylsilyl-protected **13** *via* reaction of the lithium enolate with diethyl chlorophosphate. Removal of the silyl moiety with potassium fluoride in methanol, followed by *meta*-chloroperoxybenzoic acid oxidation results in the formation of **33** as a single isomer. Although the relative stereochemistry of **33** has never been assigned, a strong analogy with the epoxidation reaction is evident, as well as in the stereochemistry of the opening products (*vide infra*).

3.3. Epoxyphosphate opening reactions

The opening reaction of epoxyphosphate **33** with alcohols and amines is shown in Scheme 14. In the reaction of epoxyphosphate with amines, an unstable aminoketone is obtained which is protected *in situ* as the allyl carbamate **35**.

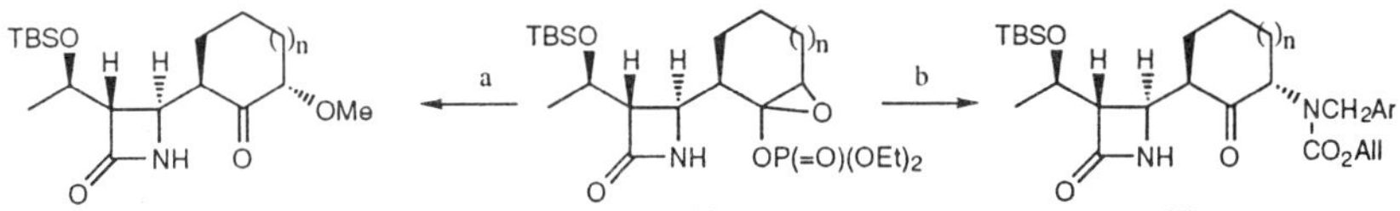

a) MeOH, 2,6-lutidine; b) $ArCH_2NH_2$, 2,6-lutidine, *then* AllOCOCl, 2,6-lutidine

Scheme 14

The epoxyphosphate opening reaction has provided a very direct method of introducing a substituent in position 4, when compared to the longer epoxide formation and opening route, thus allowing for the rapid synthesis and evaluation of a series of derivatives.

Reaction with amines, in particular benzylamines, has been extensively exploited to yield a potent series of compounds.[40]

Modification of a 4-amino side-chain has been accomplished *via* Wittig-type reaction as shown in Scheme 15;[41] reaction of epoxyphosphate **33** with 2-*tert*butyldimethylsilyloxyethylamine followed by *in situ* protection as the allyloxycarbonyl derivative gave a protected aminoketone **36** which was then transformed into the fully protected trinem **37**.

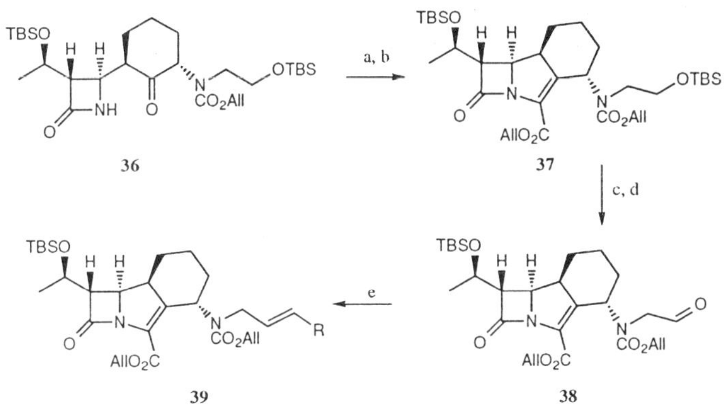

a) $ClCOCO_2All$, TEA, CH_2Cl_2; b) $P(OEt)_3$, xylene, 130°C; c) Bu_4NF, AcOH, THF; d) Swern oxidation; e) Wittig

Scheme 15

In analogy to the 4-oxygen substituted series, selective monodeprotection of the primary hydroxyl, Swern oxidation and Wittig-type olefination gave access to a potent series of compounds. In the five-membered trinem series this reaction has been exploited to synthesise the analogue of sanfetrinem.[42]

3.4. Reactions with electrophiles

The reaction of the lithium enolate of **13** with electrophiles has been exploited in the synthesis of some 4-heterosubstituted derivatives. In Scheme 16 the reaction of the lithium enolate of *N*-methylthio protected azetidinone **4** with methyl methanethiolsulphonate is shown.[43]

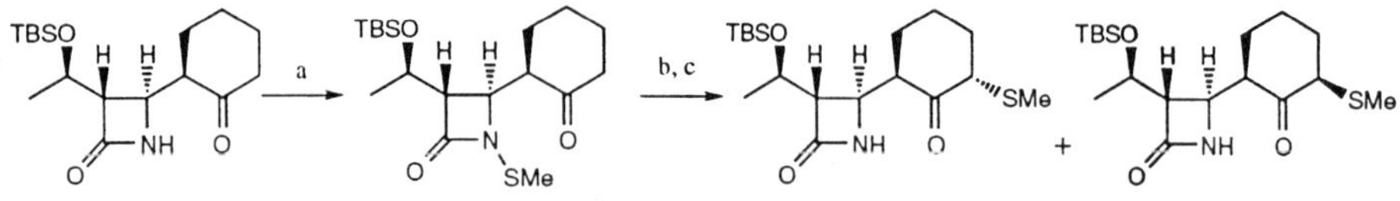

13 **40** **41** **42**

a) LHMDS, THF, *then* $MeSSO_2Me$; b) LHMDS, THF, *then* $MeSSO_2Me$; c) 2-pyridinethiol

Scheme 16

Stepwise introduction of the two methylthio groups was necessary in order to obtain a good diastereoselectivity in the enolate reaction.

When the reaction was attempted on the dianion derivative, a mixture of the two diastereomers **41** and **42** was obtained. It has been hypothesised that the mixture of isomers results from unselective reaction of the dianion, and that *in situ* protection of the azetidinone nitrogen followed by reaction at carbon was leading to isomer **41**.

Padova *et al.*[44] reported the reaction of the lithium enolate of *N*-*tert*-butyldimethylsilyl protected mixture of isomers **43** with *N*-fluorobenzenesulphonimide to give to 4-fluoro substituted trinem precursors (Scheme 17). The reaction showed complete *trans* selectivity and (4*S*, 8*S*) 4-fluorotrinem was prepared *via* the usual procedure. Unfortunately, the compound proved to be unstable in aqueous media.[45]

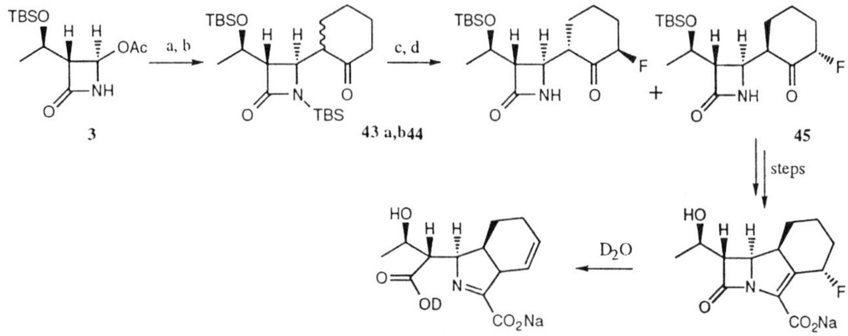

3 43 a,b44 45 47 46

a) TBSCl, Et_3N CH_2Cl_2; b) 1-trimethylsilyloxycyclohexene, $SnCl_4$, MeCN; c) LHMDS, $FN(PhSO_2)_2$, THF, -78°C; d) NH_4F, MeOH

Scheme 17

3.5. Reactions on aminoacid derivatives

In the synthesis of ureido derivatives **50**,[46] the sodium salt of the aminoacid, prepared *via* opening of the epoxide, protection as benzyl carbamate and ester, and double hydrogenation in the presence of one equivalent of sodium hydrogen carbonate, has been successfully reacted with some isocyanates in aqueous medium, as shown in Scheme 18.

The deprotected aminoacid has also been used in the synthesis of amidino derivatives **49** (Scheme 18) *via* the reaction with imidoethers in aqueous media.[47]

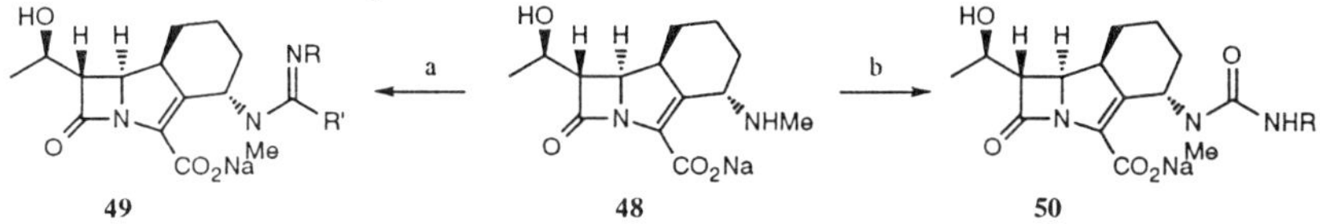

a) R'C(=NR)OEt, water; b) RN=C=O, water

Scheme 18

4. Other derivatives

A few patents describe trinem derivatives either non-substituted or with a heterocyclic third ring C.[48] Hanessian *et al.* [49] published the synthesis of the 5-methoxytrinems, shown in Scheme 19.

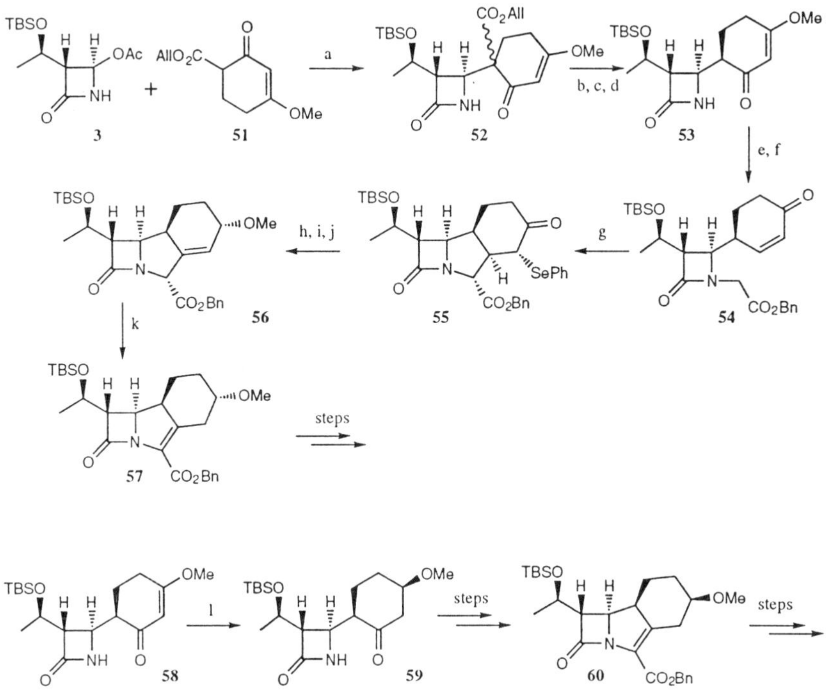

a) NaH, THF, 0°C; b) TBSCl, 2,6-lutidine, CH_2Cl_2, 0°C; c) $Pd(OAc)_2$, PPh_3, HCO_2H, AcOEt, reflux; d) AcOH, Bu_4NF, THF; e) DIBAL-H, THF, -78C, work-up with 10%HCl; f) NaHMDS, THF, *then* $BrCH_2CO_2Bn$, -78°C; g) NaHMDS, THF, -78°C, *then* $ZnCl_2$, *then* PhSeBr, -78°C; h) mCPBA, CH_2Cl_2, -78°C; i) $NaBH_4$, MeOH, -78°C; j) MeI, Ag_2O; k) DBU, CH_2Cl_2; l) H_2, Pd(C), AcOEt, *separate isomers*

Scheme 19

Condensation of the sodium salt of **51** with acetoxyazetidinone **3** afforded **52** as a mixture of isomers, which was decarboxylated by treatment with palladium acetate and formic acid to give, with high stereoselectivity, isomer **53**. Regeneration of the enone moiety and alkylation of the azetidinone moiety gave the Michael-type substrate **54**. The cyclisation was accomplished by treatment of the sodium enolate of **54** with zinc chloride and trapping of the Michael product with phenylselenyl bromide. Standard methodology led to fully protected **56** which was treated with DBU to obtain the regioisomer **57** possessing the required trinem nucleus. The final compound was obtained by application of the methodology described in the previous sections.

The stereoisomeric (5*R*) derivative **60** was obtained through non-selective hydrogenation of the double bond in **58** followed by application of standard methodology to give the fully deprotected trinem.

5. The stereoselective synthesis of sanfetrinem

The progression of sanfetrinem to Clinical Phases I and II, prompted various research groups to study alternative approaches to this molecule, and in particular to a stereoselective synthesis of advanced intermediate **61** (Figure 5).

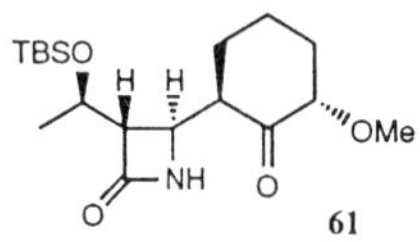

61

Figure 5

A first stereoselective synthesis of **61** could be established by taking advantage of the initial studies on the synthesis of (4α,8ß) 4-substituted trinems, through conversion of epoxide **19** into **61** in two steps.

In order to obtain a robust large-scale route towards sanfetrinem, *para*-^{t}butylbenzyl was chosen as the protecting group for the carboxylate, thus allowing a more practical deprotection procedure as well as affording crystalline intermediates, and the oxo-oxo cyclisation route discussed previously was employed.[50]

Although this route proved to be robust, and provided multikilogram quantities of the desired trinems to support early development studies, it still suffered from several drawbacks, such as; low overall chemical yields (8-9%) and a high number of stages (11 steps, 15 chemical transformations). Alternative routes were therefore explored in order to decrease the number of steps and to increase the overall chemical yield.

A convergent approach *via* the one-step formation of **61** from the 4-acetoxyderivative **3** and 2-methoxycyclohexanone has therefore been the subject of a series of studies.

Early synthetic efforts gave poor results in terms of both stereo- and regioselectivity[51] although they permitted the synthesis of sanfetrinem and a number of its isomers, albeit in low yield.

A novel method for the regioselective preparation of silyl enol ethers from α-amidocyclohexanones[52] permitted the preparation of silyl enol ether **62** with high yield and regioselectivity and the Lewis acid promoted reaction between **62** and **3** (or its *N*-protected derivatives) was studied.[53]

It was found that the reaction of optically active silyl enol ether **62** with the *N*-trimethylsilyl derivative of **3**, using tin (IV) chloride as catalyst gave isomer **61** with high diastereoselectivity (Scheme 20).

The same study showed that a number of Lewis bases, when used as additives, strongly influenced the reaction, probably through a modification of the activating power of tin tetrachloride. In particular diethyl ether and dimethyl sulfide improved the reaction yields and selectivity, (see Table 7) whereas more nucleophilic agents such as triethylamine, triphenylphosphine and dimethylsulphoxide effectively inhibited the reaction.

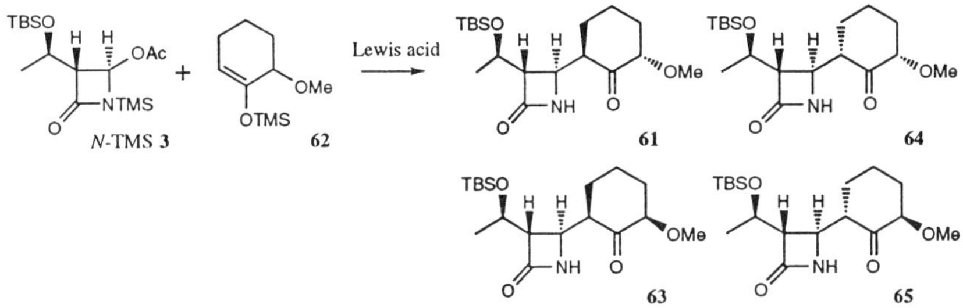

Scheme 20

Lewis acid	Temp (°C)	Yield %	Isomer ratio 61:63:64:65
TMSOTf	0	51	1.5:4:2:1
$ZnCl_2$	0-23	60	1.7:1.7:1:3.3
$TiCl_4$	0-5	decomp	-
$SnCl_4 \cdot Et_2O$	23	45	1.5:-:-:1
$SnCl_4 \cdot Me_2S$	23	50	
$SnCl_4 \cdot Me_2S$	23	50	1:-:-:-[a]

a) (6*S*)-**62** was used

Table 7

The need to conduct the reaction outlined in Scheme 20 on a large scale necessitated the development of an efficient synthesis of optically active (2*S*)-methoxycyclohexanone **69** and its silyl enol ether (6*S*)-**62**.
A collaboration between the Roberts group and Glaxo Wellcome resulted in the chemoenzymatic synthesis of ketone (2*S*)-**69** shown in Scheme 21.[54] The key step is a modification of an existing enzymatic resolution of 2-alkoxycycloalkanols.[55]
(1*S*,2*S*)-2-methoxycyclohexanol (1*S*,2*S*)-**67** was obtained in 36% yield from racemic **67** *via* an immobilised *Candida antarctica* lipase catalysed enantioselective acylation of undesired enantiomer to give **68**. The unreacted enantiomer (1*S*,2*S*)-**67** was isolated after selective extraction from the reaction solvent cyclohexane into water, thus avoiding a tedious and expensive separation by flash chromatography.
Recently a new method for the synthesis of (1*R*,2*S*)-**67** by enantioselective hydrogenation under substrate racemisation conditions of commercially available racemic **69** has been claimed by scientists at Takasago researchers.[56,57]
Other methods for the enantioselective synthesis of (1*S*,2*S*)-**67** have been reported.[58] A number of oxidizing conditions have been found to be effective in the synthesis of (2*S*)-**69** without detectable

racemisation at the stereogenic centre; among them: Swern, Jones, TEMPO, and NaClO under buffered conditions.[59]

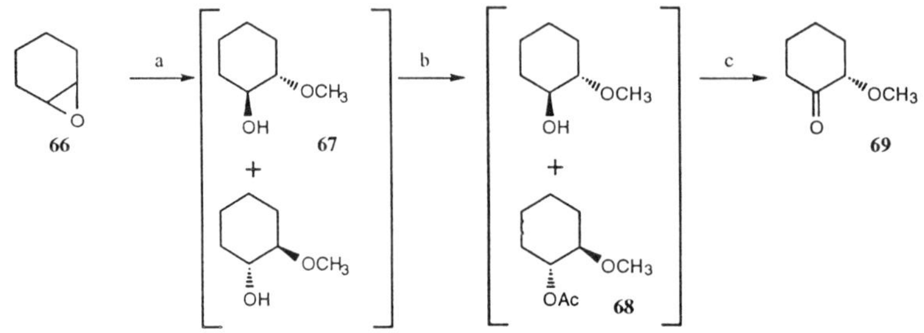

a) MeOH, *p*TSA; b) immobilised *Candida antarctica* lipase, Et_3N, cyclohexane; c) NBA, Me_2CO/H_2O *or* NaOCl/TEMPO(cat), H_2O

Scheme 21

The generation of the lithium enolate from the silyl enol ether **62**, (see Scheme 22)[60] followed by metal exchange, was the subject of further studies in this field.

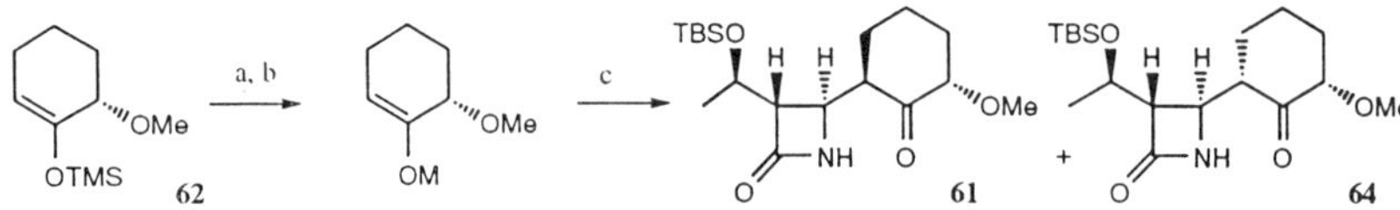

a) MeLi, THF, -78°C; b) $ZnCl_2$,*or* $ZnBr_2$,*or* Cp_2ZrCl_2,*or* $Cp_2Zr(Hex)Cl$; c) **3**

Scheme 22

Good yields and moderate stereoselectivity in the formation of the methoxyketone **61** were observed with zinc[61] and zirconium[62] enolates, although these methodologies do not appear to show significant advantage over those already reported.

In a more convergent approach, (2*S*)-**69** reacts with **3** in the presence of $SnCl_4$ and a tertiary amine (such as *N,N*-diisopropylethylamine) to give the ketoazetidinone **61** with high yield and diastereoselectivity.[63] Low temperature NMR studies of the reaction indicated the formation of a 1:1 chelate complex **70** (see Scheme 23) between $SnCl_4$ and (2*S*)-**69** which is transformed into a highly reactive chelated tin enolate upon addition of base.

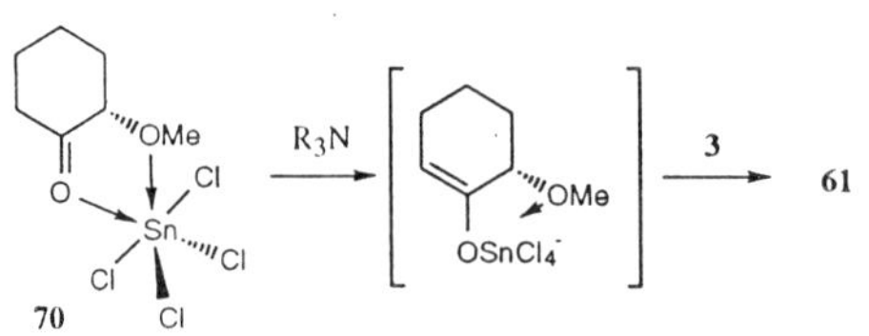

Scheme 23

The formation of complex **70** has been confirmed by single crystal X-ray analysis.

The high diastereoselectivity observed is believed to derive from an open transition state (see Figure 6) where the chelated $SnCl_4$ amplifies the stereochemical influence of the methoxy group.

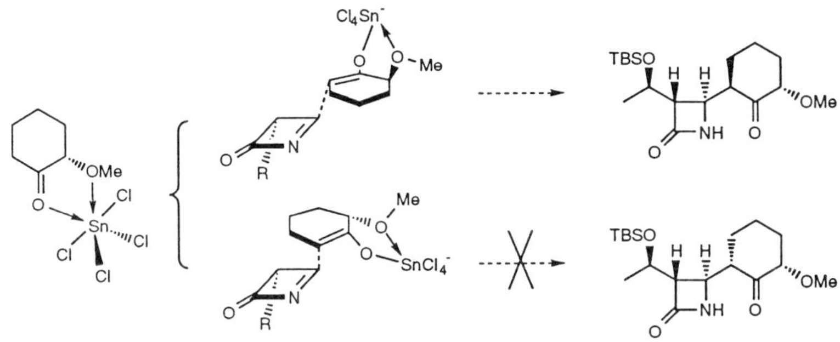

Figure 6

This reaction, which offers considerable advantages over all existing syntheses of the ketoazetidinone **5** and has been successfully scaled up in a pilot plant for the production of multikilogram amounts of **61**, is currently under evaluation for inclusion in the industrial synthesis of sanfetrinem.

Hanessian *et al.*[64] reported a stereoselective route to sanfetrinem relying on a similar strategy to that applied in his synthesis of 5-methoxy trinems. The key steps of this synthesis are reported in Scheme 24; condensation of the sodium salt of racemic **71** on acetoxyazetidinone **3** led to a mixture of all diastereomers which were *N*-protected as the *tert*-butyldimethylsilyl derivatives.

Palladium acetate catalysed decarboxylation, through a modification of an existing methodology[65] stereoselectively afforded **61** and **63**. It was found that formation of the lithium dianion and treatment with one equivalent of zinc bromide followed by diethyl malonate afforded stereoselectively the required (2'*S*)-**61** isomer upon work-up.

Deuterium incorporation studies led to the exclusion of the initial hypothesis of malonate acting as the proton source (Scheme 24, intermediate **A**), in favour of intermediate **B**, where malonate acts as a ligand and the proton is delivered through the axially coordinated diisopropylamine.

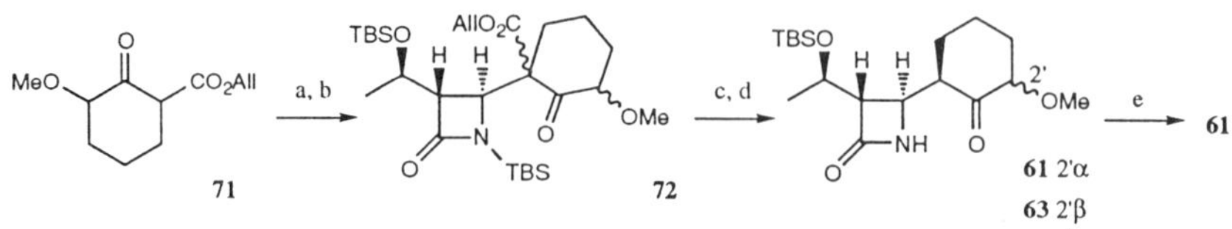

a) NaH, THF, -20°C, *then* **3**; b) TBSOTf, 2,6-lutidine, CH_2Cl_2; c) $Pd(OAc)_2$ (cat.), Ph_3P (cat.), HCO_2H, EtOAc, reflux; d) Bu_4NF, AcOH, THF; e) LDA (3eq), -10°C, THF, *then* $ZnBr_2$ (1eq), -10 to 0°C, *then* diethyl malonate (5eq), 0°C to room temp.

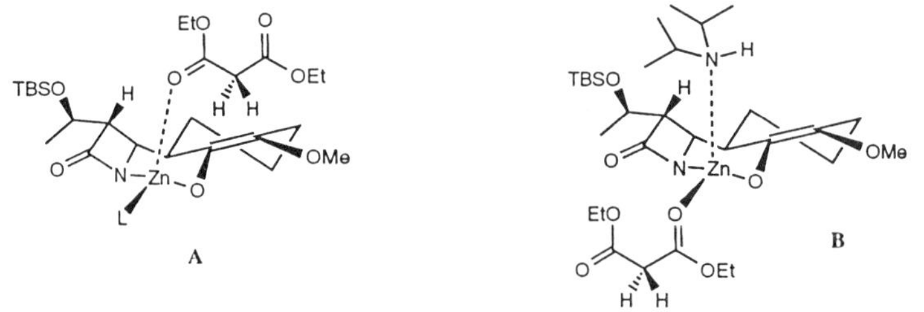

Scheme 24

Sanfetrinem was then obtained under standard conditions in moderate overall yield.

Panunzio *et al.*[66] (Scheme 25) reacted the lithium enolate of *tert*-butyl acetate with the silyl imine derived from (2-*tert*-butyldimethylsilyloxy)cyclohexyl-1-carboxaldehyde. The racemic azetidinone obtained from this enolate-imine condensation was then protected as the *N,O*-acetonide and the hydroxyethyl side-chain introduced *via* reaction of the lithium enolate of **75** with acetyltrimethylsilane followed by Brooke rearrangement. In order to obtain the enantiomerically pure trinem nucleus, a classical resolution was then carried out.

After a protection-deprotection sequence, racemic *N,O*-bis(*tert*-butyldimethylsilyl) derivative **43a** was obtained and reacted under the already described conditions with the homochiral ephedrine-derived chlorophosphate **A** to give a diastereomeric pair of enolphosphates **77** and **78**.

These compounds were separated and progressed further through oxidation followed by opening with methanol to advanced intermediates of sanfetrinem and its enantiomer.

Also in the case of the ephedrine-derived enolphosphate, the presence of the unprotected lactam nitrogen was crucial in the reaction with *meta*-chloroperoxybenzoic acid to obtain good yields and diastereoselectivity.

In another formal synthesis of sanfetrinem, the reaction of homochiral 2-methoxycyclohexanone silyl enol ether **62** with racemic unsubstituted 4-acetoxyazetidinone **81** was exploited (Scheme 26).[67] The Lewis acid catalysed aldol-type reaction proceeded with fairly good yields but poor diastereoselectivity, and a mixture of the four diastereomers was isolated.

The isomers could be separated and the hydroxyethyl side-chain introduced *via* Brooke rearrangement on the acetonide derivative **84**.

Further elaboration finally led to the synthesis of sanfetrinem methyl ester.

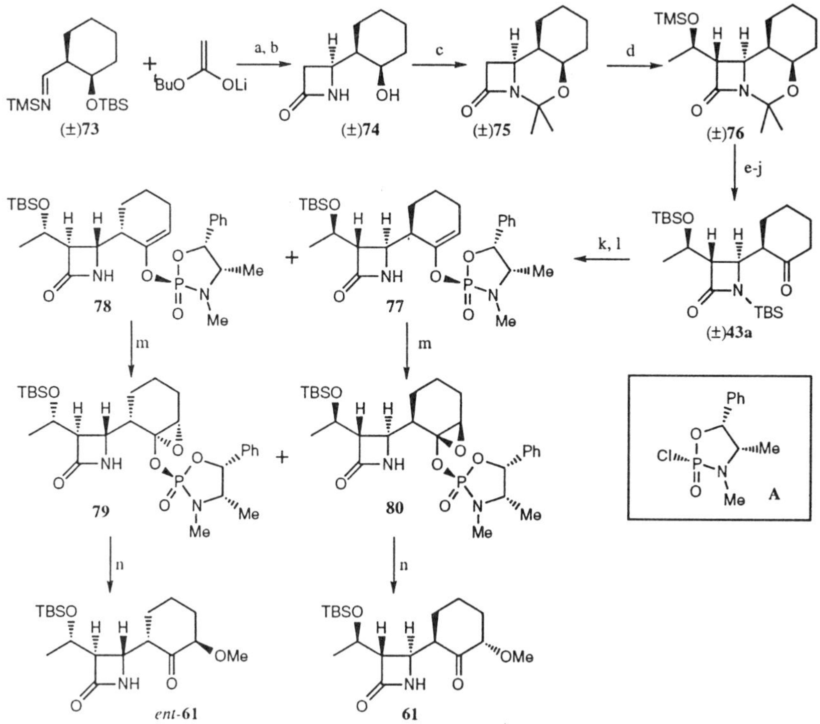

a) THF, -78°C to room temp.; b) LHMDS, Bu_4NF, THF; c) dimethoxypropane, $BF_3 \cdot Et_2O$, CH_2Cl_2; d) LDA, MeCOTMS, *then* ^tBuOK, ^tBuOH; e) HF_{aq}; f) $pNO_2BNOCOCl$, DMAP, CH_2Cl_2; g) Jones, Me_2CO; h) TBSCl, Et_3N, CH_2Cl_2; i) H_2, Pd/C, AcOEt; j) TBSCl, imidazole, DMF; k) LDA, THF, -78°C; l) KF, MeOH; m) *m*CPBA, $NaHCO_3$, CH_2Cl_2; n) MeOH, 40°C

Scheme 25

6. Other approaches to the trinem nucleus

A radical route to trinems has been reported,[68] where the phenylseleno derivative **87** was reacted with differently substituted enones **A** upon slow addition of tributyltin hydride and AIBN, as shown in Scheme 27. The formation of captodative radical **88** has been postulated, which would rearrange to the less sterically demanding system **89** in 50-60% combined yield.

Only the *cis* isomers where obtained, of which the (2'*S*, 6'*S*) 2'-*tert*-butyldimethylsilyloxy derivative was converted under standard conditions to the fully protected trinem derivative.

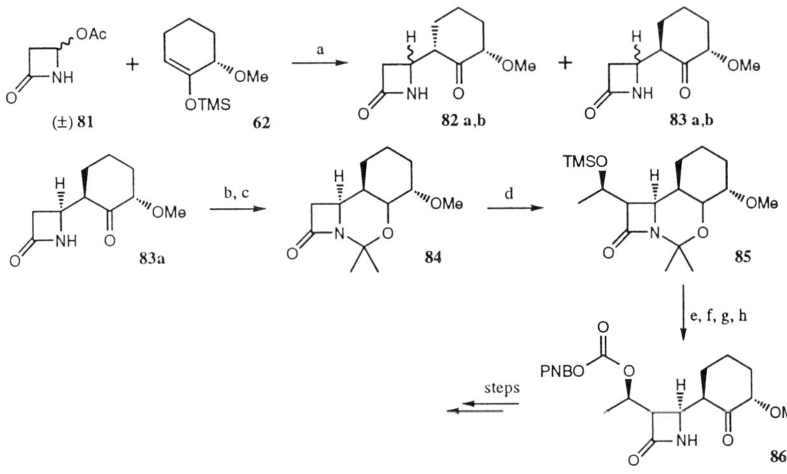

a) $ZnCl_2$, CH_2Cl_2; b) $NaBH_4$, EtOH, 0°C; c) 2,2-dimethoxypropane, $BF_3{\cdot}Et_2O$, CH_2Cl_2; d) LDA, THF, -78°C, *then* MeCOTMS, -78°C, *then* ^tBuOK, ^tBuOH, -78°C to room temp.; e) Bu_4NF, AcOH, THF; f) $pNO_2BnOCOCl$, DMAP, CH_2Cl_2; g) TFA, H_2O; h) Swern oxidation

Scheme 26

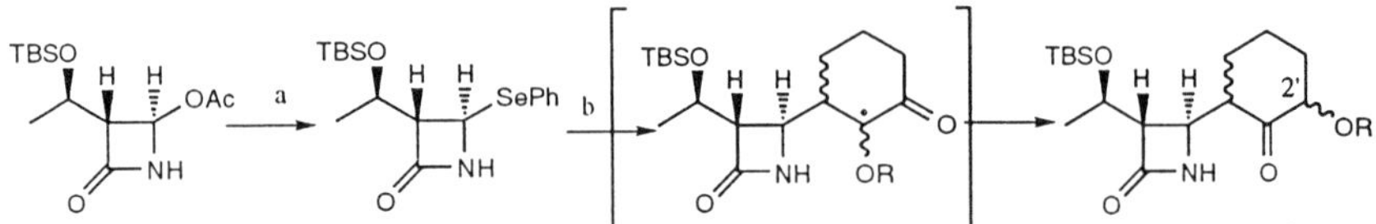

cis isomers **89a,b**
ca. 1:1

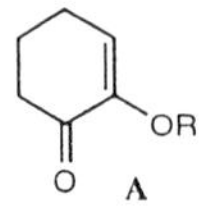

R = COMe
COPh
TBS

a) Bu_3P, PhSeSePh, NaOH, THF; b) **A**, Bu_3SnH, AIBN, THF, reflux

Scheme 27

An interesting approach, shown in Scheme 28, was attempted by Alcaide *et al*,[69] on a slightly different system.

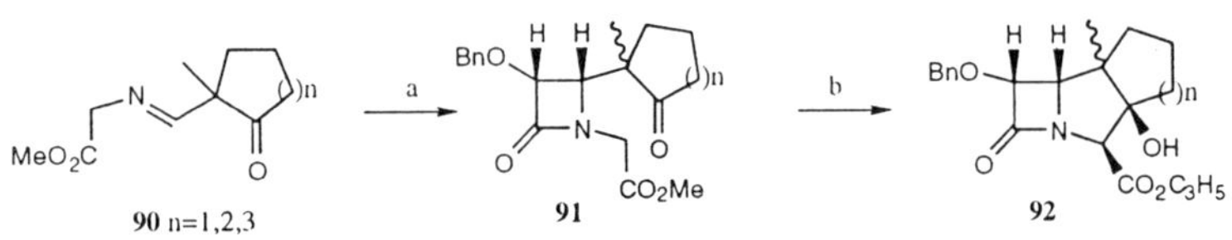

a) $BnOCH_2COCl$, Et_3N, room temp.; b) LHMDS, THF, -78°C

Scheme 28

The Staudinger reaction of ß-ketoaldimines **90** derived from cyclic ketoaldehydes afforded the *cis* substituted ß-lactam ring **91** bearing the cycloalkanone moiety and a benzyloxy side-chain (instead of the usual hydroxyethyl). Intramolecular aldol reaction afforded the tricyclic system **92** only in the cyclohexanone and cycloheptanone case. Among the limitations of this methodology are the necessity of using an α,α-disubstituted substrate **90**, and the *cis* stereochemistry of the products, which could nevertheless be overcome by base-catalysed epimerisation.

7. Conclusions

The discovery of the trinems has given us a novel and potent new class of ß-lactam antibacterials, and the structural complexity of the molecules has provided a considerable synthetic challenge for a growing number of research groups.

During the preparation of this manuscript further work on trinems has been published; among these, the aldol reaction of the enolate derived from phosphorane **7** with formaldehyde (Scheme 29): after selective cyclisation of the (2'*R*) isomer and Swern oxidation, aldehyde **94** was submitted to Wittig-type reactions to give a series of double bond derivatives.[70]

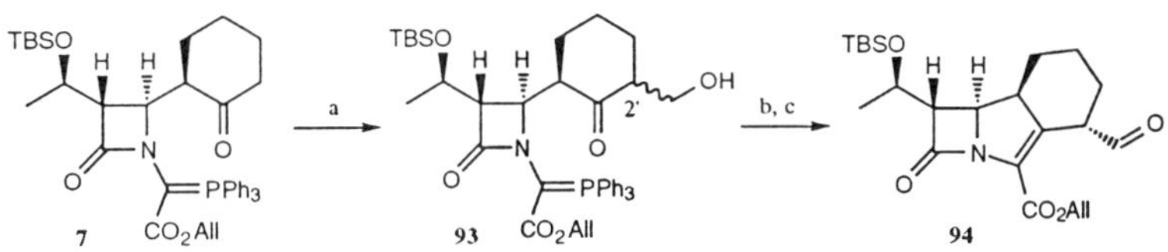

a) LHMDS (1.5eq), -78°C, then excess HCHO (*ca.* 0.47M in THF), -78°, 61%; b) toluene, 80°C; c) Swern oxidation

Scheme 29

References

1. For a descriptive book on the development of bacterial resistance, see Levy, S.; *The antibiotic paradox*, **1995**, Plenum Press, New York and London.
2. See for example; Neu, H. C.; and Cooper, R. D. G., in *The Chemistry of ß-Lactams*, **1992**, Page, M. I., Ed., Chapman and Hall, London.
3. a) Ngo, J.; Castañer, J., *Drugs of the Future*, **1996**, *21*, 1238, Prous Science Publ.; b) Spangler, S. K.; Jacobs, M. R.; Appelbaum, P. C., *Antimicrob. Agents, Chemother.*, **1997**, *41*, 148; c) Doern, G. V.; Pierce, G.; Brueggemann, A.B., *Diagn. Microbiol. Infect. Dis.*, **1996**, *26*, 39; d) Singh, K. V.; Coque, T. M.; Murray, B. E., *Antimicrob. Agents Chemother.*, **1996**, *40*, 2142; e) Wise, R.; Andrews, J. M.; Brenwald, N., *Antimicrob. Agents Chemother.*, **1996**, *40*, 1248; f) Di Modugno, E.; Erbetti, I.; Ferrari, L.; Galassi, G.; Hammond, S.M.; Xerri, L., *Antimicrob. Agents Chemother.*, **1994**, *38*, 2362.
4. a) Biondi, S.; Andreotti, D.; Rossi, T.; Carlesso, R.; Tarzia, G.; Perboni, A., *Eur. Pat. Appl.*, **1992**, EP 502464 A1; b) Gaviraghi, G.; Donati, D.; Tarzia, G.; Perboni, A.; Ursini, A., *Eur. Pat. Appl .*, **1992**, EP 502465 A1; c) Perboni, A.; Rossi, T.; Gaviraghi, G.; Ursini, A.; Tarzia, G., *PCT Int. Appl.*, **1992**, WO 9203437 A1; d) Perboni, A.; Donati, D.; Tarzia, G., *Eur. Pat. Appl.*, **1992**, EP 502468 A1; e) Tamburini, B.; Perboni, A.; Rossi, T.; Donati, D.; Andreotti, D.; Gaviraghi, G.; Biondi, S.; Bismara, C.; *Eur. Pat. Appl.*, **1991**, EP 416952 A2; f) Tamburini, B.; Perboni, A.; Rossi, T.; Donati, D.; Andreotti, D.; Gaviraghi, G.; Carlesso, R.; Bismara, C., *Eur. Pat. Appl.*, **1990**, EP 416953 A2.

5. a) Palomo, C., in *Recent Progress in the Chemical Synthesis of Antibiotics*, **1990**, Lucaks, G.; Ohno, M., Eds., Springer-Verlag, Berlin; b) Basker, M. J.; Boon, R. J.; Hunter, P. A.; *J. Antibiot.*, **1980**,*33*, 878.
6. a) Pflander, H. R.; Gosteli, J.; Woodward, R. B.; Rihs, G., *J. Am. Chem. Soc.*, **1981**, *103*, 45263; b) Ernest, I.; Gosteli, J.; Woodward, R. B., *J. Am. Chem. Soc.*, **1979**, *101*, 6310.
7. For some selected references, see a) Afonso, A.; Hon, F.; Weinstein, J.; Ganguly, A. K., *J. Am. Chem. Soc.*, **1982**, *104*, 6138; b) Battistini, C.; Scarafile, F.; Foglio, M.; Franceschi, G.; *Tetrahedron Lett.*, **1984**, *25*, 2395.
8. Hanessian, S.; Desilets, D.; Bennani, Y. L., *J. Org. Chem.*, **1990**, *55*, 3098.
9. Rossi, T., **1996**, *Chiral USA '96*, oral communication.
10. Tughan, G., *PCT Int. Appl.*, **1994**, WO 9421638.
11. Owen, M., *Chemistry in Britain*, **1996**, *32*, 46.
12. Araldi, G. L., *PCT Int. Appl.*, **1994**, WO 9421637.
13. Jeffrey, P. D.; McCombie, S. V., *J. Org. Chem.*, **1982**, *47*, 587.
14. Andreotti, D.; Biondi, S.; Di Fabio, R.; Donati, D.; Piga, E.; Rossi, T., *Bioorg. Med. Chem. Lett.*, **1996**, *6*, 2019.
15. Tranquillini, M. E.; Araldi, G. L.; Donati, D.; Pentassuglia, G.; Pezzoli, A.; Ursini, A., *Bioorg. Med. Chem. Lett.*, **1996**, *6*, 1683.
16. Perboni, A.; Tamburini, B.; Rossi, T.; Donati, D.; Tarzia, G.; Gaviraghi, G., in *Recent Advances in the Chemistry of Antiinfective Agents*, **1993**, Bentley, P.H.; Ponsford, R., Eds., The Royal Society of Chemistry, Cambridge, p. 21.
17. Tamburini, B.; Perboni, A.; Rossi, T.; Donati, D.; Andreotti, D.; Gaviraghi, G.; Biondi, S.; Bismara, C., **1991**, *Eur. Pat. Appl.*, EP 416952 A2.
18. a) Andreotti, D.; Rossi, T.; Marchioro, C., *Bioorg. Med. Chem. Lett.*, **1996**, *6*, 2589; b) Biondi, S.; Gaviraghi, G.; Rossi, T., *Bioorg. Med. Chem. Lett.*, **1996**, *5*, 525; c) Di Fabio, R.; Feriani, A.; Gaviraghi, G.; Rossi, T., *Bioorg. Med. Chem. Lett.*, **1995**, *5*, 1235.
19. For related systems, see also Dininno, F. P., **1994**, US 5372993 A.
20. Wollmann, T.; Gerlach, U.; Hörlein, R.; Krass, N.; Lattrell, R.; Limbert, M.; Markus, A., in *Recent Advances in the Chemistry of Antiinfective Agents*, **1993**, Bentley, P.H.; Ponsford, R., Eds., The Royal Society of Chemistry, Cambridge, p. 50.
21. Mukaitama, T.; Stevens, R. W.; Iwasawa, N., *Chem. Lett.*, **1982**, 353.
22. Budt, K. H.; Fischer, G.; Hoerlein, R.; Kirstetter, R.; Lattrell, R., *Tetrahedron Lett.*, **1992**, *33*, 5331.
23. Marchioro, C.; Pentassuglia, G.; Perboni, A.; Donati, D., *J. Chem. Soc. Perkin Trans. 1*, **1997**, 463.
24. Hoveyda, A. H.; Evans, D. A.; Fu, G. C., *Chem. Rev.*, **1993**, 1307.
25. Biondi, S., in *Anti-infectives: Recent Advances in Chemistry and Structure-Activity Relationships*, **1997**, Bentley, P.H.;O'Hanlon, P. J., Eds., The Royal Society of Chemistry, Cambridge, 86.
26. Bellassoued, M.; Frangin, Y.; Gaudemar, M., **1977**, *Synthesis*, *3*, 205.
27. Perboni, A.; Bismara, C.; Pentassuglia, G., *Eur. Pat. Appl.*, **1992**, EP 466509 A1.
28. Bismara, C.; Di Fabio, R.; Donati, D.; Rossi, T.; Thomas, R. J., *Tetrahedron Lett.*, **1995**, *36*, 4283.

29. a) Rossi, T.; Biondi, S.; Contini, S.; Thomas, R. J.; Marchioro, C., *J. Am. Chem. Soc.*, **1995**, *117*, 9604; b) Biondi, S.; Rossi, T.; Contini, S., *Eur. Pat. Appl.*, **1994**, EP 617017 A1.
30. See for example Ueda, Y.; Maynard, S. C., *Tetrahedron Lett.* **1985**, *26*, 6309.
31. Brown, H. C.; Bhat, K. S.; Jadhav, P. K., *J. Chem. Soc. Perkin Trans. 1*, **1991**, 2633.
32. Perboni, A.; Bismara, C.; Pentassuglia, G., *Eur. Pat. Appl.* **1991**, EP 0466509 A1.
33. a) Andreotti, D.; Biondi, S.; Di Fabio, R.; Donati, D.; Piga, E.; Rossi, T., *Bioorg. Med. Chem. Lett.*, **1996**, *6*, 2019; b) Tranquillini, M.E.; Araldi, G.L.; Donati, D.; Pentassuglia, G.; Pezzoli, A.; Ursini, A., *Bioorg. Med. Chem. Lett.*, **1996**, *6*, 1683.
34. Chini, M.; Crotti, P.; Favero, L.; Macchia, F.; *Tetrahedron Lett.*, **1991**, *32*, 4775; and Ref. 14
35. Di Fabio, R.; Rossi, T.; Thomas, R. J., *Tetrahedron Lett.*, **1997**, *38*, 3587.
36. Ref. 24.
37. Di Fabio, R.; Andreotti, D.; Biondi, S.; Gaviraghi, G.; Rossi, T., *Bioorg. Med. Chem. Lett.*, **1996**, *6*, 2025.
38. Carlesso, R.; Holman, S.; Perboni, A.; Rossi, T., *PCT Int. Appl.*, **1994**, WO 9405666 A1.
39. a) Perboni, A.; De Magistris, E., *Eur. Pat. Appl.*, **1993**, EP 586017 A1; b) Perboni, A., *Eur. Pat. Appl.*, **1992**, EP 502488 A2.
40. Perboni, A.; Sbampato, G., *PCT Int. Appl.*, **1995**, WO 9523149 A1.
41. Ref. 24
42. Biondi, S.; Gaviraghi, G.; Rossi, T., *Bioorg. Med. Chem. Lett.*, **1996**, *6*, 525.
43. Ref. 15.
44. Padova, A.; Roberts, S. M.; Donati, D.; Marchioro, C.; Perboni, A., *Tetrahedron*, **1996**, *52*, 263.
45. In house NMR analysis revealed that structure for compound **47** was reported incorrectly; Marchioro, C., personal communication.
46. a) Perboni, A.; Pentassuglia, G.; Andreotti, D.; Winders, J. A., *PCT Int. Appl.*, **1995**, WO 9513278 A1; b) Gehanne, S.; Piga, E.; Andreotti, D.; Biondi, S.; Pizzi, D., *Bioorg. Med. Chem. Lett.*, **1996**, *6*, 2791.
47. Perboni, A.; Donati, D.; Tarzia, G., *Eur. Pat. Appl.*, **1992**, EP 502468 A1.
48. a) Miwa, T.; Okonogi, K.; Nagai, K., **1996**, JP 08245628 A2; b) Miwa, T.; Okonogi, K., **1996**, JP 08053459 A2; c) Miwa, T.; Nagai, K.; Okonogi, K., **1994**, JP 06166688 A2; d) Sendai, M.; Miva, T., **1992**, *Eur. Pat. Appl.*, EP 507313 A1; e) Sendai, M.; Miwa, T., **1991**, *Eur. Pat. Appl.*, EP 422596 A2.
49. Hanessian, S.; Rozema, M. J.; Reddy, G. B.; Braganza, J. F., *Bioorg. Med. Chem. Lett.*, **1995**, *5*, 2535.
50. Tughan, G.; *PCT Int. Appl.*, **1994**, WO 9421638.
51. Andreotti, D.; Rossi, T.; Gaviraghi, G.; Donati, D.; Marchioro, C.; Di Modugno, E.; Perboni, A., *Bioorg. Med. Chem. Lett.*, **1996**, *6*, 491.
52. Rossi, L.; Pecunioso, A., *Tetrahedron Lett.*, **1994**, *35*, 5285.
53. Ghiron, C.; Piga, E.; Rossi, T.; Tamburini, B.; Thomas, R.J., *Tetrahedron Lett.*, **1996**, *37*, 3891

54. a) see Ref. 42; b) Stead, P.; Marley, H.; Mahmoudian, M.; Webb, G.; Noble, D.; Ip, Y. T.; Piga, E.; Rossi, T.; Roberts, S. M.; Dawson, M.J., *Tetrahedron: Asymmetry*, **1996**, *7*, 2247; see also c) Vedejs, E.; Daugulis, O.; Diver, S. T., *J. Org. Chem.*, **1996**, *61*, 430.

55. Hönig, H.; Seufer-Wassertahl, P., *Synthesis*, **1990**, 1137.

56. Miura, T.; *Chiral USA '96*, oral communication.

57. Ohkuma, T.; Ooka, H.; Hashiguchi, S.; Ikariya, T.; Noyori, R.; *J. Am. Chem. Soc.*, **1995**. *117*, 2675.

58. a) McCague, R.; *Eur. Pat. Appl.*, **1995**, EP 656344 A2; b) Sarco, C. R.; Guch, I. C.; DiMare, M., *J. Org. Chem.*, **1994**, *59*, 705.

59. a) Ghiron, C.; Piga, E.; Pecunioso, A., *UK Pat. Appl.*, **1995**, GB 2287709 A1; b) Rossi, T.; Zarantonello, P.; Thomas, R.J., *PCT Int. Appl.*, **1995**, WO 9526333 A1; c) McCague, R.; Chaplin, D.; Bannister, R., **1995**, *UK Pat. Appl.*, GB 2 283 971 A1.

60. House, H.O.; Crumrine, D. S.; Teranishi, A. Y.; Olmstead, H. D., *J. Am. Chem. Soc.*, **1973**, *95*, 3310.

61. Kennedy, G.; Rossi, T., Tamburini, B.; *Tetrahedron Lett.*, **1996**, *37*, 7441.

62. Giacobbe, S. A.; Rossi, T., *Tetrahedron: Asymmetry*, **1996**, *7*, 3079.

63. Rossi, T.; Marchioro, C.; Paio, A.; Thomas, R. J.; Zarantonello, P., *J. Org. Chem.*, **1997**, *62*, 1653.

64. Hanessian, S.; Rozema, M. J., *J. Am. Chem. Soc.*, **1996**, *118*, 9884.

65. Choi, W.-B.; Churchill, H. R. O.; Lynch, J. E.; Thompson, A. S.; Humprey, G. R.; Volante, R. P.; Reider, P. J.; Shinkai, I., *Tetrahedron Lett.*, **1994**, *35*, 2275.

66. a) Camerini, R.; Panunzio, M.; Bonanomi, G.; Donati, D.; Perboni, A., *Bioorg. Med. Chem. Lett.*, **1996**, *6*, 2467; b) Panunzio, M.; Camerini, R.; Pachera, R.; Donati, D.; Marchioro, C.; Perboni, A., *Tetrahedron: Asymmetry*, **1996**, *7*, 2929.

67. Jackson, P. M.; Roberts, S. M.; Davalli, S.; Donati, D.; Marchioro, C.; Perboni, A.; Provera, S.; Rossi, T., *J. Chem. Soc., Perkin Trans. 1*, **1996**, 2029.

68. Padova, A.; Roberts, S. M.; Donati, D.; Perboni, A.; Rossi, T., *J. Chem. Soc., Chem. Commun.*, **1994**, 441.

69. Alcaide, B.; Polanco, C.; Saez, E.; Sierra, M. A., *J. Org. Chem.*, **1996**, *61*, 7125.

70. Ghiron, C.; Rossi, T.; Thomas, R. J., *Tetrahedron Lett.*, **1997**, *38*, 3569.

4-AMINOQUINOLINES. A GENERAL APPROACH

Francisco Palacios*, Domitila Aparicio, Gloria Rubiales, Ana M. Ochoa de Retana and Eduardo Martinez de Marigorta.

Departamento de Química Orgánica, Facultad de Farmacia, Universidad del País Vasco, Apartado 450, 01080 Vitoria, SPAIN.

Abstract: *The synthesis, physical and biological properties of 4-aminoquinolines are surveyed.*

1. Introduction.

Quinoline was first isolated from coal tar in 1834. Shortly after the isolation of quinoline it was also recognised as a pyrolytic degradation product of *cinchonamine*, an alkaloid closely related to *quinine*, from which the name quinoline is derived; the word *quinine* **1**, in turn, derives from *quina*, a Spanish version of a local South American name for the bark of quinine-containing *cinchona* species. Several synthetic drugs are based on the quinoline nucleus.

Quinoline ring systems represent an important class of compounds[1] and have attracted a great deal of attention in recent years because they constitute the backbone of a wide group of biologically active products such as alkaloids[1,2] **2** and azasteroids[3] (Figure 1).

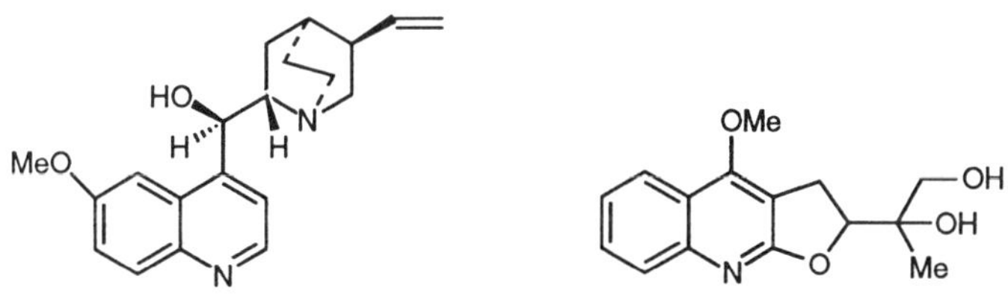

1
Quinine

2
Dubinidine

Figure 1

Quinoline compounds also provided the first photographic film sensitizer: the cyanine dye[4] *ethyl red* **3** extended photography from the blue into the green and then in 1904, with *pinacyanol* **4**, into the red (Figure 2). Subsequently, thousands of sensitizing dyes were made and tested and then quinoline-based dyes were replaced by other systems.

3
Ethyl red

4
Pinacyanol

Figure 2

New properties can appear in quinoline derivatives as consequence of the introduction of functional groups in the aromatic ring of this family of compounds. In this context, the presence of an amino group as substituent in quinolines has been recently demonstrated convincingly since such compounds have interesting pharmacological properties and are widely used in medicinal chemistry. The 2-aminoquinoline *quipazine* [5] **5**, has been reported to exhibit antidepressant activity (Figure 3). Likewise, a series of substrates with a di-, tri- or tetrapeptide attached to 3-aminoquinoline or 6-aminoquinoline have been prepared[6] for use in the quantitative detection of proteinase activity in individual cells and with application in the study of inflammatory diseases.

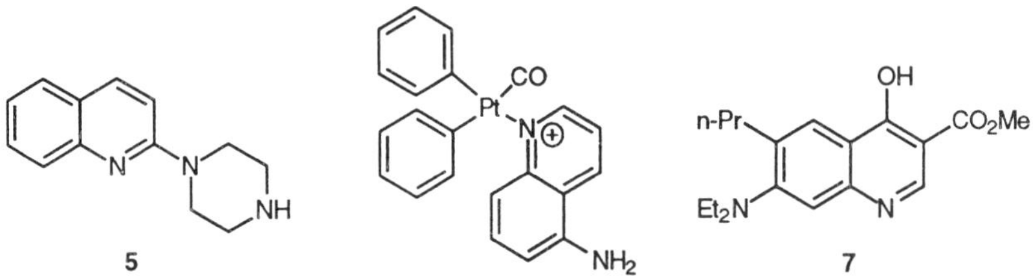

5
Quipazine

6

7
Amquinate

Figure 3

Quinoline derivatives containing an amino group in position 5 have been used as ligands in organometallic chemistry for the formation of new platinum (II) complexes **6** with luminiscence properties for biological applications[7]; likewise, carbamates derived from 6-aminoquinoline have been used for derivatization of amino acids, peptides, and proteins[8], while a 7-aminoquinoline derivative, such as *amquinate*[9] **7** has shown antimalarial activity.

Important chemotherapeutic agents such as the antimalarial drugs *pamaquine* [10], 8-(4-diethyl-amino-1-methylbutylamino)-6-methoxyquinoline **8**, and *primaquine* [11] **9**, are 8-aminoquinoline derivatives and have played a very important role for the radical cure of *plasmodium vivax* malaria. Recent efforts to develop new drugs to replace *primaquine* have focused on the search for new 8-aminoquinolines derivatives.[12,13] Furthermore, some 8-aminoquinolines like compound **10** (WR 6026) has also shown activity against visceral leishmaniasis in animal test systems[14] (Figure 4), while three 8-aminoquinolines, such as WR 6026, WR 238605 and WR 242511 were extremely effective for treatment and prophylaxis of *pneumocystis pneumonia* in rat models.[15,16] Anyway, although aminoquinolines have a broad range of applications, the purpose of this review is to focus on the synthesis, reactivity and the possible uses of 4-aminoquinoline with special emphasis on the biological activities of this family of compounds and on their use in medicinal chemistry.

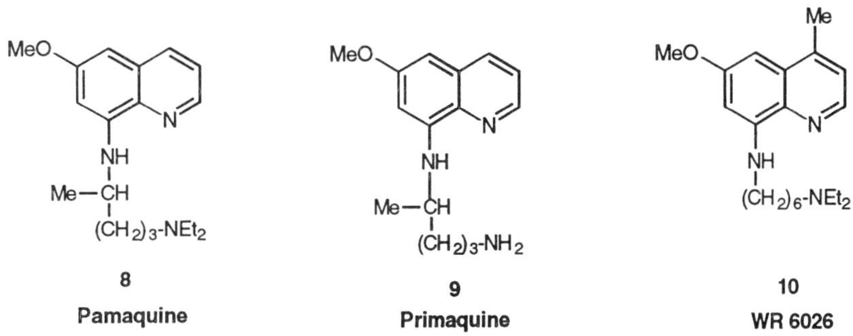

Figure 4

2. Synthesis of 4-aminoquinolines.

Retrosynthetically, the preparation of 4-aminoquinolines **11** can be envisaged by formation of the six membered nitrogen ring containing the amino group of the quinoline (Figure 5, routes a_1-a_4) or by functionalization and introduction of the amine group in the quinoline skeleton (Figure 5, route b).

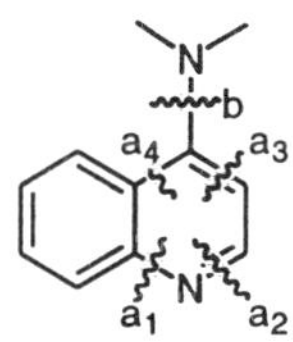

11

Figure 5

2.1. Formation of the six membered nitrogen ring carrying the amino group.

In general, the synthesis described in this section may in some aspect can show analogies to the Skraup[17] and Friedländer[18] synthesis of quinolines.[1] Four routes may be considered for the preparation of 4-aminoquinolines **11** through this approach involving heterocyclization processes for the construction of the six membered nitrogen ring (Scheme 1). Routes a_1 and a_2 involve the heterocyclic ring closure with formation of a carbon-nitrogen bond and, in both cases the amino group in the 4 position of the goal heterocycle **11** comes from the starting material. Alternatively, 4-aminoquinolines can be prepared by tandem reactions that involve both the construction of the quinoline ring and simultaneously the introduction of the amino group (routes a_3 and a_4). In both cases, the key step is the construction of the nitrogen six membered ring by means of a carbon-carbon bond formation.

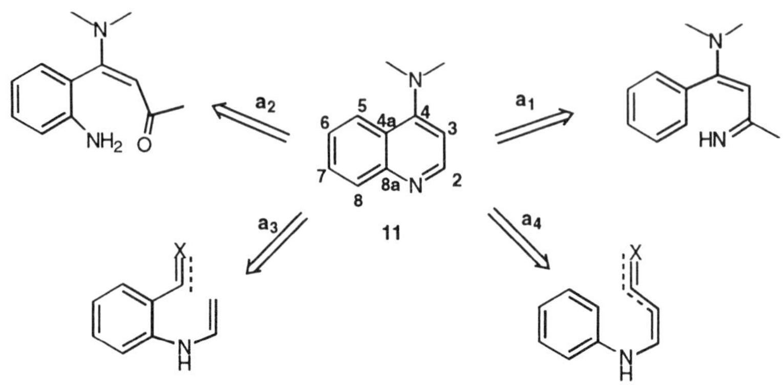

Scheme 1

2.1.1. Heterocyclization processes with formation of a carbon-nitrogen bond.

2.1.1.1. Nitrogen-aromatic carbon bond (N_1-C_{8a}, route a_1).

This strategy involves that the synthesis can be achieved through the reduction of 4-amino-3-alkenimines **12** (Scheme 2) with alkali metals[19] in the absence of protic solvents leading to the formation of a mixture of 4-aminoquinolines **13** and carbonyl **14** (X=O) or imine derivatives **14** (X=NE). The course of the process is highly affected by the nature of the alkali metal; the use of sodium or potassium in THF at room temperature afforded approximately an equimolecular mixture of acyclic **14** and cyclic compounds **13**, whereas the use of an excess of power lithium[20] gave saturated imines or ketones **14**. Formation of iminyl radicals followed by cyclization and aromatization could explain the formation of compounds **13** in accordance with other reported heterocyclization processes *via* imidyl radicals.[21,22]

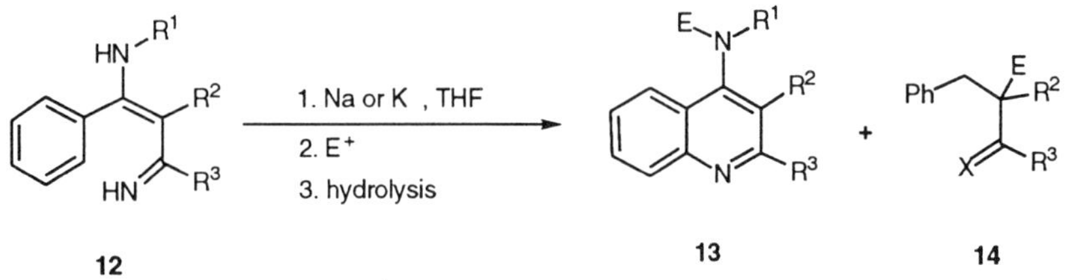

Scheme 2

2.1.1.2. Nitrogen-Carbon bond (N_1-C_2, route a_2).

Palladium catalysts have been used for carbonylative coupling of 2-aminophenylacetylenes **15** (Scheme 3) and aryl iodides **16** in dialkylamines to give 4-aminoquinolines **19**.[23] The initial carbonylative coupling of the acetylene **15** and the halide **16** gives the acetylenic ketone **17**.[24] Conjugate addition of dialkylamine to **17** produces the enaminoketone **18**,[25,26] which then undergoes intramolecular condensation of ketone and arylamine moieties affording the aminoquinoline **19**, in a similar way to that recently observed in the synthesis of pyridines.[27]

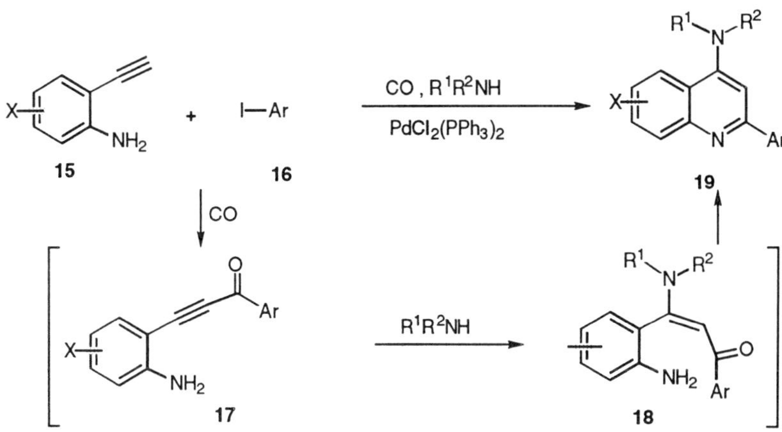

Scheme 3

Likewise, 2-alkyl (aryl) 4-aminoquinolines **22** (Scheme 4) are prepared[28,29] from functionalized isoxazoles **21**. These heterocycles can be obtained by 1,3-dipolar cycloaddition of nitrile oxides generated *in situ* from 2-nitrobenzaldoximes **20** with acetylenes. Subsequent complete reductive cleveage of the intermediate isoxazole **21** by catalytic reduction over Raney-Ni in acetic acid led to the formation of quinolines **22**.

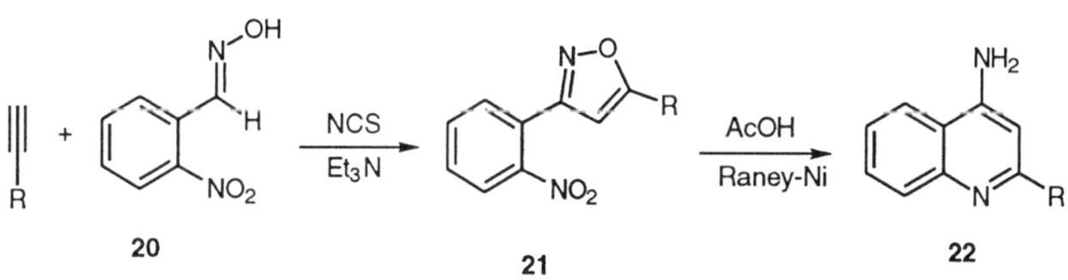

Scheme 4

2.1.2. Heterocyclization processes involving carbon-carbon bond formation.

2.1.2.1. Carbon-Carbon bond (C_3-C_4, route a_3).

Although numerous processes leading to the formation of the quinoline ring system are available, one of the most important is the reaction of *orto*-functionalized anilines **23** with a reagent that provides enamine intermediates **24** with two carbon units to complete the quinoline ring (Scheme 5). In this strategy, the key step seems to be the carbon-carbon bond formation by nucleophilic attack of enamine carbon of intermediates **24** to the electrophilic carbonyl (X = O) or nitrile (X = N) group bonded to the aromatic ring. In this context it is noteworthy that enamines are very efficient starting materials for carbon-carbon bond formation.[30-33] Enamines **24** derived from nitriles (X = N) or carbonyl (X = O) compounds (Scheme 5) are often used for this goal and these functionalized enamines **24** are easily obtained, not only by simple addition of amines to acetylenic or allenic compounds (Scheme 5, route 1), but also by condensation of amines with enolizable carbonyl compounds (Scheme 5, route 2).

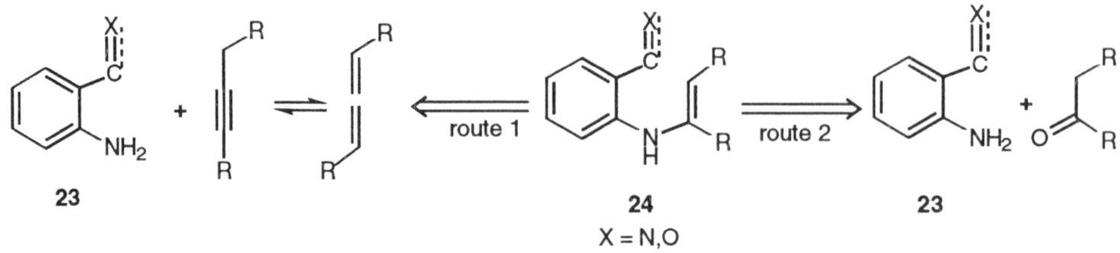

Scheme 5

In this context, an elegant method of synthesis of 4-aminoquinolines has been described by Schweizer *et al.* [34,35] by using β-enamines derived from phosphonium salts **27a** (P =$P^{+}Ph_3Br^{-}$). The synthesis of these enamines was very easily accomplished by simple addition of o-aminobenzonitrile **25** to commercialy available propargyltriphenylphosphonium bromide, although authors suggest that by the reaction conditions, the nucleophilic addition of amine **25** to allene **26a** (P=$P^{+}Ph_3Br^{-}$) can explain the formation of enamine **27a**. Intramolecular nucleophilic attack to the carbon-nitrogen triple bond[36-40] of phosphorane generated *in situ* from β-enamino phosphonium salt **27a** led to the formation of phosphazene **28**,[35] whose hydrolysis gave aminoquinoline **29a** (Scheme 6). Similarly,[41] this strategy can be used for the synthesis of 4-aminoquinolines containing a phosphoryl group in the 3 position of the nitrogen ring **29b** when allenes derived from phosphine oxide **26b** are used. Furthermore, phosphazenes **28** initially suggested[35] for the formation of quinolines **29a**, have been trapped *in situ* by aza-Wittig reaction of these phosphazenes **28** with carbonyl compounds[41] to give quinolines **30** in a similar way to that previously reported[42-44] (Scheme 6).

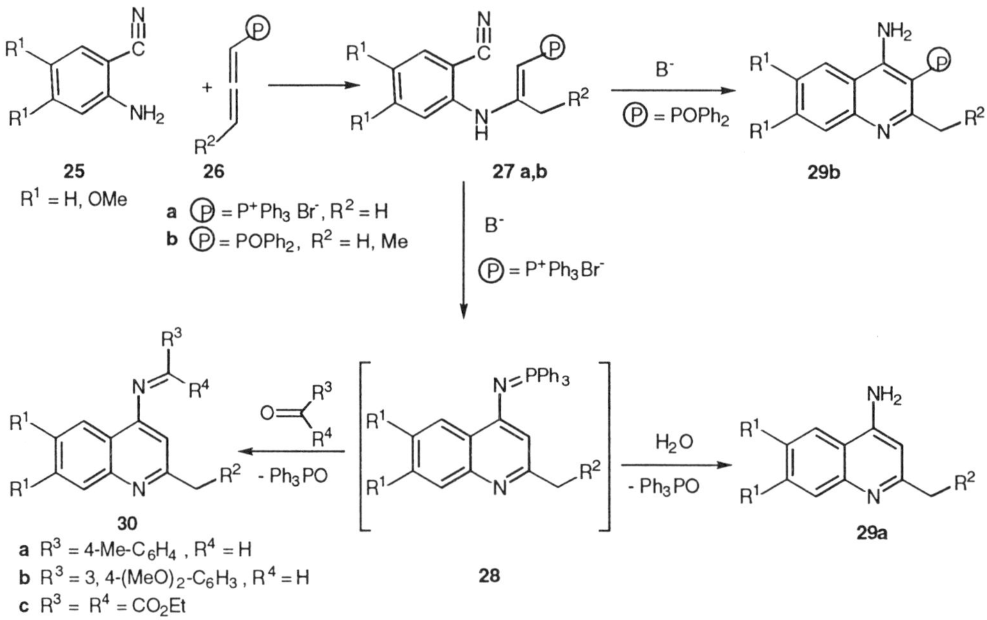

Scheme 6

A second alternative for the preparation of enamines **24** precursors of 4-aminoquinolines involves the reaction of o-aminobenzonitrile **25** with carbonyl compounds (Scheme 5, route 2). Several examples of preparation of enamines **32** (Scheme 7) by reaction of substituted aniline **25** and functionalized carbonyl compounds **31a-c** such as acetylacetone[45-47] **31a** (R^1=Me, R^2=COMe), simple ketones[48] **31b** (R^1= aryl or heteroaryl, R^2=H) and diethylacetone dicarboxylate[49] **31c** (R^1= CH_2CO_2Et, R^2=CO_2Et) have been reported. Nitrile enamines **32a-c** underwent intramolecular cyclization in the presence, not only of bases such as sodium methoxyde[45] or organolithium derivatives[48] (LDA), but also with tin (IV) chloride.[46, 47, 49] The Lewis acid aluminium chloride is often used in heterocyclization processes of functionalized enamines.[50] However, with its use in the cyclization of enamines **32a** (R^1=Me, R^2=COMe), aminoquinoline **33d** (R^1=Me, R^2=H) with loss of the acetyl group[45] was obtained instead of the expected quinoline **33a** (R^1=Me, R^2=COMe).

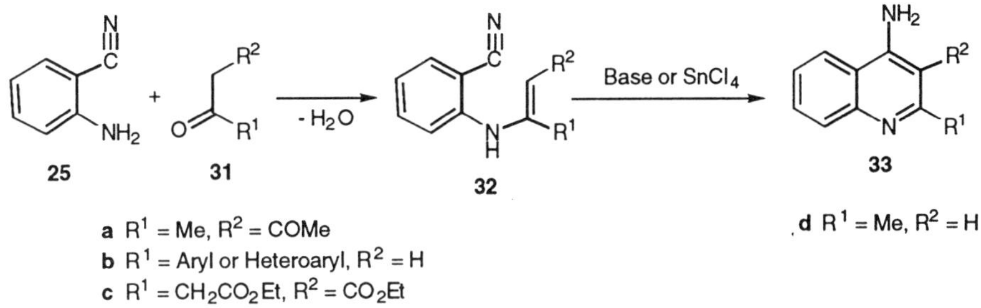

Scheme 7

Cyclization processes with formation of 4-aminoquinoline were not restricted to the use of enamine derived from nitriles **24** (Scheme 5), since aminoquinolines can also be obtained from enamines containing an oxime moiety **34**.[51] Thus, the reaction of o-aminobenzophenone oxime **34** (Scheme 8) with triphenylphosphine-carbon tetrachloride combination led to the formation of 4-aminoquinolines **35**.

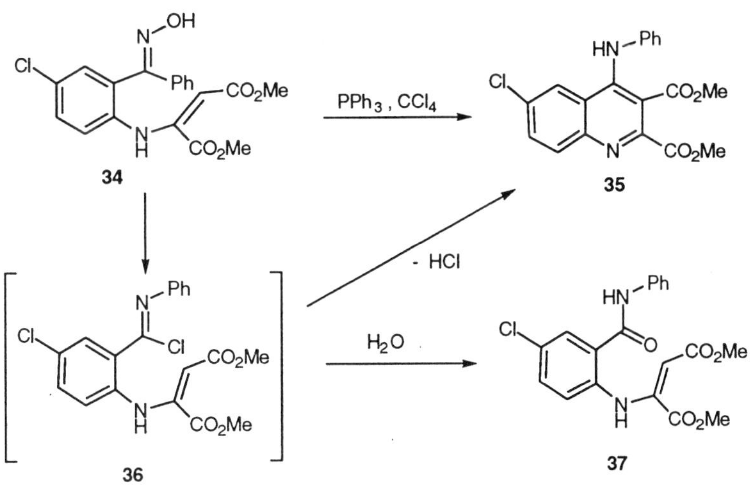

Scheme 8

The reaction of Ph_3P/CCl_4 with the oxime **34** would be expected to give the iminocarbonyl chloride **36** through a Beckman rearrangement[52,53] with elimination of triphenylphosphine oxide, in a similar way to the recently described Beckman rearrangement of oximes derived from diphosphonates.[54] Intramolecular cycloaddition with loss of HCl can afforded 4-aminoquinolines **35**. Alternatively, the iminyl choride can be trapped since it might be hydrolyzed to the amide **37** during the aqueous work-up.

A convenient synthesis of 2,4-diaminoquinolines derivatives[55] **41** based on an intramolecular cyclization of functionalized imines **39** is shown in scheme 9. In this case, ring closure is initiated *via* the central alkyl residue in the amidine moiety of compounds **39**. Reaction of anthranilonitrile **38** with triethylorthoacetate gives, in quantitative yield, the imidate **40** which, on treatment with a secondary amine, furnishes the key acetamidinonitriles **39**. Alternatively, intermediates **39** can be obtained by reaction of **38** with an imidinium chloride generated *in situ* by reaction of acetamides in the presence of phosphorus oxychloride. Cyclization of the amidine **39** with base gave 2,4-diaminoquinoline derivative **41** as the sole product. In cases where the amino residue ($-NR^1R^2$) containes a benzylic proton, base-induced cyclization did not proceed. However, ring closure could be affected by treatment of the amidine **39** with zinc chloride,[45,56,57] presumably *via* the enamine tautomer. Compounds **41** prepared by this procedure[58] has been evaluated for α-adrenoceptor affinity and antihypertensive activity. Similarly, ethyl antranilocarboxylate has been use for the synthesis of 2,4-diaminoquinolines[59] when aminocarboxylates have been treated with acetyl chloride followed by heating of the amide at 235°C in HPMT (hexametylphosphoric triamide).

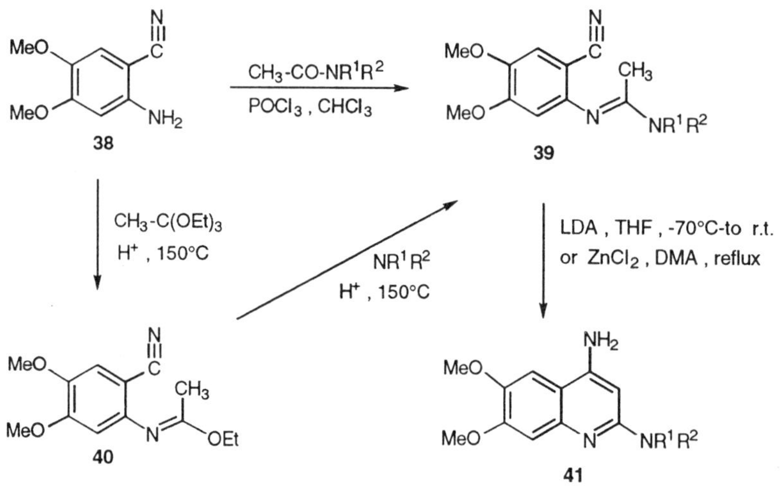

Scheme 9

Several 4-aminoquinolines **45** (Scheme 10) active against HIV-1 were prepared by a base-mediated cyclization of ketimine **42** in the presence of a lithium alkylamide or dialkylamide reagent.[60] Mechanistic studies[61-63] of these processes are consistent with the intermediates **43**. Subsequent fast addition of an amide anion followed by elimination of fluoride gives **44** which may undergo electrocyclization to afford aminoquinolines **45**. This strategy has been used to prepare 4-aminoquinolines useful as intercalators of DNA and retrovirus inhibition.[64,65]

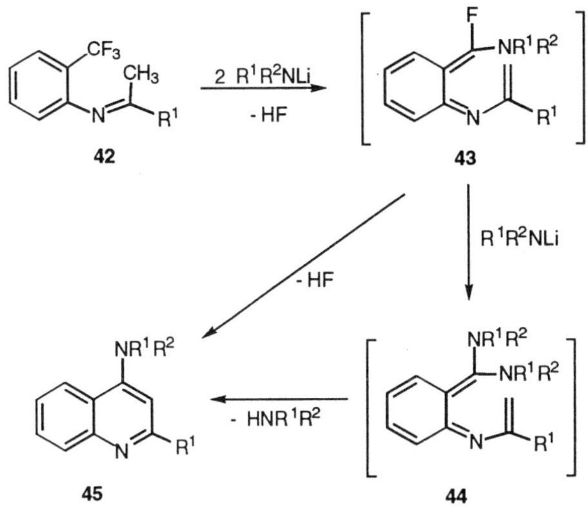

Scheme 10

Heterocycles such as triazolines **49** can also be used for the synthesis of 2,4-diaminoquinolines **51** (Scheme 11). Compounds **49** have been prepared by refluxing arylacetaldehydes **46** with the appropriate secondary amines **47** and azidonitriles **48**.[66] Thermal nitrogen elimination from triazolines **49** gave arylacetamidines **50**. These amidino intermediates **50** readily underwent base-catalysed ring closure to afford compounds **51**.

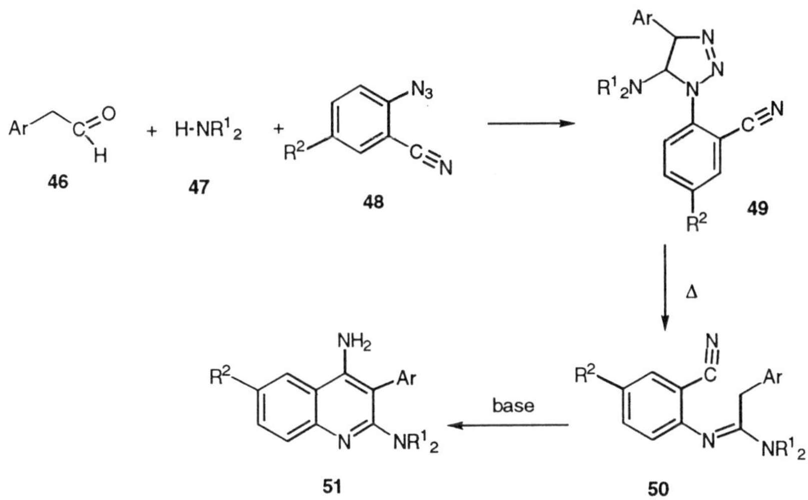

Scheme 11

2.1.2.2. Carbon-aromatic carbon bond (C_4-C_{4a}, route a_4).

A very large group of syntheses of quinolines, in which C-C bond is formed, are those in which a side chain electrophile attacks the benzene ring (Scheme 1, route a_4). These syntheses include the conventional methods of preparation of quinolines.[1] From a formal point of view these processes involve the reaction of an aromatic amine with other reagents that provide the three-carbon unit required to complete the quinoline ring.

This strategy has been reported for the synthesis of 4-aminoquinolines **54** (Scheme 12) when intramolecular cyclization of β-aryl enaminonitriles **53** has been accomplished in the presence of aluminium chloride[67] or polyphosphoric ester (PPE).[68] Likewise, a modification of the Knörr or the Conrad-Limpach synthesis of quinolines has been applied for the preparation of 2,4-diaminoquinolines **56** by heating amides **55** in HMPT as solvent[69] (Scheme 12). HMPT is know to replace potential hydroxy-groups for dimethylamino-groups[70] and carboxamides **55** may be prepared when aniline **52** is heated in refluxing HMPT with diethylmalonate. Furthermore, the same aminoquinolines **56** can be obtained from dichlorodiiminium chloride **57** by reaction, in high dilution, with arylamines **52**.[71]

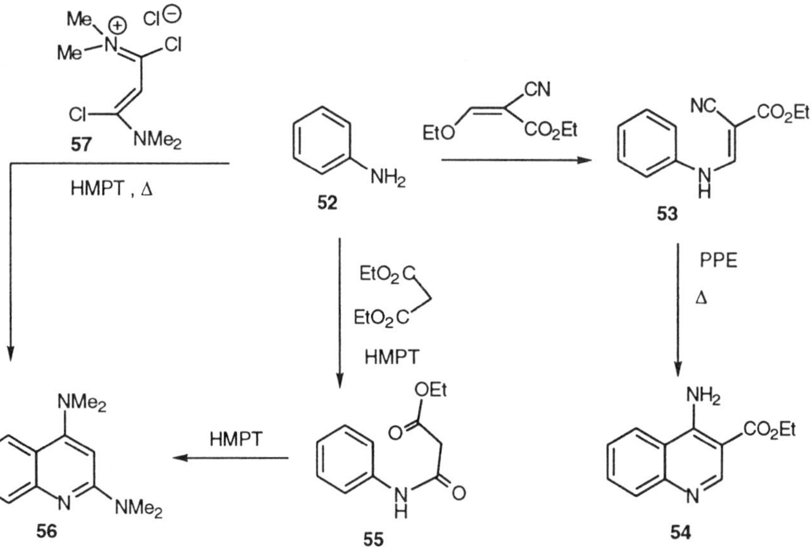

Scheme 12

Fluoro substituted 4-aminoquinolines **61** were obtained when perfluoro-2-methylpent-2-ene **58** (Scheme 13) was treated with aromatic amines **59**.[72] This reaction takes place particularly readily with aromatic amines in which the aromatic nucleus has electron-donating substituents. An isolable ketenimine **60** is formed in the reaction and electrocyclic ring closure of this heterocumulene gives the aminoquinoline **61**.

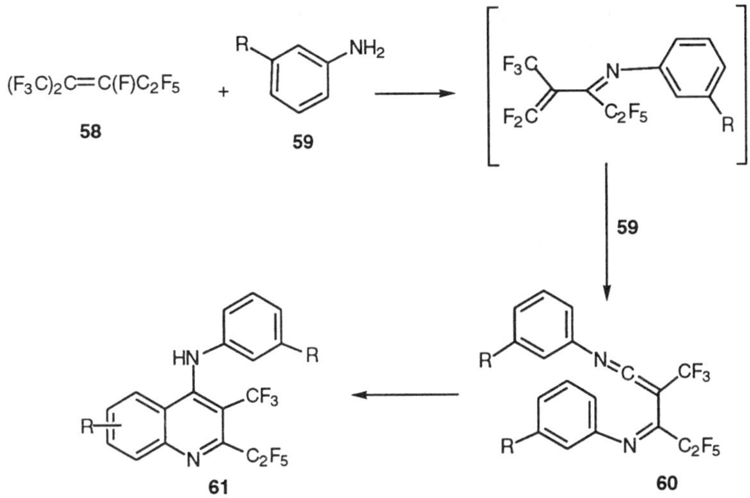

Scheme 13

Cycloaddition reactions of heterocumulenes such as N-p-tolylvinylmethylketenimine[73] with an electron-rich dienophile such as ynamines led to a method for the preparation of 4-aminoquinolines. Likewise, the reaction of heterocumulenes **62** (Scheme 14) with n-alkylaldehydes **63** led to the formation of aminoquinolines **67**. When diphenylcarbodiimide **62** (X=N-Ph) was used, the reaction with a second molecule of heterocumulene with elimination of aromatic amine gave[74] not only the 4-aminoquinolines **66**, but also pirimidines **65** (X=N-Ph). In a similar way, when phenylisocyanate **62** (X=O) was used, uraciles **65** (X=O) were obtained.[75]

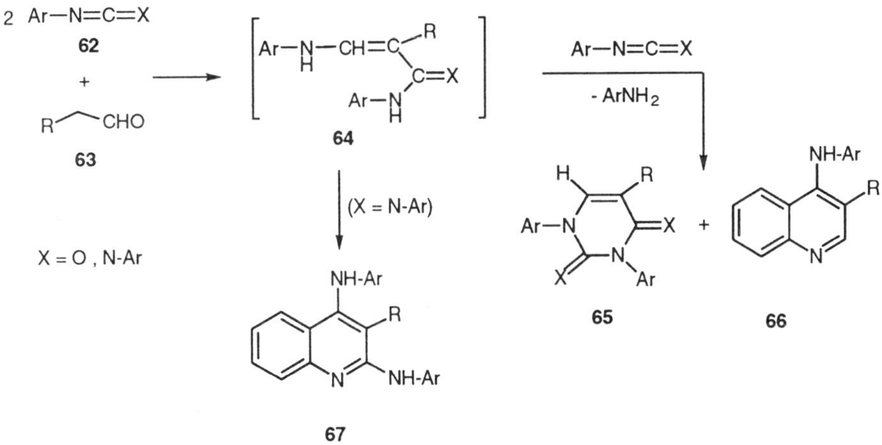

Scheme 14

Heterocumulenes **69** as intermediates in the preparation of aminoquinolines **70** were also reported when the imidoylphosphoranes **68** (Scheme 15) reacted with isocyanates. The C-keteneimine **69** then undergoes 6π-electrocyclization; subsequently aromatization gives 4-aminoquinolines **70** in good yields.[76]

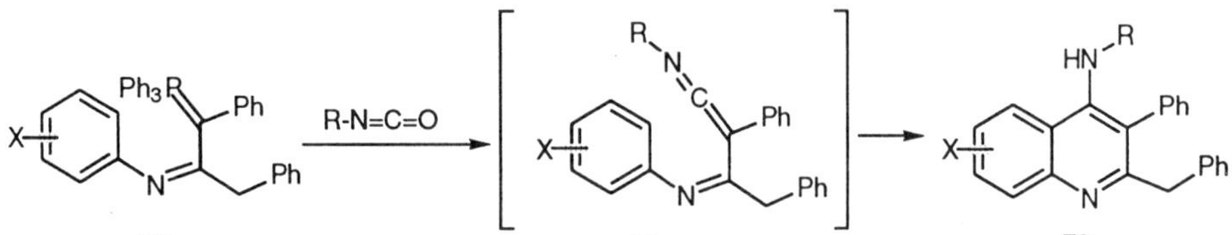

Scheme 15

Recently, an easy and efficient synthesis, of 4-aminoquinolines with a phosphine oxide, phosphine sulphide and phosphonate group **73** (Scheme 16) in the 3-position has been described.[77] Treatment of functionalized enamines **71** with phosphorus oxychloride in the presence of triethylamine led to aminoquinolines **73**. The formation of substituted quinolines **73** can be assumed to proceed *via* 6π-electrocyclization of conjugated imidoylketene imines **72** formed in a similar way to that previously reported.[74-76]

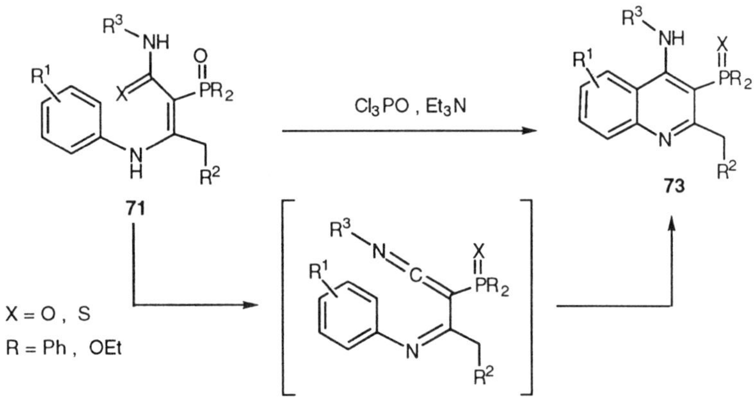

Scheme 16

Finally, a novel and elegant two step synthesis of the reversible (H^+/K^+)ATPase inhibitor SK&F 96067 **76** (Scheme 17) has been developed and evidence for the intermediacy of an imidoylketene imine has been obtained.[78] Reaction of functionalized amide **74** with Ph_3PCl_2 leads to the imidoylketene imine **75** which undergoes a thermally induced electrocyclic reaction to furnish **76**. In addition, amidine **77** has been obtained in this process when the reaction has been carried out in the presence of aniline.

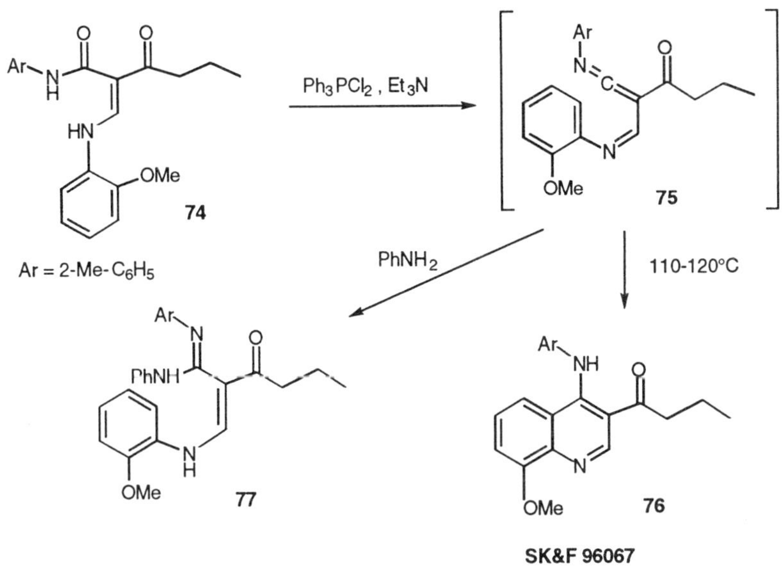

Scheme 17

2.2. Functionalization of quinoline derivatives.

2.2.1. Direct amination of the quinoline ring. (The Chichibabin reaction).

Sodium amide reacts rapidly and completely with quinoline, to give dihydro-adducts **78** and **79** with initial amide attack at C-2 (main) and C-4 (minor). The quinoline 2-adduct **78** rearranges to the more stable 4-aminated adduct at higher temperatures.[79] Oxidative trapping of the quinoline adducts provides 2- or 4-aminoquinolines[80] **80** or **81** respectively (Scheme 18).

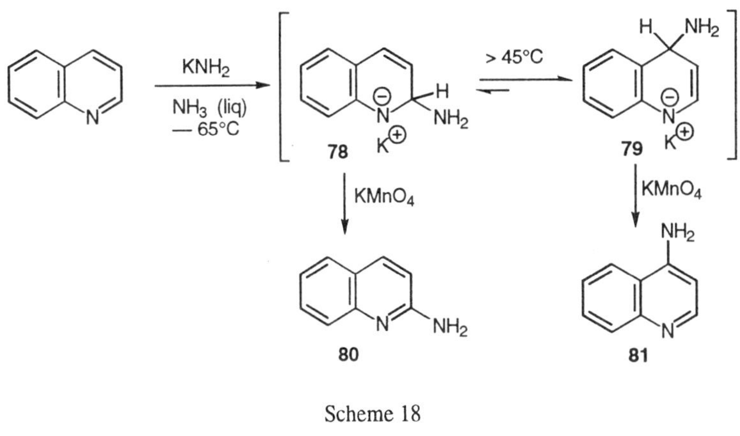

Scheme 18

2.2.2. Displacement of substituted quinolines containing good leaving groups towards nitrogen nucleophiles.

This strategy is that most often used for the synthesis of 4-aminoquinolines. Displacement of the reactive leaving group at the 4 position of quinoline **82** by means of aliphatic and aromatic amines can offer an excellent method for preparation of 4-aminoquinolines **83** (Scheme 19). Thus, 4-haloquinolines have the same susceptibility as γ-halopyridines[81] to react with amines.

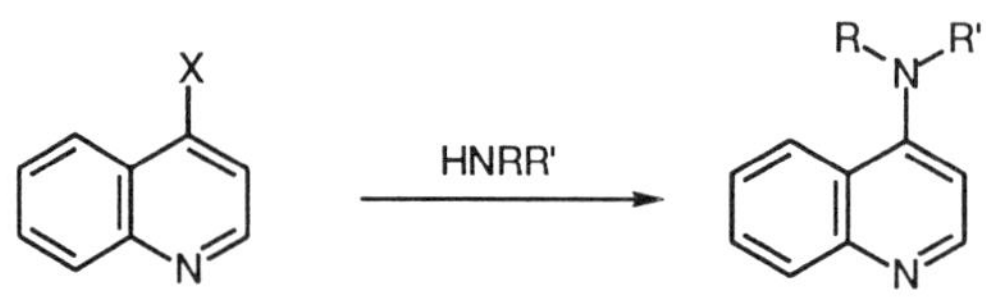

82 **83**

X = Cl , F , OR

Scheme 19

Fluoroquinolines have been used for a straightforward synthesis of quinoline derivatives **86** triplex DNA intercalators[82] by the reaction of 1,3-diamine derivatives **84** with 4-fluoro-2-(2-naphtyl)quinoline **85**, followed by deprotection of terminal nitrogen atom and formation of compounds **86**, (Scheme 20). Similarly, displacement of fluorine in heptafluoroquinoline towards ammonia in acetone gave 2-amino and 4-aminoquinolines in the ratio 57:43.[83]

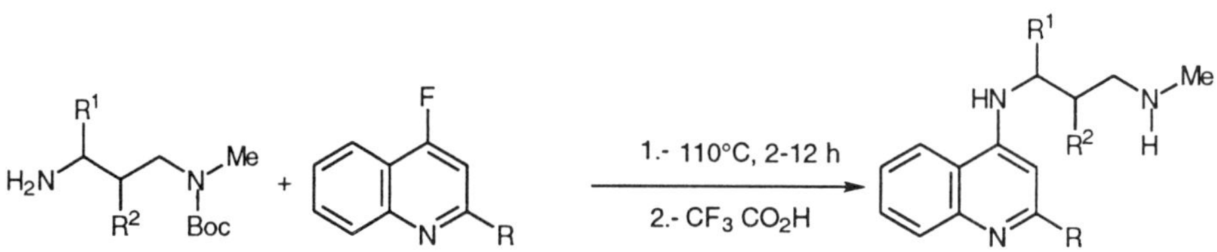

84 **85** **86**

Scheme 20

4-Chloroquinolines are also suitable starting materials and are very often used for the synthesis of 4-aminoquinolines. There are numerous examples reported in the literature in which 4- and 4,7-dichloroquinolines are the starting material for the synthesis of a great number of compounds with important biological activities, not only as antimalarial agents, but also with other pharmacological applications in medicinal chemistry. By this way the total synthesis of *amquisine*[84] **94** and *chloroquine*[85] **93** from chloroquinolines has been carried out (Scheme 21). The synthesis of aminoquinolines **93** and **94** started with one of the standard sequences for the preparation of 4-hydroxyquinolines, i.e., condensation of amines **87** with carbonyl compounds with formation of Schiff base **88**. Thermal cyclization gave quinolinones **89**; these heterocycles then spontaneously tautomerized to the enol form **90a** (R^3=CO_2Et). Saponification and decarboxilation afforded the quinolol **90b** (R^3=H). The reaction of these heterocycles with phosphorus oxychloride led to the corresponding chloride derivatives **91, 92**, and treatment of 4-chloroquinolines **91** and

92 with ammonia or amine derivatives gave *amquisisin* **93** and *chloroquine* **94** respectively. Similarly, this strategy has been used for the synthesis of a wide group of biologically active 4-aminoquinolines.[86-112]

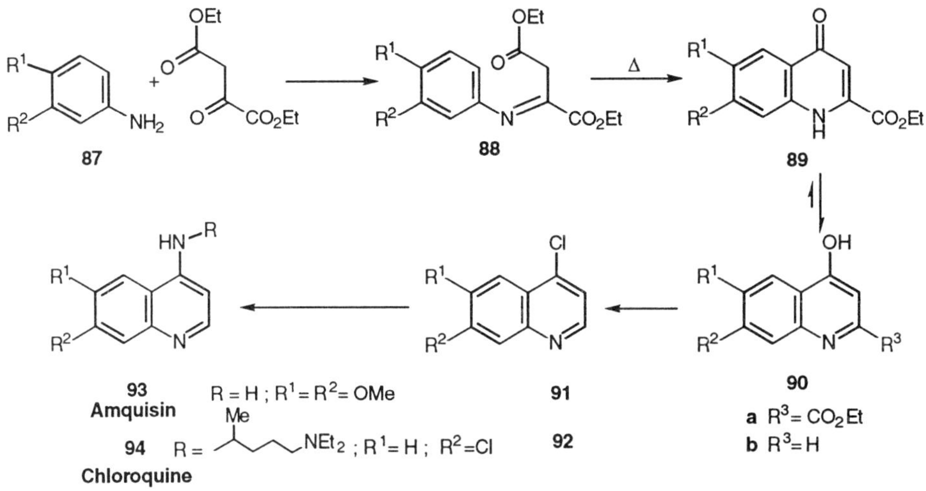

Scheme 21

Nucleophilic substitution of 4-chloroquinolines containing bulky substituents in the position 3 of the quinoline ring, did not give 4-aminoquinolines. Therefore, the use of alternative leaving groups such as phenoxy groups in the position 4 of quinoline ring instead of the halogen groups previously described have been explored.[113] The key intermediate **97** has been prepared by means of the replacement of the chlorine atom in quinoline **95** by the phenoxy group, followed by lithiation and acylation with pivalic anhydride of substituted phenoxyquinoline **96**. Subsequent displacement of the 4-phenoxy group with o-toluidine gave the aminoquinoline **98** containing a bulky carbonyl substituent in the 3 position (Scheme 22).

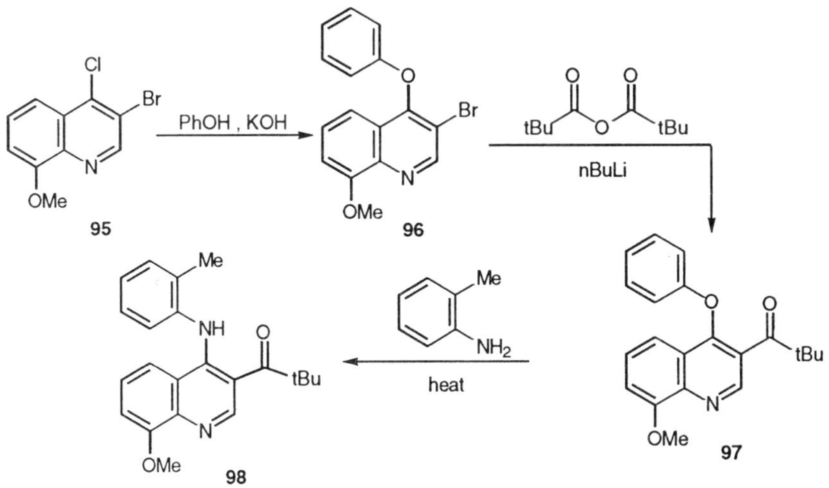

Scheme 22

Likewise, 4,6-diaminoquinolines have been obtained by substitution of 4-methoxyquinoline with amines.[114] The compounds were prepared by the procedure described in scheme 23 in which the key step is the displacement of methoxyquinoline towards nitrogen nucleophiles. The condensation of amine with ethyl acetoacetate gave functionalized enaminoester **99**. Heterocyclization of the enamine **99** followed by methylation and treatment of the methoxyquinoline **101** with ammoniun acetate led to 4,6-dimethylaminoquinolines **102**, after the deprotection of the aromatic amino group.

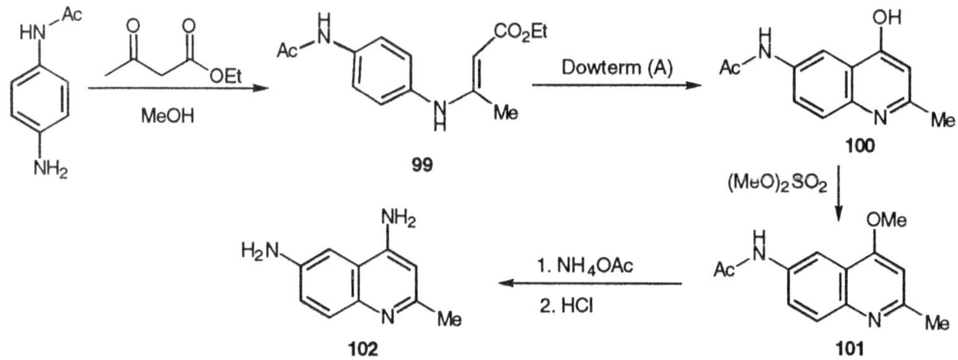

Scheme 23

Similar methoxyquinoline intermediates were used for the preparation of sulfur containing aminoquinoline disulfide derivatives.[115] Aminotiols, compounds derived from disulfides were introduced on the heterocycle of *chloroquine* (Scheme 24). In the first step the 4,7-dicloroquinoline was added to a mixture of sodium hydroxide and phenol to give phenoxyquinolines **103**. Substitution of the 4-phenoxyquinoline with amino disulfides **104** and **105** followed by reduction of disulfides with tributylphosphine gave respectively thiols **106** and **107**.

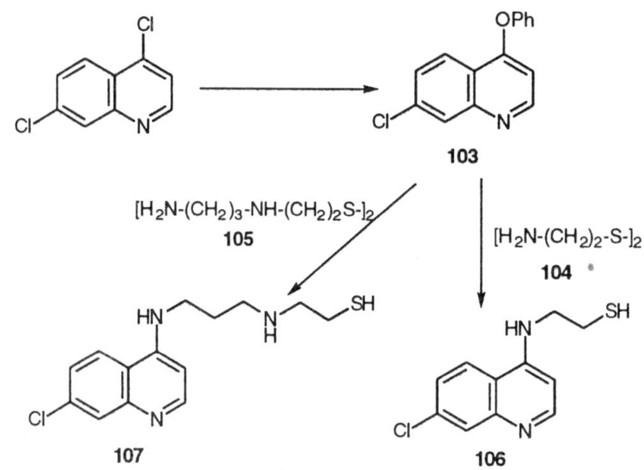

Scheme 24

An alternative strategy that attemps to palliate the problems of the resistance[116] of the malaria parasite to *chloroquine* and its derivatives, searching for a greater duration in pharmacological activity[117] and lower toxicity, has given as a result a vast number of bisquinolines, obtained from 4,7-dichloroquinoline. These bisquinolines **109** (Scheme 25) have been obtained through a displacement reaction of 4,7-dichloroquinoline, with alkanediamines **108** and triethylamine in a 2:1:2 ratio using N-methylpyrrolidinone as solvent.[118] Similar strategies[119,120] have also been described for the synthesis of other bisquinolines.

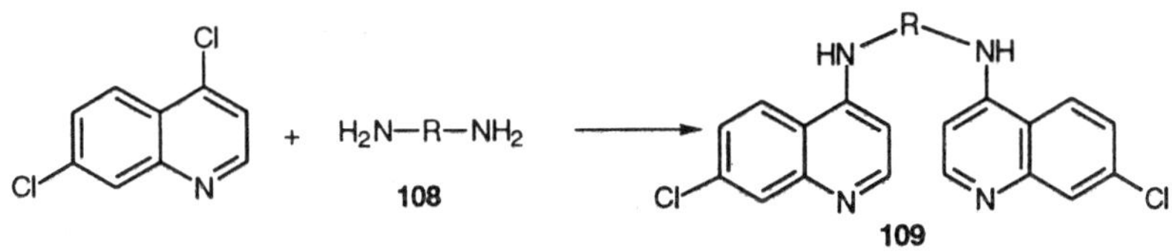

Scheme 25

3. Physical properties of 4-aminoquinolines.

3.1. Hydrophobicity.

Hydrophobicity, along with solubility and dissociation constants are important modulators of biological activity and therefore closely related to anaesthesiology, pharmacokinetics, drug design, toxicology, bioaccumulation and enzyme binding, among other areas of medicinal chemistry. Hydrophobicity may be defined as the tendency of organic molecules to partition away from water into some less polar medium, and is traditionally expressed by the logarithm of the partition coefficient of the molecule (P), determined as the concentration ratio C_2/C_1 of a single species between a non-polar solvent and water at equilibrium.[121]

A comparative study of the hydrophobicity of indoloquinoline **110** and two non-cyclized analogues such as 4-aminoquinolines **111** and **112** has been carried out. The relative hydrophobicity of these compounds, according to the log P measured by the shake-flask method and UV spectroscopy, is in the order **111>112>110**.[122]

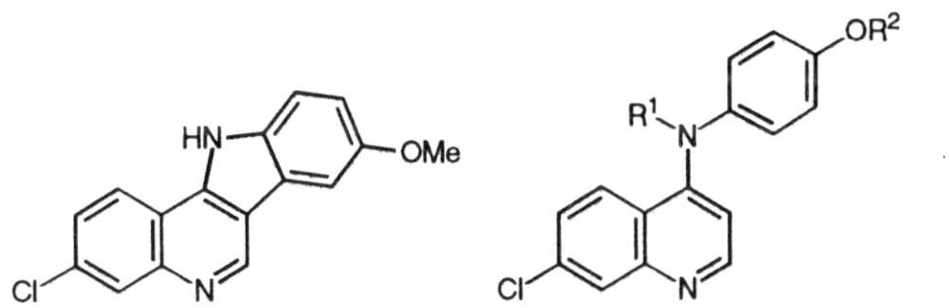

110 pKa = 3.99[122]

111 R^1 = H, R^2 = Me pKa = 7.59[122]
112 R^1 = Me, R^2 = H pKa = 6.65[122]

Figure 6. Some 4-aminoquinoline derivatives and their pKa values.

Thus, even though **110** may be considered as the ring closed analogue of **111**, the former is less hydrophobic than the latter and so **110** has greater aqueous solubility. This reduction of hydrophobicity is much greater than can be accounted for a mere loss of two non polar hydrogen atoms on ring formation, but it can be partly attributed to a reduction in molecular surface area on ring formation.

Hydrophobicity values have also been determined for several 4-amino-7-chloroquinoline derivatives and their corresponding N-oxides, by the shake-flask method and reverse-phase liquid chromatography (RPLC). It was found that the ring nitrogen, when unprotonated, was less polar than the corresponding N+—O— fragment of the N-oxide. However, when protonated, the N+—H fragment was more polar than the N+—O— fragment.[123]

3.2. Acidity-basicity.

It is well-known that 4-aminoquinolines show enhanced basicity (**113a**, Figure 8, pKa 9.17) compared to other aminoquinolines, probably due to the delocalization of the amino lone pair of electrons to the quinoline ring, thus, increasing the electron density on the quinoline nitrogen.[124]

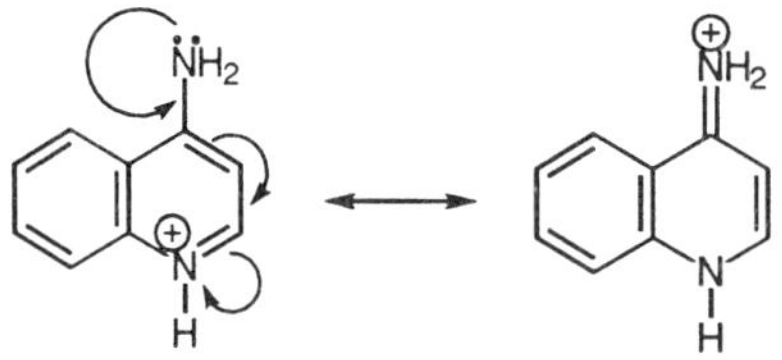

Figure 7

Furthermore, alkyl substitution of the side chain amino function in 4-aminoquinoline causes a decrease in the quinoline ring basicity, as can be see in Figure 8 for 4-methyl- and 4-dimethylaminoquinolines **113b-c**.[125] The reason seems to be the steric inhibition of conjugation caused by the bulky alkyl substituents, an effect that can also be detected in the case of quinolines **111 and 112** (Figure 6), where the more substituted amino nitrogen gives rise to a weaker base.[122]

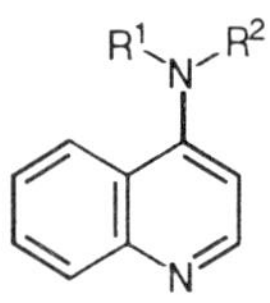

113

a $R^1 = R^2 = H$ pKa = 9.17[124]
b $R^1 = Me, R^2 = H$ pKa = 9.06[125]
c $R^1 = R^2 = Me$ pKa = 8.39[125]

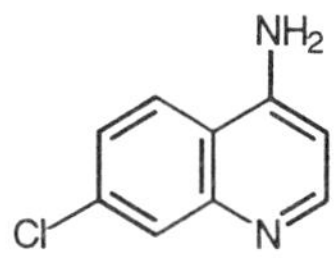

114

pKa = 8.23[126]

Figure 8. pKa values of some 4-aminoquinolines.

Aromatic substitution also causes a fall in pKa and, therefore, in basicity. In fact, compounds **111** and **112** (Figure 6) are less basic than 4-amino-7-chloroquinoline **114** (Figure 8), due to the phenyl susbstituent which competes with the quinoline ring for the amino lone pair of electrons. Surprisingly, however, the pKa of the indoloquinoline **110** (Figure 6) means that this is a much weaker base than **111** and **112**, although all of them have very similar structures.[122] It can be said, then, that the indolo nitrogen, unlike the 4-amino group in **111** or **112** does not affect the basicity of the quinoline nitrogen, which is very much a function solely of the basicity of the quinoline ring. In fact, the pKa of **110** is lower than that of unsubstituted quinoline (pKa 4.9), being closer to that of quinoline derivatives with electron withdrawing groups.[126] Anyway, it should always be kept in mind that pKa values change slightly according to the method used for their determination, and the same compound is often given different values. For instance, compound **111** (Figure 6) has pKa 6.49 according to a potentiometric titration[123] and pKa 7.59 when the solubility method was employed.[122]

On the other side, it is well-known that acid-base reactions and, therefore, pKa values are affected by changes in solvent and temperature. However, reaction heats of 4-amino-2-methylquinoline with methanosulphonic acid from 30 to 200°C in nonaqueous media show very little variation over the temperature range, thus suggesting that extrapolation of data for acid-base reaction near room temperature to higher temperatures will be straightforward.[127] Finally, the rate of the proton transfer of monocationic forms of bases like 4-aminoquinoline have been determined along with excited state pKa* values, using fluorescence titrations.[128]

3.3. Kinetic properties.

Rate constants for the hydrolysis of a series of aromatic nitrogen mustards **115** have been obtained using an HPLC technique. According to these data, the hydrolysis has a general-base-catalysed component and, when occurring on unprotonated mustard, the participation of an aziridinium ion intermediate, instead of a direct S_N2 displacement has been proposed.[129]

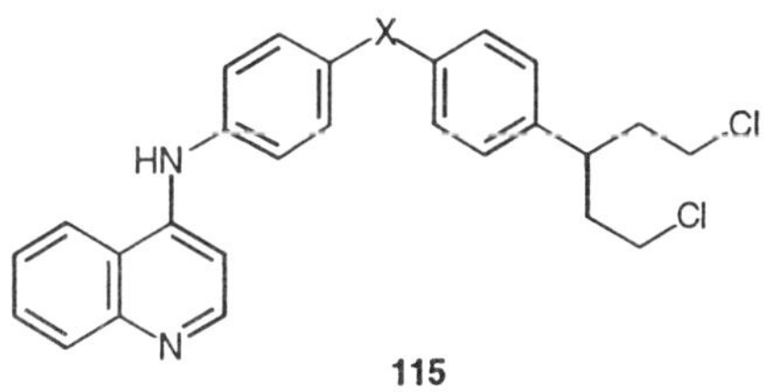

115

X = O, CH_2, CONH, S, CO, CO_2 pKa ≈ 8.4[129]

Figure 9

3.4. Molecular Orbital calculations.

Hydrophobicity along with the energy of the lowest unoccupied molecular orbital (LUMO) have been used in medicinal chemistry to propose a correlation model (QSAR-Quantitative Structure Activity Relationship) between those physical properties and the biological activity of heterocyclic compounds.[130] Therefore, it is of great importance to be able to calculate the energy of the molecular orbitals of bioactive molecules like many 4-aminoquinolines. In fact, such calculations were done some time ago for antimalarial drugs such as *chloroquine* **94** and *primaquine* **9**. In this research Hückel molecular orbital π electronic charge density and energy levels of the HOMO and LUMO were calculated and this made it possible to elucidate the contributions of the substituents to the π electronic properties of the antimalarials and to analyze the electronic aspects of their interaction with DNA.[131] More recently, semiempirical calculations using the MOPAC-AM1 method were carried out for substituted quinolines and optimal positions of strong electron-donor and electron-acceptor substituents determined. Using the same method, properties like intrinsic ground-state dipole moments and polarizabilities were also calculated.[132]

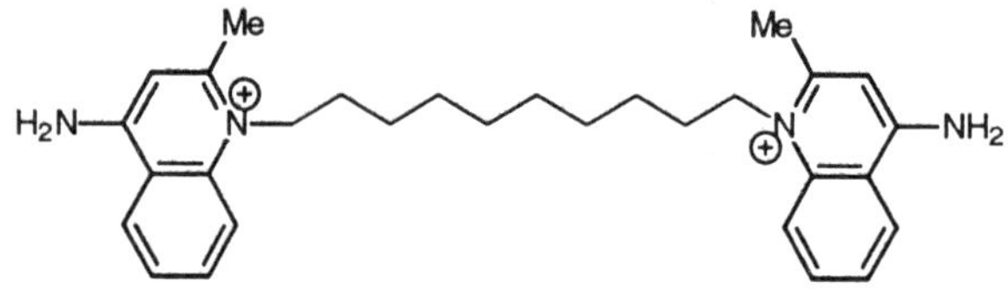

116

Dequalinium

Figure 10

Finally, a very recent report has dealt with the role of the amine group in the 4 position of *dequalinium* **116** (Figure 10), and shown that it contributes substantially to blocking potency in K^+ channels. Furthermore, a QSAR analysis on several analogues of *dequalinium* was performed using electronic indices obtained from AM1 molecular calculations on model compounds [133] The newer PM3 Hamiltonian[134] was not used, as the partial charge on the N-ring calculated with this method was positive, in contrast to results from MNDO, AM1 and *ab initio* (3-21G) calculations which indicated that the ring N-atom carries a negative partial charge. This analysis revealed that the blocking potency correlates with the partial charge in the ring N-atom, the energy of the LUMO and the energy of the HOMO. The correlation for the energy of the LUMO suggests that the higher the energy of the LUMO the more potent is the analogue.[133]

3.5. Spectroscopic properties.

3.5.1. NMR spectroscopy.

^{1}H-NMR Spectra of several 2-methyl-4-aminoquinolines were first reported by Craig and Pearson, trying to explore the tautomeric structure of 4-aminoquinolines.[135] Likewise, the ^{1}H-NMR spectrum of the antimalarial *chloroquine* has been fully assigned in several studies.[136] One of those studies was aimed to establish its spatial conformation in solution by using complexes of lanthanide ions.[136a] Some chemical shifts obtained in recent research are shown in Figure 11,[137] while the use of a long-range correlation experiment (COLOC) allowed the assignment of the four quaternary carbons, C-4, C-7, C-9 and C-10 of the ^{13}C-NMR spectrum of *chloroquine*[137] (Figure 11), some of which had been misassigned in a previous report.[138]

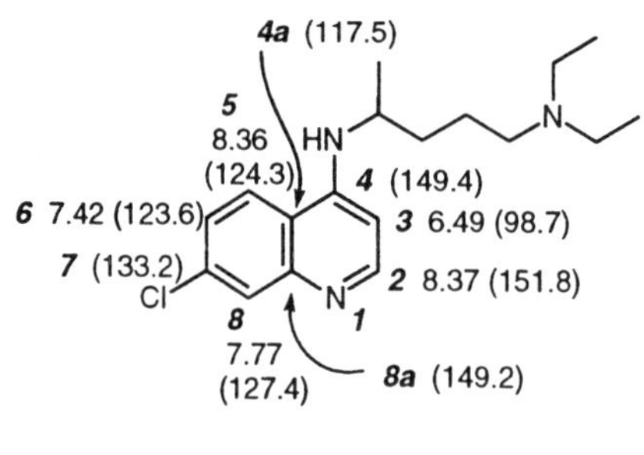

Figure 11. Some ^{1}H- and ^{13}C-NMR (in parenthesis) chemical shifts of the quinoline ring in *chloroquine*.[137]

Complete ^{1}H- and ^{13}C-NMR data for related 7-dechlorinated 4-aminoquinolines have also been reported, making use of COSY 2D, DEPT and direct heteronuclear (^{1}H-^{13}C) chemical shift correlations.[137] It is noteworthy that the ^{13}C-NMR data mentioned above are consistent with those of *hydroxychloroquine* **117** (Figure 12).[139] Likewise, it should be mentioned that ^{19}F- and ^{31}P-NMR have also been used as a diagnostic tool for fluorinated[140] and phosphorylated[77] 4-aminoquinolines, respectively.

3.5.2. Infrared and Ultraviolet spectroscopy.

The typical IR absorption bands of 4-aminoquinolines appear at 3500-3300 cm^{-1}, characteristic of the amino NH bond, 1615-1605 cm^{-1} for CN and CC bonds in the heterocyclic ring, and 1590-1570 cm^{-1} corresponding to the vibration of CC bonds in the phenyl ring.[23,77,141]

The UV absorption spectra of 4-aminoquinolines have also been reported in several publications.[125,141a,c,142] For instance, the positions of the maximum absorpion bands for the simple 4-aminoquinoline are 240, 320 and 330 nm, while the more complex *hydroxychloroquine* **117** shows absorptions at 219, 237, 256, 331 and 343 nm.[142] Protonation of the ring nitrogen causes a bathochromic shift and an increase in intensity of the band at higher wavelength in the 4-aminoquinoline. Furthermore, the introduction of one methyl group in the amine nitrogen leads to a bathochromic shift and an increase in intensity of the bands at higher wavelengths for both protonated and unprotonated compounds. Finally, the introduction of a second methyl group in the amine group causes a bathochromic shift for the ion and a hypsochromic shift for the base in the band at higher wavelength.[125]

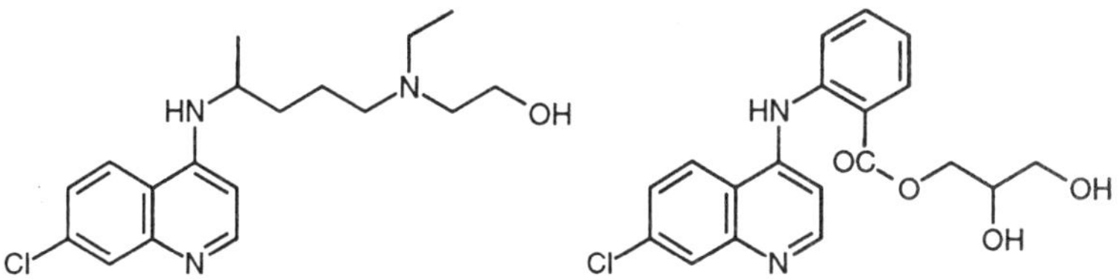

117 **Hydroxychloroquine** **118** **Glaphenine**

Figure 12

The UV-visible absorption of these molecules has been used to develop several colorimetric methods for the determination of *glaphenine* **118** in pharmaceutical formulations.[143] Furthermore, related methods employing a derivative spectrophotometric procedure, rather than a colorimetric one, have also been proposed and have shown better resolution in some instances.[144]

On the other hand, the effect of the substituents, including the amino group, on the energy of the transition of the α- and n-bands for quinolines has been studied, one of the conclusions being that the effect of the substitution on the position of the absorption band of the neutral molecule is lower than for the corresponding cations.[145] Finally, magnetic circular dichroism (MCD) of several aminoquinolines has been reported, in order to search for the effect of the aza substitution in 1- and 2-aminonaphthalenes.[146] This report includes the UV spectra of 2-methyl-4-aminoquinoline, among other related aminoquinolines.

3.5.3. Mass spectrometry.

The major fragment in the mass spectra of aminoquinolines is formed by the loss of HCN from the molecular ion. The loss of the primary amino group, however, is an insignificant fragmentation.[147] The presence of a side chain in the amine group, like in *hydroxychloroquine* **117**, produces peaks corresponding to fragmentation of this chain.[139] On the other hand, isomeric 2- and 4-N-substituted aminoquinolines can be differentiated thanks to characteristic diagnostic fragmentations in their mass spectra.[148]

3.5.4. Fluorescence.

The fluorescence of 4-aminoquinolines and their derivatives has been widely used, for instance, in kinetic studies to measure the rate of proton transfer of protonated bases,[128] or as a detection system in chromatographic methods[149] (see next section).

3.6. Chromatographic properties.

Due to the the wide use of some aminoquinolines as drugs, a number of methods for the purification of these compounds and of related metabolites have been developed to date. High pressure liquid chromatography (HPLC) is one of the most employed techniques, for example, a simple HPLC method for the determination of *chloroquine* and its two major metabolites, *desethylchloroquine* and *bidesethylchloroquine* in biological fluids has been reported. This method includes a normal phase column, a mixture of acetonitrile, methanol and ammonia as mobile phase and a fluorescence detection system, with the readily available *hydroxychloroquine* as internal standard.[149] Another method for the separation of *chloroquine* and *desethylchloroquine* makes use of both, straight- and reverse-phase chromatography, although the best performance is attained when a reverse-phase system with Nucleosil C_{18} at pH 3 (phosphate buffer) in acetonitrile with perchlorate as a counter ion.[150] The reverse-phase chromatography (RPLC) has also been used to measure the change in hydrophobicity on N-oxidation of the ring nitrogen of some 4-aminoquinoline derivatives.[123]

HPLC has also found interest in the determination of adducts of 4-acetoxyaminoquinoline 1-oxide and DNA, in an effort to evaluate the effect of the compound cited in the DNA tertiary structure,[151] as well as in the determination of rate constants for the hydrolysis of aromatic nitrogen mustards derived from 4-aminoquinolines **113**.[129] Finally, rather than being the subject of purification, some aminoquinolines and related compounds have been used to check the validity of a HPLC model which is able to predict, from experimental isocratic data, retention times of solutes under gradient conditions,[152] or also as stationary liquid phase for gas-chromatography.[153]

There are many more reports on physical properties of 4-aminoquinolines in the chemical and pharmacological literature. However, since in this review we have tried to extract the most helpful general data available, we are going to close this chapter on physical properties by mentioning that the photochemical stability of *hydroxychloroquine* **117** was investigated by subjecting it to the effect of light at a wavelength of 240-600 nm and, as a result, four decomposition products were isolated and identified by mass spectrometry and ^{13}C- and ^{1}H-NMR.[139]

4. Reactivity of 4-aminoquinolines.

4.1. The amino group.

It is well established that for most aminopyridines and their benzo derivatives the amino form **119** (Scheme 26) greatly predominates over the piridone-imine tautomer **120** and shows the normal reactivity of the amino group.

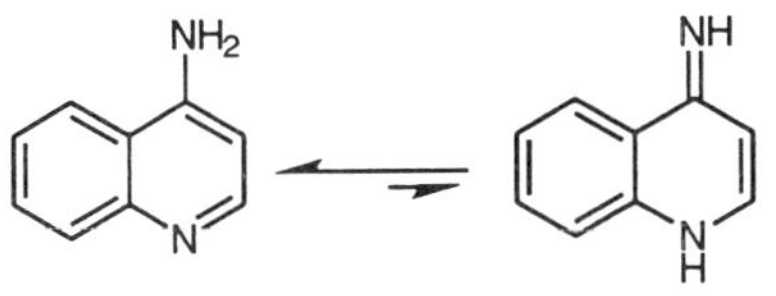

119 **120**

Scheme 26

Thus, aminoquinolines are readily acylated and tosylated[154] and they can react in the same way as aromatic amines in nucleophilic substitution reactions. This fact has been used for the preparation of aminoquinoline derivatives as agrochemical fungicides.[155] The treatment of 4-amino-8-fluoroquinolines **121** (Scheme 27) with NaH followed by the reaction with 2-chloro-4,6-dimethoxytriazine gave the active functionalized heterocycle **122**.

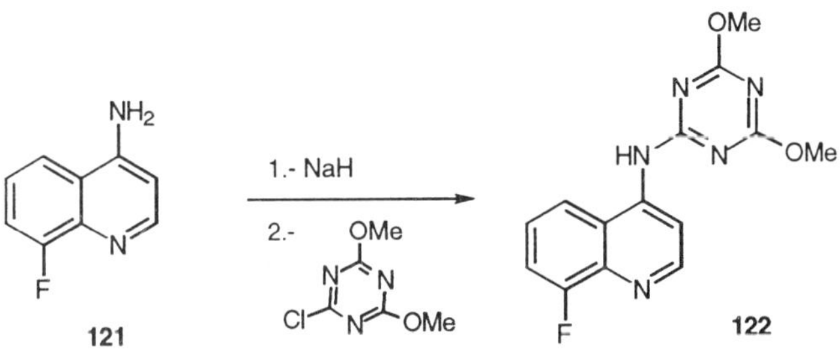

Scheme 27

Similarly, diazotation of 4-aminoquinolines **123** in the presence of sulphuric acid (Scheme 28) led to the corresponding 4-hydroxyquinolines **124**, which produced, by treatment with diazometane, alkaloids **125**.[28] This type of compounds have been also isolated from *lunasia amara* and *galipea officianalis*.

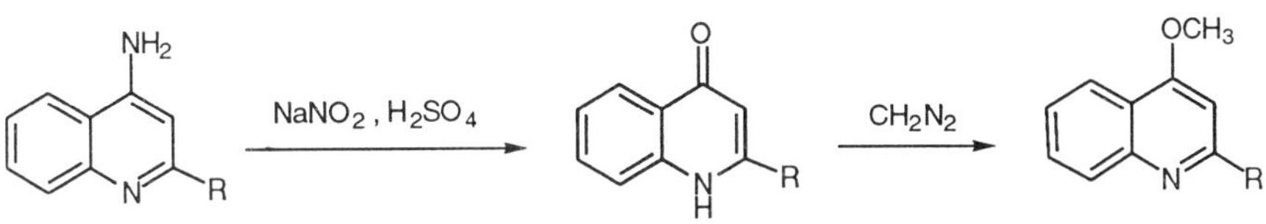

124 **125**

Scheme 28

The amino group of aminoquinolines can also form Schiff bases upon reaction with carbonyl compounds. This reaction (Scheme 29) has been used[156] for the preparation of *leniquinsin* **126** by treatment of *amquinsin* **93** with veratraldehyde.

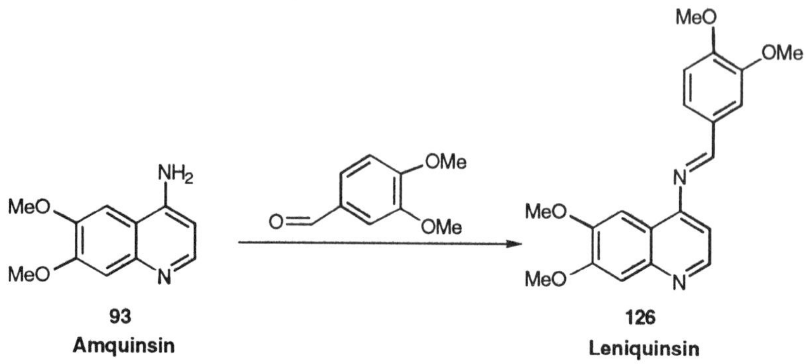

Scheme 29

Furthermore, the reactivity of the amino group in the 4 position has been used for the synthesis of other herocyclic rings[157] such as imidazoquinolines **128**, triazoloquinolines **129** and thioimidazoquinolines **130** with different biological activities. Cyclization of 3,4-diaminoquinolines **127** with formic acid, carbon sulphide or by diazotization leading to the heterocycles **128-130** is shown in the Scheme 30.

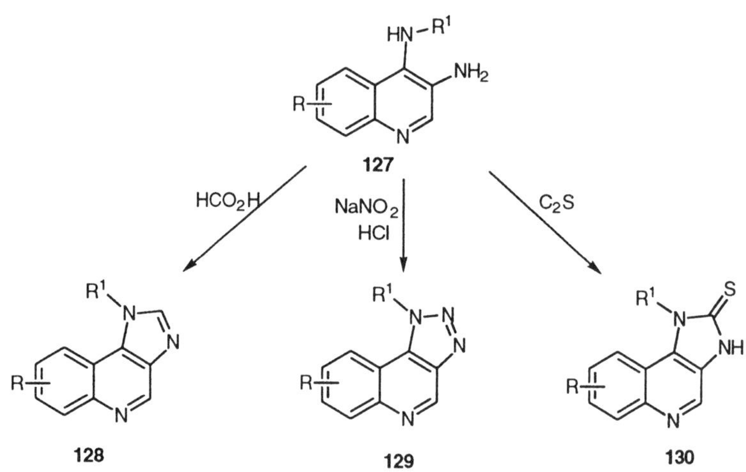

Scheme 30

The same type of reaction has been used for the synthesis of *imiquimod*.[158] In this case, the key step is the cyclization of an 3,4-aminoquinoline by means of ethyl ortoformiate to give the imidazole ring. Similarly, imidazoquinolines with biological activity[159] have been prepared by cyclization of 3,4-diaminoquinolines with orthoesters, acids, aldehydes, anhydrides, or acid halides.

4.2. The quinoline ring.

Most reactions of 4-aminoquinolines are similar to those of the 4-aminopyridine.[160] Fusion of a benzene ring to pyridine increases reactivity towards nucleophiles, and attack takes place at both 2 and 4 positions on the quinoline ring (see scheme 18). This may be attributed to a smaller loss of aromaticity in forming the initial adduct than in pyridine, and thus a corresponding decreased tendency to rearomatize is also observed.

Electrophilic substitution at carbon is, by its nature, an intrinsically unfavoured reaction for aromatic derivatives containing pyridine rings. However, it is known that the amino group in the C-4 position strongly activates the C-3 position. The nitration in C-3 position of 4-aminoquinolines can be formaly achieved to produced 3-nitro-4-aminoquinolines, but reaction can proceed *via* nitroaminoquinolines **132**.[154] These compounds **132** (Scheme 31) are the intermediates formed during the nitration of 4-aminoquinolines **131** and undergo acid-catalyzed rearrangement to nitro compounds **133**.

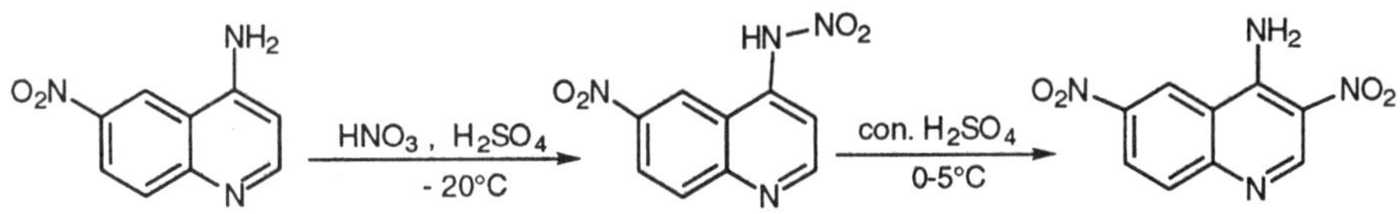

131 **132** **133**

Scheme 31

4.3. The heterocyclic nitrogen.

The lone pair on a quinoline nitrogen reacts with electrophiles under mild conditions. The ease of the reaction is controlled by the electronic effect of the substituents and, as one might expect, electron-releasing groups usually promote while electron withdrawing groups suppress reaction. Protonation of 4-aminoquinolines to form monocations occurrs on the nitrogen ring[124-129] and has been demonstrate for 4-amino-2-methylquinoline and other aromatic heterocycles.[154]

Quinolines with electron-withdrawing groups at the 4 position are very weak nucleophiles, and alkylation of the N-ring in the final quaternization step becomes problematic. On the contrary, the alkylation of N-ring in 4-aminoquinolines has been used for the synthesis of *dequalinium* analogues as K^+ channel blockers.[134,161] Compounds **135** were synthetized according to Scheme 32 when 4-aminoquinolines **134** were reacted with 1,10-diiodododecane.

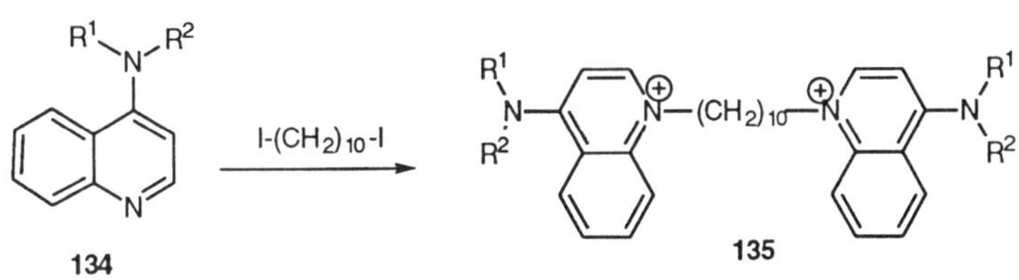

134 **135**

Scheme 32

Likewise, potential antitumor agents **137** (Scheme 33) have been synthesized[162] using *chloroquine* **94** and cis-diaminedichloroplatinum (II) **136**. The obtained complex was hygroscopic and tended to form an oil, so the *chloroquine* ligand was protonated at the N4'-position by addition of HNO_3 to give a crystalline solid **137**.

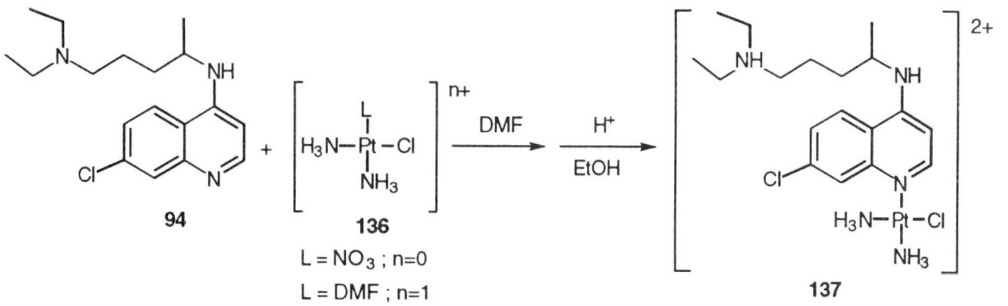

Scheme 33

5. Bilogical activity.

5.1. Antimalarials.

The aminoquinoline antimalarial agents constitute one of the earliest examples of pharmacophoric heterocyclic systems. It was recognized in the 1930's that 4-amino-7-chloroquinoline bearing an additional amino group in the side chain was often endowed with activity against the *plasmodia* that cause malaria. Most representative compounds of this series of antimalarial agents are *chloroquine*[85] **94**, and *hydroxychloroquine*[163] **117** (Figure 13). Likewise, it is known that the presence of a methyl substituent on the 1-position of the alkyl moiety of the quinoline may reduce toxicity and increase the therapeutic index.[164] The biological activity seems to be similar when a modification of the aromatic ring by the introduction of a methyl group in the 3 position has been carried out, such as has been described in the case of *sontoquine* **138**.[94] Likewise, the inclusion of an aromatic ring in the side chain sequence is apparently quite consistent with antimalarial activity as has been observed for *amodiaquine*[96] **139**, and *amopyroquine*[165] **140.**

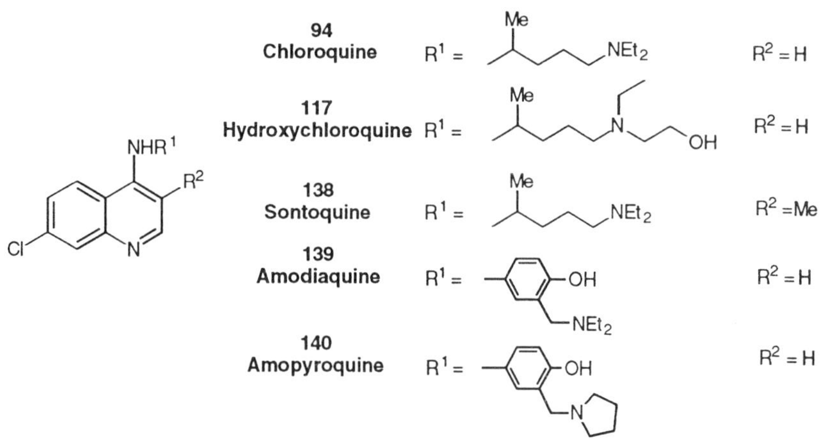

Figure 13

However, from this series of 4-aminoquinoline compounds, only *chloroquine* has found acceptance in the treatment of amebiasis.[166] *Chloroquine* was a very widely used antimalarial drug, which was synthesized as an analogue of quinine and which has remained as a major chemotherapeutic agent for over 40 years. The popularity of *chloroquine* has been due to its high efficacy and the low risk of side effects when it is taken in the correct dosage. Unfortunately, *chloroquine* resistant strains of the malaria parasite are developing and research is underway to develop new drugs to treat them. The mode of action of *chloroquine*, and other 4-aminoquinoline drugs, is not completely understood,[167] although it is known that the antimalarial action of QCDs (quinoline-containing antimalarial drugs) is rapid, and irreversible.[168-173] A photoreactive analogue of *chloroquine,* N-(4-(4-diethylamino-1-methylbutyl-amino)quinolin-6-yl)-4-azido-2-hydroxybenzamide **141** (*ASA-Q*) (Figure 14), has been recently synthesized and shown to mimic the action of *chloroquine* in possessing substantial antimalarial activity against a *chloroquine*-sensitive strain of *plasmodium falciparum.*[174] Furthermore, differences in conformational planarity of the key nuclei have little effect on activity, while the N...N+ distance between the ring and side-chain nitrogen atoms of series of 4-aminoquinoline derivatives is determinant of their activity. In fact, analogues of these compounds with a distance comparable to the N...N+ distance of *amodiaquine* and *chloroquine* demonstrated antimalarial activity.[100]

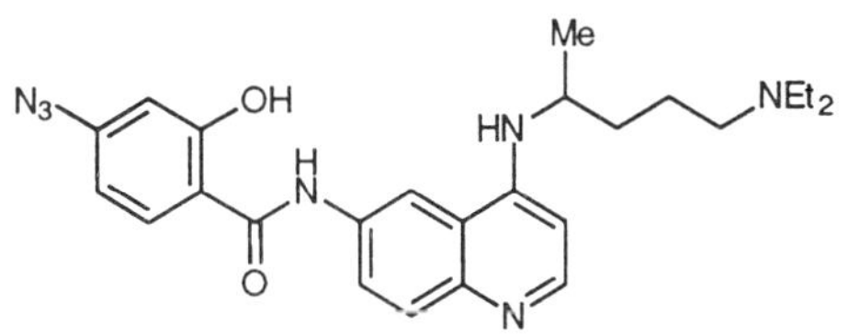

141
ASA-Q

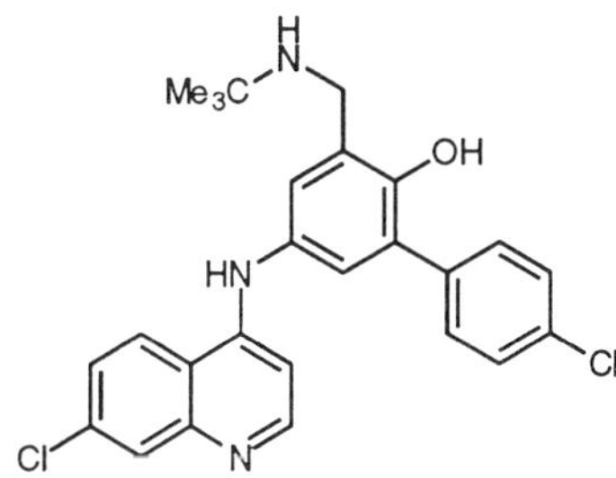

142
Tebuquine

Figure 14

Over the last 30 years, *chloroquine* resistance has become an increasingly serious problem and is now observed in all countries where malaria is endemic. The biochemical basis of resistance is not completely understood, however, it is clear that resistant parasites accumulate *chloroquine* less efficiently than sensitive isolates.[175,176] It has been suggested that the mechanism of *chloroquine* resistance has similarities with the multiple drug resistance phenotype of mammalian tumor cells, since *verapamil*, an agent which reverses multiple drug resistance of tumor cells, is also able to modulate *chloroquine* resistance.[177,178] The continuing development of resistance by the malaria parasite against chemotherapeutic agents has led to an ongoing, through sporadic, effort towards the synthesis of new chemical entities aimed at that target in the hope of finding drugs which will overcome resistance. That there is still interest in classical 4-aminoquinolines is attested to by the preparation of *tebuquine*[95] **142** (Figure 14), an hybrid compound combining structural elements of *amodiaquine* **139** and some 2-(dialkylamino)-o-cresols which have shown antimalarial activity. Similarly, (4-(aminoalkyl)amino)quinoline antimalarials for *chloroquine*-resistant pathogens have been synthesized.[98]

The effect of the introduction of fluorine atoms into the structure of this type of molecule has been recently studied and the *fluorotebuquine*-fluorine analogue of *tebuquine*- has been prepared[179] in order to know if the introduction of a fluorine atom in the structure is able to increase the biological activity of this series of drugs.[180] However, *fluorotebuquine* which was active against the *chloroquine* resistant K1 strain of *plasmodium falciparum*, seems to be less potent than both *chloroquine* and *tebuquine* against the 3D7 strain.[181] On the other hand, fluorine substitution in *amodiaquine* shows similar *in vitro* antimalarial activity to *amodiaquine* itself and makes it more resistant to oxidation and hence less likely to form toxic quinonimine metabolites *in vivo*.[182]

Recently, dimeric quinolines **143** (Figure 15), unlike their monomeric counterparts, have emerged as a small but growing class of compounds which have a wide range of pharmacological actions. Although the first synthesis of bisquinolines[183] and the study of their *in vivo* activity against marine *trypanosoma gambiense* was reported in the fifties, several 4-aminoquinolines recently prepared, including bisquinolines, exhibited potent activity against *plasmodium berghei* in mice, and therefore, these results indicate that these compounds conform to a putative receptor for quinoline antimalarials.[184] In this context, it is noteworthy that N,N'-bis(quinolin-4-yl)diamine derivatives and their salts are useful as agents for preventing and treating malaria, especially in those locations where the pathogens are resistant to *chloroquine*.[119]

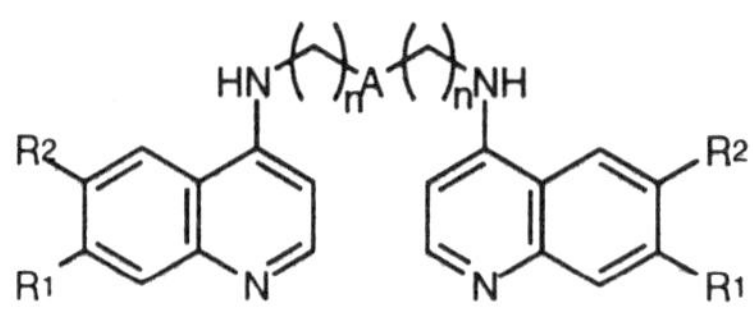

143

Figure 15

5.2. Antiinflammatories and analgesics.

Deletion of the basic nitrogen atom remote to the quinoline ring serves to abolish antimalarial activity, but makes it possible to promote other pharmacological actions. For example, N-aryl anthranilic acid derivatives ("fenamic acids") often inhibit cyclooxygenase and thus this group of derivatives can possess antiinflammatory and analgesic potency. *Glaphenine*[185] **118**, *floctafenine*[91] **144** (Figure 16), and series of fenamic acid analogues were prepared and evaluated for their analgesic, and antiinflammatory activities.[186] Other quinoline derivatives such as quinolinylguanidines **145** (Figure 16) also showed analgesic and antiinflammatory activity,[187] while substituted 4,6-diaminoquinolines have been described as inhibitors of anaphylotoxin C5a receptor binding.[114] Furthermore, 7-chloro(phenylthio)-4-phenylaminoquinolines and simple 3,4-diaminoquinolines showed interesting analgesic activity,[106,188] while N-amino functionalized aminoquinolines antagonize *bradykinin* and are useful as analgesic agents in the treatment and management of pain.[189]

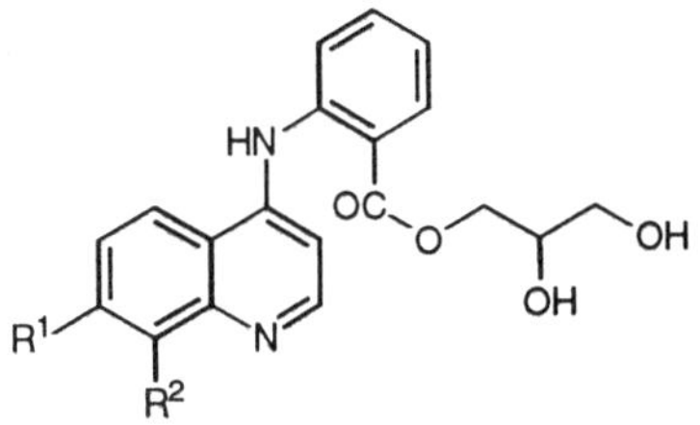

118
Glaphenine R^1 = Cl; R^2 = H

144
Floctafenine R^1 = H; R^2 = CF_3

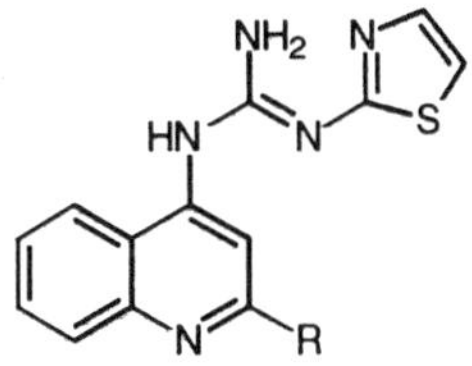

145

Figure 16

5.3. Antihypertensives.

Amquisin **93** shows hypotensive activity,[84] and so does *losulazine* [190] **146**, an orally and parenterally active agent, which prevents vasoconstriction by reducing available levels of that neurotransmitter following stimulation of adrenergic nerves. Likewise, new 7-(trifluoromethyl)-4-aminoquinolines **147, 148** (Figure 17) that are hypotensive agents and that act by a novel sympatholytic mechanism have been described,[191] while 2-methyl-4-aminoquinoline derivatives have been reported for treating hypertension or congestive heart failure.[108] Similarly, 2,4-diamino-6,7-dimethoxyquinoline[55,58] derivatives have been prepared and some of them are analogues of *prazosin* and *doxazosin* which showed exceptional α_1-adrenoceptor binding affinity.

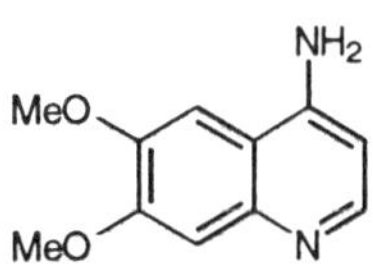

93
Amquisin

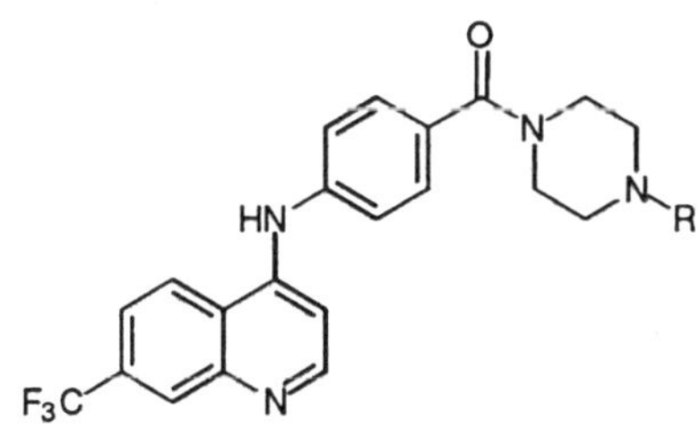

146
Losulazine R= SO_2-C_6H_4-F

147 R= C_6H_5

148 R= C_6H_4-OMe

Figure 17

5.4. Other pharmacological activities.

5.4.1. Reversible inhibitors of the (H^+/K^+)-ATPase.

Reversible (H^+/K^+)-ATPase inhibitors, as shorter acting inhibitors of gastric acid secretion, have begun to attract attention as potential therapies for acid related gastrointestinal disorders. In this context a serie of 3-carbethoxy-4-(arylamino)quinolines[157,192] **149** which showed to be reversible K^+-competitive gastric (H^+/K^+)-

ATPase inhibitors have been described. In a posterior work[113] the contribution of 3-acyl substituent has been explored and, from this series, compound **76** (*SK&F* 96067)[193] was shown to be a potent inhibitor of histamine-stimulated gastric acid secretion after oral dosing in the Heidenhain pouch dog and it was selected for further development and evaluation in man. More recently, studies showed that aminoquinoline **150** (CP-113,411) is not only a reversible inhibitor of gastric (H^+/K^+)-ATPase, but also a potent inhibitor of bone resorption[194] (Figure 18).

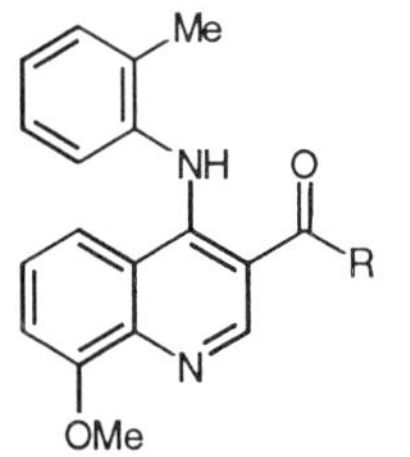

149 R = OEt

76 **(SK&F 96067)** R = OBu^t

150 **(CP-113,411)** R = OCH_2-CH_2-CO_2Et

Figure 18

5.4.2. K+ Channel blockers.

Dequalinium **116** (Figure 10), a bisquinolinium compound that has been used for many years as an antiseptic,[195] has recently been shown to be a potent and selective non-peptidic blocker of the SK_{Ca} channel.[196-198] Quantitative structure-activity relationship (QSAR) analysis of *dequalinium* analogues and the effect of 2-Me and 4-NH_2 substituents have been studied[199,161,134] in order to design more potent blockers and also to obtain structural information on the SK_{Ca} channels. Similarly, QSAR of a novel series of bisquinolinium small conductance Ca^{2+}-activated K^+ channel blockers related to *dequalinium*, in which the two quinolinium rings are linked via the 4-position to an α,ω-diamino alkylene chain and the ring N atom is quaternized with a methyl or benzylic group are described.[200]

5.4.3. Antivirals.

A new class of potential antivirals such as 4-aminoquinolines **151** (Figure 19),[201] functionalized with aryl ethers or aryl thioethers groups, were screened for their antiviral activity against tobacco mosaic virus in *nicotiana tabaccum* leaves showing significant activity. Furthermore, the classification technique of linear discriminant analysis has been applied to study the structure-activity relationship among antiviral N-quinolin-4-yl-N´-benzylidenehydrazines **152**.[202]

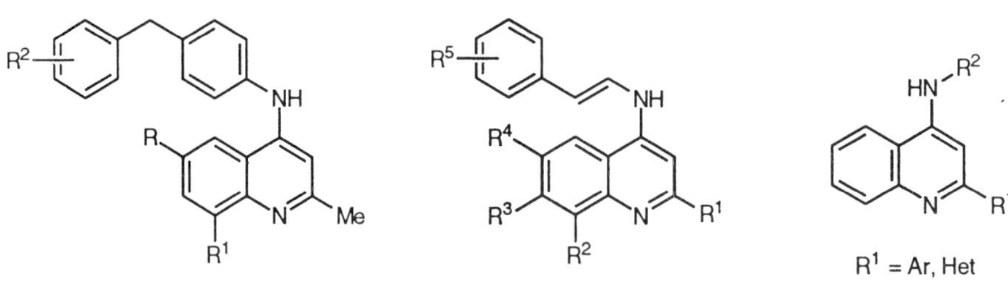

151 **152** **153**

Figure 19

Recently, a new class of non-nucleoside HIV-1 inhibitors has emerged,[203] represented by the 2-aryl (or heteroaryl)-quinoline-4-amines **153** (Figure 19). Some of these compounds have demonstrated quite high *in vitro* activity against HIV-1, in combination with low toxicity. Three-dimensional quantitative structure-activity relationships (QSAR), have been established for a novel class of non-nucleoside HIV-1 inhibitors providing reasonable guidelines to delineate a pharmacophore for this series of anti-HIV agents.[204]

5.4.4. Antitumor activity.

Organosilicon compounds derived from 4-aminoquinolines **154** (Figure 20) were prepared and tested for antitumor activity against leukemia L 1210, adenocarcinoma 755, and melanoma B_{16}.[205] On the other hand, novel organometallic complexes derived from platinum such as *cis*-$(Pt(NH_3)_2(N1\text{-}HCQ)Cl)(NO_3)_2$ have been prepared by reaction of the anticancer drug cis-diaminedichloroplatinum with the DNA intercalator *chloroquine*(CQ). These complexes have the potential to bind both covalently and intercalatively to DNA and, consequently, are potential antitumoral agents.[162] Likewise, a tetraphenyl porphine linked to a 4-amino-7-chloroquinoline was synthesized and examined as a potential photosensitizer for photodynamic therapy (PDT) of proliferative diseases.[206]

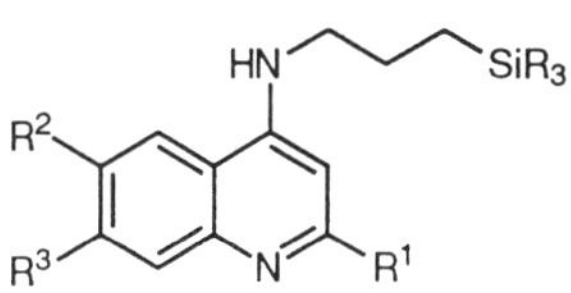

154

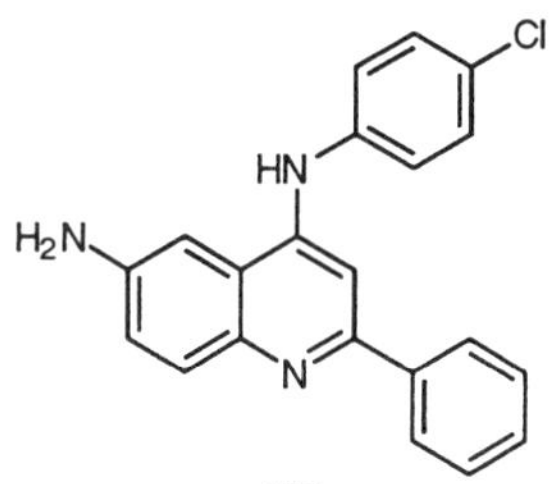

155

Figure 20

5.4.5. Immunostimulant activity.

The immunostimulant activity of 4-aminoalkylquinolines, arylquinolines[88] and 4,6-diaminoquinolines **155** (Figure 20) has been described as well as their evaluation in a mouse protection model designed to identify immunostimulant activity. These compounds represent a novel series of potent immunostimulants.[207]

5.4.6. Treatment of cardiac arrhythmia.

Intracellular recording was used to examine the effects of 4-aminoquinolines on action potentials of frog atrial myocardium depolarized by K^+,[209] while 4-amino-6,7-dimethoxy-2-(6,7-dimethoxy-1,2,3,4-tetrahydroisoquinol-2-yl)quinoline, in addition to its action as a hypertensive agent, also is useful in the treatment of cardiac arrhythmias at much lower dose levels.[209]

5.5. Other activities.

5.5.1. Pesticides and fungicides.

Interest in 4-aminoquinolines is not only restricted to the field of medicinal chemistry but is also important in the area of agrochemicals due to their behaviour as fungicides,[210,211,110,112] insecticides[112] and pesticides.[111] Furthermore, new functionalized quinolines have been prepared by condensation of 4-aminoquinolines with triazine derivatives. They are more potent than known products in controlling various plant diseases, particularly mildew of crops, showing no damage to a broad range of crops. Thus, these types of compounds, for example **122** (Figure 21), are used as substitutes for azole-based fungicides against some resistant fungi.[155]

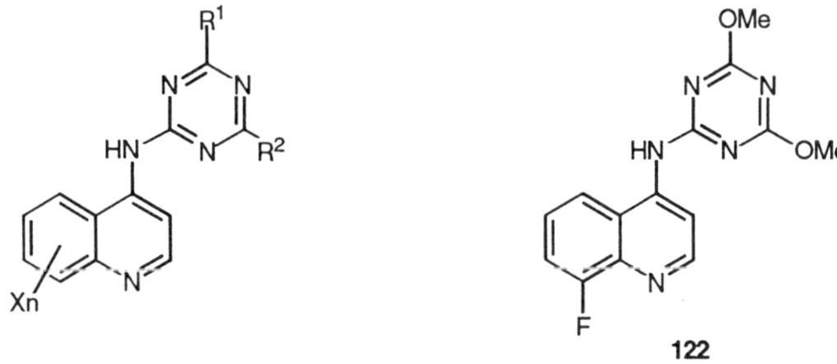

122

Figure 21

Acknowledgements.

Financial support by Dirección General de Investigación Científica y Técnica (DGICYT, Madrid; PB 93-0501) and by the Consejería de Educación, Universidades e Investigación del Gobierno Vasco (Vitoria-Gasteiz, PI 96-36) is gratefully acknowledged.

References.

1. Jones, G. In *Comprehensive Heterocyclic Chemistry*; Katritzky, A. R.; Rees, C. W., Potts, K. T., Eds.; Pergamon Press: Oxford, 1984; Vol. 2, p. 395.
2. a) Grundon, M. F. In *Alcaloids Chemical; Biological Perspectives*; Pelletier, S. W., Ed.; Wiley: New York, 1988; Vol 6, p. 339. b) Fortunak, J. M. D.; Mastrocola, A. R.; Mellinger, M.; Sisti, N. J.; Wood, J. L.; Zhuang Z. P. *Tetrahedron Lett.* **1996**, *37*, 5679.
3. Meller, J. M.; Merriman, G. D. *Steroids* **1995**, *60*, 693.
4. Hamer, F. M. In *The Chemistry of Heterocyclic Compounds*; Weissberger, A., Ed.; Wiley-Interscience, 1964, Vol. 18.
5. Rodriguez, R. 1970. German Patent 2,006, 638. *Chem. Abstr.* **1970**, *73*, 98987.
6. Brynes, P. 1990. U.S. Patent 4,897,444. *Chem. Abstr.* **1990**, *113*, 55051.
7. Scolaro, L. M.; Alibrandi, G.; Romeo, R.; Ricevuto, V.; Campagna, S. *Inorg. Chem.* **1992**, *31*, 2074.
8. Li, G.; Yu, J.; Krull, I. S. *J. Liq. Chromatography* **1995**, *18*, 3889.
9. Clark, R. L.; Patchett, A. A.; Rogers, E. F. 1968. U. S. Patent 3, 377, 352. *Chem. Abstr.* **1969**, *71*, 38821.
10. Elderfield, R. C.; Gensler, W. J.; Griffing, J. M.; Williamson, T. A.; Kupchan, S. M.; Maynard, J. T.; Kreysa, E. J.; Wright, J. B. *J. Am. Chem. Soc.* **1946**, *68*, 1584.
11. Elderfield, R. C.; Mertel, H. E.; Mitch, R. T.; Wempen, I. M.; Werble, E. *J. Am. Chem. Soc.* **1955**, *77*, 4816 .
12. Idowu, O. R.; Peggins, J. O.; Brewer, T. G. *Drug Metab. Disposition* **1995**, *23*, 18.
13. Nodiff, E. A. 1990. U.S. Patent 4,980,360. *Chem. Abstr.* **1991**, *114*, 207057.
14. Berman, J. D.; Lee, L. S. *Am. J. Trop. Med. Hyg.* **1983**, *32*, 753.
15. Bartlett, M. S.; Queener, S. F.; Tidwell, R. R.; Milhouse, W. K.; Bernan, J. D.; Ellis, W. Y.; Smith, J. W. *Antimicrob. Agents Chemother.* **1991**, *35*, 277.
16. Queener, S. F.; Barlett, M. S.; Nasr, M. *Antimicrob. Agents Chemother.* **1993**, *37*, 2166.
17. Mariske, R. H. F.; Kulka, M. *Org. React.* **1953**, *7*, 59.
18. Cheng, C. *Org. React.* **1982**, *28*, 37.
19. Barluenga, J.; Aguilar, E.; Joglar, J.; Olano, B.; Fustero, S. *J. Org. Chem.* **1989**, *54*, 2596.

20. Aguilar, E. *Doctoral Thesis*, University of Oviedo, 1991.
21. Hudson, R. F.; Lawson, A. J.; Reurd, K. A. F. *Chem. Commun.* **1974**, 484.
22. Forrester, A. R.; Gill, M.; Thompson, R. H. *Chem. Commun.* **1976**, 677.
23. Torii, S.; Xu, L. H.; Sadakane, M.; Okumoto, H. *Synlett* **1992**, 513.
24. Kobayashi, T.; Tonaka, M. *Chem. Commun.* **1981**, 333.
25. Torii, S.; Okumoto, H.; Hexu, L. *Tetrahedron Lett.* **1991**, *32*, 237.
26. Torii, S.; Hexu, L.; Okumoto, H. *Synlett* **1991**, 695.
27. Palacios, F.; Ochoa de Retana, A.; Oyarzabal, J. *Tetrahedron Lett.* **1996**, *35*, 4577.
28. Thomsen, I.; Torssell, K.B.G. *Acta Chem. Scand. Ser. B* **1988**, 309.
29. Casnati, G.; Quilico, A.; Ricca, A.; Vita-Finzi, P. *Gazz. Chim. Ital.* **1966**, *96*, 1073.
30. Cook, A.G. In *Enamines: Synthesis Structures and Reactions*; Dekker, H.; Ed.; New York, 1987.
31. Palacios, F.; Aparicio, D.; García, J. *Tetrahedron* **1996**, *52*, 9609.
32. Palacios, F.; Aparicio, D.; García, J. *Synlett* **1994**, 260.
33. Palacios, F.; García, J.; Ochoa de Retana, A.; Oyarzabal, J. *Heterocycles* **1995**, 1915
34. Schweizer, E. E.; Kim, C.S.; Labaw, C.S.; Murray, W.P. *Chem. Commun.* **1973**, 7.
35. Schweizer, E. E.; Goff, S.D.; Murray, W.P. *J. Org. Chem.* **1977**, *42*, 200.
36. Palacios, F.; Alonso, C.; Rubiales, G. *J. Org. Chem.* **1997**, *62*, 1146.
37. Barluenga, J.; Ferrero, M.; Lopez, F.; Palacios, F. *J. Chem. Soc., Perkin Trans.1* **1989**, 615.
38. Palacios, F.; Alonso, C.; Rubiales, G. *Tetrahedron* **1995**, *51*, 3693.
39. Ciganek, E. *J. Org. Chem.* **1970**, *35*, 3631.
40. Meyers, A. I.; Sircar, J. C. In *The Chemistry of the Cyano Group*; Patai S., Ed.; Wiley: London, 1970, p.341.
41. Palacios, F.; Aparicio, D.; García, J. Unpublished results.
42. Palacios, F.; Perez de Heredia, I.; Rubiales, G. *J. Org. Chem.* **1995**, *60*, 2384.
43. Palacios, F.; Rubiales, G. *Tetrahedron Lett.* **1996**, *37*, 6379.
44. Palacios, F.; Aparicio, D.; De los Santos, J. *Tetrahedron* **1996**, *52*, 4857.
45. Schäfer, H.; Sattler, K.; Gewald, K. *J. Prakt. Chem.* **1979**, *321*, 695.
46. Veronese, A. C.; Callagari, R.; Morelli, C. F. *Tetrahedron* **1995**, *51*, 12277.
47. Veronese, A. C.; Callagari, R.; Salah, S. A. A. *Tetrahedron Lett.* **1990**, *31*, 3485.
48. Strekowski, L.; Kong, S.B.; Cegla, M. T.; Harden, D.B. *Heterocycles* **1989**, *29*, 539.
49. Charvat, T.; Potacek, M.; Marek, *J. Monatsh. Chem.* **1995**, *126*, 333.
50. Barluenga, J.; Palacios, F.; Fustero, S.; Gotor, V. *Synthesis* **1981**, 200.
51. Hester, J. B. *J. Org. Chem.* **1974**, *39*, 2137.
52. Appel, R.; Warning, K. *Chem. Ber.* **1975**, *108*, 1437.
53. Appel, R. *Angew. Chem.* **1975**, *87*, 863.
54. Yokomatsu, T.; Yoshida, Y.; Nakabayashi, N.; Shibuya, S. *J. Org. Chem.* **1994**, *59*, 7562.
55. Campbell, S. F.; Hardstone, J. D.; Palmer, M.J. *Tetrahedron Lett.* **1984**, *25*, 4813.
56. Moore, J. A.; Kornreich, L. D. *Tetrahedron Lett.* **1963**, 1277.
57. Schroeder, H. E.; Rigby, G. W. *J. Am. Chem. Soc.* **1949**, *71*, 2205.
58. Campbell, S. F.; Hardstone, J. D.; Palmer, M. J. *J. Med. Chem.* **1988**, *31*, 1031.
59. Pedersen, E. B. *Tetrahedron* **1977**, *33*, 217.
60. Strekowski, L; Patterson, S. E.; Janda, L.; Wydra, R. L.; Harden, D. B.; Lipowska, M.; Cegla, M. T. *J. Org. Chem.* **1992**, *57*, 196.
61. Strekowski, L; Wydra, R. L.; Cegla, M. T.; Czarny, A.; Harden, D. B.; Patterson, S. E.; Battiste, M. A.; Coxon, J. M. *J. Org. Chem.* **1990**, *55*, 4777.
62. Strekowski, L; Wydra, R. L.; Harden, D. B.; Honkan, V. A. *Heterocycles* **1990**, *31*, 1565.
63. Wydra, R. L.; Patterson, S. E.; Strekowski, L. *J. Heterocycl. Chem.* **1990**, *27*, 803.
64. Strekowski, L.; Mokrosz, J. L.; Honkan, V. A.; Czarny, A.; Cegla, M. T.; Wydra, R. L.; Patterson, S. E.; Schinazi, R. F. *J. Med. Chem.* **1991**, *34*, 1739.
65. Strekowski, L.; Wydra, R. L.; Patterson, S. E.; Schinazi, R. F. 1994. US 5,304,554 U.S. *Chem Abstr.* **1994**, *121*, 50080.
66. Beccalli, E. M.; Erba, E.; Gelmi, M. L.; Pocar, D. *J. Chem. Soc., Perkin Trans.1* **1996**, 1359.
67. Schäfer, H.; Gewald, K. *Monatsh. Chem.* **1978**, *109*, 527.
68. Okomura, K.; Adachi, T.; Tomie, M.; Kondo, K.; Inoue, I. *J. Chem. Soc., Perkin Trans.1* **1972**, 173.
69. Pedersen E. B. *Acta Chem. Scand. Ser. B* **1976**, *30,* 133.
70. Pedersen, E. B.; Lawesson, S. O. *Tetrahedron* **1974**, *30*, 875.
71. Viehe, H. G.; de Voghel, G. J.; Smets, F. *Chimia* **1976**, *30*, 189.
72. Flowers, W. T.; Haszeldine, R. N.; Owen, C. R.; Thomas, A. *Chem. Commun.* **1974**, 134.
73. Sovenaux, E.; Ghosez, L. *J. Am. Chem. Soc.* **1973**, 5417.
74. Yamamoto, I.; Gotoh, H.; Minami, T.; Ohshiro, Y.; Agawa, T. *J. Org. Chem.* **1974**, *39*, 3516.

75. Yamamoto, I.; Furukawa, T.; Nakajima, H.; Gotoh, H. *J. Chem. Soc., Perkin Trans.1* **1976**, 1597.
76. Saito, T.; Nakane, M.; Miyazaki, T.; Motoki, S. *J. Chem. Soc., Perkin Trans.1* **1989**, 2140.
77. Palacios, F.; Aparicio, D.; García, J. *Tetrahedron* **1997**, *53*, 2931.
78. Andrews, I. P.; Bannister, R.; Etridge, S. K.; Lewis, N. J.; Mullane, M. V.; Wells, A. S. *Tetrahedron Lett.* **1995**, *36*, 7743.
79. Zoltewicz, J. A.; Helmick, L. S.; Oestreich, T. M.; King, R. W.; Kadetzki, P.E. *J. Org. Chem.* **1973**, *38*, 1947.
80. Tondys, H.; van der Plas, H. C.; Wozniak, M. *J. Heterocycl. Chem.* **1985**, *22*, 353.
81. Bulton, A. J.; Mc Killop, A. In *Comprehensive Heterocyclic Chemistry*; Katritzky, A. R.; Rees, C. W.; Potts, K. T., Eds.; Pergamon Press: Oxford, 1984; Vol. 2, p. 59.
82. Strekowski, L; Gulevich, Y.; Van Aken, K.; Wilson, D. W. *Tetrahedron Lett.* **1995**, *36*, 225.
83. Brooke, G. M.; Chambers, R. D.; Drury, C. J.; Bower, M. *J. Chem. Soc. Perkin Trans.1* **1993**, 2201.
84. Ebetino, F. F.; Wright, G. C. 1965. French Patent 1,388,756. *Chem. Abstr.* **1965**, *63*, 589.
85. Surrey, A. R.; Hammer, H. F. *J. Am. Chem. Soc.* **1946**, *68*, 113.
86. Gerster, J. F. 1985. Eur. Pat. Appl. 145,340. *Chem. Abstr.* **1985**, *103*, 19609.
87. Andre, J. D.; Lagain, D. 1991. US 4,988,815 U.S. *Chem. Abstr.* **1991**, *114*, 228916.
88. Moyer, M. P.; McFarland, J.W. 1993. PCT Int. Appl. WO 93 03,030. *Chem. Abstr.* **1993**, *119*, 49248.
89. Moyer, M. P.; Weber, F. H.; Gross, J. L. *J. Med. Chem.* **1992**, *35*, 4595.
90. Ferenc, K.; Laszlo, F.; Andras, B.; Istvan, F. 1993. HU 61,294. Hung. Teljes. *Chem. Abstr.* **1993**, *119*, 49247.
91. Allais, A. *Chim. Ther.* **1973**, *8*, 154.
92. McCall, J. M. 1979. US 4,167,567 U.S. *Chem. Abstr.* **1980**, *92*, 555.
93. Andersen, K. E.; Lundt, B. F.; Jørgensen, A. S.; Braestrup, C. *Eur. J. Med. Chem.* **1996**, *31*, 417.
94. Steck, E. A.; Hallock, L. L.; Holland, A. J. *J. Am. Chem. Soc.* **1946**, *68*, 380.
95. Werbel, L. M.; Coock, P. D.; Elslager, E. F.; Hung, J. H.; Johnson, J. L.; Kesten, S.J.; Mc Namara, D. J.; Ortwein, D. F.; Worth, D. F. *J. Med. Chem.* **1986**, *29*, 924.
96. Burckhalter, J. H.; Tendrick, F. H.; Jones, E.M.; Holcomb, W. F.; Rawling, A. L. *J. Am. Chem. Soc.* **1948**, *70*, 1363.
97. O'Neill, P. M.; Harrison, A. C.; Storr, R. C.; Hawley, S. R.; Ward, S. A.; Park, B. K. *J. Med. Chem.* **1994**, *37*, 1362.
98. Hofheinz, H.; Jaquet, C.; Jolidon, S. 1995. CA 2,133,620. Can. Pat. Appl. *Chem. Abstr.* **1995**, *123*, 313786.
99. Tripathi, R. C.; Saxena, M.; Chandra, S.; Saxena, A. K. *Indian J. Chem., Sect. B* **1995**, *34B*, 164.
100. Koh, H. L.; Go, M. L.; Ngiam, T. L.; Mak, J.W. *Eur. J. Med. Chem.* **1994**, *29*, 107.
101. Dubey, R.; Abuzar, S.; Sharma, S. *Indian J. Chem., Sect. B* **1985**, *24B*, 408.
102. Abuzar, S.; Dubey, R.; Sharma, S. *Eur. J. Med. Chem.* **1986**, *21*, 5.
103. Chauhan, P.M.S.; Pratap, R.; Sharma, S. *Indian J. Chem., Sect.* B **1985**, *24B*, 1154.
104. Abuzar, S.; Dubey, R.; Sharma, S. *Indian J. Chem., Sect. B* **1985**, *24B*, 848.
105. Agrawal, V. K.; Sharma, S. *Indian J. Chem., Sect. B* **1987**, *26B*, 550.
106. Bethegnies, G.; Dupuis, M. H.; Marcincal, L.; Trupin, N.; Brunet, C. *Farmaco, Ed. Sci.* **1986**, *41*, 471.
107. Klein, J. T.; Davis, L.; Olsen, G. E.; Wong, G. S.; Huger, F. P.; Smith, C. P.; Petko, W. W.; Cornfeldt, M.; Wilker, J. C.; Blitzer, R. D.; Landau, E.; Haroutunian, V.; Martin, L. L.; Effland, R. C. *J. Med. Chem.* **1996**, *39*, 570.
108. Koh, K.; Tauzuki, K.; Taniguchi, M.; Kushida, H.; Ozawa, K. 1993. PCT Int. Appl.WO 92 22,533. *Chem. Abstr.* **1993**, *118*, 233898.
109. Dreikorn, B. A.; Kaster, S. V.; Kirby, N. V.; Suhr, R. G.; Thoreen, B. R. 1995. PCT Int. Appl. WO 94 04,526. *Chem. Abstr.* **1995**, *122*, 187600.
110. Schaper, W.; Maerkl, M.; Preus, R.; Klein, R.; Salbeck, G.; Braun, P.; Knauf, W.; Sachae, B.; Waltersdorfer, A. 1995. Ger. Offen. DE 4,308,014. *Chem. Abstr.* **1995**, *122*, 9884.
111. Dreikorn, B. A.; Suhr, R. G.; Johnson, P. L. 1995. PCT Int. Appl. WO 94 04,527. *Chem. Abstr.* **1995**, *122*, 56046.
112. Coghlan, M. J.; Dreikorn, B. A.; Jordan, G. P.; Suhr, R. G. 1994. US 5,296,484 U.S. *Chem. Abstr.* **1994**, *121*, 35356.
113. Ife, R. J.; Brown, T. H.; Keeling, D. J.; Leach, C. A.; Meeson, M. L.; Parsons, M. E.; Reavill, D. R.; Theobald, C. J.; Wiggall, K. J. *J. Med. Chem.* **1992**, *35*, 3413.
114. Lanza, T. J.; Durette, P. L.; Rollins, T.; Siciliano, S.; Cianciarulo, D. N.; Kobayashi, S. V.; Caldwell, C. G.; Springer, M. S.; Hagmann, W. K. *J. Med. Chem.* **1992**, *35*, 252
115. Demonchaux, P.; Laayoun, A.; Demeunynck, M.; Lhomme, J. *Tetrahedron* **1989**, *45*, 6455
116. LeBras, J.; Deleron, P.; Charmot, G. *Lancet* **1983**, *1*, 73.

117. Benazet, F. *Ann. Soc. Belge. Méd. Trop.* **1965**, *45*, 459.
118. Vennerstrom, J. L.; Ellis, W. Y.; Ager, A. L.; Andersen, S. L.; Gerena, L.; Milhous, W. K. *J. Med. Chem.* **1992**, *35*, 2129.
119. Hofheinz, W.; Leupin, W. 1996. PCT Int. Appl. WO 95 35,287.*Chem. Abstr.* **1996**, *124*, 260860.
120. Hofheinz, H.; Masciadri, R. 1996. PCT Int.Appl. WO 95 35,288. *Chem. Abstr.* **1996**, *124*, 260861.
121. a) Taylor, P. J. In *Comprehensive Medicinal Chemistry*; Hansch, C., Ed.; Pergamon Press: Oxford, 1990, Vol. 4, p. 241; b) Leo, A. J. In *Comprehensive Medicinal Chemistry*; Hansch, C., Ed.; Pergamon Press: Oxford, 1990, Vol. 4, p. 295.
122. Go, M. L.; Ngiam, T. L. *Heterocycles* **1989**, *29*, 2353.
123. Go, M. L.; Ngiam, T. L. *Chem. Pharm. Bull.* **1988**, *36*, 1393.
124. Joule, J. A.; Mills, K.; Smith, G. F. In *Heterocyclic Chemistry*; Chapman Hall, 1995, p. 130.
125. Girault, G.; Coustal, S.; Rumpf, P. *Bull. Soc. Chim. Fr.* **1972**, 2787.
126. Perrin, D. D. In *Dissociation Constants of Organic Bases in Aqueous Solutions*; Butterworths, 1965.
127. Hughes, D. L.; Arnett, E. M. *J. Am. Chem. Soc.* **1983**, *105*, 4157.
128. Kelly, R. N.; Schulman, S. G. 1988. ASTM Spec. Tech. Publ. 1009, 75. *Chem. Abstr.* **1989**, *111*, 38655.
129. O'Connor, C. J.; Denny, W. A.; Fan, J.-Y.; Gravatt, G. L.; Grigor, B. A.; McLennan, D. J. *J. Chem. Soc., Perkin Trans.2* **1991**, 1933.
130. Debnath, A. K.; Lopez de Compadre, R. L.; Debnath, G.; Shusterman, A. J.; Hansch, C. *J. Med. Chem.* **1991**, *34*, 786.
131. Singer, J. A.; Purcell, W. P. *J. Med. Chem.* **1967**, *10*, 754.
132. Ryu, U.; Choi, D.; Kim, N.; Lee, Y. S. *J. Korean Chem. Soc.* **1993**, *37*, 62. *Chem. Abstr.* **1993**, *118*, 233411.
133. Galanakis, D.; Calder, J. A. D.; Ganellin, C. R.; Owen, C. S.; Dunn, P. M. *J. Med. Chem.* **1995**, *38*, 3536.
134. Stewart, J. J. P. *J. Comput. Chem.* **1989**, *10*, 209.
135. Batterham, T. J. In *NMR Spectra of Simple Hetrocycles*; John Wiley: New York, 1973; p. 303.
136. a) Angerman, N. S.; Danyluk, S. S.; Victor, T. A. *J. Am. Chem. Soc.* **1972**, *94*, 7137, b) Moreau, S., Perly, B.; Biguet, J. *Biochimie* **1982**, *64*, 1015.
137. Moreau, S.; Veignie, E. *Magn. Reson. Chem.* **1990**, *28*, 377.
138. Atouk, A.; Viala, A.; Samat, A. *Spectrosc. Lett.* **1985**, *18*, 1.
139. Tønnesen, H. H.; Grislingaas, A. L.; Woo, S. O.; Karlsen, J. *Int. J. Pharm.* **1988**, *43*, 215.
140. Brooke, G. M.; Chambers, R. D.; Drury, C. J.; Bower, M. J. *J. Chem. Soc., Perkin Trans.1* **1993**, 2201.
141. a) Ayachit, N. H.; Rao, K. S.; Shashidhar, M. A. *Indian J. Pure Appl. Phys.* **1984**, *22*, 303. b) Pouchert, C. J. In *The Aldrich Library of FTIR Spectra;* Aldrich Chemical: Milwaukee, 1985; Vol. 2, p. 863C. c) Grasselli, J. G.; Ritchey, W. M. In *Atlas of Spectral and Physical Constants for Organic Compounds;* CRC Press, 1975; Vol. IV, p. 459.
142. Grasselli, J. G.; Ritchey, W. M. In *Atlas of Spectral and Physical Constants for Organic Compounds*, CRC Press, 1975; Vol. IV, p 458-460 and 465.
143. Daabees, H. M. G. *Anal. Lett.* **1991**, *24*, 1571.
144. Hassib, S. T.; El-Sayed, N. M.; Romeih, F. A.; ELZaher, A. A. *Anal. Lett.* **1994**, *27*, 887.
145. Vysotskii, Y. B. *Opt. Spektrosk.* **1978**, *44*, 1025. *Chem. Abstr.* **1978**, *89*, 120046.
146. Souto, M. A.; Michl, J. *J. Am. Chem. Soc.* **1978**, *100*, 6853.
147. Brown, E. V.; Plasz, A. C.; Mitchell, S. R. *J. Heterocycl. Chem.* **1970**, *7*, 661.
148. Moeller, J.; Pedersen, E. B. *Chem. Scr.* **1982**, *19*, 185.
149. Chaulet, J. F.; Mounier, C.; Soares, O.; Brazier, J. L. *Anal. Lett.* **1991**, *24*, 665.
150. Bergqvist, Y.; Olin, A. *Acta Pharm. Suec.* **1982**, *19*, 161. *Chem. Abstr.* **1982**, *97*, 119986.
151. Menichini, P.; Fronza, G.; Tornaletti, S.; Galiegue-Zouitina, S.; Bailleul, B.; Loucheux-Lefebvre, M. H.; Abbondandolo, A.; Pedrini, A. M. *Carcinogenesis (London)* **1989**, *10*, 1589.
152. Marengo, E.; Baiocchi, C.; Gennaro, M. C.; Bertolo, P. *Chromatographia* **1989**, *27*, 19.
153. Sharif'yanova, L. N.; Vigdergauz, M. S.; Trufanova, A. I.; Shitova, E. N. 1984. U.S.S.R. SU 1,111,103. *Chem. Abstr.* **1984**, *101*, 183214.
154. Uff, B.C. In *Comprehensive Heterocyclic Chemistry*; Katritzky, A. R.; Rees, C. W.; Potts, K. T., Eds; Pergamon Press: Oxford, 1984; Vol. 2, p. 340.
155. Ishikawa, K.; Yooshikawa, Y.; Ishii T.; Tanigawa, H.; Maeda, S.; Kawashima, H.; Shimotori, H.; Yanese, J.; Tomura, N. 1994. JP 06 49,062, Jpn. Kokai Tokkyo Koho.*Chem. Abstr.* **1994**, *121*, 134158.
156. Winterstein, A. 1966. US 3,272, 806. U.S. *Chem. Abstr.* **1966**, *65*, 18567.
157. Brown, T. H.; Ife, R. J.; Keeling, D. J.; Laing. S. M.; Leach, C. A.; Parson, M. E.; Price, C. A.; Reavill, D. R.; Wiggall, K. J. *J. Med. Chem.* **1990**, *33*, 527.

158. Gester J. F. 1985. Eur. Pat. Appl. 145340. *Chem. Abstr.* **1985**, *103*, 196090.
159. a) Berenyi, D. M.; Budai, Z. M.; Petocz, L.; Grasser, K.; Szirt Kiszelly, E.; Knoll, J.; Furst, Z.; Kovacs, A. HU 34,479 Hung. Teljes. *Chem. Abstr.* **1986**, *105*, 115064. b) Andre, J.D.; Lagain, D. US 4,988,815. *Chem. Abstr.* **1991**, *114*, 228916.
160. Scriven, E. F. V. In *Comprehensive Heterocyclic Chemistry*, Katritzky, A. R.; Rees, C. W.; Potts, K. T. Eds.; Pergamon Press: Oxford, 1984; Vol. 2, p. 170.
161. a) Galanakis, D.; Davis, C. A.; Del Rey Herrero, B; Ganellin, C. R.; Dunn, P. M.; Jenkinson, D.H. *Biorg. Med. Chem. Lett.* **1995**, *5*, 559. b) Galanakis, D.; Davis, C. A.; Ganellin, C. R.; Dunn, P. M. *J. Med. Chem.* **1996**, *39*, 3592. c) Campos, J. R.; Galanakis, D.; Ganellin,C. R.; Dunn, P. M. *J. Med. Chem.* **1996**, *39*, 4247.
162. Sundquist, W. I.; Bancroft, D. P.; Lippard, S. J. *J. Am. Chem. Soc.* **1990**, *112*, 1590.
163. Surrey, A. R.; Hammer, H. F. *J. Am. Chem. Soc.* **1950**, *72*, 1814.
164. Peters, W. In *Chemotherapy and Drug Resistance in Malaria*; Academic Press: London, 1970.
165. Verdier, F.; Paussard, E.; Clavier, F.; Le Bras, J.; Gaudebout, C. *Eur. J. Clin. Pharmacol.* **1989**, *33*, 409.
166. Dujoune, C. A.; Azarnoff, D. L.; Huffman, D. H.; Pentikainen, P; Hurwitz, A.; Shoeman, D. W. *Clin. Pharmacol. Ther.* **1976**, *19*, 352.
167. Ginsburg, H.; Krugliak M. *Biochem. Pharmacol.* **1992**, *43*, 63.
168. Yayon, A.; Wande Waa, J. A.; Yayon, M.; Geary, T. C.; Jensen, J. B. *J. Protozool.* **1983**, *30*, 642.
169. Peters, W.; Howells, R. E.; Portus, J.; Robinson, B. L., Thomas, S.; Warhurst, D. C. *Ann. Trop. Med. Parasitol.* **1977**, *71*, 407.
170. O'Brien, R. L.; Olenick, J. G.; Hahn, F. R. *Proc. Natl. Acad. Sci. USA* **1966**, *55*, 1511.
171. O'Brien, R. L.; Hahn, F. R. *Antimicrob. Agents. Chemother.* **1965**, 315.
172. Meshnick, S. R. *Parasitol. Today* **1990**, *6*, 77.
173. Chou, S. C.; Conkli, K. A.; Levy, M. R.; Warhust, D. C. *Antimalarial Drugs* **1984**, *1*, 281.
174. Foley, M.; Deady, L. W.; Ng, K., Cowman, A. F.; Tilley, L. *J. Biol. Chem.* **1994**, *269*, 6955.
175. Fitch, C. D. *Proc. Helminthol. Soc. Washington* **1972**, *39*, 265.
176. Yayon, A.; Cabantchik, Z. I.; Ginsburg, H. *Proc. Natl. Acad. Sci. US.A.* **1985**, *82*, 2784.
177. Krogstad, D. J.; Gluzman, I. Y.; Kyle, D. E.; Odoula, A. M.; Martin, S. K.; Milhous, W. K.; Schlesinger, P. H. *Science* **1987**, *238*, 1283.
178. Martin, S. K.; Oduola, A. M. J.; Milhous, W. K. *Science* **1987**, *235*, 899.
179. O'Neill, P. M.; Hawley, S. R.; Storr, R. C.; Ward, S. A.; Park, B. K. *Bioorg. Med. Chem. Lett.* **1996**, *6*, 391.
180. Wilch, J. T.; Eswarakrishman, S. In *Fluorine in Biorganic Chemistry*; Wiley: New York, 1991.
181. Charmot, G.; Goujon, C. *Bullet. Soc. Phatol. Exotiq.* **1987**, *80*, 266.
182. O'Neill, P. M.; Harrison, A. C.; Storr, R. C.; Shaun, S. R.; Stephen, A.; Park, B. K. *J. Med. Chem.* **1994**, *37*, 1362.
183. Schock, R. U. *J. Am. Chem. Soc.* **1957**, *79*, 1672.
184. Ismail, F. M.; Dascombe, M. J.; Carr, P.; North, S. E. *J. Pharm. Pharmacol.* **1996**, *48*, 841.
185. Lednicer, D.; Mitscher, L. A. In *The Organic Chemistry of Drug Synthesis;* John Wiley and Sons: New York, 1977; Vol. 1, p. 342-343.
186. El-Azzouny, A. A.; El-Shabrawy, O. A.; El-Azzouny, M. M.; Ebeid, M. Y.; Lehmann, *J. Sci. Pharm.* **1995**, *63*, 81.
187. Quadro, G.Eur. Pat. Appl. EP 1984. 123,146. *Chem. Abstr.* **1985**, 102, 95553.
188. Berenyi, D.; Budai, Z.; Petocz, L.; Grasser, K.; Szirt Kiszelly, E.; Knoll, J.; Furst, Z.; Kovacs, A. 1985. Hung. Teljes HU 34,3479. *Chem. Abstr.* **1986**, 105, 115064.
189. Mobilio, D.; Musser, J. H. 1993. U. S. US 5,216,165. *Chem. Abstr.* **1994**, 120, 106792.
190. McCall, J.M. 1979. U. S. Patent 4, 167,567. *Chem. Abstr.* **1980**, 92, 6, 555.
191. McCall, J. M.; TenBrink, R. E.; Kamdar, B. V.; Skaletzty, L. L.; Perricone, S. C.; Piper, R. C.; Delehanty, P. J. *J. Med. Chem.* **1986**, *29*, 133.
192. Leach, C. A.; Brown, T. H.; Ife, R. J.; Keeling, R. J., Laing, S. M.; Parsons, M. E.; Price, C. A.; Wiggall, K. J. *J. Med. Chem.* **1992**, *35*, 1845.
193. Keeling, D. J.; Malcom, R. C.; Laing, S. M.; Ife, R. J.; Leach C. A. *Biochem. Pharmacol.* **1991**, *42*, 123.
194. Sarges, R.; Gallagher, A.; Chambers, T. J.; Yeh, L.-A. *J. Med. Chem.* **1993**, *36*, 2828..
195. Babbs, M.; Collier, H. O. J.; Austin, W. C.; Potter, M. D.; Taylor, E. P. *J. Pharm. Pharmacol.* **1956**, *8*, 110.
196. Castle, A.; Haylett, D. G.; Morgan, J. M.; Jenkinson, D. H. *Eur. J. Pharmacol.* **1993**, *236*, 201.
197. Dunn, P. M. *Eur. J. Pharmacol.* **1994**, *252*, 189.
198. Dunn, P. M.; Davis, C. M.; Ganellin, C. R.; Haylett, D. G.; Morgan, J. M.; Jenkinson, D. H. Br. *J. Pharmacol.* **1991**, *104*, 165.

199. Galanakis, D.; Davis, C. A. ; Del Rey Herrero, B.; Ganellin, C. R.; Dunn, P. M.; Jenkinson, D. H. *J. Med. Chem.* **1995**, *38*, 595.
200. Galanakis, D.; Davis, C. A.; Ganellin, C. R.; Dunn, P. M. *J. Med. Chem.* **1996**, *39*, 359.
201. Misra, V. S.; Dhar, S. *Ind. J. Chem., Sect. B.* **1979**, *18B*, 262.
202. Gombar, V. K. *Arzneim.-Forsch.* **1985**, *35*, 1633.
203. Strkowski, L.; Mokrosz, M. J.; Harden, D. H.; Mokrosz, J. L.; Wilson, W. D. In *Adv. Chemother. AIDS*; Pergamon: New York, 1990.
204. Kireev, D. B.; Chrétien, J. R.; Raevsky, O. A. *Eur. J. Med. Chem.* **1995**, *30*, 395.
205. Lukevics, E.; Zidermane, A.; Dauvarte, A.; Lapina, T. V.; Khokhlova, L. N.; Segal, I. D. *Khim. Farm. Zh.* **1978**, *12*, 62.
206. Morliere, P.; Momenteau, M.; Candide, C.; Simonin, V.; Santus, R.; Maziere, J.-C.; Dubertret, L.; Goldstein, S.; Hüppe, G. *J. Photochem. and Photobiol., B. Biology* **1990**, *5*, 49.
207. Moyer, M. P. ; Weber, F. H.; Gross, J. L.; Isaac, J. W.; Saint Fort, R. *Bioorg. Med. Chem. Lett.* **1992**, *2*, 1589.
208. Guerrero, S.; Zacharias, *J. Gen. Pharmacol.* **1986**, *17*, 37.
209. Davey, M. J. 1989. Eur. Pat. Appl. EP 303,357. *Chem. Abstr.* **1989**, 111, 209164 .
210. Mukherjee, P.; Pathak, B. *Ind. J. Chem., Sect. B.* **1983**, *22B* , 837.
211. Coghlan, M. J.; Dreikorn, B. A.; Suhr, R. G.; Jourdan, G. P. 1989. Eur. Pat. Appl. EP 326,328. *Chem. Abstr.* **1990**, *112*, 55907.

STEREOSELECTIVE SYNTHESIS OF FUNCTIONALIZED FIVE-, SIX- AND SEVEN-MEMBERED HETEROCYCLIC SYSTEMS *VIA* INTRAMOLECULAR CYCLOADDITION OF C-ALKENYL NITRONES AND NITRILE OXIDES

Ugo Chiacchio,[a] Antonio Rescifina[a] and Giovanni Romeo[b]
[a]*Dipartimento di Scienze Chimiche, Università di Catania, Viale A. Doria 6, 95125 Catania, Italy*
[b]*Dipartimento Farmaco-Chimico, Università di Messina, Viale Annunziata, 98168 Messina, Italy*

Abstract. *The most recent advances on synthetic applications of the intramolecular cycloaddition of nitrones and nitrile oxides towards the stereoselective construction of functionalized five-, six- and seven-membered heterocycles have been surveyed.*

1. Introduction

1,3-Dipolar cycloadditions offer one of the most powerful synthetic methodologies for a valuable entry towards functionalized heterocyclic systems:[1] the reaction usually proceeds regio- and stereoselectively, *via* a [4+2] addition process, so involving the formation of a heterocyclic ring.

Nitrones and nitrile oxides, as easily accessible compounds and important members of many known 1,3-dipoles, have greatly contributed to this field for a long time.[2,3] The synthetic potential of these 1,3-dipoles is based primarily on two factors. One is their high reactivity with a wide range of dipolarophiles such as electron-deficient, electron-rich or non activated species, to afford five-membered 1,2-*N,O*-heterocyclic systems. The second feature is that the resulting heterocycles such as 2-isoxazolines or isoxazolidines have various synthetically useful functionalities masked in the rings, which can be released through ring cleavage giving easy access to a variety of differently functionalized open chain derivatives.

For instance, nitrile oxide-olefin cycloadditions serve as useful starting materials for bifunctional systems such as amino alcohols, keto alcohols, β-diketones or unsaturated ketones, *via* cleavage of the *N-O* bond of the primarily formed isoxazoline (Scheme 1).[4]

Analogously, isoxazolidines populates different reaction channels, following cleavage of *N-O*, *C-N* and/or *C-O* bonds, alternative and competitive, according to the nature of the reagent and the experimentally adopted conditions (Scheme 2).[5]

Furthermore, one of the key features of the strategy based on the ring-opening of five-membered *N,O*-heterocycles resides on the regioselectivity and stereo-specificity of the cycloaddition that transports the stereochemical informations of the olefin and of the dipole to the new stereocentres of the heterocyclic ring .[6]

Since the reaction creates several contiguous stereogenic centres in a well-defined manner, in recent years the intramolecular version of the 1,3-dipolar cycloaddition has received a very considerable synthetic and mechanistic interest, as an important stereogenic step, especially in the synthesis of the intriguing carbon frameworks occurring in natural products and other stereochemically complicated molecules.[6,7]

In fact, when both reactants are part of the same molecule, the intramolecular nature of this synthetic scheme leads to a simultaneous double annulation, with formation of two rings and allows, *via* the cleavage of the isoxazoline or isoxazolidine ring, a direct access to highly functionalized systems in a regioselective and stereocontrolled manner. One ring would be heterocyclic or carbocyclic, the second an isoxazoline or isoxazolidine ring, which could be opened to produce stereoselectively positioned functional groups.

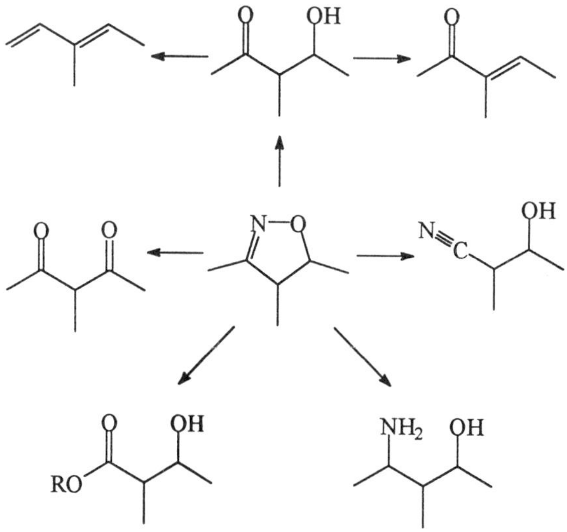

Scheme 1

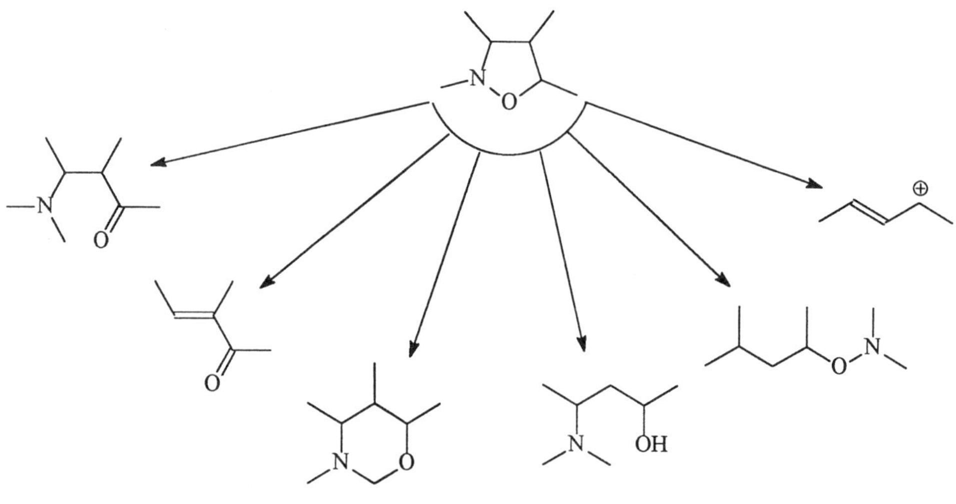

Scheme 2

While HOMO-LUMO interactions mainly direct intermolecular nitrile oxide and nitrone addition to 1-alkenes to produce regioselectively 5-substituted isoxazolines or isoxazolidines, intramolecular addition appear to behave differently with formation of 4-substituted *N,O*-heterocycles, due to geometrical constraints, or 5-substituted derivatives, due to favourable frontier molecular orbital interactions (Scheme 3).

A significant body of information regarding the regiochemistry and the stereochemistry of the process has been collected. In the case of nitrile oxide intramolecular cycloaddition (INOC), the obtainment of 4-substituted

2-isoxazolines indicates that generally geometric factors can force the reaction to occur in the opposite manner from that normally encountered in the intermolecular process. However, the INOC is function of chain length between the dipole and dipolarophile:[7] in fact, macrocarbocycles originate from the normal regiochemical directing effects, provided by the matching of HOMO-LUMO interactions, as a result of ring strains and transanular steric effects (Scheme 4).[8]

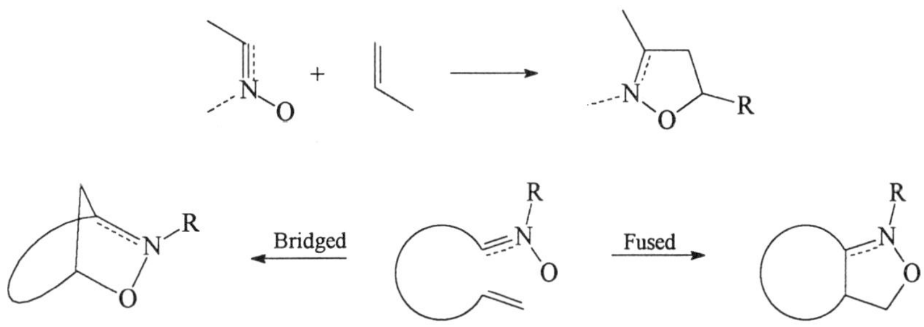

Scheme 3

On the contrary, the intramolecular cycloaddition of nitrones (INC) can be viewed as a competitive process between the bridged and fused modes of cycloaddition, controlled by a suitable interplay of factors such as alkene polarity, ring strain and other non-bonded interactions.[8]

A comprehensive report on nitrile oxide and nitrone intramolecular cycloadditions covers the literature up to 1991.[9] The present article reviews on the most recent advances on the synthetic applications of these dipoles towards the stereoselective constructions of functionalized heterocyclic systems and covers the literature till 1996.

Scheme 4

2. Nitrones

LeBel fundamental studies on the regiochemistry of intramolecular cycloaddition of acyclic alkene nitrones are instructive but not conclusive.[10]

A preference was found for the formation of fused bicyclic compounds, but bridged adducts appear when bond formation between the β-position of the alkene and the α-position of nitrone was hindered by substitution in those positions. Thus, 5-hexen-1-imino-*N*-oxides **1** react with high regio- and stereoselectivity to give the fused 3-oxa-2-azabicyclo[3.3.0]octanes (*cis*) **2** as exclusive products; however, when both R_3 and R_5 are CH_3, essentially equal amounts of **2** and **3** are formed, suggesting that steric effects at the reacting carbon termini may operate against fused-product regiochemistry (Scheme 5).[11]

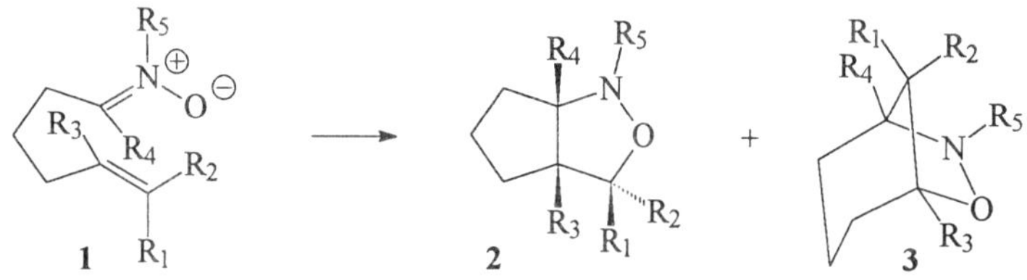

Scheme 5

The regiochemical outcome of the intramolecular cycloaddition of 6-alkenyl aldonitrones depends on the substitution at the alkene carbons. Then, 6-hepten-1-imino-*N*-oxides **4** give predominantly fused products **5**, 8-oxa-7-azabicyclo[4.3.0]nonanes (*cis* and *trans*), when C_6 is unsubstituted (R_3 = H), but the bridged 7-oxa-8-azabicyclo[4.2.1]nonanes **6** are exclusively formed when R_3 is an aryl group (Scheme 6).[12]

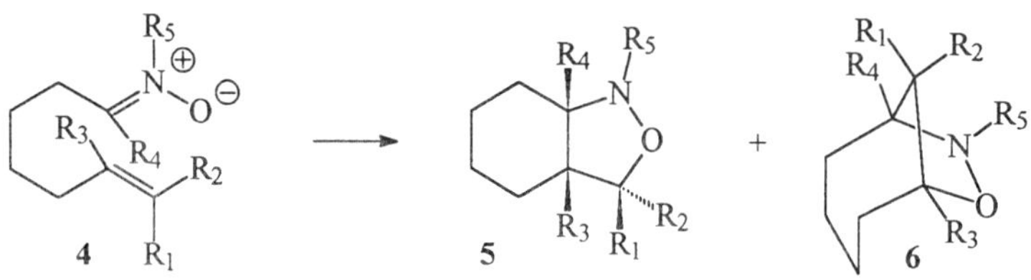

Scheme 6

The stereochemical features guiding the intramolecular cycloadditions are still now not too well understood in spite of the pioneering work of LeBel. While the stereochemical informations present in the alkene moiety are maintained in the obtained cycloadducts, the stereochemistry of the ring fusion (*cis* or *trans*) depends on the nitrone configuration and the geometries of the more stable **TS**.[13]

A chiral centre in the starting nitrone could cause an asymmetric induction giving rise to the formation of new chiral centres with definite configuration in the cycloadduct [6,7]. In particular, stereoselection at C_3 of the acyclic substrate is reported to give the best control of the new formed stereogenic centres (Scheme 7).[14]

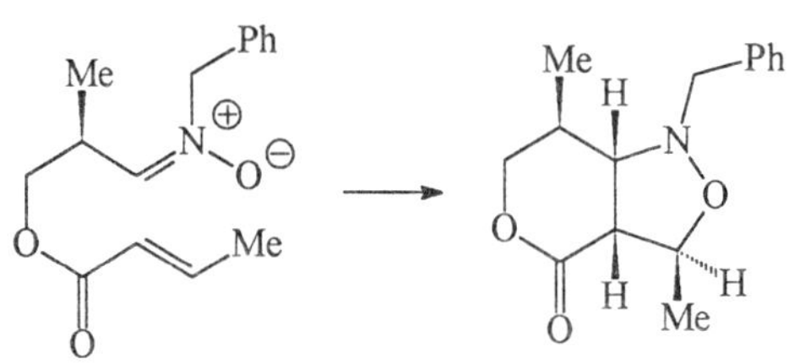

Scheme 7

Introduction of a heteroatom in the tether between the nitrone and the alkene groups considerably extends the synthetic potential of the reaction. Such reactions assume great utility in the synthesis of pyrrolidines, pyrrolizidines, furans and thiophenes. In particular, the intramolecular cycloaddition of several differently substituted 5-hexenyl-1-imine-*N*-oxides with nitrogen, sulphur and oxygen at the 3-position has been investigated as a tool for the diastereoselective or enantioselective synthesis of stereochemically defined heterocycles (Scheme 8).[15]

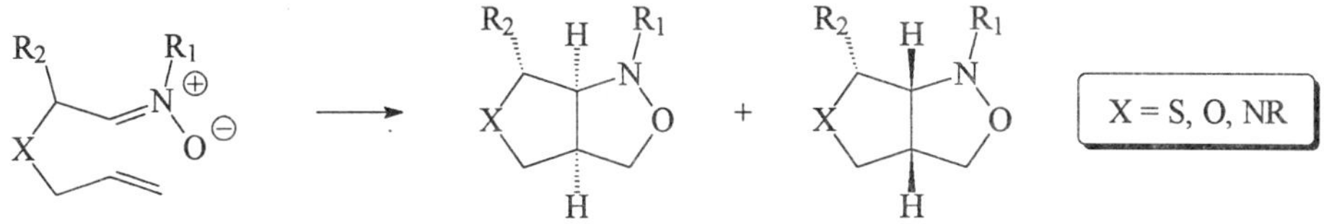

Scheme 8

2.1 Five-membered rings

The synthetic versatility of the intramolecular cycloaddition of nitrones containing additional oxygen, nitrogen or sulfur atoms has been exploited in a wide series of examples. Applications of this methodology are seen in the synthetic routes towards functionalized furan, pyrrole and thiophene derivatives, in some cases starting from homochiral educts.

Factors controlling the stereochemical outcome of the reaction have been elucidated on the basis of a kinetic control of the process and the stability of the different transition states leading to different diastereo-mers.

2-Oxa- and 2-thio-4-pentenyl nitrones **7** react in a regioselective and stereoselective manner to give cis fused furans and thiophenes **8** exclusively;[16] selective cleavage of isoxazolidine ring affords 3-amino-4-hydroxymethyl furans and thiophenes **9**. Removal of the sulfur atom from **9** leads to 1,3-aminoalcohols **10** with a controlled stereochemistry (Scheme 9).

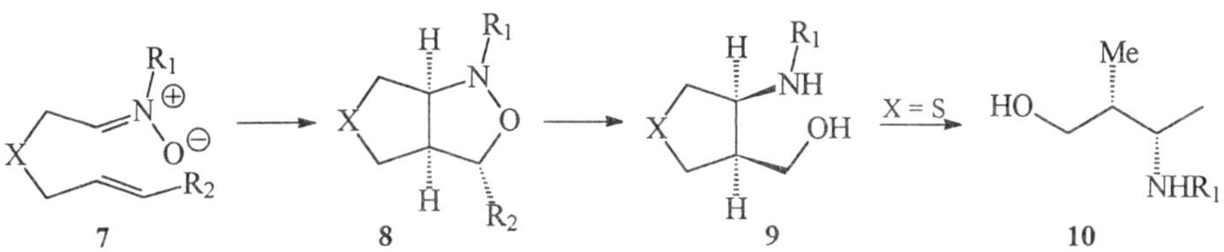

R_1 = Me, *t*-Bu, *p*-*t*-BuC_6H_4, PhCH_2; R_2 = H, Me, CO_2Me; X = O, S

Scheme 9

On the same basis, the conversion of a variety of enantiomerically pure α-hydroxy esters to enantiomerically pure bicyclic furans **12**, *via* the corresponding nitrones **11**, has been reported[17,18] (Scheme 10).

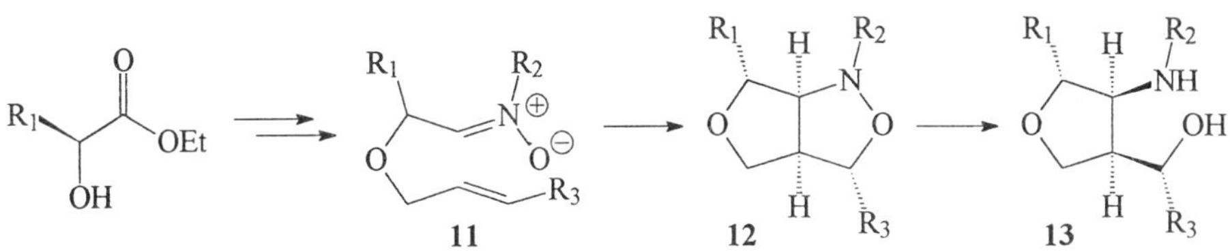

Scheme 10

The intramolecular cycloaddition proceeded with complete diastereoselectivity: the stereogenic centre originating from the α-hydroxy esters gave high asymmetric induction resulting in the formation of 4 contiguous stereogenic centres in the obtained 3,7-dioxa-2-azabicyclo[3.3.0]octanes **12**. Opening of the isoxazolidine ring of **12** by various reductive methods yielded the highly substituted tetrahydrofuran derivatives **13** with a γ-aminoalcohol group: the ^1H NMR spectra reveal a diastereomeric purity of at least 98%.

Fluoro analogues of these bicyclic systems, of potential biological interest, have been also synthesised (Scheme 11).[19] Thus, nitrone **15**, prepared from chiral β-ketosulphoxide **14**, was converted into the enantiomerically pure 3-amino-4(hydroxyalkyl)tetrahydro-furan **17** in 61% yield *via* the reductive cleavage of the isoxazolidine cycloadduct **16**.

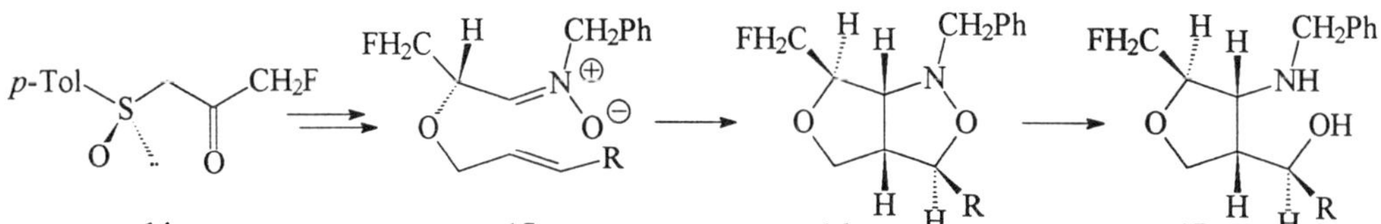

14 **15** **16** **17**

Scheme 11

A novel general approach to functionalized γ-lactams has been reported starting from a series of unsaturated α-amido aldehydes (Scheme 12).[20]

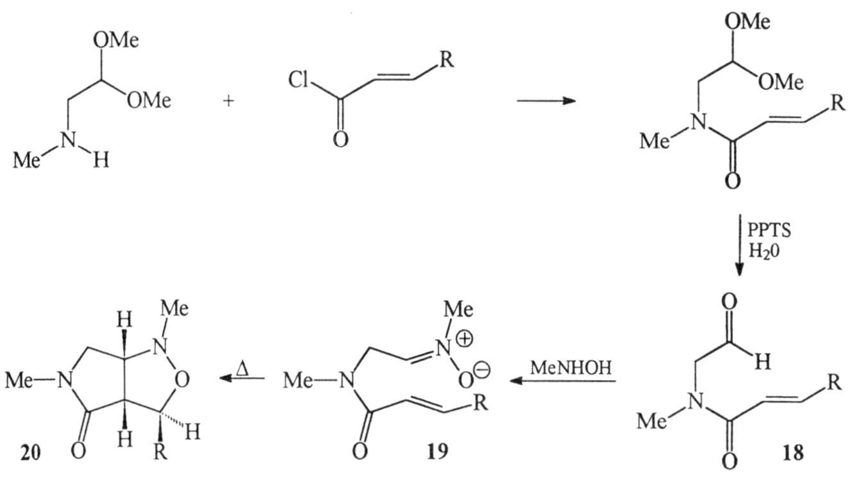

20 **19** **18**

Scheme 12

Nitrones **19**, prepared in situ from the related aldehydes **18** with *N*-methyl hydroxylamine, reacts in a regioselective and stereoselective manner to give *cis* fused 3-oxa-4-substituted-2,7-diazabicyclo[3.3.0]octan -6-ones **20**, exclusively.

A stereocentre located in position α to the nitronic functionality completely controls the stereochemical course of the intramolecular cycloaddition, which proceeds diastereoselectively, furnishing homochiral compounds **22** from homochiral nitrones **21** with simultaneous introduction of four stereocentres (Scheme 13).[21]

The exclusive formation of **22** is supported by PM3 calculations. Analysis of these data point out that **TS** involving the nitrone **21a** in *E* and *Z* forms is energetically favoured with respect its conformers **21b** which, instead, would lead to the formation of compounds **23**.

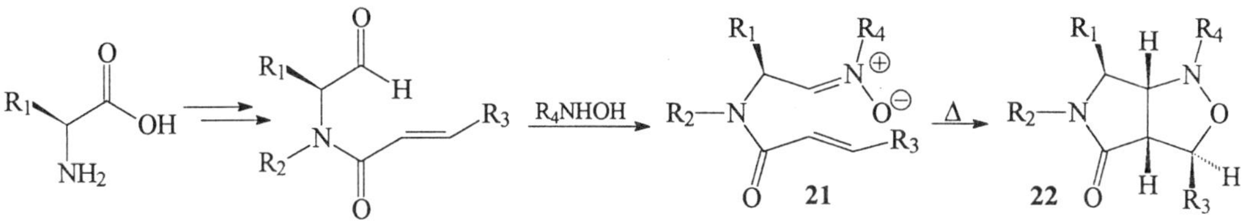

Scheme 13

Moreover, according to the Boltzman equation, the conformational equilibrium is *E* **21a** = 88% and *Z* **21a** = 12%, while *E* **21b** and *Z* **21b** are absent. On this ground, therefore, compounds **22** are the only products of the reaction (Scheme 14).

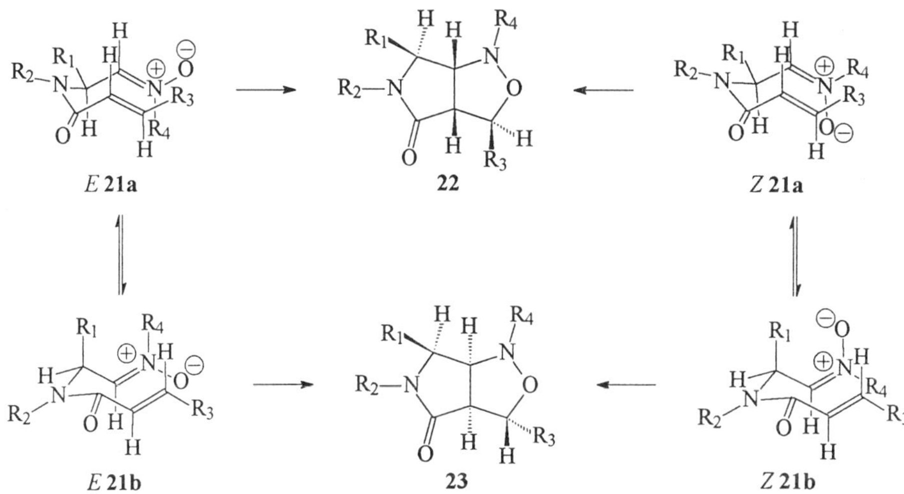

Scheme 14

Selective and reductive ring cleavage of the isoxazolidine ring represents an easy synthetic access towards homochiral functionalized pyrrolidine and pyrrolizidine systems, starting from homochiral aminoacids (Scheme 15).

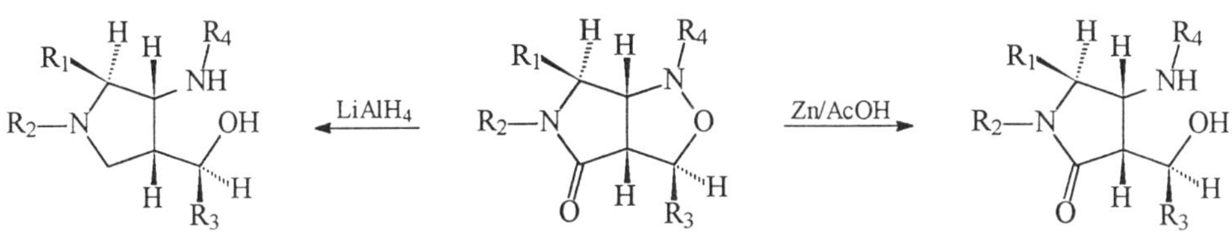

Scheme 15

The amino and alcoholic functions present in the obtained compounds offer the possibility of usefully synthetic manipulations directed towards the synthesis of natural alkaloids.

Some sultams **25** have been prepared in similar way starting from unsaturated sulphonamide aldehydes **24** (Scheme 16).[22]

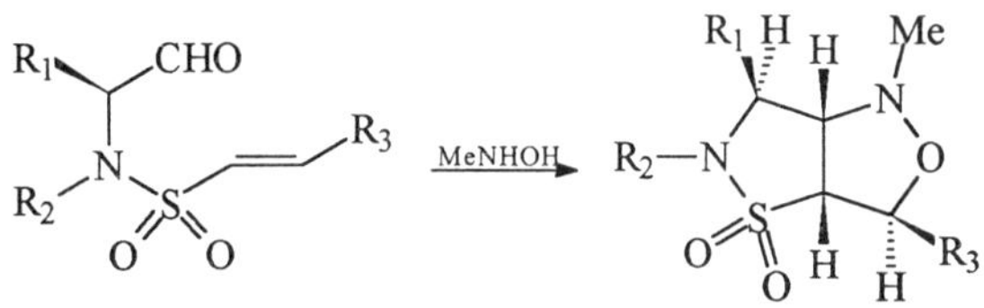

24 **25**

Scheme 16

Homochiral sulphonamides, containing a stereocentre located in α to the aldehyde function, yield homochiral bicyclic compounds **25**. These products can be used as Oppolzer's sultams in enantioselective syntheses.[23]

Analogously, nitrones **27**, obtained from the corresponding α-amino-acids,[24,25] yielded the bicyclic compounds **28** (Scheme 17).

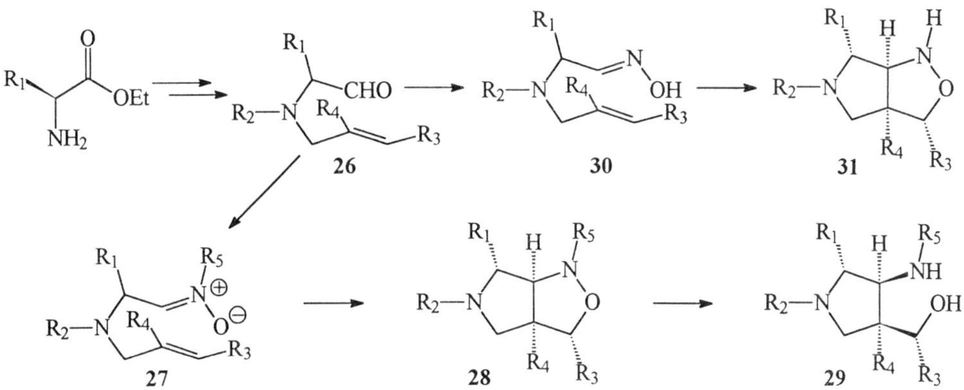

Scheme 17

The intramolecular cycloadditions proceed diaste-reoselectively furnishing homochiral compounds **28** from homochiral starting material.

Further reductive cleavage of the isoxazolidine ring afforded stereochemically defined pyrrolidines **29**.

Pyrrolidines **31** have been also obtained from oximes **30** by heating in toluene at 110 °C.

Compounds **28** and **29** were tested as catalysts for the reaction of diethylzinc with benzaldehyde and found to be moderately active, with an observed enantiomeric excess not more than 61%.

The enolate compounds **33**, generated by treatment of nitrone **32** with LDA, can act as dipolarophiles in intramolecular nitrone cycloaddition, yielding the bicyclic compounds **34** diastereomerically pure. Nitrone **32** were synthesised by nucleophilic addition of *N*-benzylalaninol with 3-buten-2-one, followed by Swern oxidation of the resulting compound and subsequent reaction with RNHOH (Scheme 18).[26]

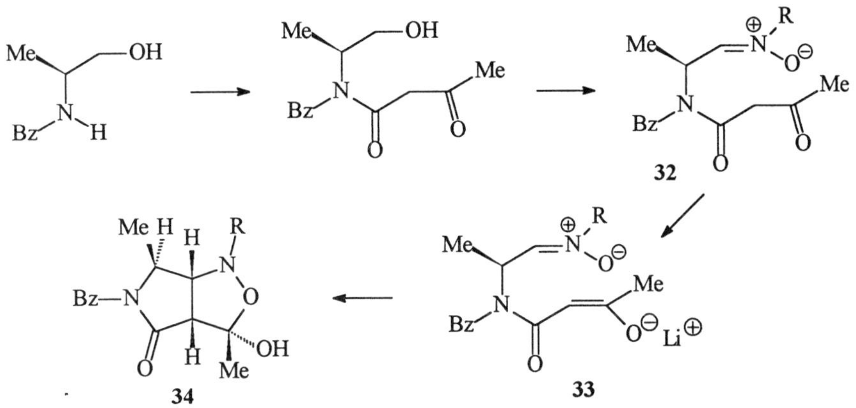

Scheme 18

The high diastereoselectivity observed in these reactions can be explained by the attack of nitrone group to *Re* side of the alkene moiety because of a minor steric interaction between oxygen and hydrogen in the transition state *a* (Scheme 19).

On the contrary, the attack on the *Si* side would cause a strong steric interaction between the oxygen and the CH_3 group at position 2 (see **TS** *b*).

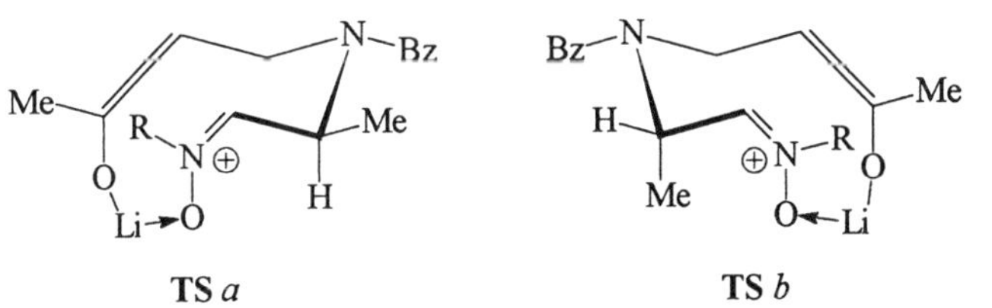

Scheme 19

According to the results obtained for oxa- and aza-nitrones,[18,24] thioalkenyl nitrones **35**, obtained from chiral mercapto alcohols, underwent spontaneously intramolecular cycloaddition to give the diastereomerically pure bicyclic compounds **36** with 4 contiguous chiral centres (Scheme 20).[27]

The process occurs with an high asymmetric induction by the chiral centre at C_2; however, during the reaction course partial racemization has been observed. In contrast, nitrones **37**, in which the chiral centre is adjacent to the nitrogen atom, afforded a mixture of the two diastereomeric cycloadduct **38** and **39**: the change from 1,2- to 1,3-induction caused a dramatic loss of diastereoselectivity.

Shift of sulfur atom from position 3 to position 4 in the thioalkenyl nitrones conjugates the advantage of the high asymmetric induction by a chiral centre at C_2 with the reduction of the susceptibility to racemization by the *S* atom: nitrones **40** possessing even two chiral centres, gave bicyclic compounds **41**, with 5 contiguous chiral centres, diastereomerically and enantiomerically pure (Scheme 20).

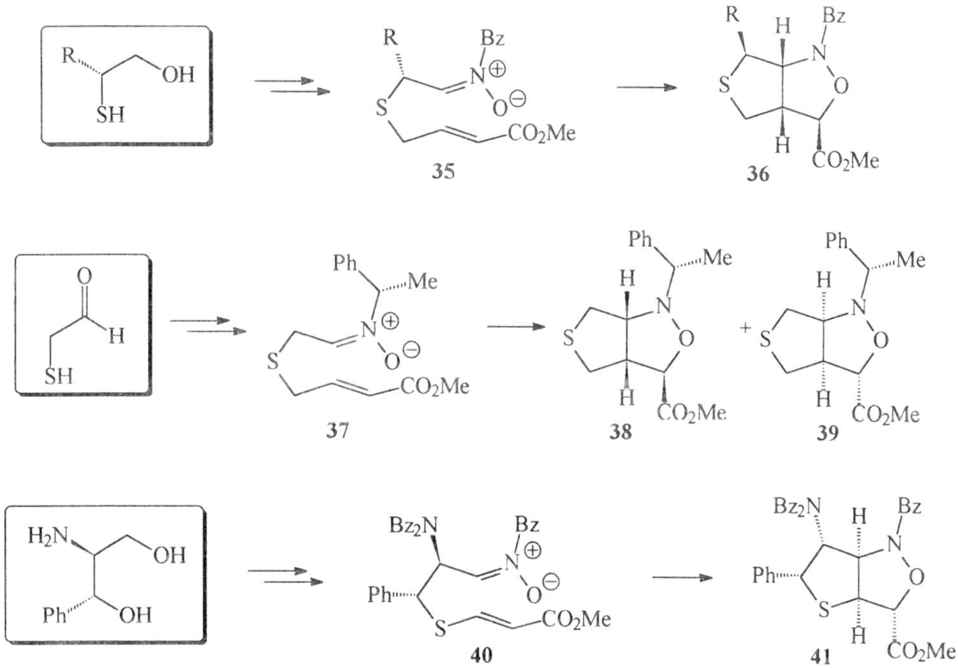

Scheme 20

The 4-cyano-substituted 3-oxa-7-thia-2-azabicyclo [3.3.0]octanes **42** allow a new entry to β-lactams **45**. The action of LDA on the *N*-arylsubstituted compounds **42**, possessing an *exo*-hydrogen atom at 4-position, easily abstractable by sterically hindered bases, afforded the bicyclic lactams **45**. The process is rationalizable on the basis of the ring-opening of the carbanion **43**, initially formed by reaction with LDA, to afford the amide anion **44**, which, through intramolecular substitution at the acyl cyanide group, leads to **45** with loss of the cyanide ion. The aryl substituents at the *N* atom play a decisive role on the rearrangement pathways by stabilizing the negative charge which develops at the nitrogen atom (Scheme 21).[28]

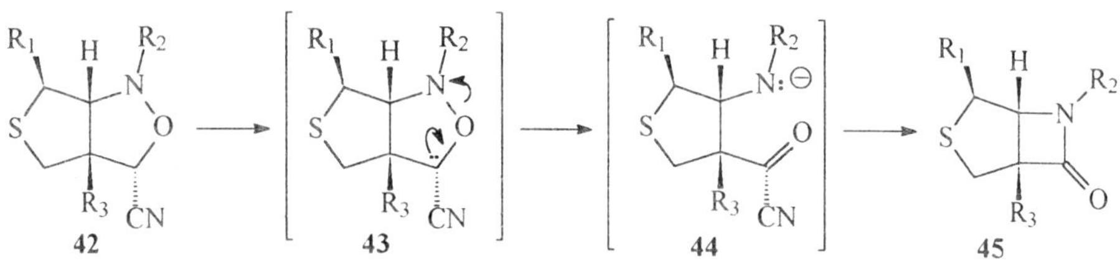

Scheme 21

The influence of an allylic stereocentre, located outside the tether connecting dipole and dipolarophile units, has been investigated in a series of intramolecular nitrone cycloadditions to chiral allylic esters; the stereochemical outcome, exerted by several factors, including the nature of the substituents at the stereocentre, the alkene geometry and steric and electronic features of the double bond, have been evaluated.

The allylic stereocentres control the steric course of the cycloaddition more effectively than a chiral auxiliary located at the *N* atom, too far away from the core of the transition state.[29]

Nitrone **46** and **47** (*Z*, *E* isomers) have been obtained from the corresponding alcohols by α-bromo acetylation, treatment with $AgNO_3$, reduction of nitrate esters with DMSO and AcONa and condensation of the resulting glyoxylate with *N*-alkylhydroxylamines. The subsequent intramolecular cycloaddition gave a mixture of diastereomeric furoisoxazolidinones **48**, **49** and **50**, **51** respectively.

The best stereoselection, observed with glyceroaldehyde derived substrate (R_2, R_3 = $-OC-[(CH_2)_5]-OCH_2-$; **48**:**49** = 86:14 and **50**:**51** = 97:3) was rationalised on the basis of a stronger donor ability of the CH_2OR group compared to an alkyl group. This effect is promoted by a lone-pair participation of the homoallylic oxygen or by a direct trough-space interaction of the homoallylic oxygen lone-pair with the alkene π-bond (Scheme 22).[30,31]

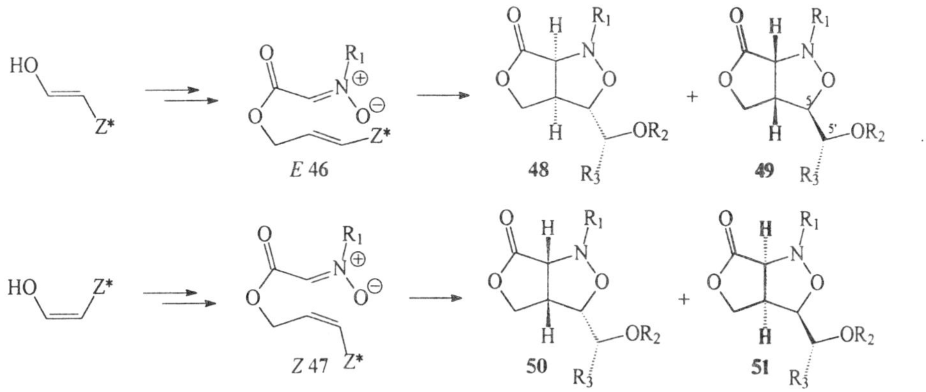

Scheme 22

(*Z*)-Alkenyl nitrones undergo more stereoselective cycloadditions than their *E*-counterparts. The stereochemical results can be explained by the inspection of the relative transition state **TS**. In fact, with *Z*-olefins, **TS** *a* appears to be more stable than *b* because of the lack of repulsive interactions between the nitrone oxygen and the alkoxy group. For *E*-alkenes stereoelectronic effects mainly rule the process: inside alkoxy conformations are preferred (**TS** *c* over **TS** *d*) because nitrone oxygen attacks away from the alkyl group **R** located in the *anti* position (Scheme 23).

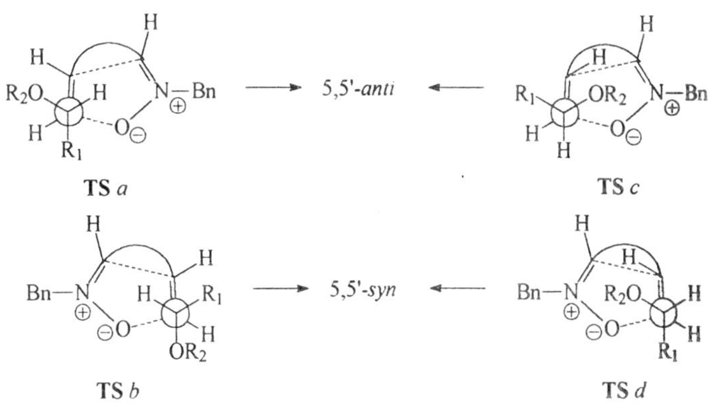

Scheme 23

α-Allyloxycarbonyl nitrones **54** have been also prepared, as transient intermediates, in one step: treatment of α-methoxycarbonylnitrones **52** with allylic alcohols **53** in the presence of $TiCl_4$ and molecular sieves 4A causes tandem transesterification and *E*/*Z* isomerization of the nitrone moieties. Compound **54** undergoes spontaneous intramolecular cycloaddition to afford bicyclic or tricyclic compounds **55** (Scheme 24).[32,33]

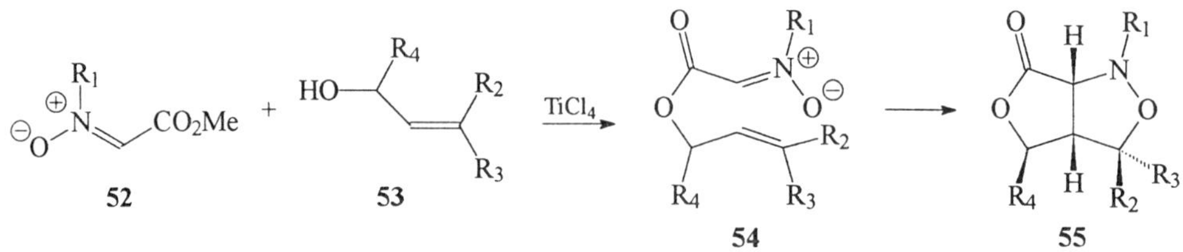

Scheme 24

In the cases employing cyclic allylic alcohols [R_2, R_4 = $-(CH_2)_4-$], the reaction could control four relative stereogenic centres in one step. In the absence of $TiCl_4$, the reaction gave the expected epimeric mixture of cycloadducts **56**, having the opposite regiochemistry, originated from the intermolecular cycloaddition process (Scheme 25).

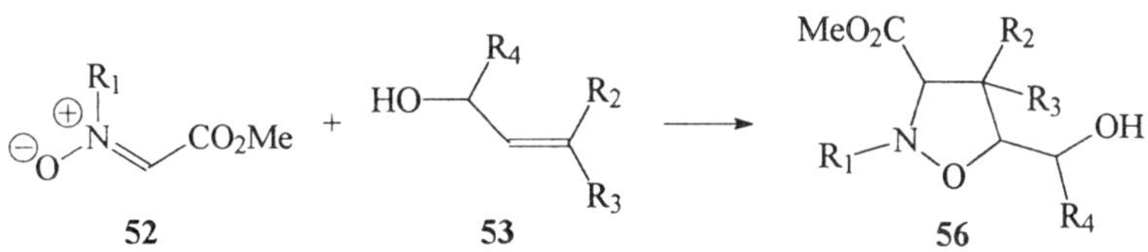

Scheme 25

On this basis, the mechanism reported on the scheme 26 has been proposed.

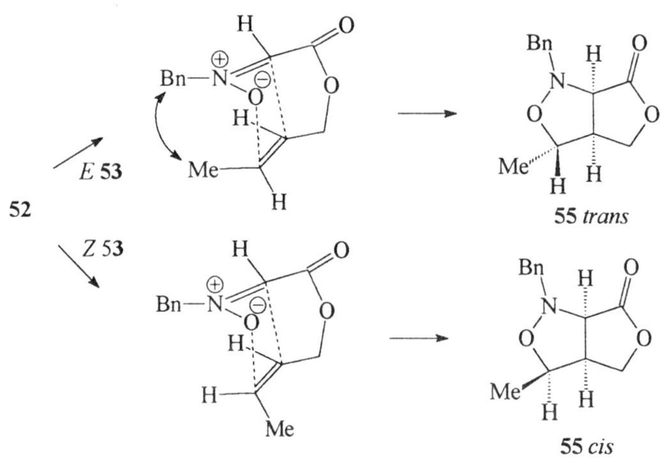

Scheme 26

Possibility of intramolecular cycloaddition from E nitrone is neglected due to the high steric strain in the **TS**. Furthermore, it was found that the process is characterised from a geometry differentiated cycloaddition: reaction of Z allyl alcohols **53** with **52** is faster than those of E allylic alcohols. The reduced reactivity with E allylic alcohol is due to the steric hindrance between benzyl and methyl group.

A synthetic route towards butenolide systems, versatile synthons fore the obtainment of β-lactams has been designed. Racemic 2-amino-5H-furanones have been prepared by intramolecular cycloaddition of the glyoxylate of allylic alcohols, containing a stereogenic centre at position 3 of the tether connecting the dipole and dipolarophile moieties.[34] The treatment of the epimeric mixture of furo[3,4-c] isoxazol-6-ones **57** with NaH affords diastereoselectively **58**, by an aldehyde elimination from the corresponding carbanion generated by H abstraction at C_{6a} of furoisoxazole ring (Scheme 27).

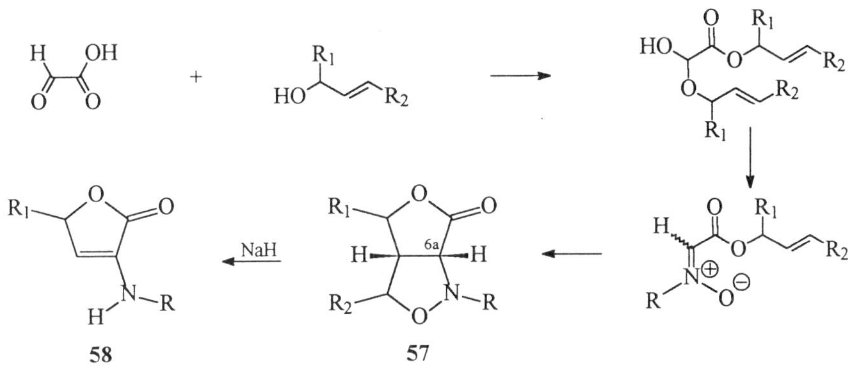

Scheme 27

The same reactions performed with homochiral allylic alcohols afforded enantiomerically pure compounds.

Convenient routes to the pyrrolo[1,2-a]indole ring system **60**, which is contained in numerous natural products, like as antitumor antibiotic mitomycin and isoborreverine has been envisioned starting from *N*-allylindole-2-carbaldehydes **59** with *N*-methylhydroxyl-amine (Scheme 28).[35-37]

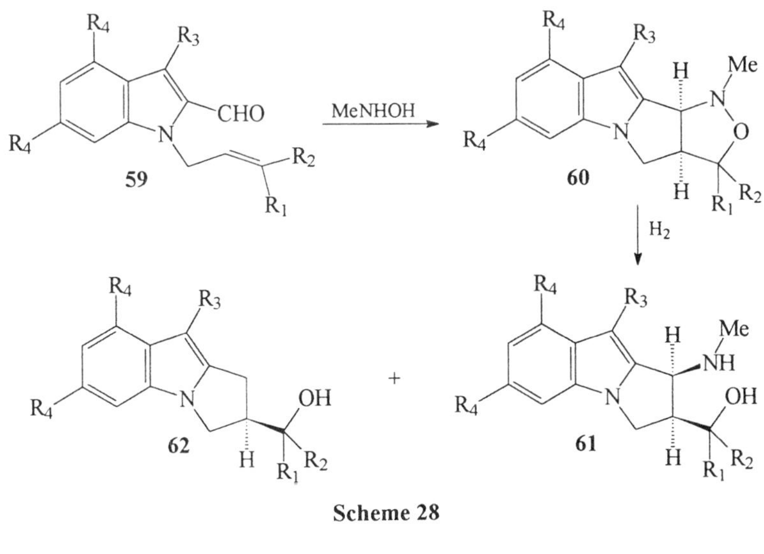

Scheme 28

Ring opening of *N-O* bond by hydrogenolysis give rise the amino alcohols **61** together with traces of the alcohols **62**.

The tandem intramolecular 1,3-dipolar cycloaddition methodology has been exploited in a synthetic scheme towards nucleosides with a 2'- or 3'-spiro- unit, possessing anti-HIV-1 activity, starting from *C*-alkenyl-nitrone derivatives.[38]

Thus, an isomeric pair of tricyclic *cis*-fused-spiro-isoxazolidine nucleosides **64** and **66** have been prepared regio- and diastereospecifically through 2'- or 3'-*N*-methylnitrones **63** and **65**, generated in situ from the parent ketones (Scheme 29).

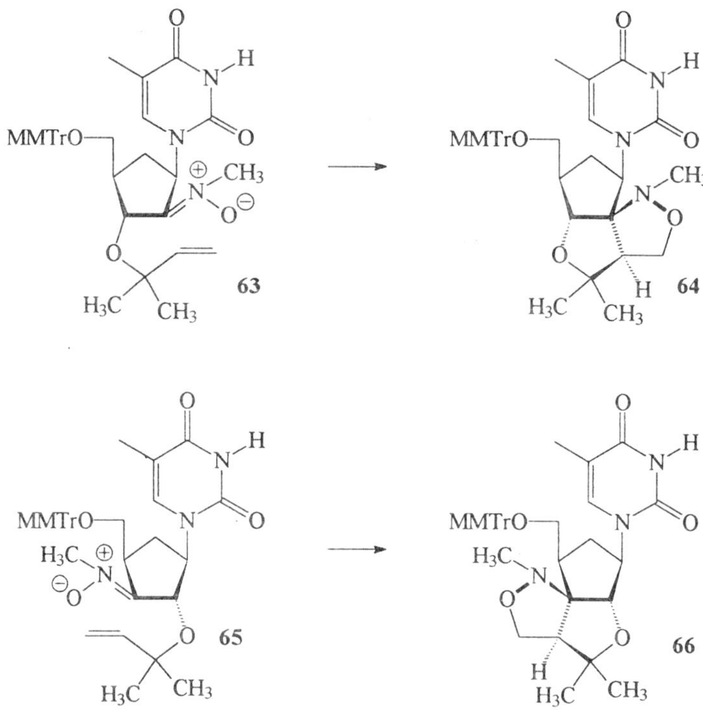

Scheme 29

Functionalized spiropyrrolidines **69** have been synthesised by intramolecular cycloaddition of nitrones **68** originated from the reduction of the corresponding nitroketones **67** (Scheme 30).[39]

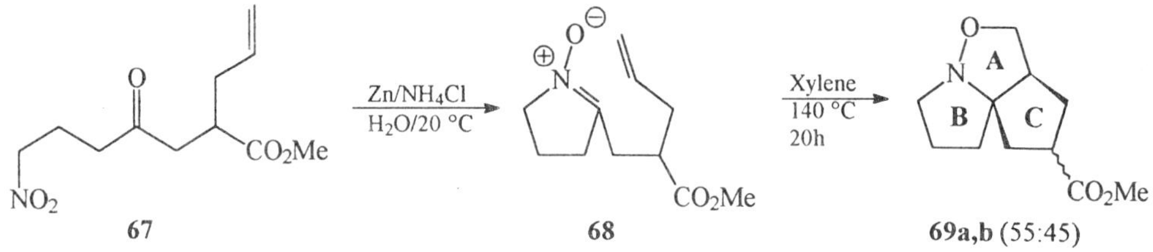

Scheme 30

Two diastereomers **69a,b** with exclusive *cis* fusion of rings **A** and **B** have been obtained. The homologue of **67** gives a mixture of annulated and bridged cycloadducts.

Non racemic 2,6-dioxa-3-azatricyclo[5.3.1.0] undecanes **73** and 2-oxa-3,6-diazatricyclo[5.3.1.0] undecanes **74** have been prepared by intramolecular cycloaddition of δ-unsaturated nitrones in which the unsaturated moiety is part of ring system (Scheme 31).[40]

Starting from (*S*)-ethyl lactate and (*S*)-*N*-benzyl-alanol **70**, 3-oxanitrones **71** and 3-azanitrones **72** respectively have been obtained as diastereomeric mixtures (Scheme 31).

Further spontaneous cycloaddition gives rise to diastereomer cycloadducts **73a** and **74a** (starting from **71**) and **73b** and **74b-c** (from **72**).

The configuration at C_4 determines the side from which the cyclohexene ring is attacked by nitrone group: thus, the *cis*-relationship of protons H_1, H_{11}, H_7 and H_4 in the cycloadducts **73a/74a** and **73b/74b** results from an *exo*-approach of the *Z*-nitrone group.

The unusual *cis*-relationship[41] of H_4 and H_5 in compounds **74a/74b** is caused by the dominating effect of the stereogenic centres in the cyclohexene ring. Finally, the *trans*-annulation of the two five-membered ring in **74c** is assumed to result from the attack of the *E*-nitrone group; an *endo* approach of the *Z*-nitrone group appears to be severally strained.

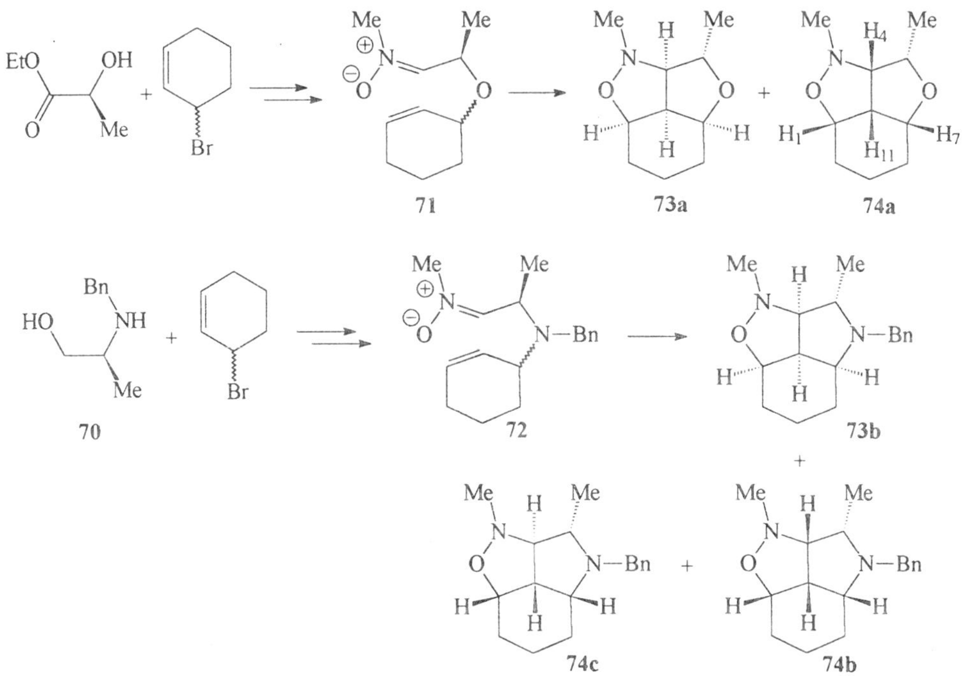

Scheme 31

Tetracyclic compounds **75**, which are similar to **73** and **74**, insofar as they contain the same tricyclic frame but an additional oxygen atom in the six-membered ring, have been prepared starting from glucal by standard procedures (Scheme 32).[42]

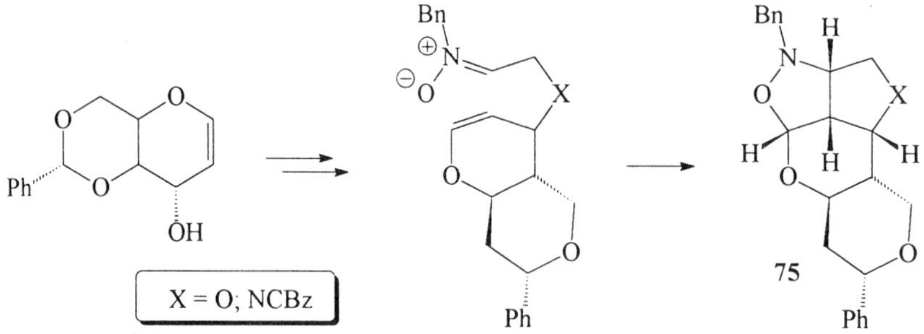

Scheme 32

The intramolecular cycloaddition of oximes has been investigated: these ring closures lead to stereospecific introduction of two or more stereochemical centres, and proceed *via* four tandem nitrone generation cycloaddition processes, as recently rationalised by Grigg.[43]

Route **a** (1,2-prototropy) proceeds *via* a thermally promoted tautomeric 1,2-proton shift to the intermediate nitrone species 77, which undergoes cycloaddition reaction to **78** (Scheme 33).

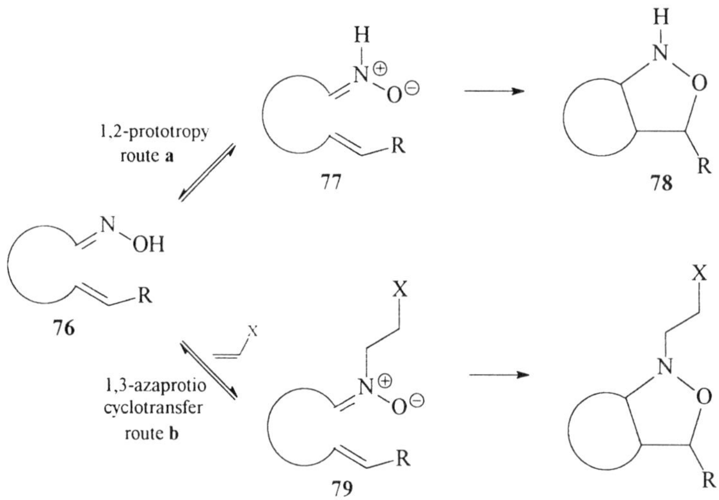

Scheme 33

In routes **b** and **c**, nitrones **79** and **80** are generated by addition of oximes to activated and non-activated double bonds, in a inter- or intramolecular way, respectively, through a concerted ene-like mechanism which involves a $2\pi + 2\sigma + 2n$ electron process (n = lone pair electrons) and a 5-membered transition state (1,3-azaprotio cyclotransfer reaction) (Scheme 33 and 34).

Route **d** involves nucleophilic attack on 3-membered heterocycles *via* the *N* lone pair of the oxime to generate the nitrone **82** which in turn cyclizes to **83** (Scheme 34).

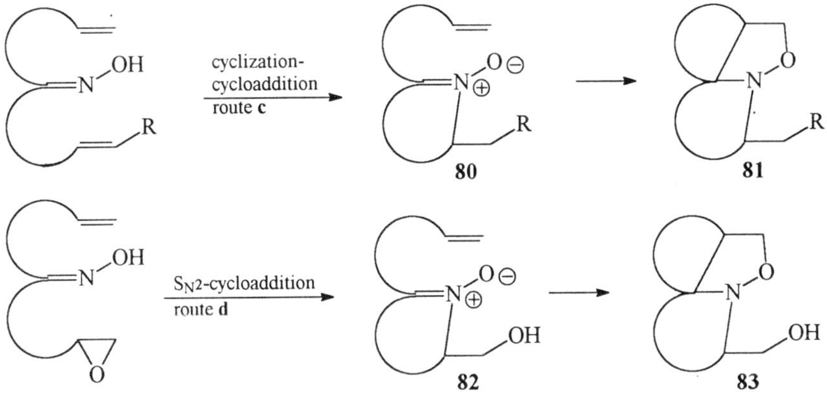

Scheme 34

Hassner has used the 1,2-prototropic cycloaddition pathway (route **a** - IOOC) in the stereospecific formation of pyrrolidines **85** containing 3 or 4 stereocentres, starting from oximes **84** which possess an amine function (Scheme 35).[44]

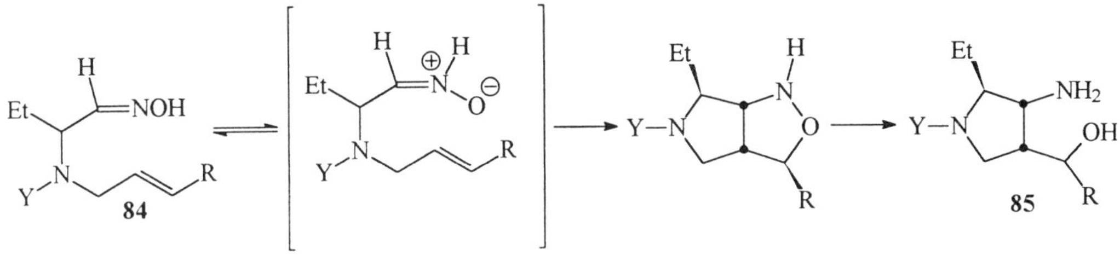

Scheme 35

The presence of an allylic electronegative centre in the non-activated alkene, such as nitrogen or oxygen, results in a beneficial lowering of the alkene LUMO energy as well as in the creation of a more favourable geometry for cycloaddition, as a result of the smaller C-X-C bond angle.[45,46]

Furthermore, this strategy has been exploited in the stereoselective synthesis of pyrrolizidines **87** and **91**, related to necine alkaloids, as well as indolizidine **89** (Scheme 36 and 37).[47]

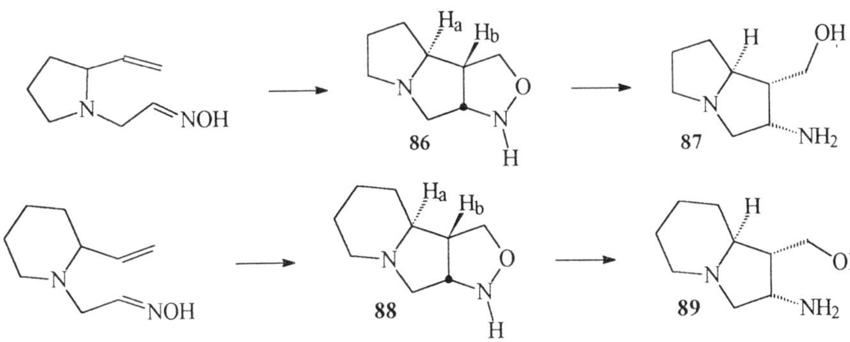

Scheme 36

Closure to 5-membered rings fused to the isoxazolidine, regardless whether part of a pyrrolizidine or of an indolizidine system, led mainly to the *cis-anti* (H_a/H_b) isomers: IOOC products **86**, **88** and **90** were obtained stereochemically pure, while **92** is obtained as a 75:25 mixture of *anti-syn* (H_a/H_b) isomer.(Scheme 37).

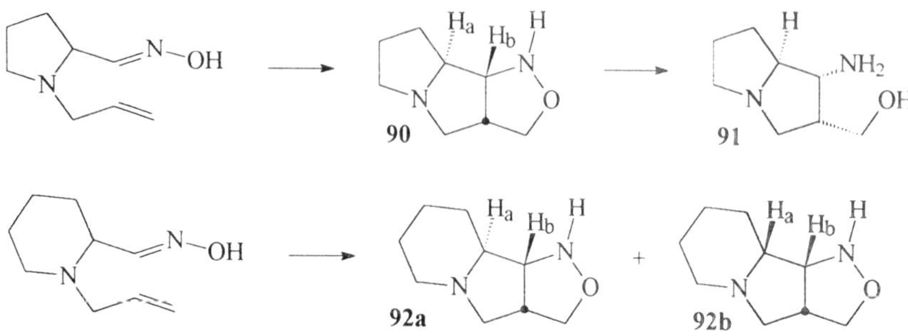

Scheme 37

By means of molecular mechanics calculations it has been possible to predict the more stable isomer formed as well as to correlate or predict NMR coupling constants.[48]

A further interesting application of IOOC approach towards natural alkaloids is afforded by the synthesis of (-)-supinidine **93** and its regioisomers **94** from L-proline (Scheme 38).[49]

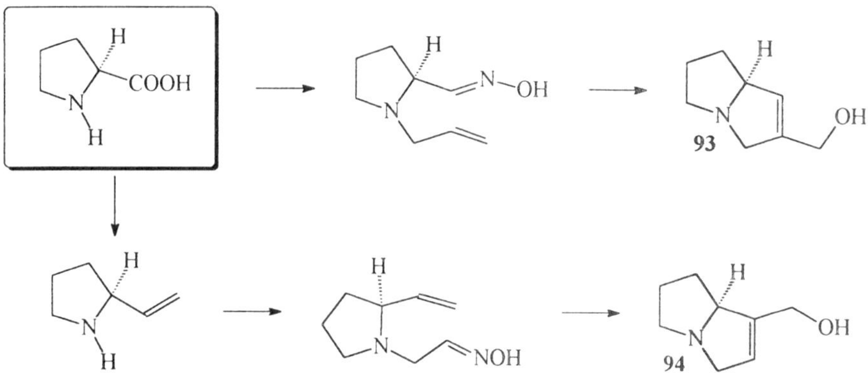

Scheme 38

A route towards asymmetric five-membered aza-sugars, starting from L-serine and utilizing an intra-molecular oxime-olefin cycloaddition, has been developed (Scheme 39). Pyrrolidines **95** show a selective inhibition of α-glucosidase.[50]

Conformational preferences for oxadiazabicyclo [3.3.0]octanes **97**, obtained from IOOC of α-allylamino aldoximes and ketoximes have been evaluated by means of NMR (Scheme 40).[48]

It has been found that some substituents in **97** restrict the conformation mobility of these five-membered rings to the extent that only one conformer was detected at 20 °C. In other cases an equilibrium between two major conformers was revealed by NMR. An alkyl group at C_4 prefers a pseudoequatorial orientation which predisposes the molecule to conformation **97a**. On the other hand, an alkyl substituent at C_7 favours conformer **97b**.

In cases where both conformers were produced, equilibrium measurements indicated a free energy conversion of 13.4-13.2 Kcal/mol, which probably is due to the energy barrier for *N*-inversion around the *N-O* bond. NMR studies also show that *NH* proton in isoxazolidines **97** prefers an axial orientation (Scheme 41).

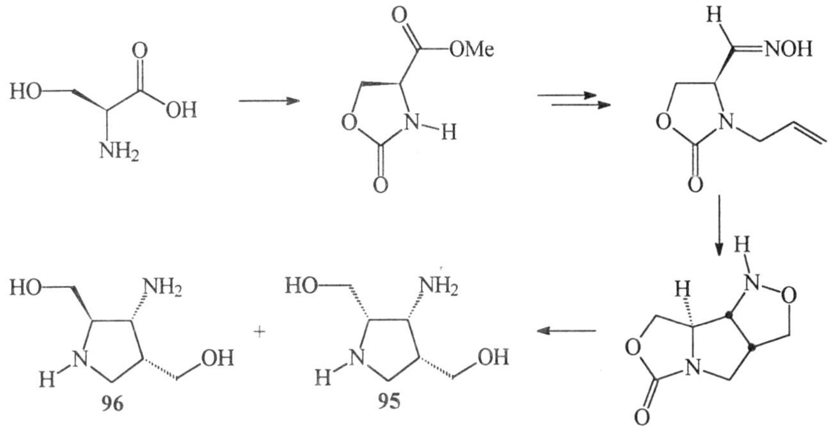

Scheme 39

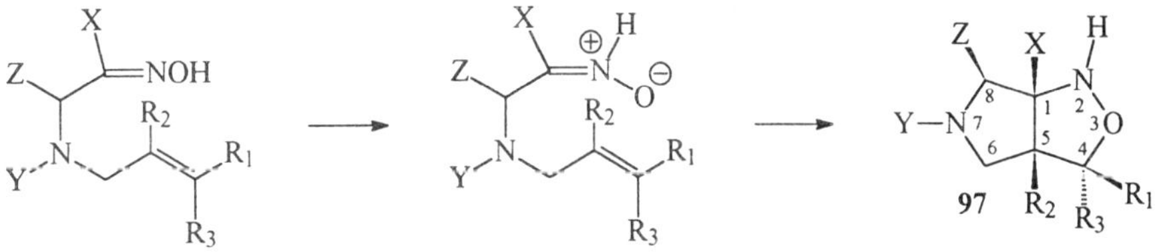

Scheme 40

R = alkyl, favours conformer **97a** (> 10:1); Z = alkyl, favours conformer **97b** (> 10:1) Z and R = H, give 1:2 to 2:1 mixtures of **97a**:**97b**.

Scheme 41

δ-alkenyl oxime **98** gave, on heating at 80 °C in benzene, the tetrahydrofuran derivative **100** in 30% yield (Scheme 42).[51]

The failure of **99** to cyclize reflects the larger atomic radius of sulfur, the greater C-S-C angle bond and the smaller electronegativity difference between S and C.

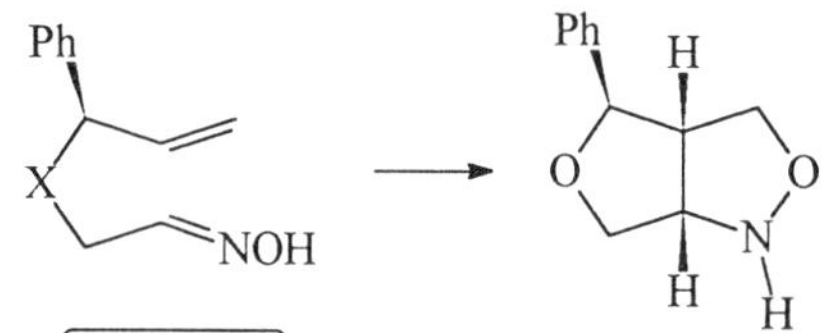

X = O; **98**
X = S; **99**

100

Scheme 42

The intramolecular cycloaddition of homochiral oximes **101**, carrying an unsaturated amido group, has been reported;[52] the process shows a complete regio- and stereoselectivity. Subsequent reduction of the obtained bicyclic compounds **102a-c** affords homochiral functionalized pyrrolidinone and pyrrolizidinone systems **103a-c**. Moreover, reduction of **102a-c** with $LiAlH_4$ affords the enantiomerically pure pyrrolidines **104a,b** and pyrrolizidine **104c** in high yields (Scheme 43).

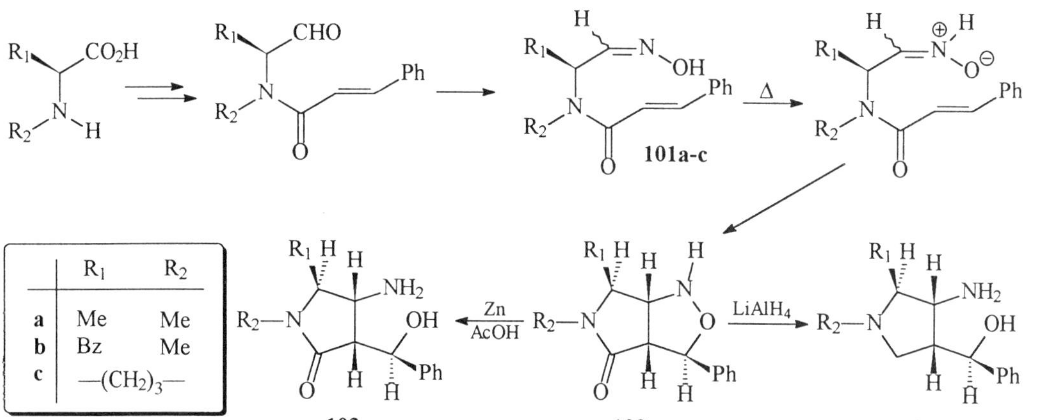

	R_1	R_2
a	Me	Me
b	Bz	Me
c	—$(CH_2)_3$—	

101a-c **103a-c** **102a-c** **104a-c**

Scheme 43

Noteworthy, reduction with LAH in THF at reflux gives the pyrrole **105** (Scheme 44).[20]

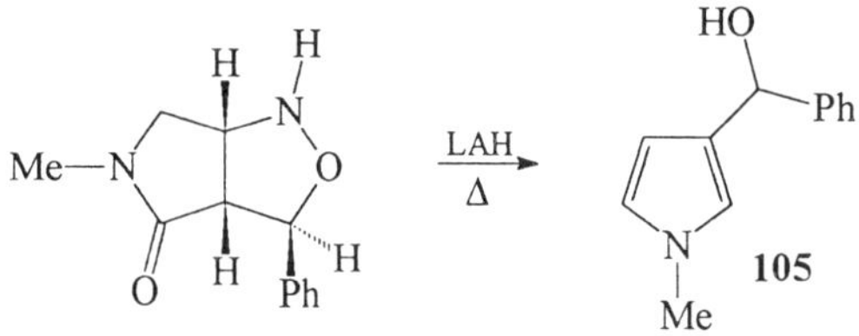

105

Scheme 44

The oxime of *N*-allylated pyridine-2-carboxyalde-hyde **106** underwent the tandem 1,2-prototropic-cycloaddition sequence to give an unstable cycloadduct **107** which was converted in the indolizine **108** by treatment with cold NaOH. Reduction of **107** with sodium borohydride afforded the tetrahydroderivative **109** (Scheme 45).[53]

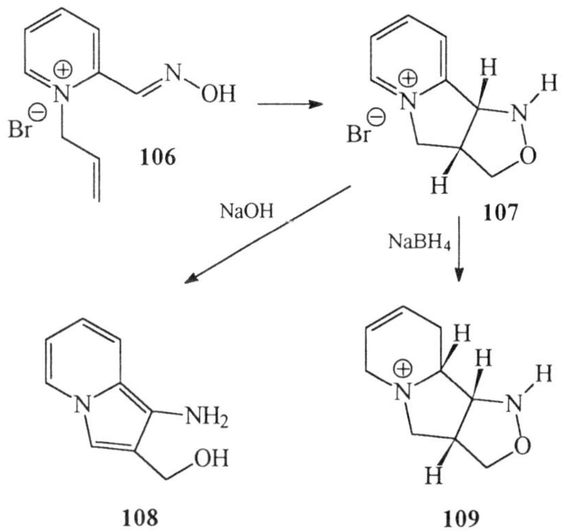

Scheme 45

Oximes of allyl esters of α-ketoacids **110** undergo cycloadditions *via* 1,3-azaprotio cyclotransfer pathway (Scheme 33, route **b**) to give *cis*-fused bicyclic lactones **111**;[54] no bridged adducts **112** were detected (Scheme 46).

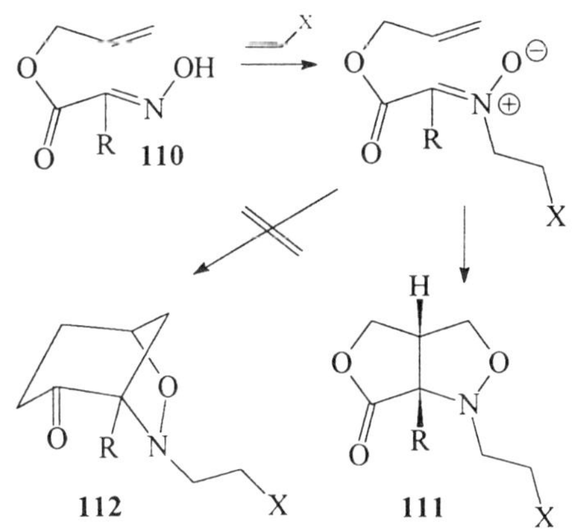

Scheme 46

The reaction can be rationalised by considering that the alternative 1,2 prototropic reaction scheme, leading to the intramolecular attack of oxime to the tether double bond, is a higher energy process than the 1,3-azaprotio cyclotransfer. Moreover, when the oxime α-carbon atom is disubstituted, the 1,3-azaprotio cyclotransfer is the preferred reaction course.[55-57]

The bicyclic isoxazolidine **111** could arise from the *Z*-nitrone *via* an *exo*-transition state **113** or from the *E*-

nitrone *via* an *endo* transition state **113** (Scheme 47).[54]

*Z-exo***113** *E-endo***114**

Scheme 47

The 2:1 mixture of *E*- and *Z*-isomers of *N*-allylpyr-role oxime **115** reacts with benzyl acrylate to give isoxazolidine **116** in 30% yield: no bridged ring isomer **117** was isolated (Scheme 48).[54] At lower temperature (60 °C), *Z*-(**115**) affords a mixture of isoxazolidines **116** and **117** (12%) together with the nitrone **118** (39%). By heating in xylene, a 1:1.6 mixture for **115** and **116** is obtained in quantitative yield.

A similar reaction of *Z*-(**115**) with phenyl vinyl sulphone and methyl vinyl ketone affords a nearly equimolar mixture of **116** and **117** in 85% yield.

The stereochemical results are explained by considering that the *trans*-fused isomer of **116** would originate from an *exo*-nitrone transition state where, however, the restriction of the conformational flexibility of the linking chain between oxime and dipolarophile, exerted by the pyrrole ring, prevents the correct alignment of the four reacting centres. Furthermore, the rigidity and angular disposition of the ring substituents, imparted by the pyrrole nucleus, raise the fused transition state energy so allowing the bridged transition state to compete effectively.

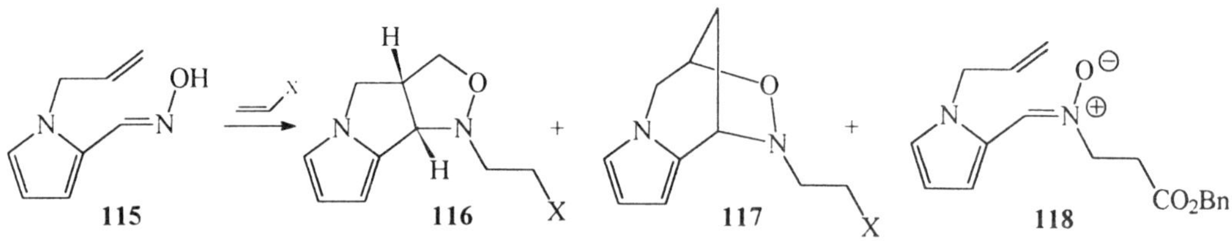

Scheme 48

The tandem process, consisting of nitrone generation via 1,3-azaprotio cyclotransfer reaction and subsequent intramolecular cycloaddition of the resulting nitrone to both activated and non-activated dipolarophiles (Scheme 33, route **b**), involves oximes carrying an alkene moiety.

An interesting paper deals with substrates where alkene moieties are positioned in each of two substituent of a ketoxime **119** or of a branched aldoxime **120** (Scheme 49).[58]

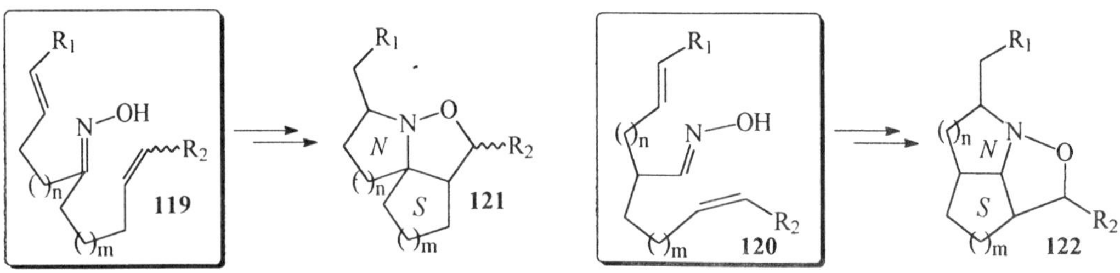

Scheme 49

In each case tricyclic spiro- **121** or fused ring-product **122** are formed stereospecifically and these are classified in terms of the size of the two variable rings, the nitrone (*N*) ring, and subsidiary (*S*) ring created in the cycloaddition step.

Electrophilic induced oxime-alkene reactions as a source of intramolecular tandem nitrone function-cycloaddition protocols have been also exploited (Scheme 50).

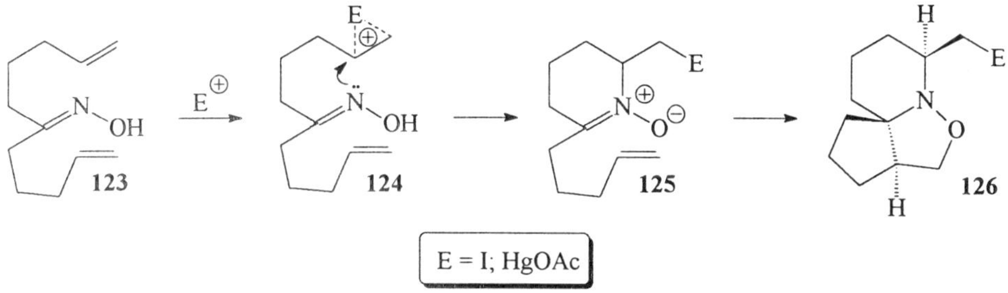

Scheme 50

The reaction involves the electrophilic attack of I_2 or $Hg(OAc)_2$ to alkene moiety of oxime **123** to produce the activated intermediate **124** which readily cyclizes by nucleophilic attack of oxime lone pair to nitrone **125**.

Analogously, oximes derived by bis(alk-γ,δ-enyl)- **127** and (alk-δ,δ-enyl)ketones **128** are regiospecifically cyclized by treatment with phenylselenyl bromide to the corresponding cyclic nitrones **129** and **130**, which undergo stereospecific cycloaddition to spirocyclic products **131** and **132** respectively (Scheme 51).[59,60]

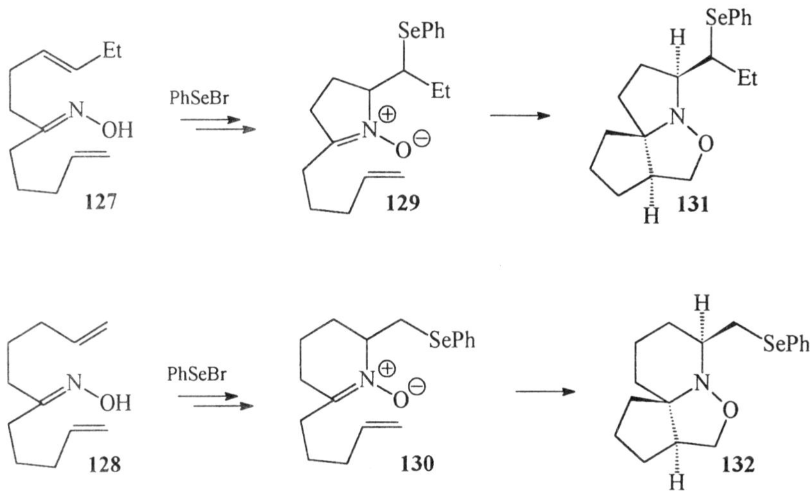

Scheme 51

The methodology of nitrone formation by nucleophilic attack on 3-membered heterocycles *via* the lone pair of oxime nitrogen atom and subsequent 1,3-dipolar cycloaddition sequence (Scheme 34, route **d**) has been exploited in the regiochemical conversion of oxime **133** in the spiroadduct **134** (Scheme 52).[61]

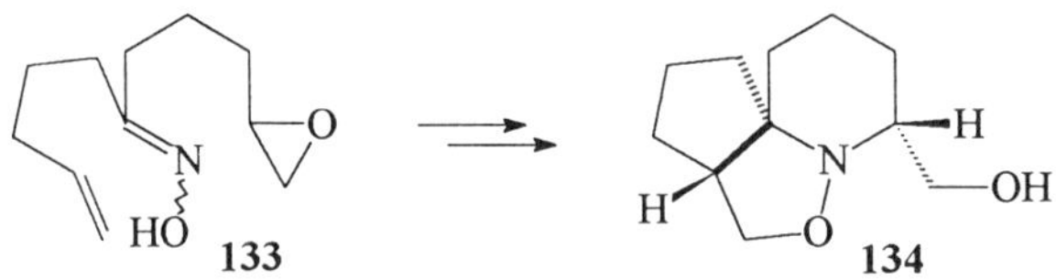

Scheme 52

2.2 Six- and seven-membered rings

Intramolecular nitrone cycloaddition strategy has been proved to be particularly valuable in the synthesis of six-membered heterocycles, as carbohydrate derivatives and alkaloids.

As an example, relatives of amino sugars **137**, lacking the anomeric hydroxyl function, but possessing many of the other key structural and stereochemical features of the natural compounds, can be elaborated by intramolecular 1,3-dipolar cycloaddition of nitrones **135**, via the bicyclo[4.3.0]nonanes **136** (Scheme 53).

Such compounds can be recognised as amino sugars and, then, interfere with processes as cell-wall assembly for which the presence of the anomeric OH group should not be important.

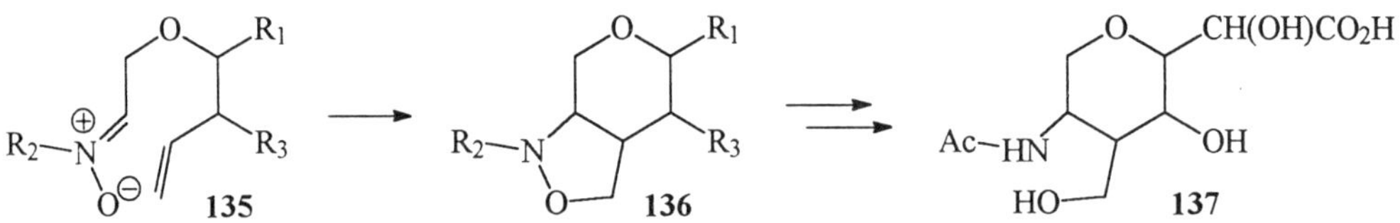

Scheme 53

Thus, nitrones **140** and **141**, derived from omoallylic alcohols **138** and **139**, afford diastereoselectively and regioselectively compounds **142-144** and **145-147** respectively. The "all-*cis*" diastereomer **142** is the predominant compound in the cycloaddition starting from **140**, while **141** gives diastereomer **146** as major compound (Scheme 54).[62]

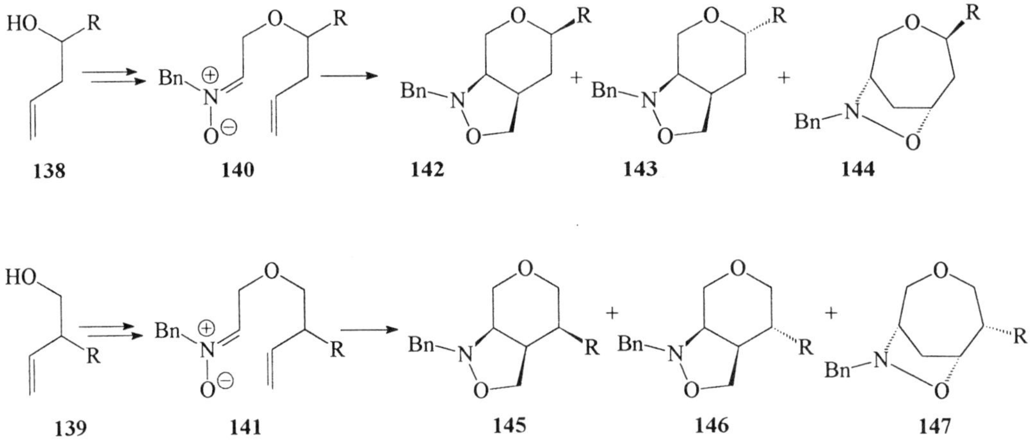

Scheme 54

The stereoselectivity of the cyclization process is controlled by an "equatorial" positioning of the substituent and by the preference for a less eclipsed orientation of the reaction centres, as in **148**, with respect to the alternative boat-like conformation **149** (Scheme 55).

Scheme 55

Nitrones formed from a sugar unit provide a novel strategy for the synthesis of chiral oxygen heterocycles *via* an intramolecular cycloaddition to the alkene double bond of an *O*-allyl residue in the sugar molecule.

In fact, according to Scheme 56, the oxepanoisoxazolidines **152** and **153**, as pure enantiomers, have been synthesised from D-glucose by an INC process starting from the corresponding nitrones **150** and **151**.[63,64] Reductive cleavage of isoxazolidine **153** leads to the chiral oxepane derivative **154**. This compound is potentially useful as a precursor for naturally occurring oxepanes, like zoapatanol[65] which has generated immense interest as a reproductive medicine.

The factors responsible for the reversal of regioselectivity, with respect to other INC processes involving *O*-allyl residues, have been not ascertained.

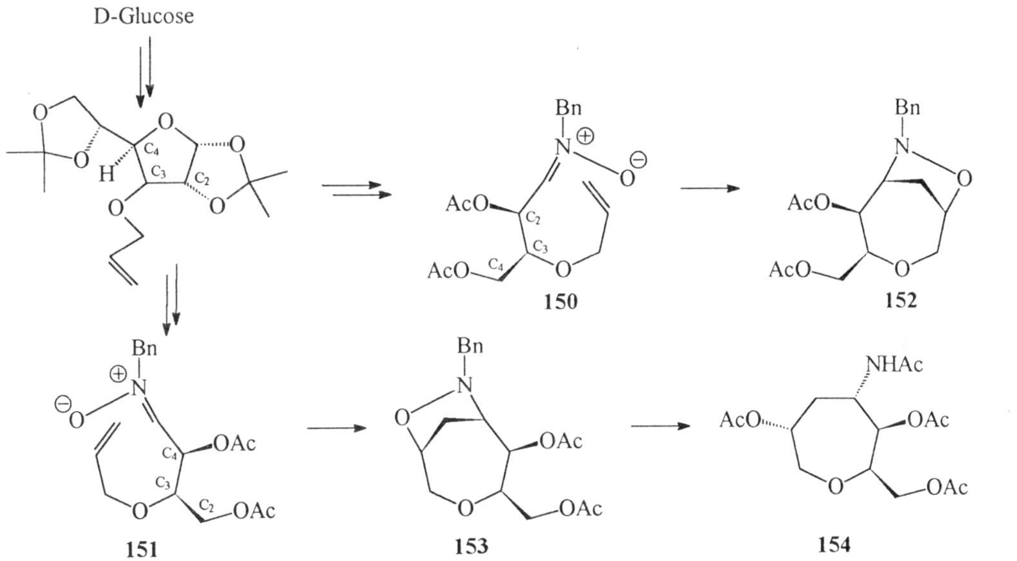

Scheme 56

Starting from 6-aldehyde glucopyranoside allyl ethers, pyrano-pyran and pyrano-oxepan systems have been synthesised. Treatment of aldehyde **155** with *N*-benzyl hydroxylamine leads to nitrone **156**, which subsequently added to the 4-*O*-allyl group to give, after deallylation, the enantiomerically pure fused isoxazolidine diols **157** and the bridged isoxazolidine diols **158**.[66]

Removal of the *N*-benzyl protecting group and reductive isoxazolidine ring cleavage by H_2/Pd gave the highly functionalized chiral pyrano-pyran **159** and pyrano-oxepan **160** derivatives, which may be converted, by deamination and partial deoxygenation, into compounds quite closely related to rings of the marine toxin brevatoxin B (Scheme 57).

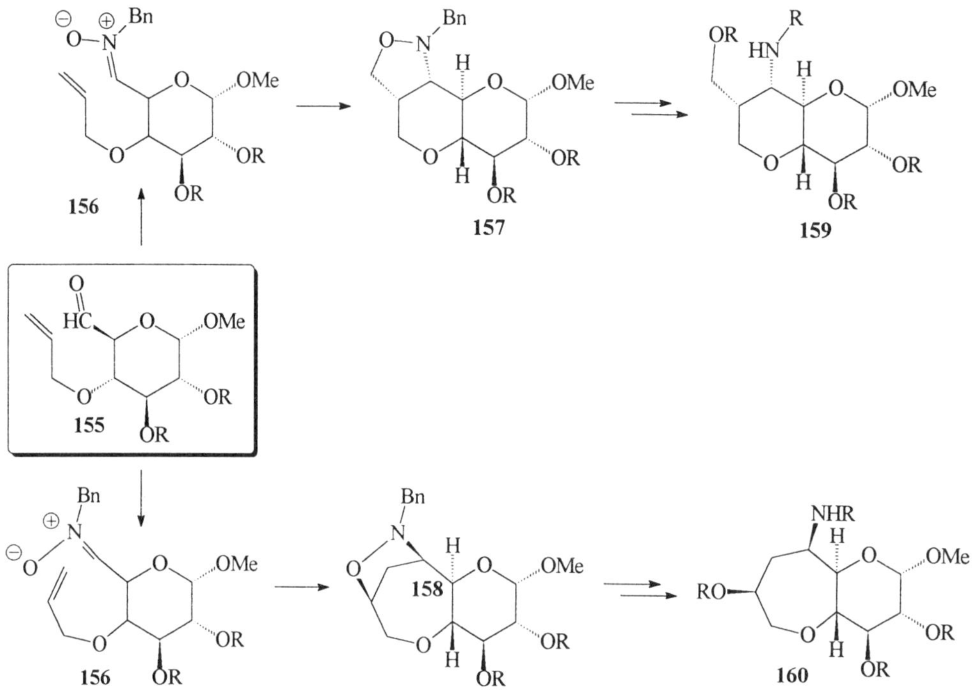

Scheme 57

Aldehydes **161** and **162** gave, by treatment with methyl hydroxylamine in refluxing ethanol, a mixture of adducts **163**, **164** (1:1.4 ratio) and the oxepans **165** respectively (Scheme 58).[67]

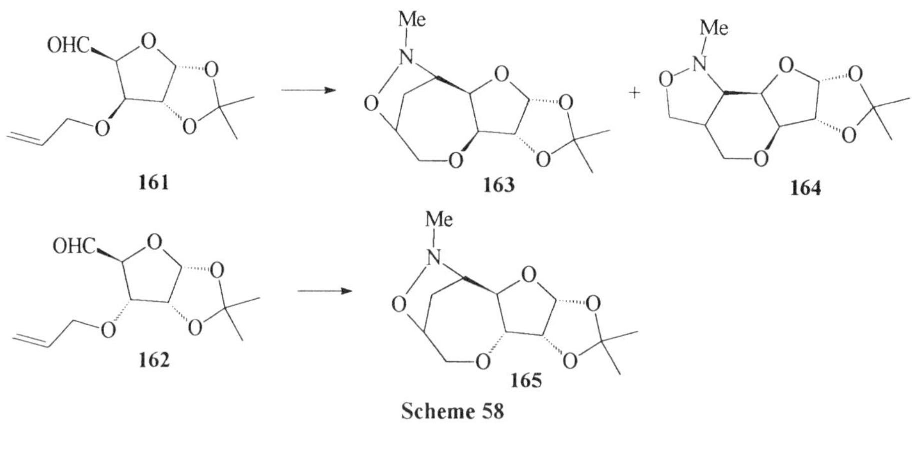

Scheme 58

A different approach to homochiral oxepans or tetrahydropyrans involving a INC process is shown in scheme 59.

Introduction of a methyl substituent at β position with respect to the aldehyde moiety altered the ring-selectivity from oxepan to tetrahydropyran. Thus, INC of **166** afforded the tetrahydropyran **167** as the only isolable cycloadduct. The methyl substituent induces steric interactions with the *Z*-nitrone moiety in the **TS 168**, leading to bridged compounds, and, therefore, the alternative *E*-nitrone-1-*trans*-allyl ether **TS** is preferred, furnishing the fused cycloadduct **169**.

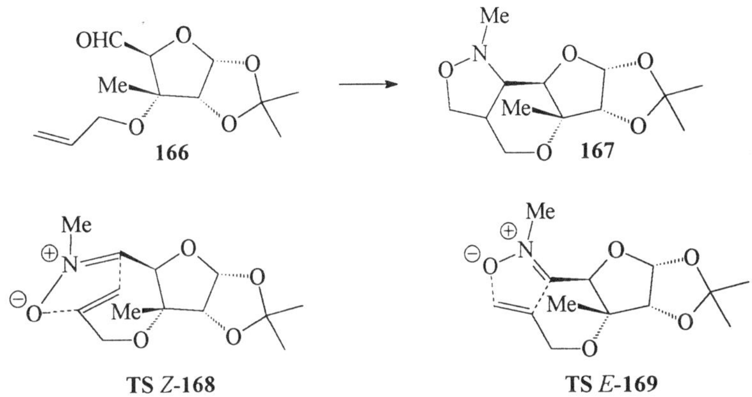

Scheme 59

4-Thia- and 4-oxa-6-heptenal **170**, by reaction with substituted hydroxylamines, are converted into fused and bridged bicyclic compounds **171** and **172** according to the substituents. As expected, ring cleavage of **171** (X = O) gives rise to tetrahydropyrans **173**, while reduction of **172** yields the hexahydrothiepine and oxepane derivatives **174** and **175** respectively (Scheme 60).[16]

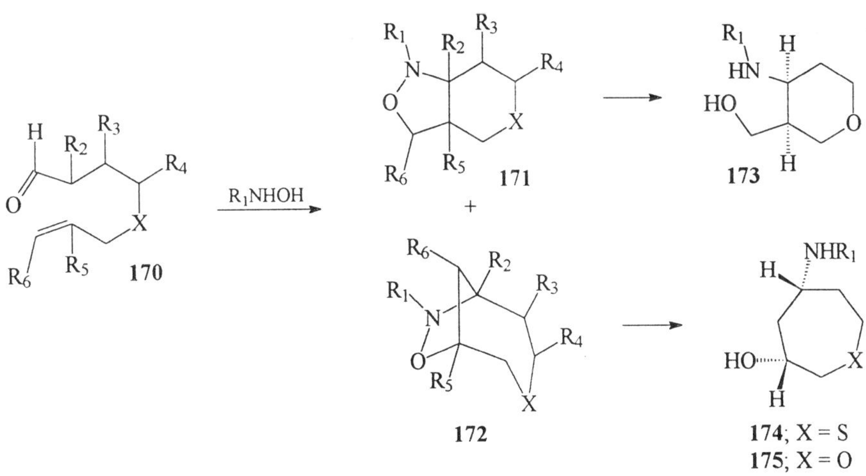

Scheme 60

The intramolecular nitrone cycloaddition at the periphery of a 2-oxo-2H-1-benzopyran system has been investigated: nitrone **177**, obtained from 4-(allylbenzyl-amino)-2-oxo-2H-1-benzopyran-3-carbaldehyde **176**, reacts with complete regio- and stereoselectivity to give isoxazolidines **178** in good yields. The stereochemical features of the obtained adducts suggest that the nitrone with *Z*-configuration interacts with the olefinic dipolarophile in a *exo*-approach manner (Scheme 61).[68]

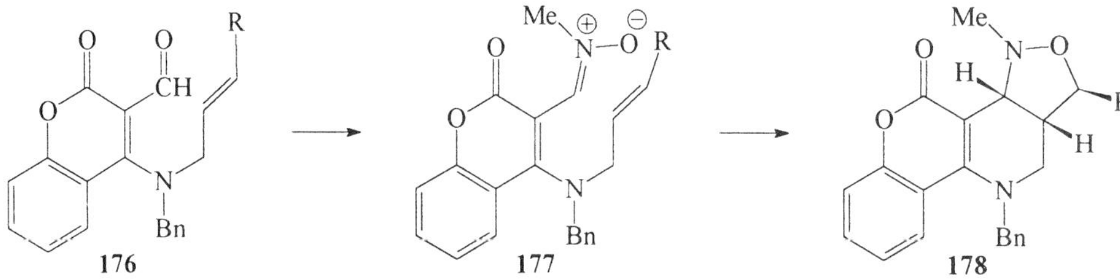

Scheme 61

In a similar way, the 5-(allylsulphenyl)-pyrazole-4-carbaldehyde **179** was transformed into fused **181** or bridged **182** tricyclic compounds by reaction with methyl hydroxylamine *via* the not isolated nitrone intermediate **180** (Scheme 62).[69]

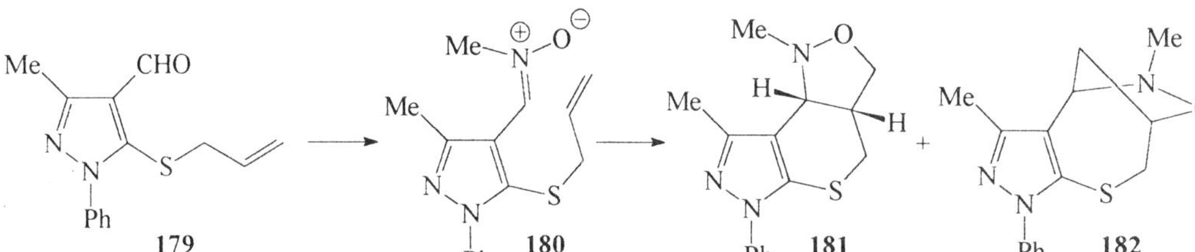

Scheme 62

The presence of an amido group in the tether connecting the dipole and dipolarophile has been investigated. Starting from *E*-enamide propanal derivatives **183**, by reaction with *N*-substituted hydroxylamines, the corresponding 8-oxa-3,7-diazabicyclo[4.3.0]nonan- 2-ones **185** *via* not isolated nitrones **184** have been synthesised, as exclusive products.

Treatment of α-amido aldehydes **187** with *N*-methyl hydroxylamine afforded a mixture of *cis* and *trans* 8-oxa-3,9-diazabicyclo[4.3.0]nonan-4-ones **188**, and **189** (Scheme 63).[70]

The stereochemical outcome of this latter reaction appears of some interest, because the nature of the ring fusion is linked to the substitution pattern at the alkene carbon. Thus, when R is an hydrogen atom, only *trans* adducts are obtained, on the contrary *cis* adducts are the prevalent one when R is methyl (*cis/trans* ratio = 70:30) or phenyl (*cis/trans* ratio = 99:1).

The obtained fused δ-lactams have been then transformed by LAH treatment in substituted piperidines **186**, **200**.

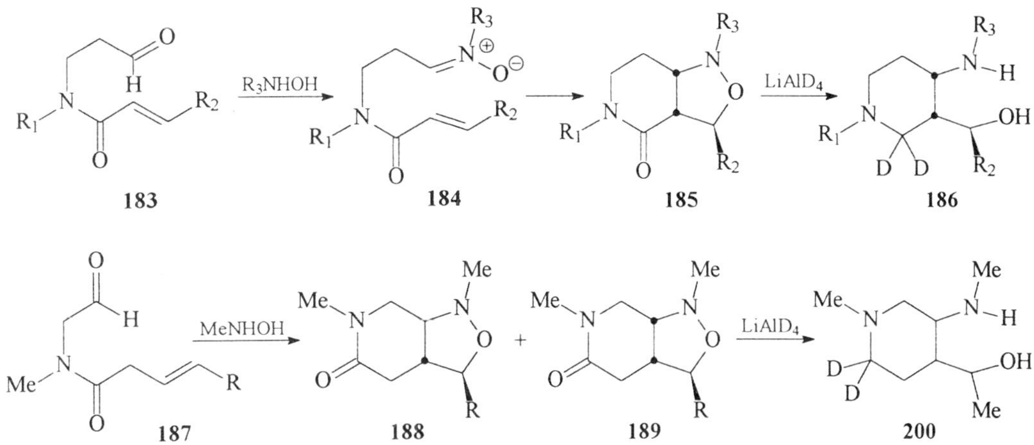

Scheme 63

The enhanced reactivity and stereoselectivity of INC with respect to the intermolecular version has been exploited in a intramolecular cycloaddition process starting from the nitrones carrying the normally poorly reactive allene group.

The reaction of *O*-(1,2-propadienyloxy)benzaldehy-de **201** with phenyl hydroxylamine proceeds smoothly to give dioxaazabicyclo[3.2.1]octane **202** originated from the attack of the intermediate, not isolated *N*-phenylnitrone to the C_1-C_2 π-bond of the allene (Scheme 64).[71]

The regiochemistry of the process, with exclusive formation of the bridged adduct, is consistent with a transition state possessing both better orbital overlap and fewer nonbonded interactions.

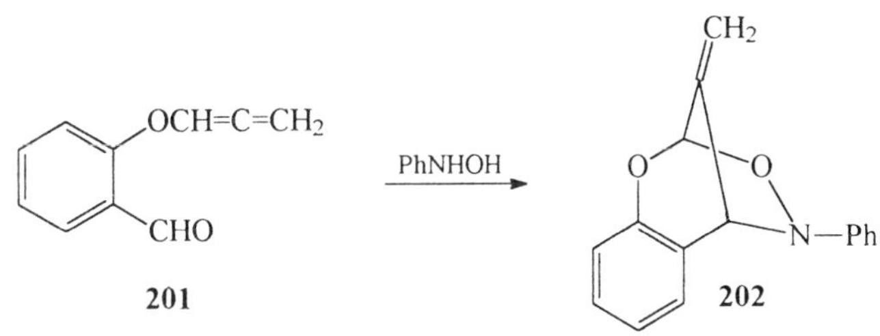

Scheme 64

The isomeric acetylenic nitrones **203** give rise to the transient 4-isoxazolines **204** which evolve according to the substituents present on the triple bond. Thus, **204a** (R = Ph) afforded a single rearranged product, the benzopyran **206a**, *via* the diradical **205a** and subsequent 1,2-hydrogen shift.

Derivative **204b** (R = H) furnished a 1:5 mixture of compounds **206b** and **207**: formation of **207** is explainable on the basis of the series of the reactions outlined in scheme 65.[72]

In a similar intramolecular nitrone-alkyne cycloaddition, nitrone **208**, where an oxygen atom is present in the tether chain, afforded isoxazoline **209** as stable product in moderate yields. By further reduction, 3-hydroxy-3-pyrrolin-2-ones **210** have been obtained, while oxidation of **209** leads to α-keto-β,γ-unsaturated esters **211**

(Scheme 66).[73]

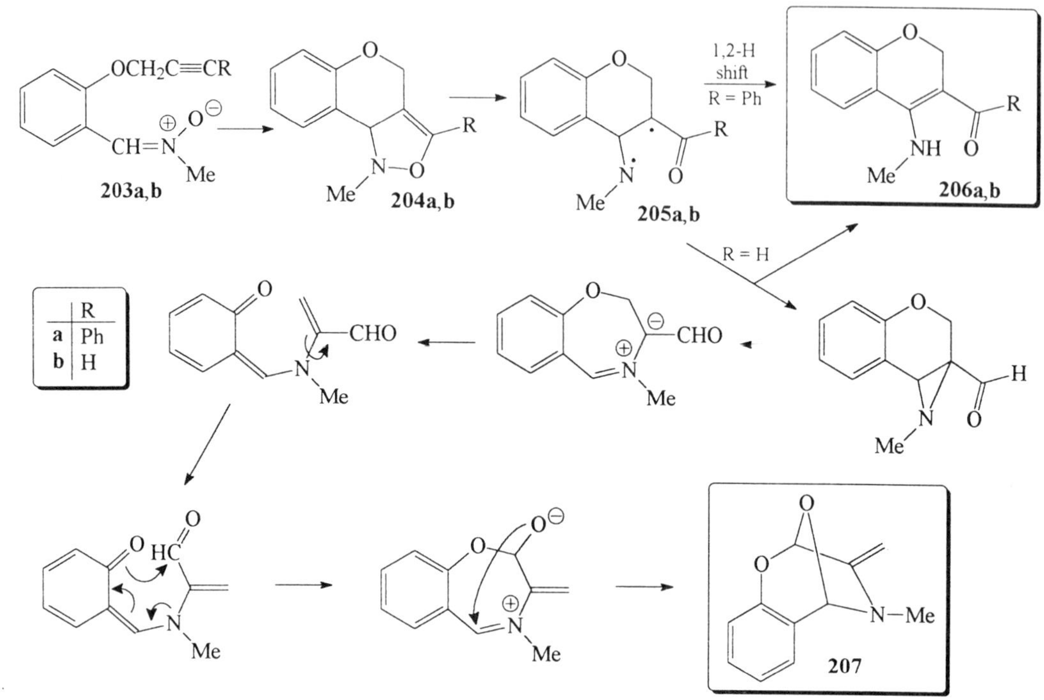

Scheme 65

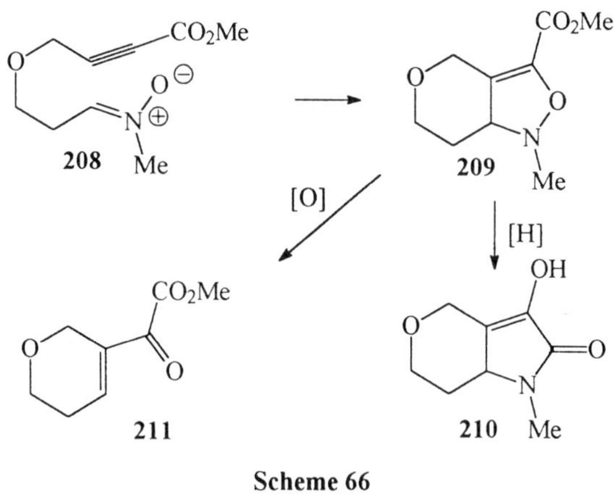

Scheme 66

By taking advantage of the reduced rotational freedom of the cyclohexene ring, which should favour a diastereofacial discrimination, the intramolecular cycloaddition of 2-(cyclohexen-3-yloxy)benzaldehyde **212** has been investigated in an attempt to obtain an absolute stereochemical control of the process. Treatment of **212** with homochiral 1-phenylethylhydroxyl-amine, as chiral auxiliary, afforded a mixture of cycloadducts **213-216**

(Scheme 67). The process appears to be highly stereoselective: no cycloadducts with a *trans* junction between two hexatomic rings have been obtained. Moreover, the predominance (90%) of diastereoisomers with a fully *cis* fused policyclic skeleton, **213** and **214**, reflects a preference for the less crowded *exo*-like approach by the *Z*-nitrone configuration.[74]

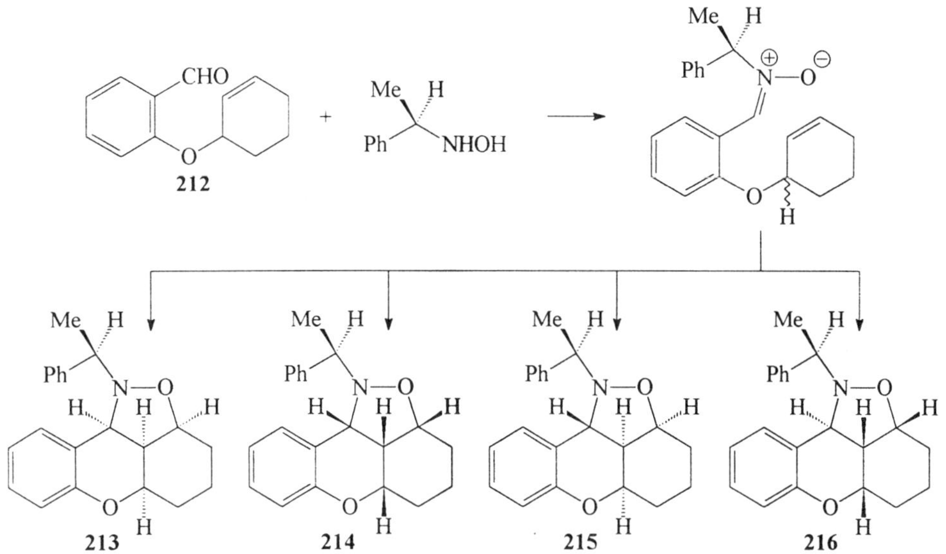

Scheme 67

Hydrogenolysis of the isolated cycloadducts **213-216** afforded the enantiomerically pure 1,2,3,4,-4a,9a-hexahydro-9H-xanthen-1-ols **217-220** (Scheme 68).

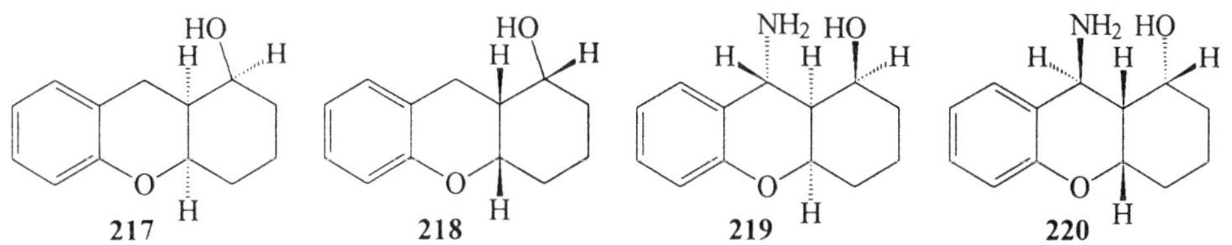

Scheme 68

According to the consideration that the selective introduction into a substrate of fluorine atoms leads to an enhancement of the biological activity in many natural products, INC strategy was exploited in a completely stereoselective synthetic approach to trifluoro-methylsubstituted isoxazolidino[4,3-c]-chroman-4-ones **221** and -quinolin-4-ones **222** (Scheme 69).[75]

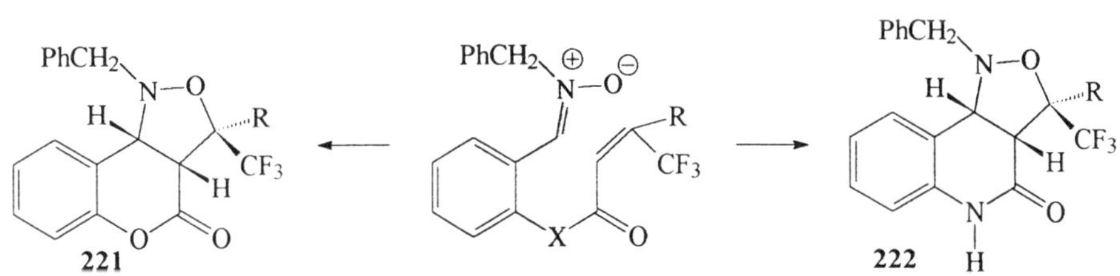

Scheme 69

The influence of the fluorine atom on the stereoselectivity of the process has been evaluated in the intramolecular cycloaddition of homochiral nitrones bearing one fluorine atom at the quaternary α-position. Thus, homochiral nitrones **223** afford two diastereomeric bicyclic isoxazolidines **224** and **225** (3:1 ratio).[76] The stereochemical outcome is considerably affected by the presence of the fluorine, which causes an unexpected stereoselectivity, reverse to that observed in the INC process performed on the analogoues not fluorinated nitrones.[77] The results can be explained by an examination of transition state involved in the reaction: **TS** *a* is more stable than **TS** *b* because the electronic effects of the fluorine overcome the steric repulsions between the oxygen atom of the nitrone and the methyl group at C_2 (Scheme 70).

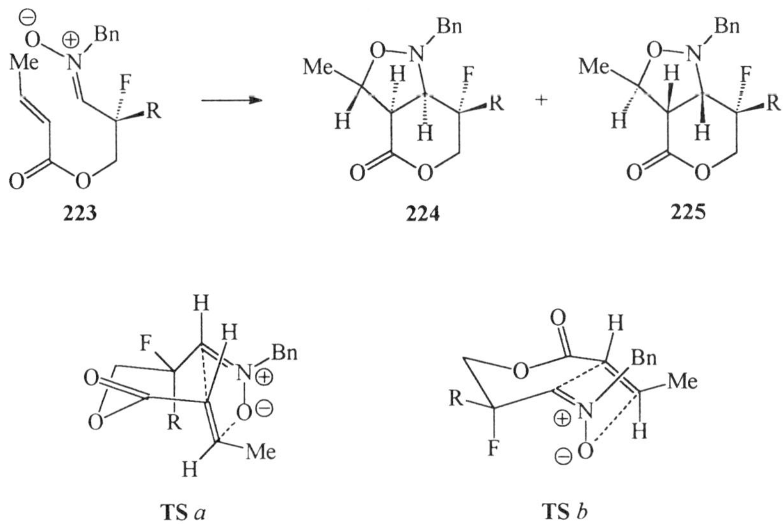

Scheme 70

The IOOC approach towards the synthesis of six- and seven-membered system has proven to be synthetically useful. Thus, through thermal IOOC, oxime olefin **226** and **227** proceeds, *via* N-H nitrone tautomer (Scheme 33, route **a**), giving rise to indolizidine **228** and quinolizidine **229** respectively (Scheme 71). Noteworthy, the IOOC products were obtained stereochemically pure with no evidence in the NMR spectra of crude product of any diastereomers. While the ring closure to 5-membered rings fused to the oxazolidine led mainly to the *cis-anti* isomers, as reported in scheme 37, ring closure to six membered fused system produces only the *cis-syn* stereoisomer, as supported by molecular mechanic calculations and NMR data.[44,47]

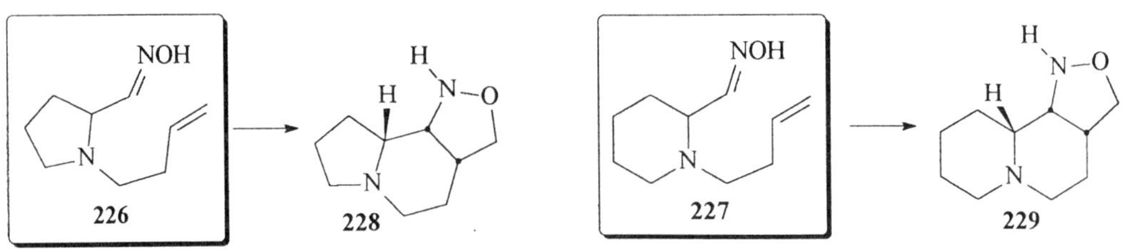

Scheme 71

Oximes of 4-(alk-2-enylamino)-1,6-dimethyl-2-oxo-1,2-dihydropyridine-3-carbaldehydes **230** and 2-(alk-2-enylamino)-4-oxo-4H-pyrido[1,2-a]pyrimidine-3-carb- aldehydes **231** underwent the IOOC reactions under very mild conditions (EtOH, reflux) to give fused isoxazolidine derivatives **232** and **233** respectively (Scheme 72).[78]

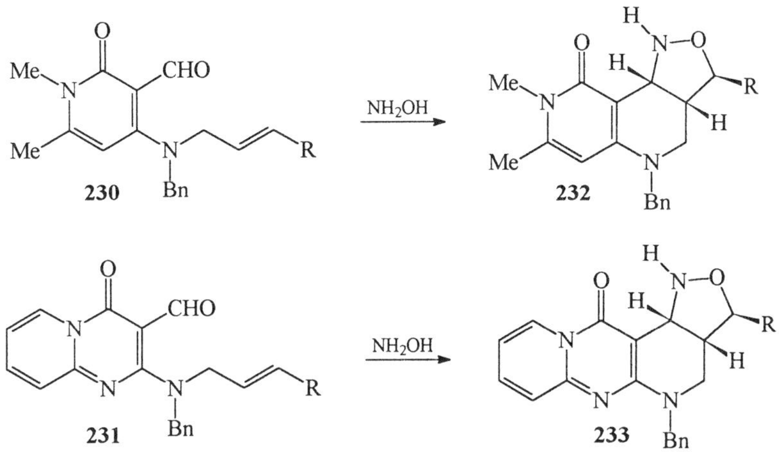

Scheme 72

The feasibility of the reaction is linked to the facile oxime-nitrone isomerization due to a mechanistic feature in which the alkenylamino nitrogen and/or carbonyl groups play an intramolecular catalyst role for this isomerization.

Another strategy to accelerate the rate of the cycloaddition relies on a steric buttressing, in order to reduce the available conformational space to the reactants, so optimising their spatial orientation and favouring their mutual approach. Thus, the selected conformation constraints, achieved by the incorporation of *o*-1,3-dipole substituents in 1-allyloxybenzenes forces the group into close proximity and hence allow cycloadditions to proceed that otherwise do not. Thus, oxime **234** reacts cleanly to give the cycloadduct **235**, whereas its isomer **236** is relatively inert (Scheme 73).[79,80]

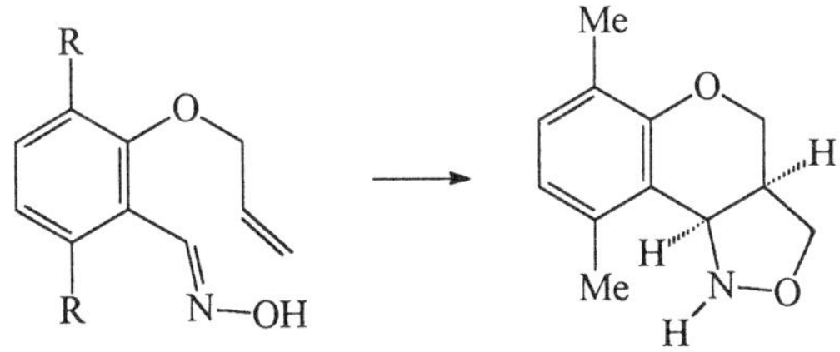

234; R = Me
235; R = H

236

Scheme 73

In a similar way the cycloaddition of **237** leads to **238**, while, in complete contrast, the isomeric oxime **239** is quantitatively converted into oxime **240** *via* para-Claisen's rearrangement (Scheme 74).[55]

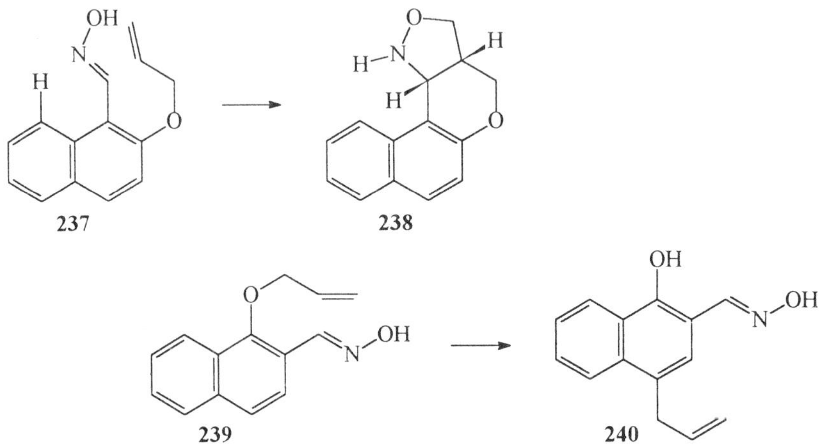

Scheme 74

In the case of compound **240**, the buttressing effect, arising from steric interaction of H_8 (peri-H) and oxime moiety, assists the attainment of the transition state geometry.

The 1,3-azaprotio cyclotransfer process (Scheme 33, route **b**) has been also exploited in the synthesis of oxygenated six- and seven-membered rings. It was found that *N*-methyl-*C*-(5-hexenyl)nitrone **241**, obtained from the corresponding oxime by reaction with benzyl acrylate, furnishes in boiling xylene a 3:1:5:2 mixture of **242**, **243**, **244** and **245** (Scheme 75).[54]

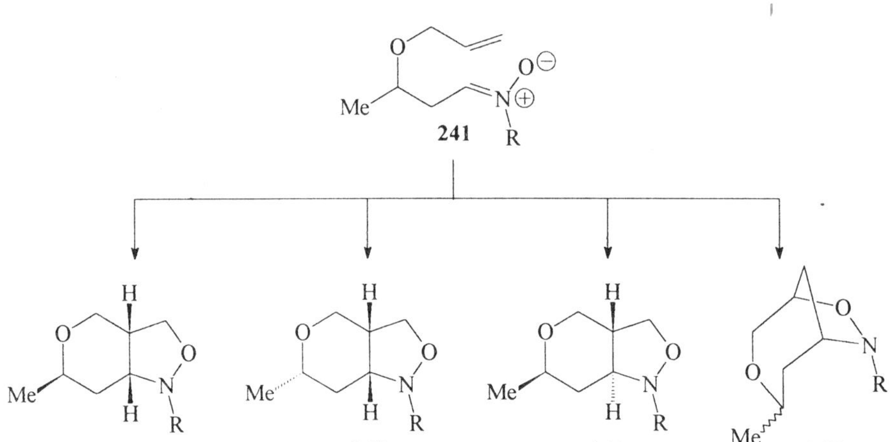

Scheme 75

3. Nitrile oxide

The intramolecular version of the nitrile oxide cycloaddition (INOC), exploited since the mid-1970s,[81] has received recently a great deal of attention for its synthetic versatility.[3] Typically, INOC reactions are run on nitrile oxides generated in situ, from oximes, chloroximes or nitroalkanes, in mild conditions because of the presence of dipolarophile.

INOC and INC reactions can be considered as alternative synthetic procedures in a number of further transformations; furthermore, isoxazolines can be converted to isoxazolidines and vice versa.[82] The use of either or two different approaches on the synthetic exploitations relies on the consideration that cycloaddition transition states are rather different for these two dipoles: nitrile oxides are linear and opposite to nitrones which are planar with an approximately 120° C-N-O bond angle. As a consequence, the nitrones usually can offer more promise of stereoselective cycloaddition, while nitrile oxides can offer more promise of regiospecific cycloaddition, especially in forming 5,6-, 5,7- and larger fused bicyclic systems.[7,8]

3.1 Five-membered rings

Substituted and functionalized tetrahydrofurans, structures of considerable importance in natural products chemistry, can be synthesised in a stereoselective manner.

A series of unsaturated oximino ethers **248**, synthesised by reaction of allylic alcohols **247** with α-bromo alkanal *O*-(trimethylsilyl)oxime **246** in the presence of TBAF, were converted in fused tetrahydrofurano[3,4-c]isoxazoles **250** by treatment with NaClO *via* transient nitrile oxides **249** (Scheme 76) and transformed into functionalized tetrahydrofuran derivatives **251** by Raney-Ni opening of the isoxazoline ring.[51]

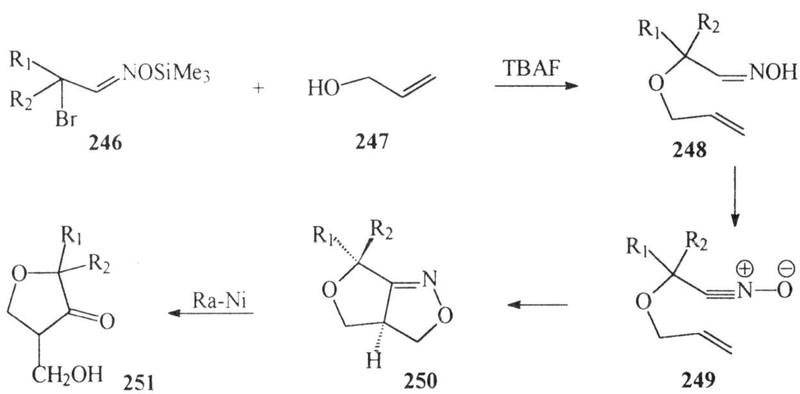

Scheme 76

It is interesting to note that the reaction which leads to phenyl-substituted tetrahydrofurans (R_2 = Ph) occurs with exclusive formation of the *trans* isomer, while the methyl analogue is obtained in a 2.4:1 ratio of *trans:cis* isomers.

This different diastereoselectivity was explained by use of MM2 calculations, assuming that the relative energy differences of the two lowest energy conformations of the diastereomeric cycloadducts parallels the energy differences in their transition states. In fact, the calculations reveal a 0.96 Kcal/mol difference between the two diastereomeric transition states for the phenyl case, but only a 0.38 Kcal/mol difference for the methyl system. In both cases, the lower energy isomer corresponds to the *trans* isomer.

The replacement of the oxygen atom with sulfur deprimes the overall yield of the process, because of the sensitivity of the sulfide function to the reagents employed (NaClO or Chloroamine T). These synthetic difficulties have been overcome by using a nitrile oxide obtained by dehydration of unsaturated nitro sulphide according to Mukaiyama method.[83]

In fact when aldehydes **252** were allowed to react with nitromethane **253** in the presence of allyl mercaptan

254, the unsaturated sulphides **255** were isolated and successively converted by INOC in tetrahydrothiopheno[3,4-c]isoxazoles **257** and **258**, in a *trans-cis* mixture, with moderately good degree of stereoselectivity (Scheme 77).[84]

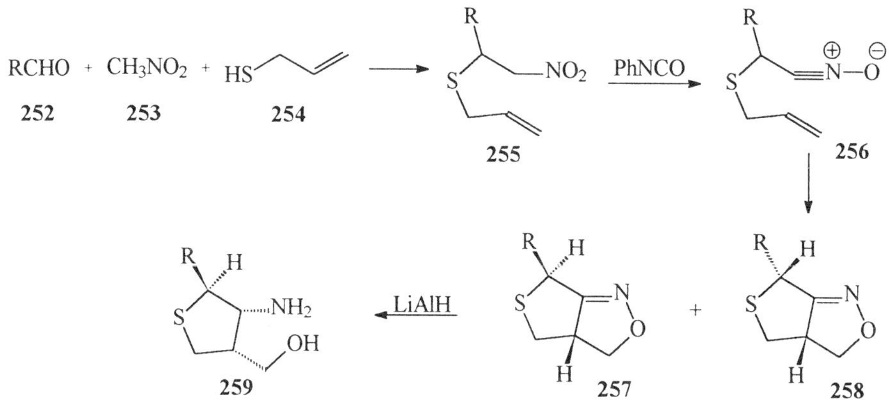

Scheme 77

Apparently, the steric factor plays a role during the cycloaddition. An equimolar mixture of stereoisomers **257** and **258** was generated from the alkyl-substituted systems **255a-c** (R = Me, Et, *i*-Pr), while, a 3:2 ratio was found with R = *t*-Bu or Ar. With the extremely hindered β-nitrosulfide **255k** (R = 2,4,6-$Me_3C_6H_2$), a 5:2 ratio was obtained. Further reduction of **257** with LAH led to tetrahydrothiophenes **259** possessing stereoselectively placed amino and alcoholic functions.

The disubstituted 6,6-dialkyl tetrahydrothiopheno-[3,4-c]isoxazoles **262** were also successfully prepared starting from the corresponding ketones **260** (Scheme 78).

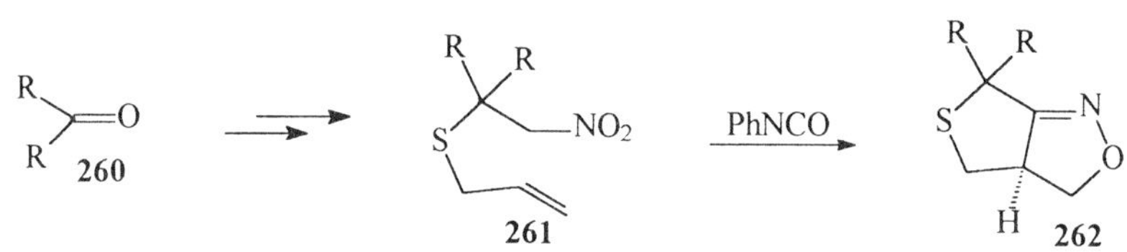

Scheme 78

An investigation of the relative asymmetric induction and double diastereoselectivity in the INOC was further studied with nitrile oxide intermediates similar to **249** and **256**.[85]

The nitroethers **263a-c** (X = O) were prepared by Michael addition of lithioallyloxide to the appropriate nitroolefin; sulphides **263d-f** (X = S) were prepared in the same way starting from allyl mercaptane. Phenyl isocianate treatment, then, produced diastereomeric heterocycles **265** and **266** in the ratio indicated in scheme 79.

This series of experiments established that, in the case of X = O, a stereogenic centre in the nitrile oxide-olefin affects the olefin face selectivity: increasing of the face selectivity with the increasing of the steric bulk.

Moreover, the effect of the substitution pattern on the olefinic moiety has been evaluated (Scheme 80): 1,1-disubstitution causes the cycloaddition to proceed with complete olefin face selectivity, while a *trans* 1,2-

olefin results in a slightly decreased face selectivity (6:1 from **268** versus 8.8:1 from **263c**).

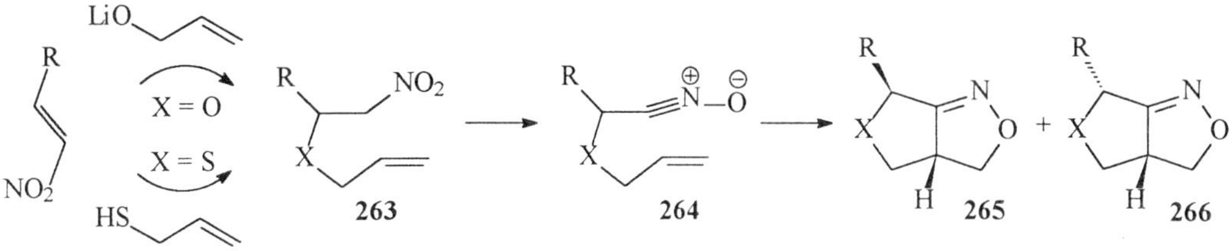

	X/R	265:266
a	O/*i*-Pr	4.0:1
b	O/*t*-Bu	8.8:1
c	O/Ph	9.0:1
d	S/*i*-Pr	1.2:1
e	S/*t*-Bu	1.6:1
f	S/Ph	2.0:1

Scheme 79

A more dramatic stereocontrol is observed when the stereogenic centre is in the allylic position (**269**): in this case only one epimer is formed (**270**) (Scheme 80). This total stereoselection is confirmed by using Allinger's MM2 program which indicates that the epimer of **270** is less stable of 3.3 Kcal/mol.

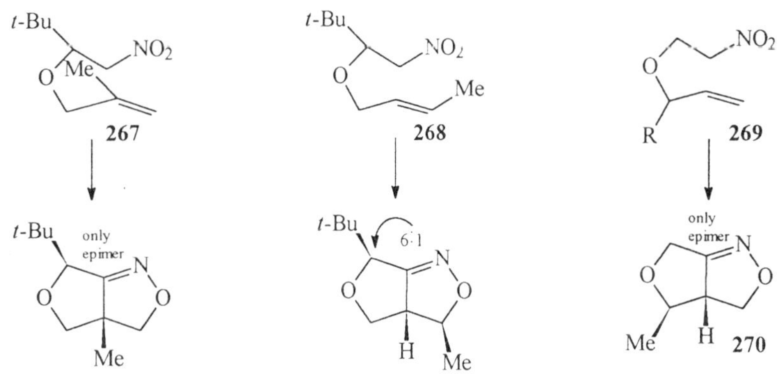

Scheme 80

The results are in sharp contrast to what obtained with the carbon analog nitrile oxides **271a-c** (Scheme 81) which give rise to marginal face selectivity.

	cis:trans ratio
a	1.5:1
b	1.0:1
c	2.5:1

Scheme 81

The ^{dd}INOC starting from nitrodiene **273** has been exploited to probe whether an α stereogenic centre in the nitrile oxide is able to differentiate the two diastereotopic olefins, as well as the two diastereotopic faces of each olefin (Scheme 82). Formation of heterocycles **274** and **275** in a 1:7.2 ratio supports a complete face selectivity and a 1:7.2 olefin selectivity.

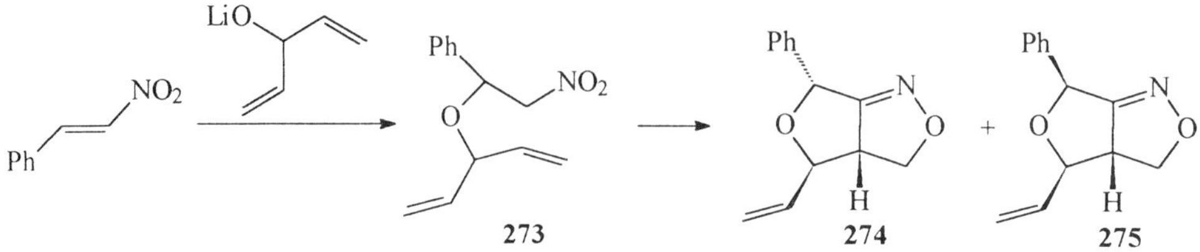

Scheme 82

In analogous substrates (Scheme 83), the cycloaddition step sets two new stereocentres: a) the geometry of the internal C=C bond provides stereochemical control at C_{3a} and C_3; b) the C_4 stereocentre and to a lesser extent the C_6 stereocentre dictate the stereofacial selectivity.

The reaction scheme has been functionalized towards the synthesis of cyclic furans by electrophilic cyclization sequence through the use of ICl.[86]

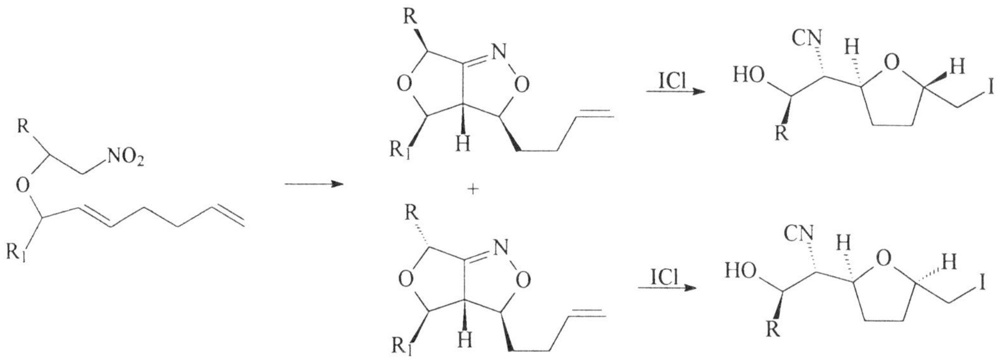

Scheme 83

Oximino ethers containing furfuryl and thiophenyl moiety as well as the cylohexene group (**276-278**) have been used to give, respectively, the tricyclic systems **279**, **280** and **281** (Scheme 84). In particular, the ring closure to **281** proceeds with stereospecific introduction of three stereocentres, the fourth stereocentre being introduced as 1:1 mixture of diastereomers.[51]

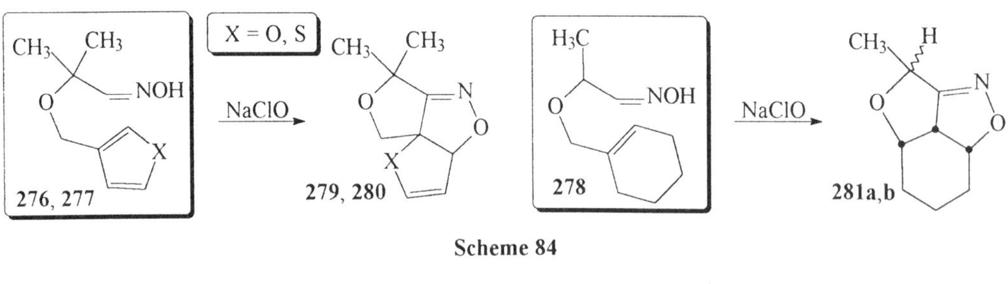

Scheme 84

A complete stereocontrol has been obtained on a INOC process starting from glucal **282** and leading enantiospecifically to tetracyclic compounds **283** (Scheme 85).[42]

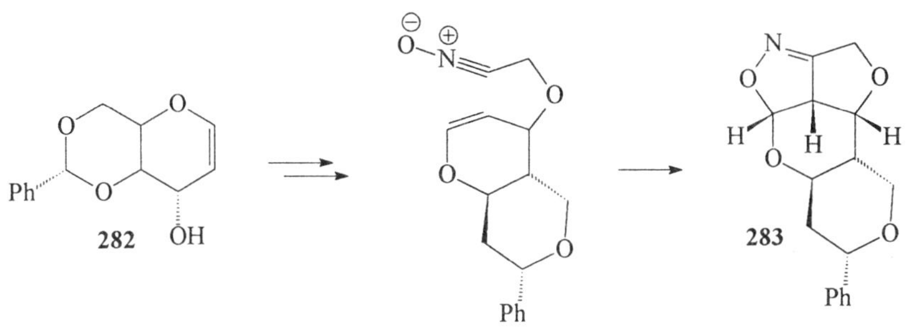

Scheme 85

Tricyclic compounds **285**, as a 5:1 mixture of stereoisomers, have been also prepared from not isolated intermediate nitrile oxide, generated by the method of Mukaiyama[83] from nitroderivative **284** (Scheme 86).[87]

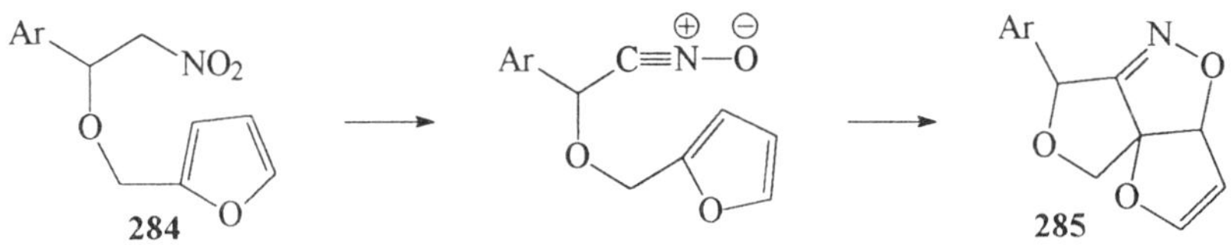

Scheme 86

The stereochemistry of isomers **285** was not determined.

Similar reactions have been performed with nitroderivatives **286-288**,[88] which, in accord with other INOC, occur with complete regiochemical control, but with low degree of diastereoselectivity (Scheme 87).

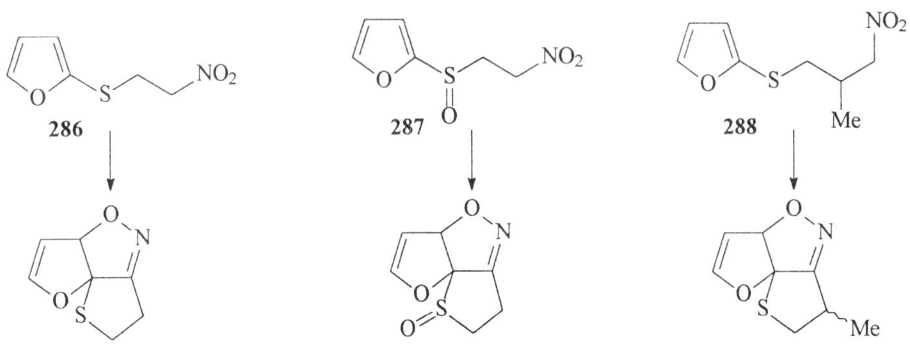

Scheme 87

α-Bromo alkanal *O*-(trimethylsilyl)oxime **246** has been also reacted with propargyl alcohol to lead, *via* acetylenic oximes, to fused tetrahydrofurano[3,4-c]iso-xazoles **289** (Scheme 88).[51]

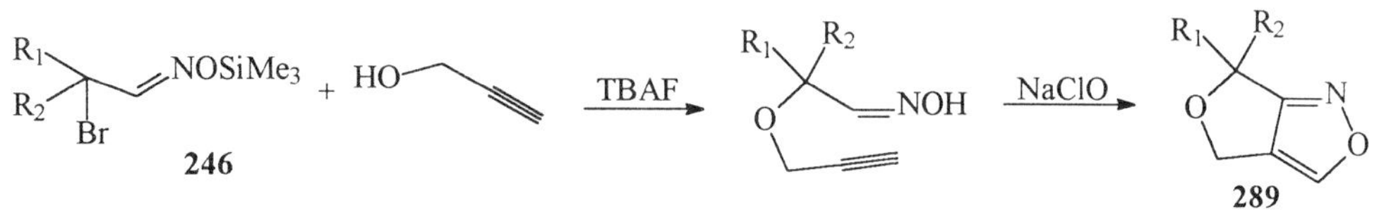

Scheme 88

In similar way, the alkanal *O*-(trimethylsylil)oximes **246** have been reacted with substituted α,β-unsatura-ted acids **290** and transformed in the oximino esters **291**, which, by subsequent oxidation with NaClO, *via* nitrile oxide intermediate **292**, were directly converted in tetrahydrofuro[3,4-c]isoxazolo-4-ones **293** (Scheme 89).[89]

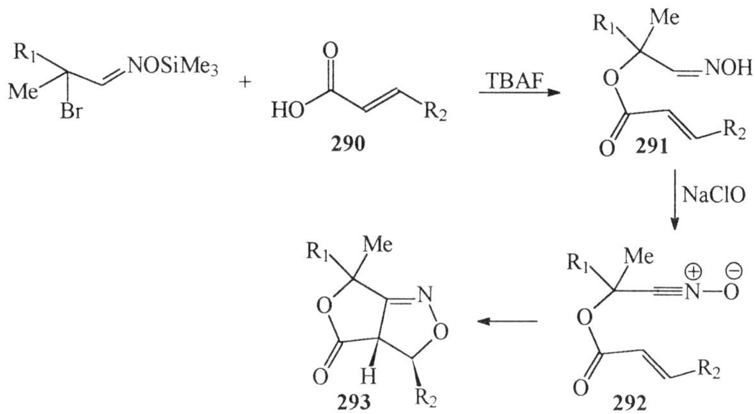

Scheme 89

However, the reaction suffers of severe limitations linked to the substitution pattern. Lack of INOC has been observed when R_1, R_2 or R_3 are hydrogen atoms, and rationalised by help of AM1 calculations. In fact, assuming that the *s-cis* conformation of corresponding nitrile oxides is the one responsible for the intramolecular cycloaddition, while the *s-trans* conformation is responsible for the oligomerization process (Scheme 90), it was found that, when R_1, R_2 or R_3 are hydrogen atoms, the energies of nitrile oxide intermediates associated to *s-trans* conformation are sharply more stable with respect to *s-cis* conformation. On the contrary, when R_1 is a methyl group and R_3 are alkyl or phenyl, the *s-cis* conformation of nitrile oxide intermediates is more stable of *s-trans* conformation.

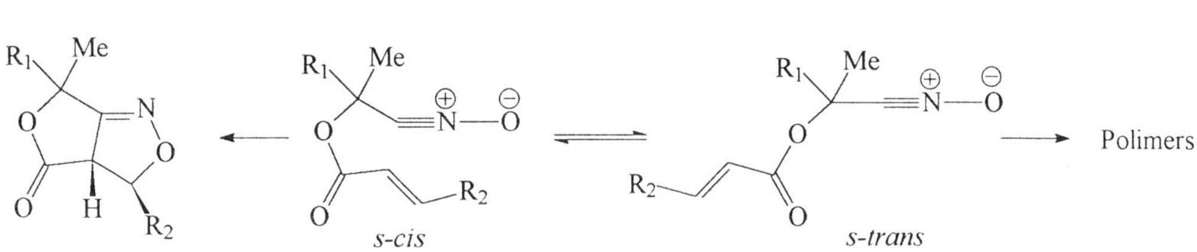

Scheme 90

The insertion of a nitrogen atom in the fused bicyclic system, as in the pyrrolidino[4,3]isoxazolines, has been achieved starting from *N*-tosylallylamines **294**, which were alkylated with dibromoalkanes and converted into unsaturated nitroderivatives **295**: INOC process leads to ring fused isoxazolines **297** (Scheme 91).[90]

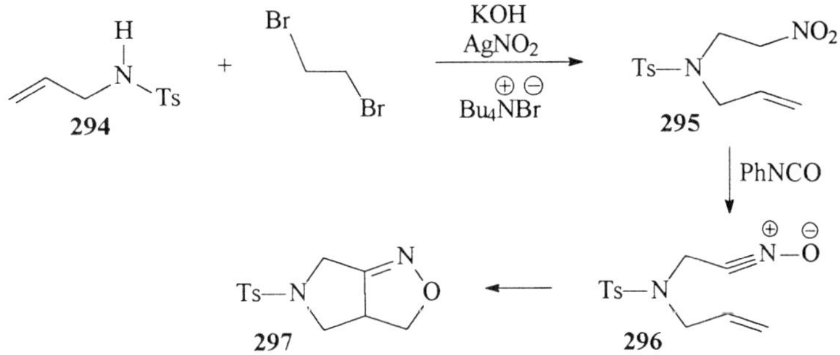

Scheme 91

A synthetic approach to 3,4-dehydropyrrolidin-2-one and 1,2-dehydropyrrolozidin-3-one systems has been designed, starting from the corresponding enamido-oximes, by reaction with NaClO.[91]

Derivatives **298a-c** led to bicyclic 2H-3-oxa-4-substituted-7-methyl-2,7-diazabicyclo[3.3.0]oct-1,5-en-6-ones **301a-c** in moderate to good yields (Scheme 92), while the analogous reaction of the homochiral derivatives **302a-d**, gives rise to a mixture of diastereoisomeric 3-oxa-4-phenyl-2,7-diazabicyclo[3.3.0]oct-1-en-6-ones **304a-d** and **305a-d** in enantiomeric pure forms (Scheme 93). The degree of olefin face selectivity in the latter reactions increases with the steric requirements of the substituent in α to the 1,3-dipole: when R is a proline group, the *trans/cis* ratio increases to 4:1.

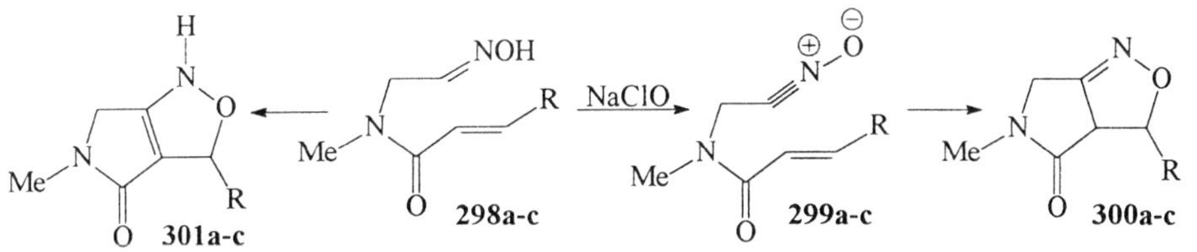

	R
a	Me
b	Et
c	Ph

Scheme 92

The observed stereocontrol parallels the results previously reported for similar INOC processes.

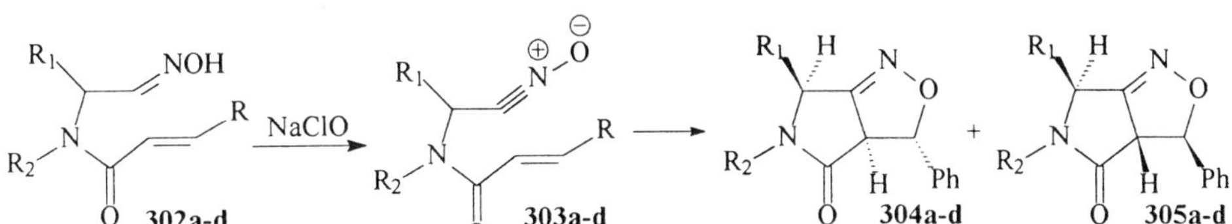

	R_1	R_2	**304:305** Ratio
a	Isopropyl	Me	3:1
b	Isobutyl	Me	1:1
c	Benzyl	Me	1:1
d	$-(CH_2)_3-$		4:1

Scheme 93

Raney-nickel reduction of cycloadducts **301c**, **304c** and **305c,d** afforded, then, the highly functionalized pyrrolydones and pyrrolizidinones **306**, **309c**, **310c,d**; oxidation of compound **301** with chloranil, followed by reduction with molybdenum hexacarbonyl, furnishes the 4-amino-pyrrolidinone **308** (Scheme 94).

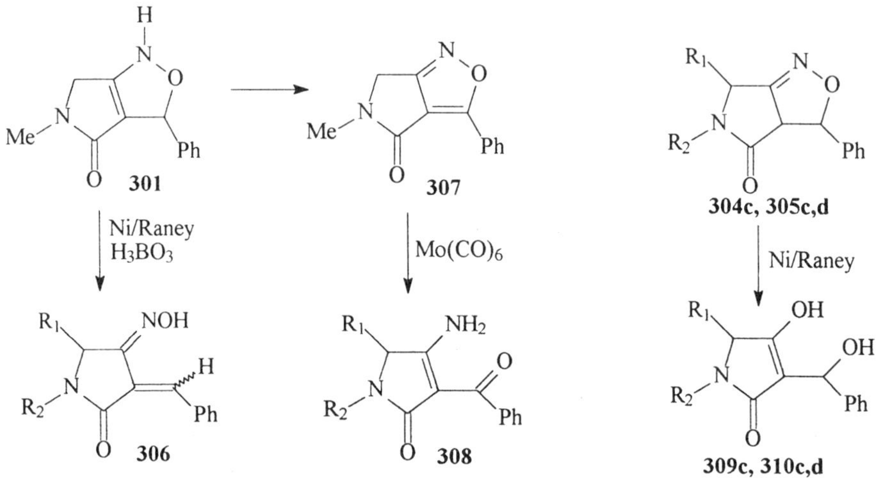

Scheme 94

Noteworthy, with respect to oximino esters INOC reactions, occurring only when two methyl groups are present in α to the nitrile oxide function, enamido oximes react smoothly, according to a transition state characterised by a better orbital overlap and low nonbonded interactions.

Following a similar procedure, tetrahydroisoxazoles **311,** analogues of Mitomycin have been prepared in high yield starting from 2-formyl-3-methyl-*N*-allyl-indoles **59** and hydroxylamine, followed by hypochlorite oxidation (Scheme 95).[35] The presence of methyl group at C_9 in these compounds enhances its importance due to the possibility of side chain manipulations at C_{10} nucleophilic bio-reductive site.

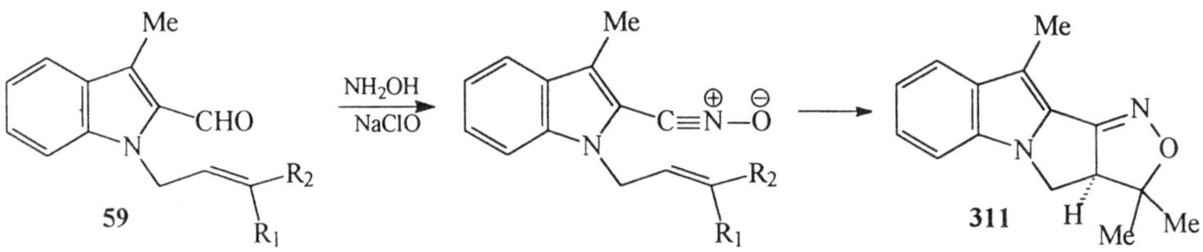

Scheme 95

3.2 Six- and seven-membered rings

Stereoselectively substituted and functionalized tetrahydropyrans, analogues of carbohydrates, have been synthesised in two different ways, starting from unsaturated oxime ethers or nitro alkyl ethers respectively (Scheme 96). The same procedure has been extended to unsaturated nitrosulfides with formation of tetrahydrothiopirans derivatives.[84] The stereoselective formations of *cis* **(312)** over *trans* isomers **(313)** is opposite to that observed in the tetrahydrofuran series. Accordingly, MM2 calculations predict an energy difference of 1.5 Kcal/mol in favour of the *cis* isomer when R = Ph.[51,92]

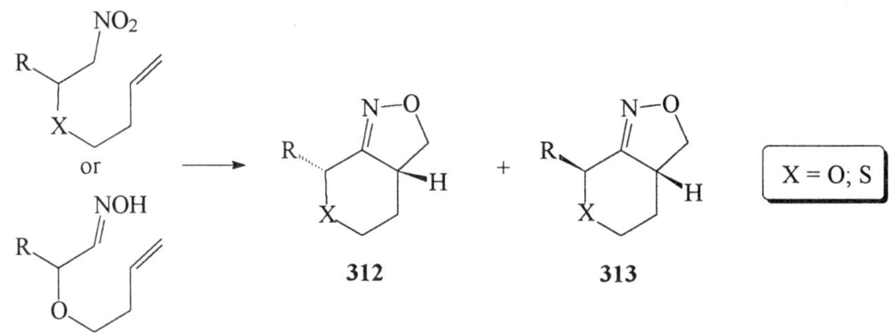

Scheme 96

The presence of a stereocentre in β position with respect to nitrile oxide function, as in compound **314**, improves the stereoselectivity: the reaction proceeds with complete face selectivity giving only the *cis* epimer **315** (Scheme 97). Clearly, a chair-like transition state conformation, which places the R-group of the nitrile oxide intermediate equatorial, is preferred in these transformations.[85]

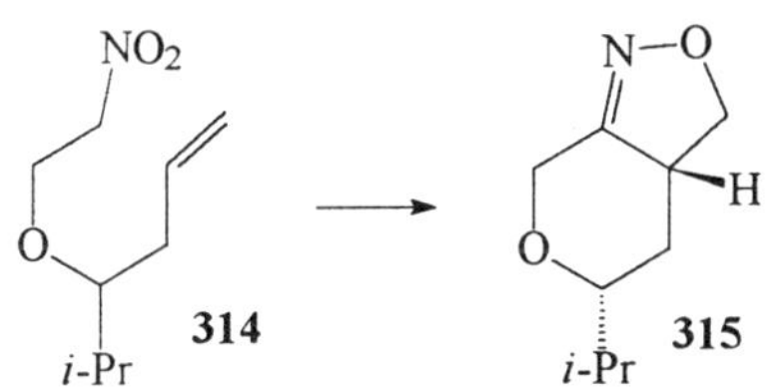

Scheme 97

The [dd]INOC reaction of **316** shows the same degree of stereoselectivity: two adducts **317** and **318** are obtained in a ratio of 11:1 respectively. The major product has phenyl, allyl and CH_2O-N moieties all situated equatorially about the tetrahydropyran ring (Scheme 98).[85]

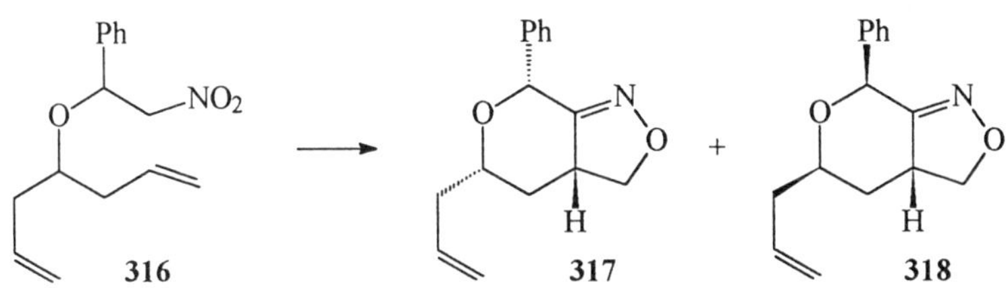

Scheme 98

Homochiral tetrahydropyrans **322-324** have been obtained by INOC of oximes **319-321** derived from homochiral 3-*O*-allyl-1,2-*O*-isopropylidene-α-D-gluco-furanose (Scheme 99).[67]

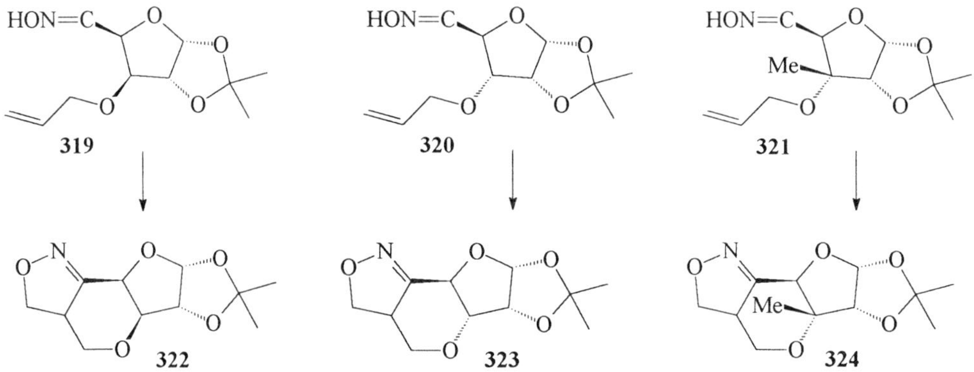

Scheme 99

Noteworthy, the analogous INC process affords a mixture of tetrahydropyrans and oxepans (Scheme 58). Molecular models indicate that, in the INOC reaction, the maximum overlap of orbitals could only be achieved in the *S-trans*-allyl ether **TS 325**, thus leading to the exclusive formation of the tetrahydropyran ring system (Scheme 100).

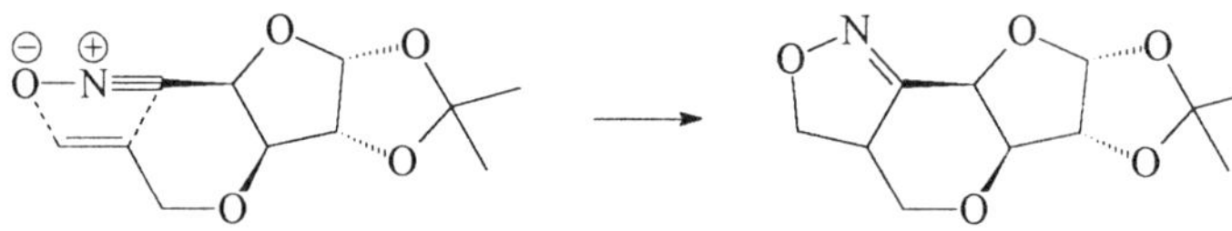

Scheme 100

Nitrile oxides derived from α-allyloxyaldehydes with β-quaternary centre cyclized to give six- instead of seven-membered *O*-heterocycles exclusively, where-as analogous nitrile oxides derived from a γ-allyloxy aldehydes cyclize to yield an oxepane ring as the sole product.

Thus, variously functionalized 4-allyl(benzylamino)-oximes **326** gave exclusively fused pyrido-isoxazole derivatives **327** and **328** by treatment with aqueous NaClO at 0 °C (Scheme 101).[68,78]

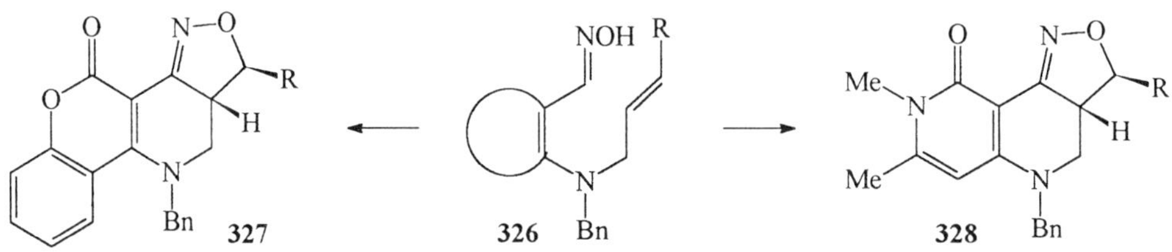

Scheme 101

Analogously, pyrazoleoximes **329** were converted via INOC into fused dihydroisoxazoles **330** (Scheme 102).[69]

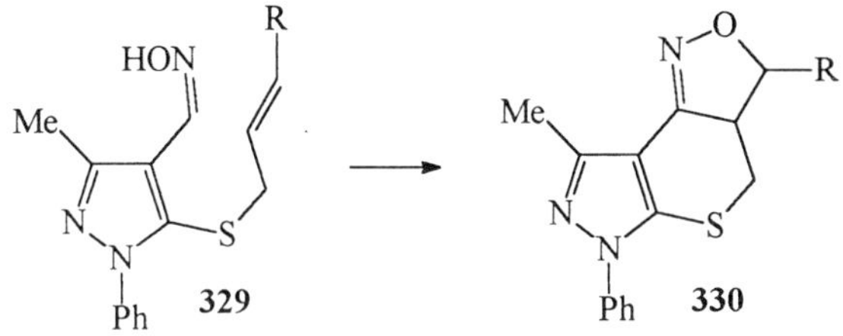

Scheme 102

According to arguments above reported, 6-(*N*-benzyl-but-3-enylamino)-4-methoxy-5-oximinomethyl-2-pivaloylaminopyrimidine **331** was transformed into the pyrimidoazepinone derivative **333** by treatment with *N*-chlorosuccinimide at 20 °C and subsequent reduction of corresponding isoxazoline **332** (Scheme 103).

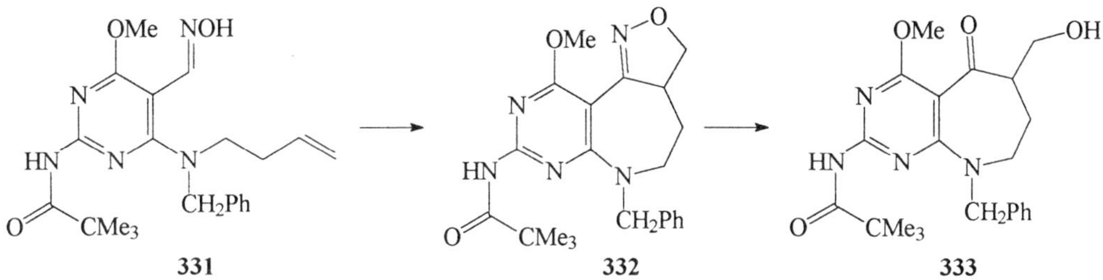

Scheme 103

Compound **333** is a pivotal intermediate for the synthesis of pyrimidoazepino based folic acid analogues.[93]

The utility of this methodology has been tested in the synthesis of various biologically interesting compounds.

The antiulcer and antisecretory tricyclic 3(2H)-pyridazinone **335** has been prepared through a five step sequence, involving an INOC of the precursor **334** (Scheme 104).[94]

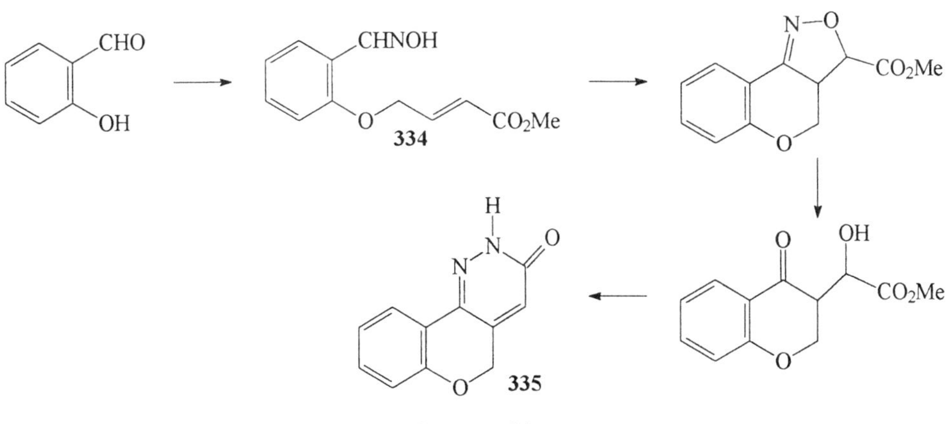

Scheme 104

A 1,8-naphtyridin-5-one **338** structurally related to potent antibacterial agents, nalidixic acid an enoxacin, has been synthesised by a INOC process between a pyridine-3-nitrile oxide **336** and an alkene dipolarophile tether to 2-position of the pyridine ring (Scheme 105).[95]

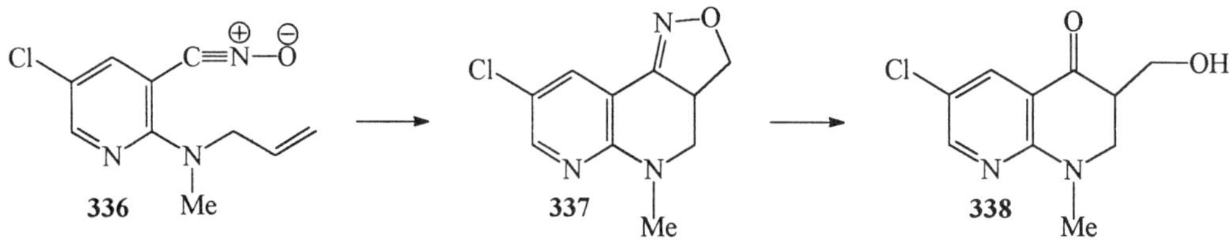

336 **337** **338**

Scheme 105

A similar strategy was employed to prepare a pyridopyrimidinone derivative **339**, a potentially useful intermediate to 5-deaza- and 5,10-dideaza-5,6,7,8-tetra-hydrofolic acid analogs (Scheme 106).[96]

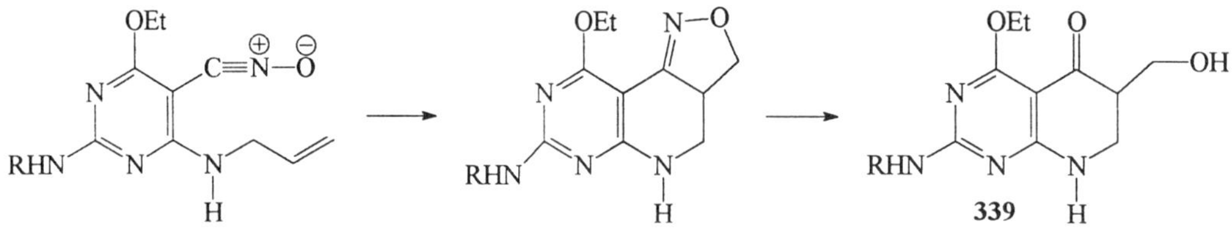

339

Scheme 106

According to the previously reported considerations, a synthetic pathway has been envisaged towards oxepane **342**, containing two stereogenic centres, a key intermediate for the synthesis of the antifertility agent zoapatanol **343** (Scheme 107).[93]

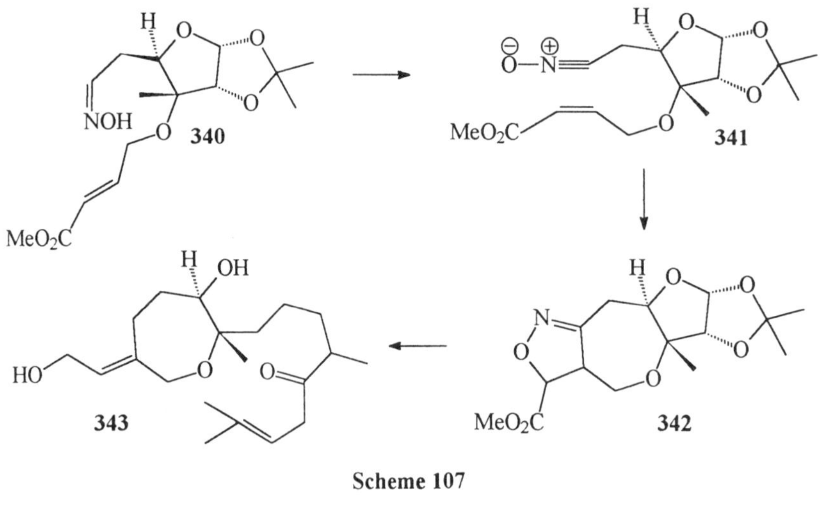

340 **341** **342** **343**

Scheme 107

The fourteen membered ring of crassin acetate **347**, which show antibiotic and antineoplastic activity against a KB and PS cell lines, has been elaborated. The INOC process starting from nitroderivate **344** and the subsequent isoxazoline ring cleavage furnish the lactone **346** with the right stereocentres at C_1 and C_{14} (Scheme 108).[95]

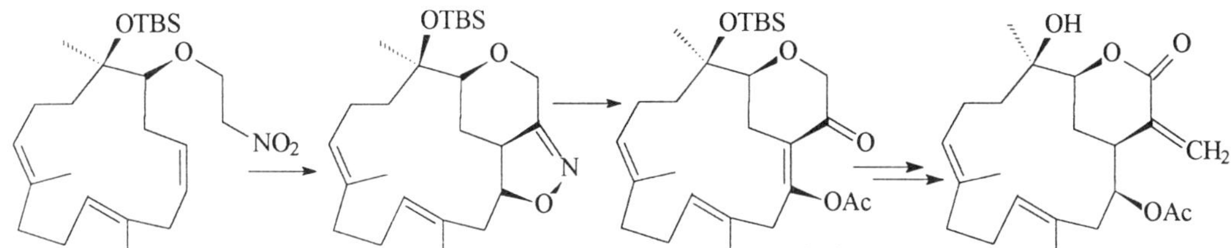

344 **345** **346** **347**

Scheme 108

INOC performed on functionalized furans **348** occurs with complete regiochemical control, but low diastereoselectivity. The presence of a sulphur atom in the obtained cycloadducts allows the enantioselective sulphide oxidation to give enantiomerically enriched products (e.e. > 95%) (Scheme 109).[88]

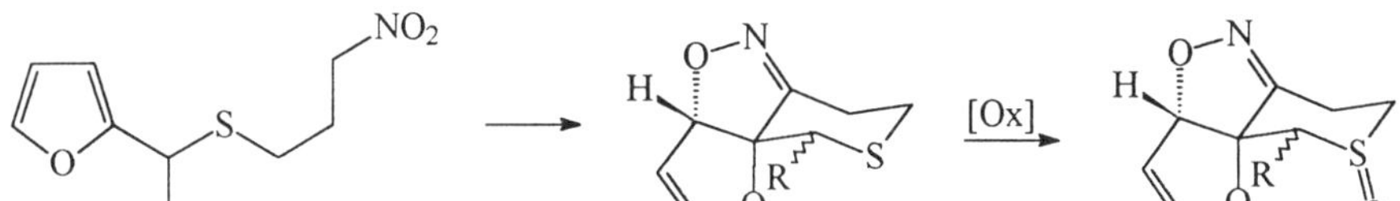

348

Scheme 109

N-Alkyl-pyrrolenitrile oxide, with n = 1 fails to give the five-membered ring annulated pyrrole and only furoxan is isolated. Starting from the higher homologues **349**, the ring closure take place with formation of carbamates **351** in moderate yields. Pyrrole **350** can not be isolated because of their great tendency toward rearomatization (Scheme 110).

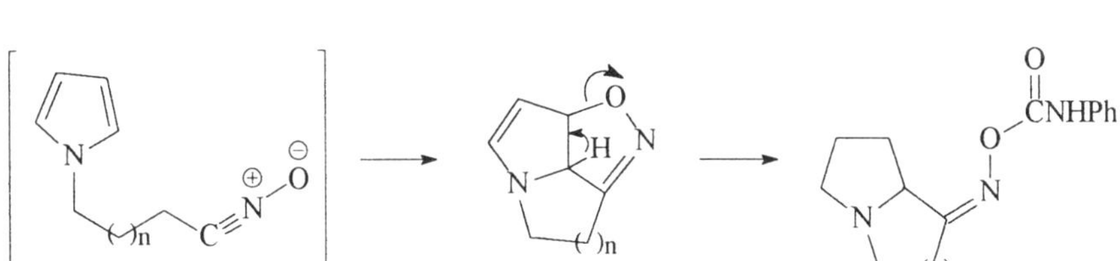

349 **350** **351**

Scheme 110

Cycloadducts of the indole on the contrary has proved to be more stable, but not for n = 1 which has not been obtained even working at higher temperatures or at higher dilution to prevent dimerization.

Cycloaddition of **352** to seven-member compounds **353** require the conditions of refluxing benzene and high dilutions. Under these conditions a partial rearomatization of cycloadduct to the corresponding oxime **354** is observed (Scheme 111).[87]

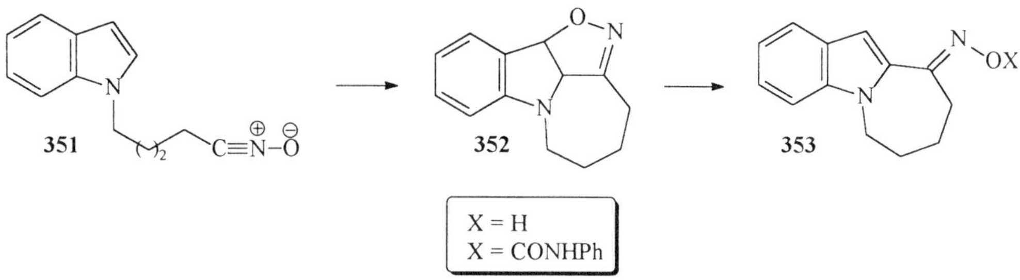

Scheme 111

4 Conclusions

The intramolecular 1,3-dipolar cycloaddition of nitrones and nitrile oxides constitutes a valid synthetic approach towards the formation of five-, six- and seven -membered heterocyclic systems in a highly stereoselective manner.

Altough the synthetic routes *via* the 1,3-dipolar cycloaddition process have been widely exploited, the versatility of this strategy is still very promising, representing in the next future an interesting field of research for the synthesis of natural and stereochemically complex molecules with a defined stereochemistry.

References

1 Padwa, A. In *1,3-Dipolar Cycloaddition Chemistry*, Padwa, A. Ed., Wiley Interscience: New York, **1984**, Voll. 1 and 2. Katritzky, A. R.; Rees, C. W. In *Comprehensive Heterocyclic Chemistry*, Pots, K. T. Ed., Pergamon Press: Oxford, **1984**, Voll. 1-8. Carruthers, V. In *Cycloaddition Reactions in Organic Synthesis*, Baldwin, J. E. Ed., FRS & Magnus, P.D., Tetrahedron Organic Chemistry Series, **1990**, Vol. 8, p 269.

2 Tufariello, J. J. *Acc. Chem. Res.* **1979**, *12*, 396. Confalone, P. N.; Huie, E. M. *Org. React.* **1988**, *36*, 1. De Shong, P.; Lander, S. W. Jr.; Leginus, J. M.; Dickens, M. *In Advances in Cycloaddition*; Curran, D. P. JAI Press Inc.: Greenwich, **1988**. Vol. 1, p. 87. Torssell, K. B. G. In *Nitrile Oxides, Nitrones and Nitronates in Organic Synthesis*, Feuer, H. Ed., VCH Publisher: New York, **1988**. Chiacchio, U.; Di Bella, M.R.; Rescifina, A.; Romeo, G.; Uccella, N. *Heterocycles* **1993**, *36*, 2209. Casuscelli, F.; Chiacchio, U.; Liguori, A.; Romeo, G.; Sindona, G.; Uccella, N. *Tetrahedron* **1993**, *49*, 5147. Chiacchio, U.; Casuscelli, F.; Corsaro, A.; Rescifina, A.; Romeo, G.; Uccella, N. *Tetrahedron* **1994**, *50*, 6671. Casuscelli, F.; Chiacchio, U.; Di Bella, M. R.; Rescifina, A.; Romeo, G.; Romeo, R.; Uccella, N. *Tetrahedron*, **1995**, *51*, 8605. Casuscelli, F.; Chiacchio, U.; Rescifina, A.; Romeo, G.; Tommasini, S.; Uccella, N. *Tetrahedron*, **1995**, *51*, 2979. Chiacchio, U.; Gumina, G.; Rescifina, A.; Romeo, R.; Uccella, N.; Casuscelli, F.; Piperno, A.; Romeo, G. *Tetrahedron*, **1996**, *52*, 8889.

3 Breuer, E. In *The Chemistry of Amino, Nitroso and Nitro Compounds and their Derivatives*, Patai, S. Ed., John Wiley, **1979**, p. 459. Tennant, G. In *Comprehensive Organic Chemistry*, Barton, D. H. R., Ollis, W. D., Ed., Pergamon Press: New York, **1979**, Vol. 2, p. 500. Caramella, P.; Grünanger, P. In *1,3-Dipolar Cycloaddition Chemistry*; Padwa, A. Ed., Wiley Interscience: New York, **1984**, Vol. 1, p.

291. Tufariello, J. J. In *1,3-Dipolar Cycloaddition Chemistry*; Padwa, A. Ed., Wiley Interscience: New York, **1984**, Vol. 2, p. 83. Murahashi, S. I.; Mitsui, H.; Shiota, T.; Tsuda, T.; Watanabe, S. *J. Org. Chem.* **1991**, *55*, 1736. Ballistreri, F. P.; Chiacchio, U.; Rescifina, A.; Tomaselli, G.A., Toscano, R.M. *Tetrahedron* **1992**, *48*, 8677. Grünanger, P.; Vita-Finzi. In *The Chemistry of Heterocyclic Compounds, in Isoxazoles*, Taylor, E.C. Ed., Wiley Interscience: New York, **1991**. Boyd, G.V. In *Progress in Heterocyclic Chemistry*; Suschitzky, H.; Scriven, E. F .V.; **1991**; Vol. 3, 166. Boyd, G.V. In *Progress in Heterocyclic Chemistry*; Suschitzky, H.; Scriven, E. F .V.; **1991**; Vol. 4, 150.

4 Kozikowski, A. P. *Acc. Chem. Res.* **1984**. *17*, 410. Kanemasa, S.; Tsuge, O. *Heterocycles* **1990**, *30*, 719. Desimoni, G.; Faita, G. In *Seminars in Organic Synthesis, XVI Summer School* "A. Corbella", Polo Editoriale Chimico Ed., **1991**, p 4.1.

5 Chiacchio, U.; Romeo, G.; Uccella, N. In *Trends in Heterocyclic Chemistry*; Menon, J.; Ed.; Council of Scientific Information: Trivandrum, India, 1995, Vol. 4, p. 261.

6 Annunziata, R.; Cinquini, N.; Cozzi, F.; Raimondi, L. *Gazz. Chim. Ital.* **1989**, *119*, 253. Cinquini, N.; Cozzi, F. In *Methods in Organic Chemistry: Stereoselective Synthesis*. Hellmchen, G.; Hoffman, R. W.; Mulzer, J.; Schaumann, E. Ed., Thieme, G. Stuttgart: New York, **1995**, Vol. E21c, chapt. 1.6.1.2.

7 Padwa, A.; Schoffstall, A. M. In *Advances in Cycloaddition*, Curran, D.P. Ed., JAI Press Inc.: Greenwich, **1990**, Vol. 2, p. 2. Hassner, A. In *Heterocycles in Bio-Organic Chemistry*, Pergamon J. Ed., Royal Soc. of Chem., Cambridge, **1991**, p. 130.

8 Kozikowski, A. P.; Mugrage, B. B. *J. Chem. Soc.; Chem. Commun. 1988*, 188. Asoaka, M.; Abe, M.; Takai, H. *Bull. Chem. Soc. Jap.* **1985**, *58*, 2145. Confalone, P.N.; Koss, T. Tetrahedron Lett. **1984**, *25*, 947.

9 Wade, P. A. In *Comprehensive Organic Synthesis*, Trost, B.M.; Fleming, I.; Semmelhack, M.F. Ed., Pergamon Press: Oxford, **1991**, Vol. 4, p. 1111.

10 LeBel, N. A.; Whang, J. J. *J. Am. Chem. Soc.* **1959**, *81*, 6634. LeBel, N. A. *Ann. N. Y. Acad. Sci.* **1965**, *27*, 858.

11 LeBel, N. A.; Banucci, E..G. *J. Org. Chem.* **1971**, *36*, 2440.

12 Baldwin, S. W.; Wilson, J. D.; Aubbé, J. *J. Org. Chem.* **1985**, *50*, 4432.

13 Raban, M.; Jones, F. B.; Karlton, E.H.; Banucci, E.; LeBel, N. A. *J. Org. Chem.* **1970**, *35*, 1496. Black, D. C.; Crozier, R. F; Davis, V. C. Synthesis **1975**, 205.

14 Ihara, M.; Takahashi, M.; Fukumoto, K.; Kametani, T. *J. Chem. Soc.; Chem. Commun.* **1988**, 9.

15 Hoffmann, R. W.; Endesfelder, A. *Liebigs Ann. Chem.* **1986**, 1832. Kozikowski, A. P., Stein, P. D. *J. Am. Chem. Soc.* **1985**, *107*, 2569. Padwa, A.; Chiacchio, U.; Dean, D. C.; Shoffstall, A.; Hassner, A.; Murthy, K. S .K. Tetrahedron Lett. **1988**, 4169.

16 Aurich, H. G.; Boutahar, M.; Köster, H.; Möbus, K. D.; Ruits, *L. Chem. Ber.* **1990**, *123*, 1999.

17 Aurich, H. G.; Biesemeier, F. *Synthesis* **1995**, 1171.

18 Aurich, H. G.; Biesemeier, F.; Boutahar, M. *Chem. Ber.* **1991**, *124*, 2329.

19 Bandiera, P.; Bravo, P.; Bruché, L; Zanda, M.; Arnone, A. *Gazz. Chim. It.* **1996**, *126*, 773.

20 Chiacchio, U.; Buemi, G.; Casuscelli, F.; Procopio, A.; Rescifina, A.; Romeo, R. *Tetrahedron* **1994**, *50*, 5503.

21 Chiacchio, U.; Casuscelli, F.; Corsaro, A.; Librando, V.; Rescifina, A.; Romeo, R.; Romeo, G. *Tetrahedron* **1995**, *51*, 5689.

22 Chiacchio, U.; Corsaro, A.; Rescifina, A.; Romeo, G. *Tetrahedron*, in press.

23 Hoppolzer, W.; Bochet, C. G.; Merifield, E. *Tetrahedron Lett.* **1994**, *35*, 7015.

24 Aurich, H. G.; Frenzen, G.; Gentes, C. *Chem. Ber.* **1993**, *126*, 787.

25 Aurich, H. G.; Gentes, C.; Harms, K. *Tetrahedron* **1995**, *51*, 10497.

26 Aurich, H. G.and Köster, H. *Tetrahedron* **1995**, *51*, 6285.

27 Aurich, H. G. and Quintero, J. R. *Tetrahedron* **1994**, *50*, 3929.

28 Aurich, H. G. and Quintero, J. R. *Tetrahedron* **1994**, *50*, 3943.

29 Annunziata, R.; Cinquini, M.; Cozzi, F.; Raimondi, L. *J. Org. Chem.* **1990**, *55*, 1901.

30 Annunziata, R.; Cinquini, M.; Cozzi, F.; Gennari, C.; Raimondi, L. *J. Org. Chem.* **1987**, *52*, 4674.

31 De Amici, M.; De Micheli, C.; Ortiz, A.; Gatti, G.; Gandolfi, R.; Toma, L. *J. Org. Chem.* **1989**, *54*, 793.

32 Tamuta, O.; Okabe, T.; Yamaguchi, T.; Gotanda, K.; Noe, K.; Sakamoto, M. *Tetrahedron* **1995**, *51*, 107.

33 Tamuta, O.; Okabe, T.; Yamaguchi, T.; Kotani, J.; Gotanda, K.; Sakamoto, M. *Tetrahedron* **1995**, *51*, 119.

34 Chiacchio, U.; Corsaro, A.; Piperno, A.; Rescifina, A.; Romeo, R.; Romeo, G. *Tetrahedron Asym.*, in press.

35 Bhuyan, B. J.; Boruah, R. C.; Sandhu, J. S. *Tetrahedron Lett.* **1989**, *30*, 1421.

36 Black, D. C.; Craig, D. C.; Deb-Das, R. B.; Kumar, N. *Aust. J. Chem.* **1993**, *46*, 603.

37 Black, D. C.; Keller, P. A.; Kumar, N. *Aust. J. Chem.* **1993**, *46*, 843.

38 Rong, J.; Roselt, P.; Plavec, J.; Chattopadhyaya, J. *Tetrahedron* **1994**, *50*, 4921.

39 Zschiesche, R. and Reißig, H. *Liebigs. Ann. Chem.* **1989**, 551.

40 Aurich, H. G.; Gheiger, M.; Gentes, C.; Köster, H. *Tetrahedron Lett.* **1996**, *37*, 841.

41 Bernet, B. and Vasella, A. *Helv. Chim. Acta* **1979**, *62*, 2411.

42 Hewson, A. T.; Jeffery, J.; Szczur, N. *Tetrahedron Lett.* **1995**, *36*, 7731.

43 Grigg, R.; Hadjisoteriou, M.; Kennewell, P.; Markandu, J.; Thornton-Pett, M. *J. Chem. Soc.; Chem. Commun.* **1992**, 1388.

44 Hassner, A.; Maurya, R. *Tetrahedron Lett.* **1989**, *30*, 5803.

45 Hassner, A.; Maurya, R.; Mesko, E. *Tetrahedron Lett.* **1988**, *29*, 5313.

46 Grigg, R.; Scott, R.; Stevenson, P. *J. Chem. Soc.; Perkin Trans. 1* **1988**, 1357.

47 Hassner, A.; Maurya, R.; Padwa, A.; Bullock W.H. *J. Org. Chem.* **1991**, *56*, 2775.

48 Hassner, A.; Maurya, R.; Friedman, O.; Gotlieb, H.E.; Padwa, A.; Austin, D. *J. Org. Chem.* **1993**, *58*, 4539.

49 Hassner, A.; Singh, S.; Sharma, R.; Maurya, R. *Tetrahedron* **1993**, *49*, 2317.

50 Hassner, A.; Falb, E.; Nudelman, A.; Albeck, A.; Gotlieb, H.E. *Tetrahedron Lett.* **1994**, *35*, 2397.

51 Hassner, A.; Murthy, K. S. K.; Padwa, A.; Chiacchio, U.; Dean, D. C.; Schoffstall, A. M. *J. Org. Chem.* **1989**, *54*, 5277.

52 Chiacchio, U.; Corsaro, A.; Pistarà, V.; Rescifina, A.; Romeo, G.; Romeo, R. *Tetrahedron* **1996**, *52*, 7875.

53 Grigg, R.; Heaney, F.; Markandu, J.; Surendrakumar, S.; Thornton-Pett, M.; Warnock, W. J. *Tetrahedron* **1991**, *47*, 4007.

54 Armstrong, P.; Grigg, R.; Heaney, F.; Surendrakumar, S.; Warnock, W. J. *Tetrahedron* **1991**, *47*, 4495.

55 Grigg, R.; Markandu, J.; Perrior, T.; Surendrakumar, S.; Warnock, W. J. *Tetrahedron* **1992**, *48*, 6929.

56 Grigg, R.; Heaney, F.; Surendrakumar, S.; Warnock, W. J. *Tetrahedron* **1991**, *47*, 4477.

57 Grigg, R.; Dorroty, M. J.; Heaney, F.; Malone, J. F.; Rajviroongit, S.; Sridharan, V.; Surendrakumar, S. *Tetrahedron* **1991**, *47*, 8297.

58 Grigg, R.; Markandu, J.; Surendrakumar, S.; Thornton-Pett, M.; Warnock, W. J. *Tetrahedron* **1992**, *48*, 10399.

59 Grigg, R.; Hadjisoteriou, M.; Kennewell, P.; Markandu, J. *J. Chem. Soc.; Chem. Commun.* **1992**, 1537.

60 Grigg, R.; Hadjisoteriou, M.; Kennewell, P.; Markandu, J.; Thornton-Pett, M. *J. Chem. Soc.; Chem. Commun.* **1993**, 1340.

61 Grigg, R. and Markandu, J. *Tetrahedron Lett.* **1989**, *30*, 5489.

62 Gravestock, M. B.; Knight, D. W.; Lovell, J. S.; Thornton, S. R. *J. Chem. Soc.; Perkin Trans. 1* **1994**, 1661.

63 Bhattacharjya, A.; Chattopadhyay, P; McPhail, A. T.; McPhail, D. R. *J. Chem. Soc.; Chem. Commun.* **1990**, 1508.

64 Datta, S.; Chattopadhyay, P; Mukhopadhyay, R.; Bhattacharjya, A. *Tetrahedron Lett.* **1993**, *34*, 3585.

65 Kocienski, P.; Love, C.; Witby, R. *Tetrahedron* **1989**, *45*, 3839. Sparks, M. A.; Williams, K. W.; Lukacs, C.; Schrell, A.; Priebe, G.; Spaltenstein, A.; Whitesides, G. M. *Tetrahedron* **1993**, *49*, 1.

66 Collins, P. M.; Ashwood, M. S.; Eder, H.; Wright, S. H. B.; Kennedy, D. J. *Tetrahedron Lett.* **1990**, *31*, 2055.

67 Shing, T. K. M.; Fung, W. C.; Wong, C. *J. Chem. Soc.; Chem. Commun.* **1994**, 449.

68 Gotoh, M.; Mizui, T.; Sun, B.; Hirayama, K.; Noguchi, M. *J. Chem. Soc.; Perkin Trans. 1* **1995**, 1857.

69 L'abbé, G.; Emmers, S.; Dehaen, W.; Dyall, L. K. *J. Chem. Soc.; Perkin Trans. 1* **1994**, 2553.

70 Chiacchio, U.; Rescifina, A.; Casuscelli, F.; Piperno, A.; Pisani, V.; Romeo, R. *Tetrahedron* **1996**, *52*, 14311.

71 Padwa, A.; Meske, M.; Ni, Z. *Tetrahedron Lett.* **1993**, *34*, 5047.

72 Padwa, A.; Meske, M.; Ni, Z. *Tetrahedron* **1995**, *51*, 89.

73 Kang, H.; Cho, Y. S.; Koh, H. Y.; Chang, M. H. *Tetrahedron Lett.* **1991**, *32*, 2779.

74 Broggini, G.; Folcio, F.; Sardone, N.; Zecchi, G. *Tetrahedron* **1996**, *52*, 11849.

75 Arnone, A.; Bruché, L.; Garanti, L.; Zecchi, G. *J. Chem. Research (S)* **1995**, 282.

76 Ihara, M.; Kawabuchi, T.; Tokunaga, J.; Fukumoto, K. *Heterocycles* **1995**, *40*, 97.

77 Ihara, M.; Takahashi, K.; Fukumoto, K.; Kametani, K. *J. Chem. Soc.; Perkin Trans. 1* **1989**, 2553.

78 Gotoh, M.; Sun, B.; Hirayama, K.; Noguchi, M. *Tetrahedron* **1996**, *52*, 887.

79 Orlek, B. S.; Sammes, P. G.; Weller, D. J. *J. Chem. Soc.; Chem. Commun.* **1993**, 1412.

80 Bishop, J. E.; Flaxman, K. A.; Orlek, B. S.; Sammes, P. G.; Weller, D. J. *J. Chem. Soc.; Perkin Trans. 1* **1995**, 2551.

81 Fusco, R.; Garanti, L.; Zecchi, G. *Chim. Ind. Milan* **1975**, *57*, 16.

82 Vasella, A. *Helv. Chim. Acta* **1977**, *60*, 1273. Barco, A.; Benetti, S.; Pollini, G. P.; Baraldi, P. G. *Synthesis* **1977**, 837.

83 Mukaiyama, T. and Hoshino, T. *J. Am. Chem. Soc.* **1960**, *82*, 5339.

84 Hassner, A.; Dehaen, W. *J. Org. Chem.* **1990**, *55*, 5505.

85 Kim, H. R.; Kim, H. J.; Duffy, J. L.; Olmstead, M. M.; Ruhlandt-Sengwe, K.; Kurth, M. J. *Tetrahedron*

Lett. **1991**, *32*, 4259.

86 Beebe, X.; Chiappari, C. L.; Kurth, M. J.; Schore, N. E. *J. Org. Chem.* **1993**, *58*, 7320.

87 Dehaen, W. and Hassner, A. *J. Org. Chem.* **1991**, *56*, 896.

88 Annunziata, R.; Cinquini, M.; Cozzi, F.; Raimondi, L. *Tetrahedron Lett.* **1989**, *37*, 5013. Annunziata, R.; Cinquini, M.; Cozzi, F.; Raimondi, L.; Licini, G. *Tetrahedron* 1991, *47*, 3869.

89 Buemi, G.; Chiacchio, U.; Corsaro, A.; Romeo, G.; Rescifina, A.; Uccella, N.; Hassner, A. *Heterocycles* **1993**, *36*, 2005.

90 Rai, K. M. L.; Hassner, A. *Heterocycles* **1990**, *30*, 817.

91 Chiacchio, U.; Corsaro, A.; Librando, V.; Rescifina, A.; Romeo, R.; Romeo, G. *Tetrahedron* **1996**, *52*, 14323.

92 Hassner, A.; Dehaen, W. *Chem. Ber.* **1991**, *124*, 1181.

93 Miller, M. L. and Ray, P. S. *Tetrahedron* **1996**, *52*, 5739.

94 Baraldi, P. G.; Bigoni, A.; Cacciari, B.; Caldari, C.; Manfredini, S.; Spalluto, G. *Synthesis* **1994**, 1158.

95 Read, M. W. and Ray, P. S. *J. Heterocyclic Chem.* **1995**, *32*, 1595.

96 Taylor, E. C. and Ray, P. S. *Heterocycles* **1991**, *32*, 1327.

97 Shing, T. K. M.; Wong, C. H. *Tetrahedron Asym.* **1994**, *5*, 1151. Shing, T. K. M.; Wong, C. H.;. Yip, T. *Tetrahedron Asym.* **1996**, *7*, 1323.

98 Tius, M. A. and Reddy, N. K. *Tetrahedron Lett.* **1991**, *32*, 3605.

PHOTOCHEMICAL ARYL-ARYL COUPLING IN FURAN, THIOPHENE, AND PYRROLE DERIVATIVES

Maurizio D'Auria

Dipartimento di Chimica, Università della Basilicata, Via N. Sauro 85, 85100 Potenza, Italy

Abstract. *There are two approaches described in the literature on the photochemical formation of the Ar-Ar bond. The first requires the presence of an halogen atom (Ar-Hal → Ar-Ar'), the second is related to the cyclization reaction of cis-stilbenes. In this review all the possible approaches in the field of furan, thiophene, and pyrrole are reported.*

1. Introduction

This review has been thought in order to give an overview on the photochemical aryl-aryl coupling methods used in the heterocyclic chemistry of furan, thiophene, and pyrrole. There are two approaches described in the literature on the photochemical formation of the Ar-Ar bond. The first requires the presence of an halogen atom (Ar-Hal → Ar-Ar'), the second is related to the cyclization reaction of *cis*-stilbenes. In this review all the possible approaches are described.

2. Arylation

2.1. An Introduction

Photochemical arylation of halogenoaromatic compounds was studied from a long time. Iodobenzene, irradiated in benzene, gave biphenyl in 65% yield.[1] The irradiation of pentafluoroiodo-benzene gave the arylation product in 89% yield,[2] while bromobenzene gave biphenyl in 40% yield.[3]

However, in the presence of substituents, some problems appeared: thus, 4-iodobenzoic acid gave the corresponding biphenyl in 80% yield[1] and ethyl 4-chlorobenzoate gave the same reaction in 48% yield,[4] but the irradiation in benzene of 4-chloro and 4-bromoacetophenone gave the corresponding biphenyl derivative in 15 and 23% yields, respectively.[3,4] Furthermore, 4-chlorobenzaldehyde gave only unidentified products[4] and 4-bromophenyl *n*-propyl ketone in toluene gave only the dehalogenation product.[5] Nevertheless, this reaction has been extensively used in the synthesis of natural compounds.[6]

Photoarylation of halogenoheterocyclic compounds is not a very useful synthetic reaction. It was used on few heterocyclic rings and there are few examples on molecules bearing functional groups.

On pyridine, the arylation product can be obtained in high yields by the irradiation of an oxime ester in benzene.[7] The irradiation of 2-iodopyridine gave different results.[8,9] These results did not indicate a high reactivity. On the contrary, the arylation of 4-iodopyridine can be carried out with high efficiency.[2] Furthermore, the reaction can be intramolecular.[10] However, when the reaction was performed in acetonitrile in the presence of indole a complex reaction mixture was obtained.[11] The reaction with thiophene gave a mixture of two products, too.[12]

Photochemical arylation was attempted on pyrimidine nucleus. 5-Iodo-2-chloropyrimidine, irradiated in the presence of thiophene, gave 5-(2-thienyl)-2-chloropyrimidine.[13] When the arylation was carried out on uracil derivatives, the iodine substituted compounds gave the expected products in good yields,[14] while the bromo substituted derivative gave a mixture of isomeric products.[15] When the reaction is carried out with 2-methylfuran the arylation product was obtained only in 28% yield.[14] Few data are available on purine derivatives. The reaction was tested when iodine was present in C-2 and C-6 giving interesting results.[16,17]

2.2. Arylation of Furans, Thiophenes, and Pyrroles

2-Iodothiophene can be arylated in good yields when irradiated in aromatic solvents.[1,18] Also 2,5-diiodothiophene gave the corresponding arylation product if irradiated in the presence of thiophene.[19] On the contrary, 3-iodothiophene gave the arylation product only in low yields.[20] This conversion probably occurs through the formation of the corresponding thienyl radical by the homolytic cleavage of the carbon-iodine bond. The subsequent reaction of this radical with the solvent gives the product.

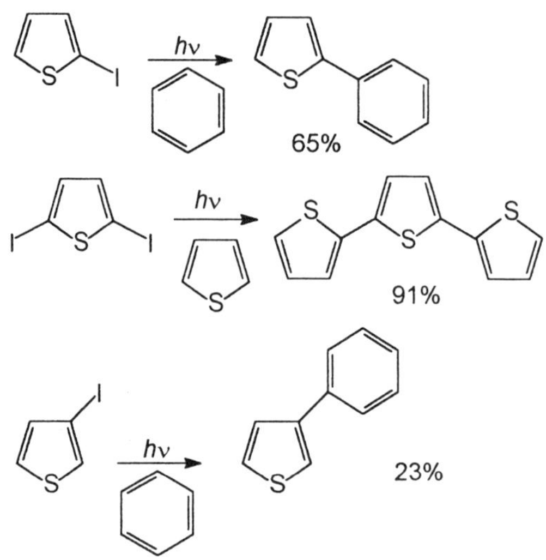

The irradiation of 5-bromofuran-2-carbaldehyde in benzene gave the corresponding 5-phenyl derivatives in good yields.[21] By using 3-bromofuran-2-carbaldehyde as starting material we obtained the corresponding 3-phenyl derivative in 31% yields. With polyhalogenofuryl derivatives, such as 4,5-dibromo-furan-2-carbaldehyde, the reaction selectively occurred in C-5: thus, we obtained 5-aryl-4-bromofuran-2-carbaldehyde. 4-Bromofuran-2-carbaldehyde was found to be completely unreactive under the same conditions. Also the reaction of 5-bromo-2-furyl-*n*-decyl ketone with furan allowed us to obtain the corresponding coupling product.[22]

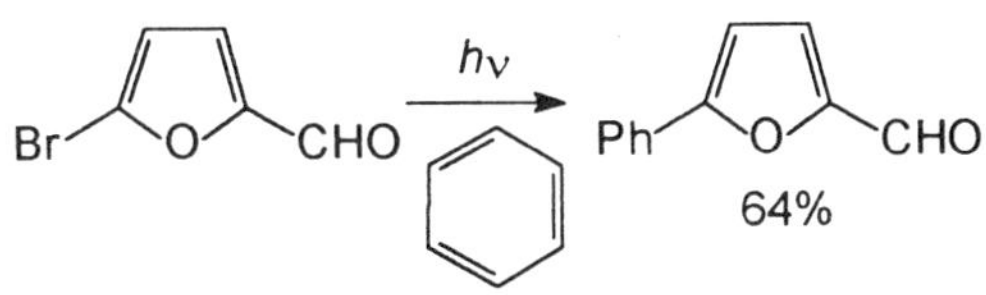

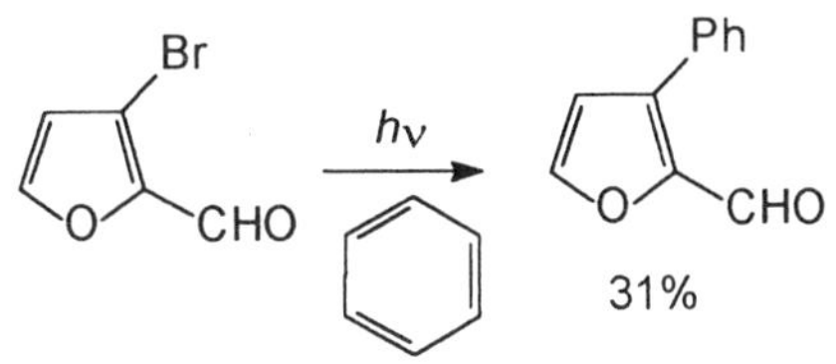

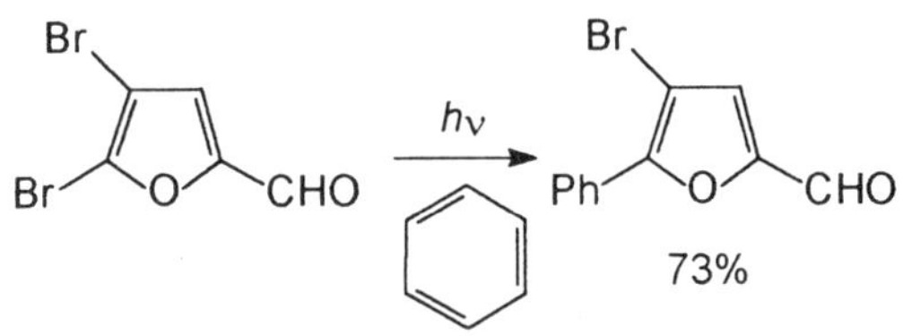

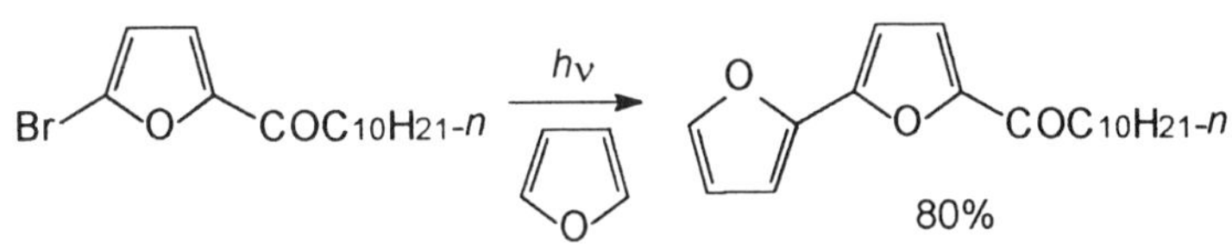

In a completely different approach to the photoarylation (more related to the cyclization of stilbenes), also the irradiation in methanol of a furyl anilide derivatives gave the corresponding arylation product.[23]

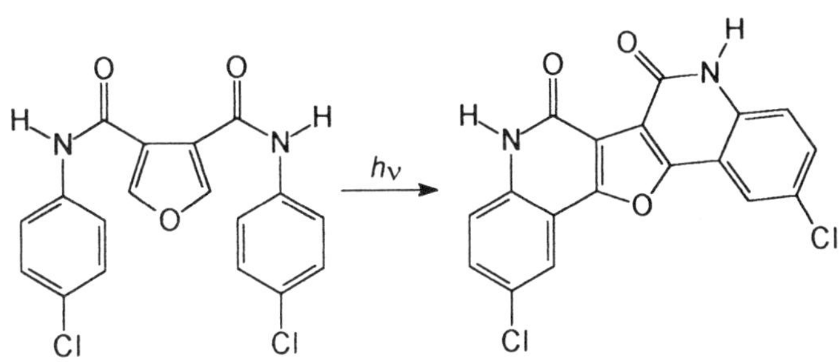

The extension of the photochemical arylation of halofuran derivatives to thiophenes showed the same trend of reactivity observed with the furans.[25] 5-Bromothiophene-2-carbaldehyde gave the corresponding 5-phenyl derivative in 54% yield.

3,5-Dibromothiophene-2-carbaldehyde gave 5-phenyl-3-bromothiophene-2-carbaldehyde in 43% yield. On the contrary, 5-bromothiophene-2-carbonitrile gave no reaction.[25] The same behaviour was observed using methyl 5-bromo-3,4-dimethylpyrrole-2-carboxylate.[26]

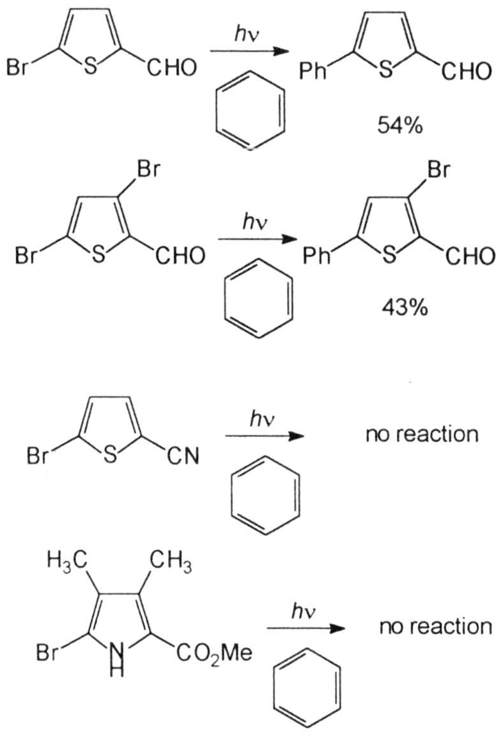

Better results can be obtained by using iodine substituted heterocycles. In fact, 5-iodofuran-2-carbaldehyde,[27] 5-iodothiophene-2-carbaldehyde,[24] 3,5-diiodothiophene-2-carbaldehyde,[24] 4,5-diiodo-2-acetyl-thiophene,[24] 3-iodothiophene-2-carbaldehyde,[24] methyl 5-iodothiophene-2-carboxylate,[25] methyl 5-iodo-3,4-dimethylpyrrole-2-carboxylate,[28] 4,5-diiodopyrrole-2-carbaldehyde,[28] and 5-iodo-2-nitrothiophene[29-31] gave the corresponding phenyl derivatives in very good yields.

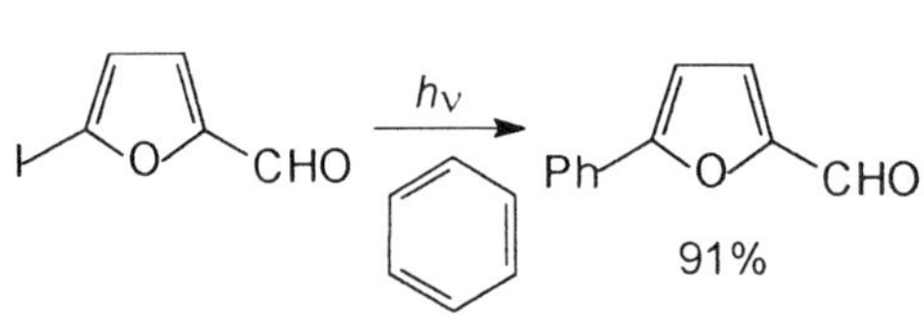

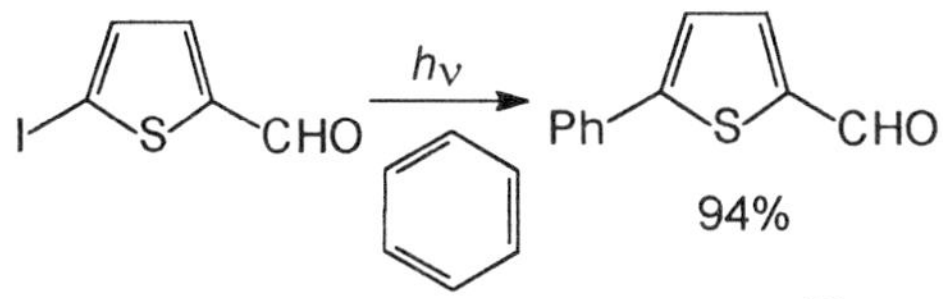
I
S
CHO
hν
Ph
S
CHO
94%

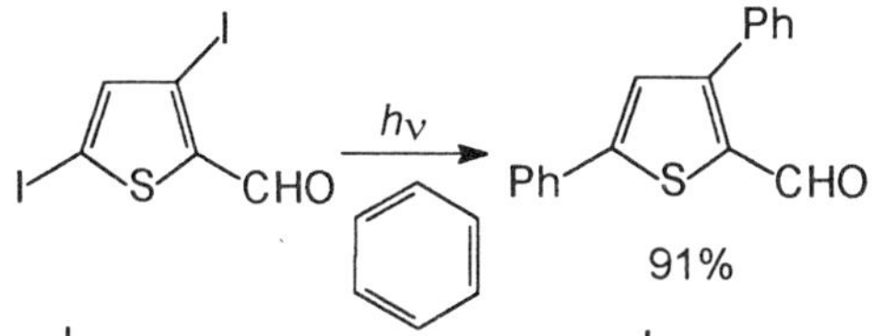
I
I
S
CHO
hν
Ph
Ph
S
CHO
91%

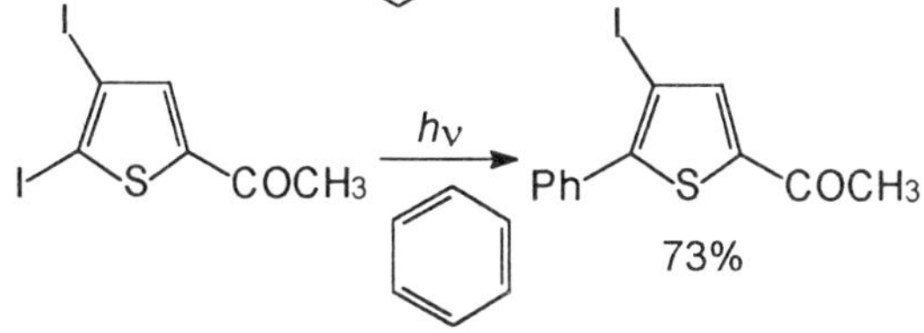
I
I
S
COCH3
hν
I
Ph
S
COCH3
73%

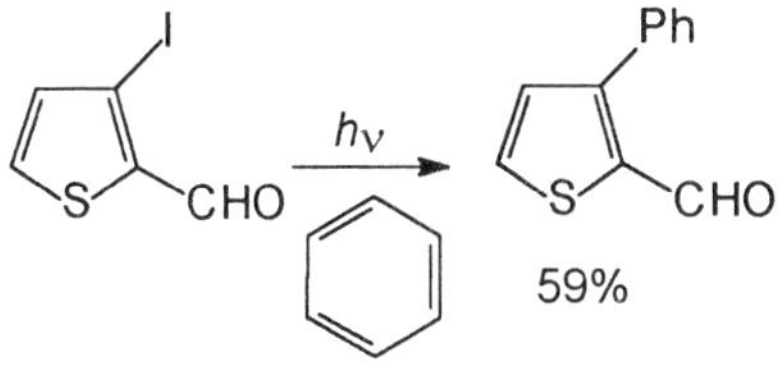
I
S
CHO
hν
Ph
S
CHO
59%

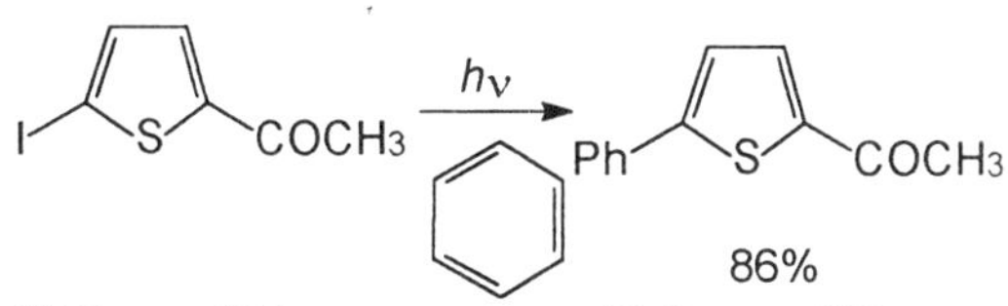
I
S
COCH3
hν
Ph
S
COCH3
86%

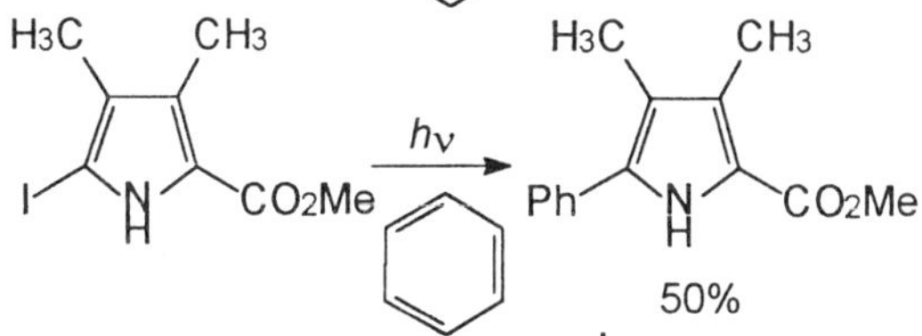
H3C
CH3
I
N
H
CO2Me
hν
H3C
CH3
Ph
N
H
CO2Me
50%

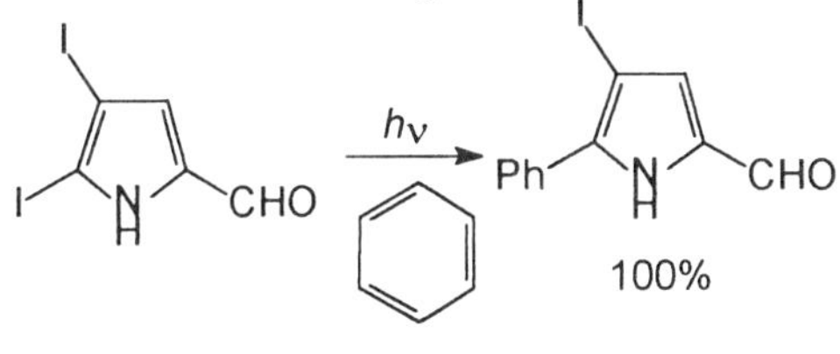
I
I
N
H
CHO
hν
I
Ph
N
H
CHO
100%

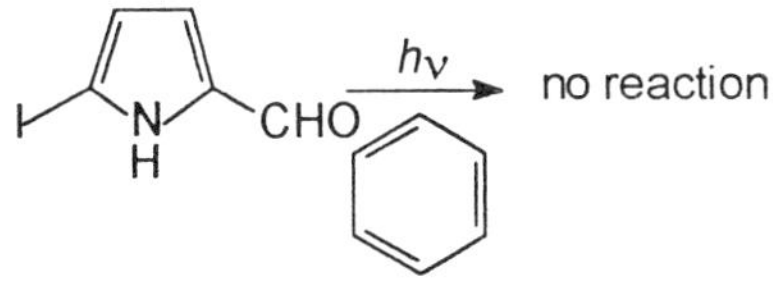
I
N
H
CHO
hν
no reaction

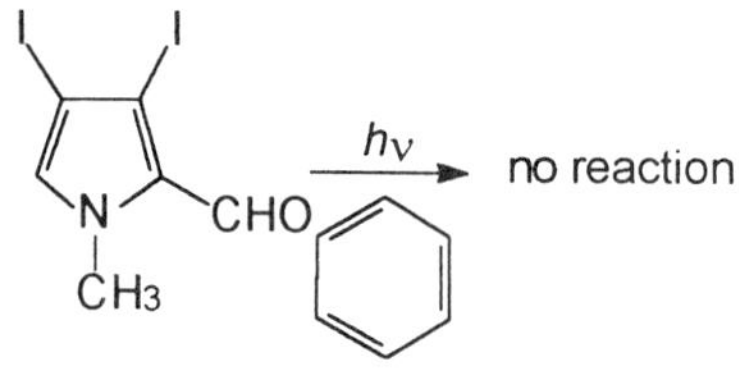

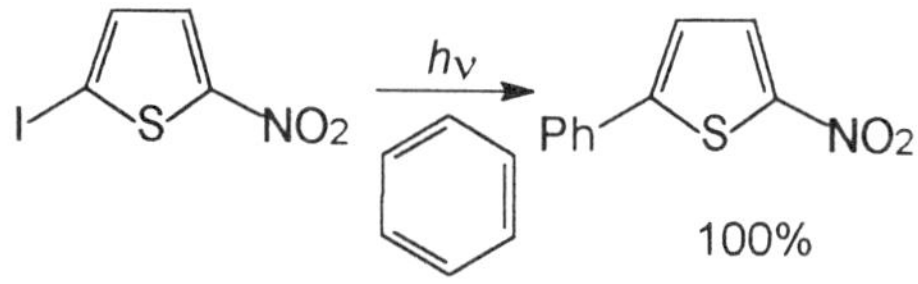

Only the pyrrole derivatives gave some different behaviours. Methyl 5-iodo-3,4-dimethylpyrrole-2-carboxylate gave a mixture of both the arylation and the dehalogenation product,[28] while 5-iodopyrrole-2-carbaldehyde and 3,4-diiodopyrrole-2-carbaldehyde did not give any reaction.[28] Furthermore, 5-iodo-thiophene-2-carbonitrile gave only the dehalogenation product.[24] Clearly, the dehalogenation and the arylation are competitive reactions and until now we do not know the factors determining either behaviour.

All the reactions described before show a limitation of this methodology: the need for an aromatic compound as solvent limits our choice of substrates to liquid, low-boiling ones only. We found that acetonitrile is the best solvent for this reaction. This way, we could synthesise some interesting products. Thus, 5-iodofuran-2-carbaldehyde could react with naphthalene to give only the β-naphthyl derivative and with anisole to give 2-methoxyphenylfuran-2-carbaldehyde in 60% yield.[27]

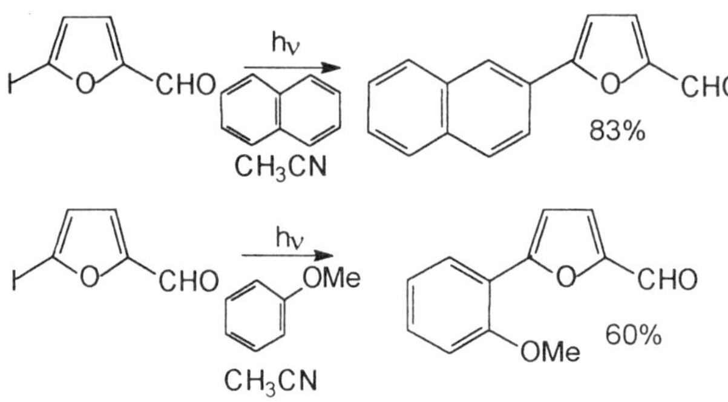

The same type of results could be obtained irradiating 5-iodo-2-acetylthiophene or 5-iodothiophene-2-carbaldehyde with halogenothiophenes,[32] 2-thienylacetic acid derivatives,[33] and substituted thiazoles,[34] or irradiating methyl 5-iodothiophene-2-carboxylate and 5-iodo-2-nitrothiophene in the presence of naphthalene or thiophene.[25,27,30]

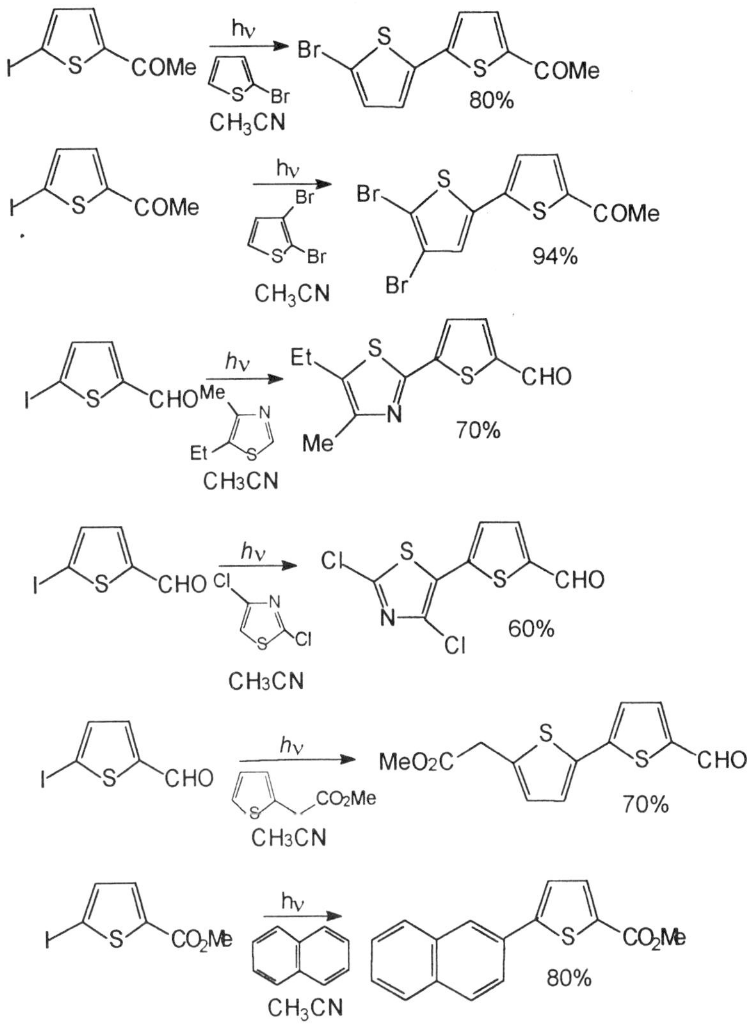

On the contrary, this type of reaction failed irradiating all the pyrrole derivatives in the presence of benzene in acetonitrile,[28] or irradiating 5-iodothiophene-2-carbaldehyde with 2-thienyl-methanol or 2-thienylacetonitrile.[33]

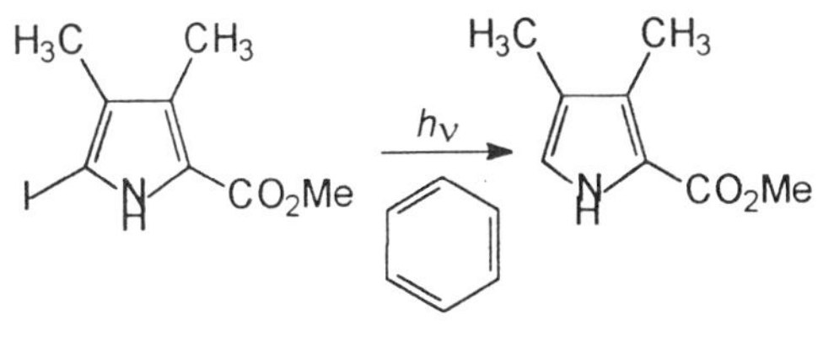

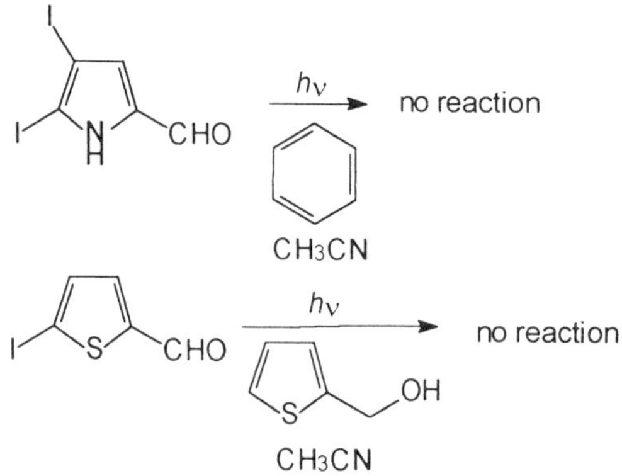

Several studies have been carried out in order to elucidate the mechanism of the above described photochemical arylation. The mechanism of this reaction has been studied recently showing some interesting data. The excitation of the substrate gave a n,π* triplet state, but this excited state is unable to give the dissociation of the carbon-iodine bond. This assumption was demonstrated considering that, if this n,π* triplet state was sensitised by chrysene, it did not produce coupling products. Then, probably the reaction occurred in an excited σ,σ* triplet state mainly localised on the carbon-iodine bond. Furthermore, the interaction between this triplet state of the substrate and aromatic compounds led to the homolytic cleavage of the C-I bond with the formation of both a 5-thienyl radical and a complex between the aromatic compound and the halogen atom. The formation of this complex was demonstrated by the presence of a short-lived transient with $\lambda_{max} = 510$ nm, showing a second order decay kinetics and a half-life of *ca.* 0.4 μs in laser flash photolysis. The thienyl radical thus formed reacted rapidly with the aromatic compound to form the corresponding arylation product.[35]

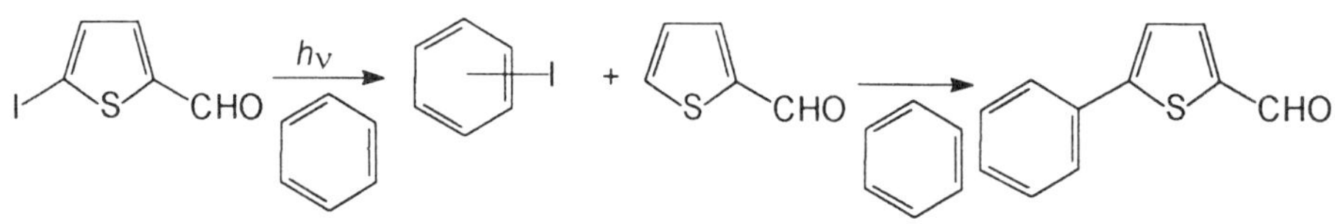

These data seem to be not in agreement with our previous reported hypothesis on the mechanism of these reaction: in fact, our kinetic data appeared to fit the hypothesis that the reaction occurs through a single electron transfer process from the aromatic compound (the donor) to the heteroaromatic derivative (the acceptor).[23,36] In particular, we followed the reaction kinetics in four solvents (benzene, chlorobenzene, toluene, and anisole) obtaining this increasing order of reactivity: benzene < anisole < chlorobenzene < toluene. This order of reactivity clearly showed that the aromatic substrates are involved in the key photochemical step. Successively, we found that this behaviour can be explained on the basis of an electron transfer process: in fact, with the exception of anisole, all our kinetic data fit the equation $\ln\left(\frac{1}{\phi}-1\right)=\beta\bullet IP$ where IP is the ionisation potential of the donor. On the basis of the above spectroscopic evidences on the mechanism, these data can be explained considering that the reaction occurs only when a complex between iodine atom and the aromatic compounds can be formed. The capability of the aromatic compounds to give these complexes can be estimated on the basis of their ionisation potential (or electrochemical properties).

Furthermore, it is noteworthy that in our experiments no transposition products were recovered in appreciable yields. The photochemical isomerisation of the thienyl derivatives is a well known reaction.[37]

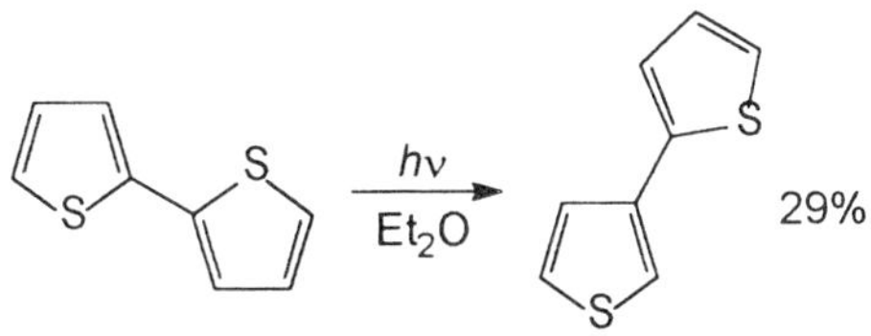

However our results can be understood considering that a) the above described isomerisation was obtained after 6 h irradiation while our reactions used shorter irradiation time; b) in ether, Kellogg *et al.* reported $\phi = 0.07$ but also observed that in ethanol ϕ was ≤ 0.01 and that in benzene isomerisation did not work.[38] Furthermore, thiophenes bearing electron-withdrawing substituents, such as NO_2, CHO, COMe, did not rearrange photochemically.[37]

The most important synthetic applications of the above described reactions were the synthesis of several naturally occurring bithienyl compounds. Bithiophenes and terthiophenes were isolated in several members of the *Compositae*. These compounds, and in particular α-terthienyl, have been extensively studied for their biological properties. In fact, they show a nematocidal activity that is enhanced in the presence of sunlight (UV-A); similarly, in the presence of UV-A they show antibiotic, ovicidal, algicidal, larvicidal, and antifeedant properties. Furthermore, they inhibit the germination of some plants and are phototoxic for some aquatic organisms. They can produce hemolysis and phototoxic dermatitis; however, they are not able to induce chromosome chromosome damage. The presence of UV-A to have biological activity can be explained by considering that this type of compounds are photosensitizers of singlet oxygen and that singlet oxygen quenchers inhibit the enzymatic inactivation by α-terthienyl.[39,40]

Because of their biological properties, these compounds are potentially useful against infesting insects.[41] Furthermore, it is known that several plants containing thiophenes are used in the traditional medicine of China, India, Pakistan, Guyana, and Malaysia against conjunctivities, mumps, sore eyes, and in the treatment of skin infections.[42] For these reasons the development of a general synthetic method was a significant target.

We was able to synthesise some naturally occurring bithienyls isolated in *Coreopsis grandiflora*,[32] *Arctium lappa*,[25,43] *Tagetes erecta*,44 *Tagetes minuta*,[32,45] *Eclipta erecta*,[43] and *Bidens radiata*.[43]

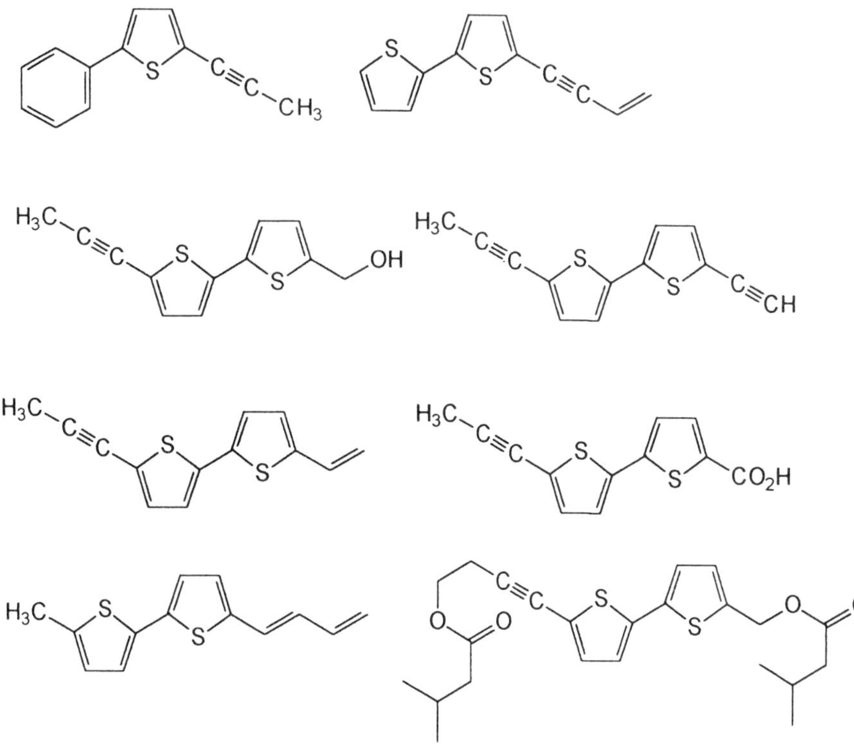

A further application of the photochemically obtained bithiophenes is given by the fact that the thienyl rings can show several different chemical reactivities. Thus, it is known that thiophenes can be reduced to alkenes through the reaction with lithium in ethylamine.[46,47] The application of this methodology to the bithienyls gave a product were only one heterocyclic ring has been reduced. Furthermore, the second ring was completely reduced and fragmented. Probably, this anomalous reactivity can be attributed to the capability of this type of compounds to give dianions.[47]

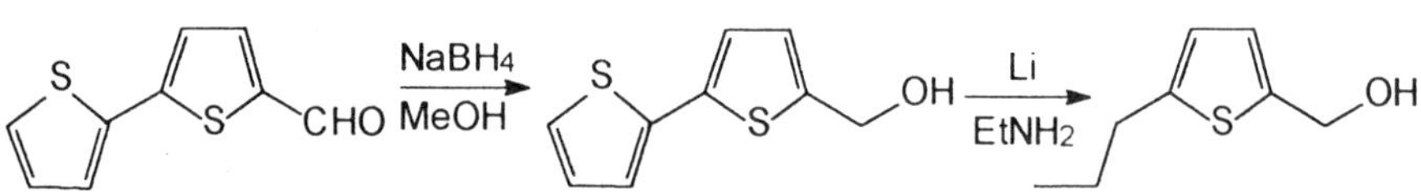

One of the problem connected with this type of research was: how do halogenoheterocyclic compounds react when they are in presence of a substrate containing functionalities able to react with them? To answer this question we have tested the reactivity of halogenoheterocyclic derivatives with arylalkenes and arylalkynes. In fact, it is well known that these compounds could give photoaddition reactions on the carbonyl group.[49] The irradiation of both halofuran and thienyl derivatives in the presence of arylalkenes gave good yields of a *cis-trans* mixture of the corresponding photosubstitution products on the alkenes.[50]

When the reaction is carried out on β-alkyl substituted compounds, the same behaviour was observed.[51] On the contrary, when an α-alkyl substituent is present, a mixture of two products was obtained.[52]

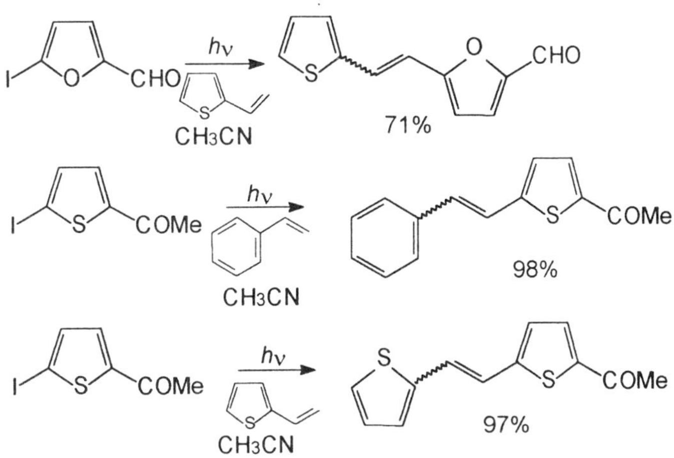

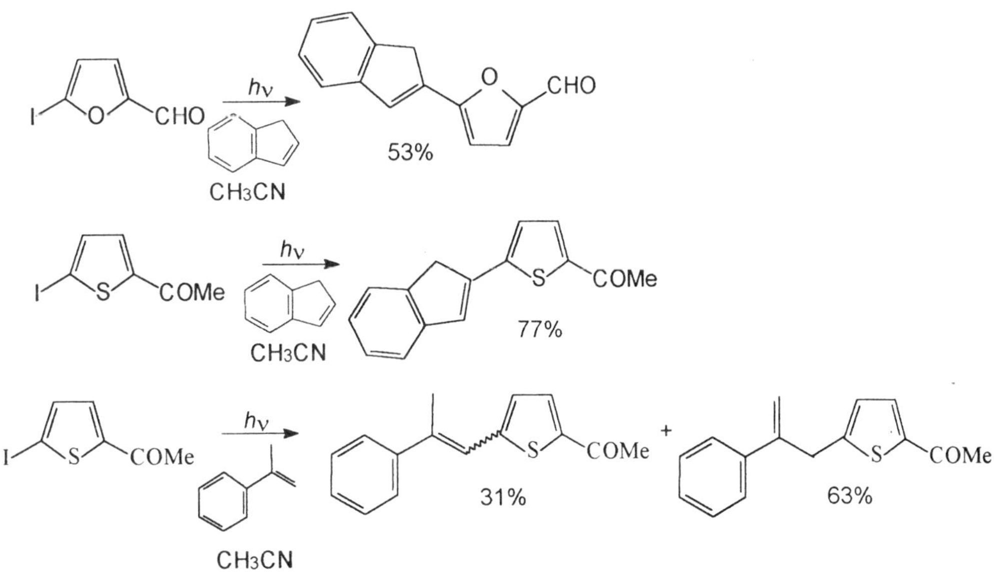

The irradiation of furyl or thienyl derivatives in the presence of arylalkenes bearing functional groups induced different behaviours. In the presence of an allylic alcohol no reaction occurred, while, in the presence of the corresponding acetate the coupling reaction occurred on the aromatic ring.[33] These bithienyls can be used in the synthesis of polyhydroxylated bithienyl compounds, as pointed out in a research devoted to synthesise polar compounds able to interact with DNA.[31] When 5-iodo-thiophene-2-carbaldehyde was irradiated in the presence of methyl 2-furylacrylate only dimers of this compound were obtained.[53-55]

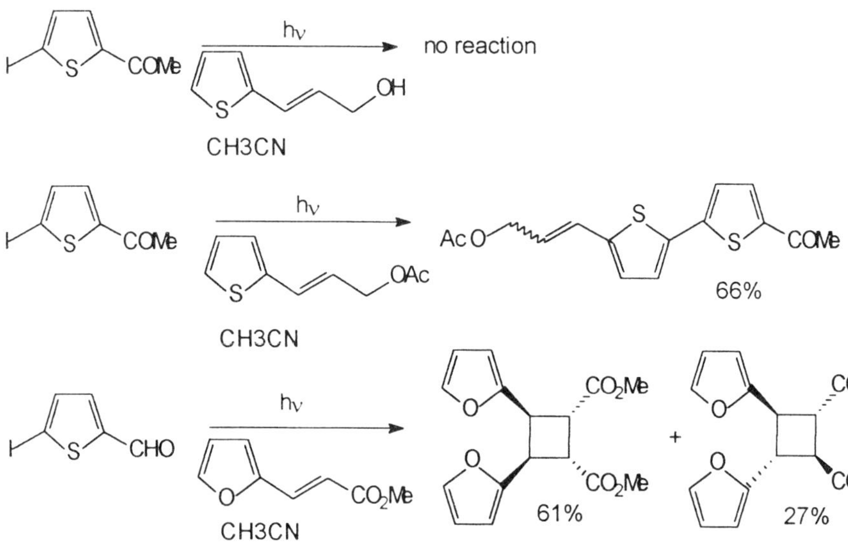

In the above described reactivity the reasons of the different behaviour between the alcohol and the corresponding acetate are not clear. However, it seems that the presence of an electron-withdrawing substituent on the alkene lowers the reactivity of both the double bond and all the molecule. The dimerization reaction occurred also with 2-furylacrolein or with the corresponding methyl ketone. Instead of 5-iodo-thiophene-2-carbaldehyde, benzophenone can be used as sensitizer. This reaction did not work with methyl cinnamate but, if methoxy substituted cinnamic derivatives were used as starting material, the dimerization reaction occurred in good yields.[56] When furylidenetetralones were used as substrates a transposition reaction occurred.[57] The mechanism of the photochemical dimerization of methyl 2-furyl-acrylate was carefully studied. Laser flash photolysis experiments showed that this reaction occurred *via* an energy transfer process from the sensitizer.[54] The regiochemistry of this reaction can be justified on the basis of the interaction between the LSOMO of the triplet state and the HOMO of the other molecule. The stereochemistry can be explained considering that the obtained compounds are the most stable ones.[55]

Different behaviours were observed also when the double bond is a part of a heterocyclic ring. The reaction of 5-iodothiophene-2-carbaldehyde with indole gave a completely different product[58]. This behaviour was in agreement with the previously reported reactivity of the indole with the carbonyl group.[59] In this case, a cycloaddition between the double bond C2-C3 of the indole and the carbonyl group was proposed. While this hypothesis was in agreement with the evidence that, when 5-iodo-2-nitrothiophene was used as substrate, the coupling product was obtained, the reactivity of aromatic and heteroaromatic aldehydes with pyrrole to give the same type of product is not in agreement with a cycloaddition reaction.[60] In this case, then, an electron transfer mechanism was proposed to account for the observed reactivity. On the contrary, benzofuran gave the expected photosubstitution product.[50] When benzimidazole was used as substrate the photosubstitution occurred but on the benzenic ring of the molecule.[61] It is difficult to explain these different behaviours. Probably, in the case of benzofuran and benzimidazole, the side of attack could be determined by the different reactivity of the rings within the molecules.

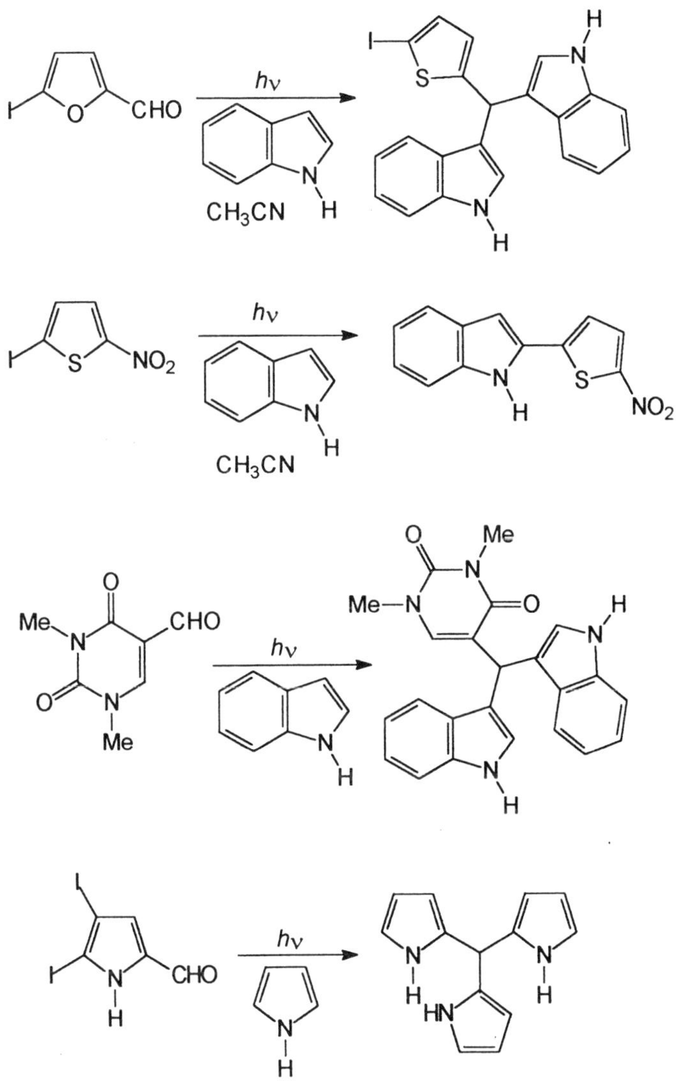

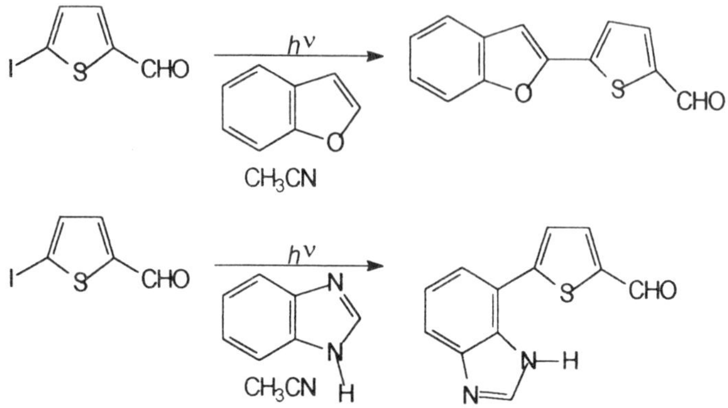

A completely different behaviour was reported for the photochemical interaction between halogenonitrothiophenes and arylalkenes. When 5-iodo-2-nitrothiophene was irradiated in the presence of styrene we observed the formation of a mixture of nitrones. The same behaviour was observed by using nitroarenes, such as nitrothiophene or nitrobenzene.[62] The reaction can be explained on the basis of a work published in 1955 where an electron transfer mechanism was proposed. The radical thus formed can give the product of addition of the nitro group to the double bond that, then, can be converted into the product.[63]

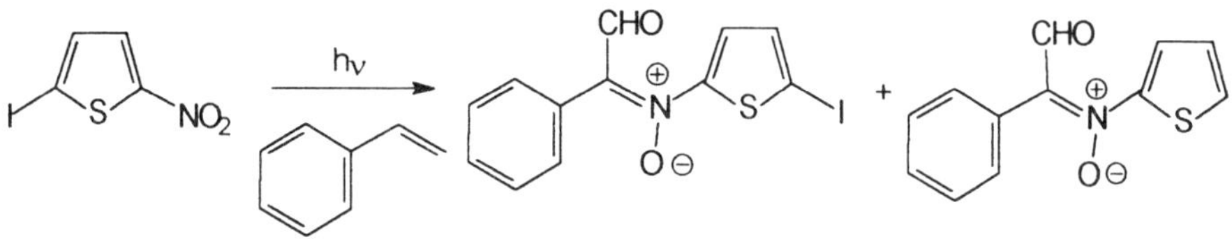

A similar behaviour was observed when the same substrates reacted with 1,1- diphenylethylene. In this case, benzophenone was obtained as the main product, together the formation of nitrones or products deriving from an unusual substitution of the nitro group.[64]

Trans-stilbene showed a low reactivity and the main product was benzaldehyde.[64] Indene showed a very complex behaviour. When it reacts with 5-iodo-2-nitrothiophene the main product was the unusual product deriving from the substitution of the nitro group.[65] However, in the presence of nitrobenzene, the main product is due to an oxidative fission of the double bond of indene.[64] These different behaviours can be explained considering that nitrobenzene and nitrothiophene show different triplet states (n,π* and π,π*, respectively) and the interaction between the LSOMO of the triplet state of the nitroarene with the HOMO of indene can be different. A further factor that affects the reactivity of the substrates is that styrene does not show a dipole moment while indene shows a dipole moment that selects the way of the approaching of the nitro derivatives.[64]

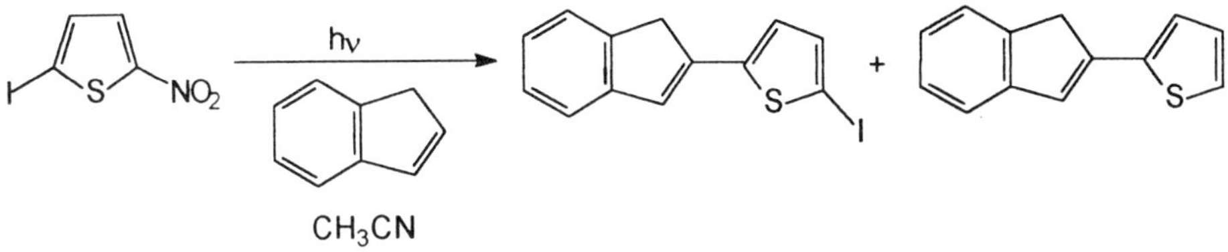

Finally, when nitrothiophene reacts with β-methoxystyrene a mixture of two products was obtained. The first was an oxidized dimer of the starting material while the second was an amide.[66]

The irradiation of halogenothienyl derivatives with phenylacetylene gave the corresponding coupling products on the alkyne in 52-54% yields.[50] The product thus obtained was a naturally occurring thiophene isolated in several *Bidens* plants[67] and in *Coreopsis grandiflora*.[68]

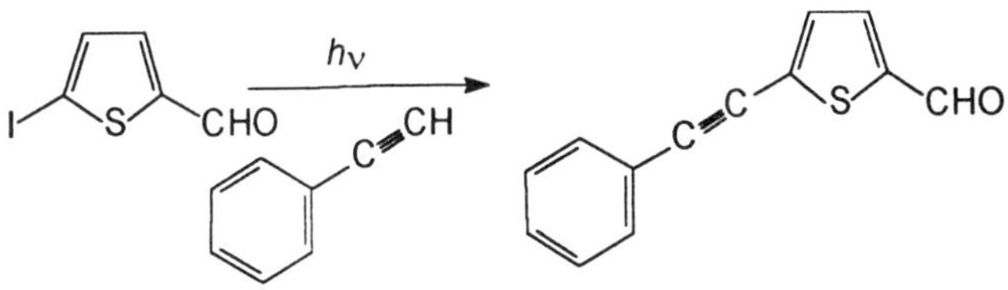

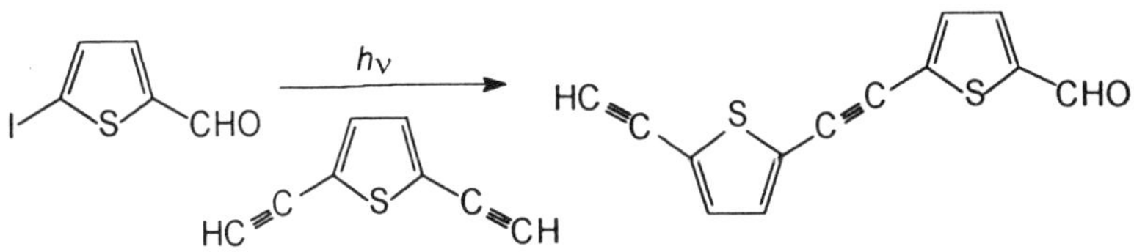

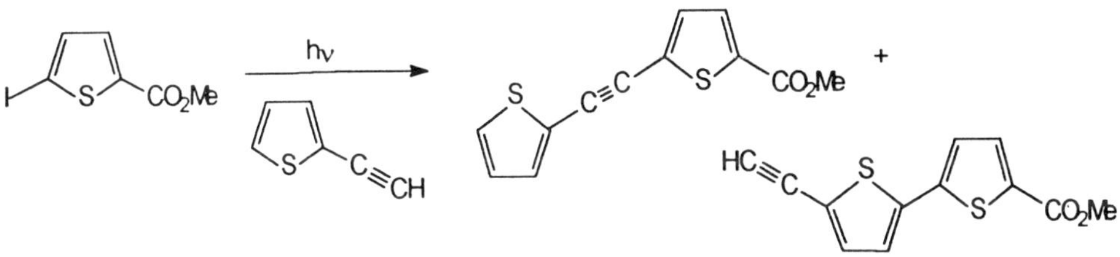

The same behaviour was observed using 2,5-diethynylthiophene as substrate.[69] Unfortunately, the extension of this reaction to other arylalkynes such as 2-thienylacetylene or 2-furyl-acetylene gave a mixture of two products.[50]

By using our usual reaction conditions we obtained the substitution product on the alkyne in 22% yield and the arylation product in 72% yield. Interestingly, we have found that using more diluted reaction conditions we can obtain the substitution product on the alkyne in 68% yield in the presence of 20% of the other product.[70] This compound is a naturally occurring thiophene isolated in *Berkheya* species.

In order to avoid tedious separation of isomers and to increase the scale of the reaction (we have to use very diluted reaction mixture to increase the formation of the substitution product on the alkyne), we tested whether protected alkynes can be used. We have used 2-ethynyl-5-trimethylsilylthiophene. The irradiation of 5-iodothiophene-2-carbaldehyde in the presence of this compound gave only the expected substitution product in 65% yield. The coupling product can be converted into the naturally occurring thiophene in acidic medium in 85% yield.[71]

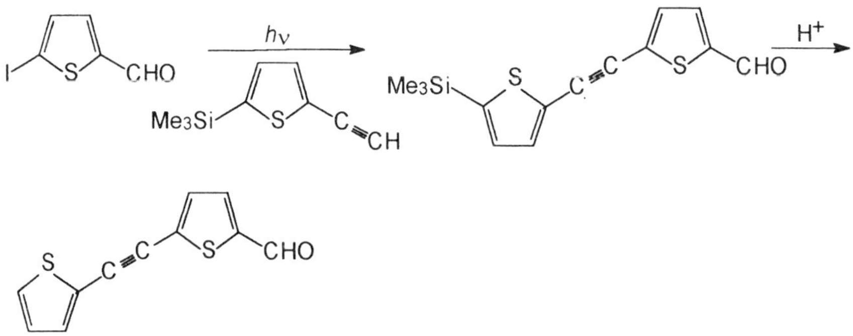

The naturally occurring thienylacetylene thus obtained can be converted into other natural thienylacetylenes. The aldehydic group can be reduced to the alcohol or converted into the corresponding acetylene. The acetylide can react with trimethylborane and iodine to give the methylated compound.[70] All these compounds are naturally occurring thiophenes isolated in *Berkheya* species.

The above described synthetic sequence using trimethylsilyl derivatives can be used in the synthesis of other naturally occurring thienylacetylenes. In fact, the irradiation of 2-acetyl-5-iodothiophene in the presence of 2-ethynyl-5-trimethylsilylthiophene gave the coupling product in 81% yield. The deprotection reaction gave a naturally occurring thienylacetylene in 91% yield.[71] Furthermore, the irradiation of 2-hydroxyacetyl-5-iodo-thiophene in the presence of 2-ethynyl-5-trimethylsilylthiophene gave the coupling product. The subsequent deprotection reaction and reduction gave the diol[71]. This is a naturally occurring thiophene isolated in *Centaurea sphaerocephala*[72].

On the contrary, using substituted arylalkynes we could not obtain interesting results. The reaction of 5-iodothiophene-2-carbaldehyde with 2-propynylthio-phene gave a mixture of products[33]. Also the reaction 2-thienyl-trimethylsilylacetylene with 5-iodo-thiophene-2-carbaldehyde did not give any interesting result. In fact, with the expected coupling product, we obtained a product deriving from the desilylation and from the isomerisation of the acetylene. In fact, the irradiation of 2-thienyl-trimethylsil-ylacetylene gave 3-thienyl-trimethylsilylacetylene.

The presence of electron-withdrawing substituents on the triple bond decreases the reactivity of the molecule. Using 2-thienyl-1-propyn-3-ol acetate as substrate, a photochemical reaction occurred, giving the product of photoarylation.[33]

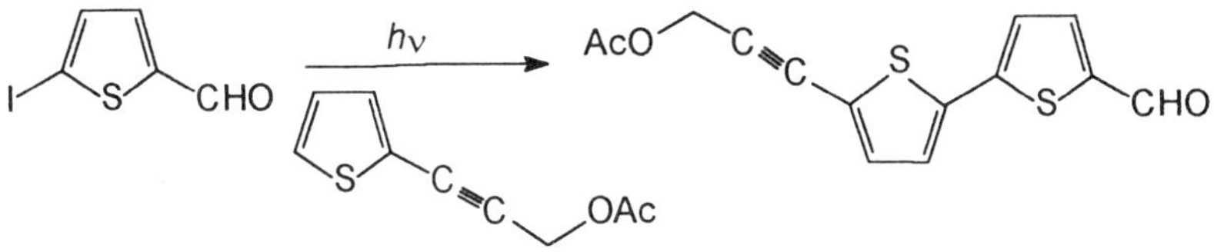

The above reported data show that the regiochemical behaviour of the photochemical coupling is a real question. In the presence of an alkene the reaction occurs on the alkene, usually. However, if a CH_2OAc group is present the reaction occur on the aromatic ring. In the presence of an alkyne the reaction gives a mixture of product. Only when a CH_2OAc group is present the arylation occurs selectively. The regiochemical behaviour of these reactions can be understood on the basis of the dipoles of the reagents. The assumptions that a reagent approaches the other on parallel planes and that the prevalent interaction is between the SOMO of the thienyl radical and the LUMO of the other reagent have been accepted. This way, the interaction between these two orbitals oriented by the dipoles of the molecules explained the regiochemical behaviour of the reaction.[73]

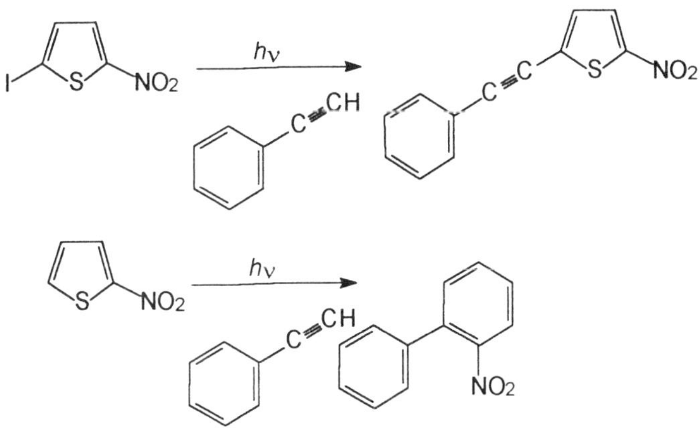

Furthermore, we have to note a different chemical behaviour when phenylacetylene is used as substrate. The reaction of 2-iodo-5-nitrothiophene with phenylacetylene gave the corresponding substitution product.[29] On the contrary, 2-nitrothiophene, 2-acetylthiophene, and thiophene-2-carbaldehyde, irradiated in the presence of phenylacetylene, gave the corresponding diphenyl derivative *ortho* substituted. The reaction is probably a photo Diels-Alder reaction with extrusion of sulphur from the adduct to give the observed product.[74]

In an approach intermediate between photochemical arylation and stilbene cyclization the reaction of chlorothienylanilides has been described.[75-80] It is noteworthy that this reaction can be selectively performed.[80]

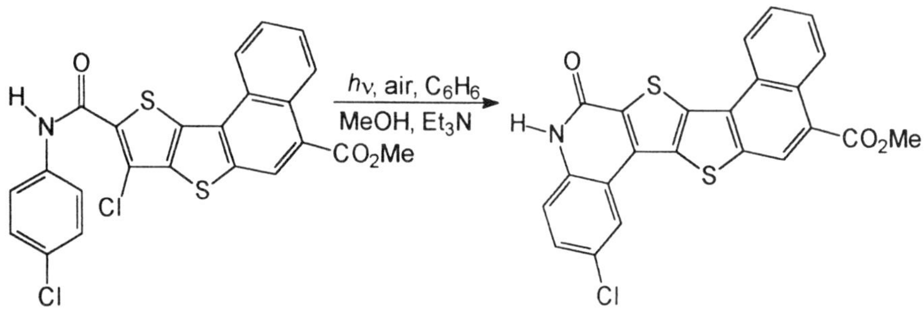

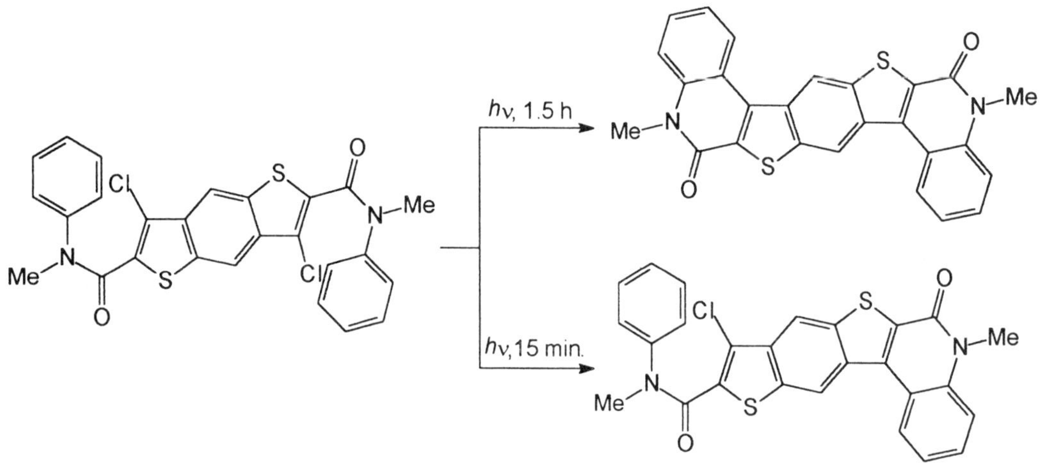

3. Intramolecular Arylation

The photocyclization of 1,2-diarylethylenes is a well known reaction and several reviews appeared in literature on this field.[81-83] This type of reaction has been used also on pyrrole derivatives. 2,3-Di-phenylpyrrole gave the indole derivative when irradiated in cyclohexane in the presence of iodine.[84] In the substrate not only the ethylenic part of the pyrrole moiety but also pyrryl ethylenes were used as starting materials. In a project devoted to the synthesis of coenzime methoxatin the key-step for the skeleton synthesis was the cyclization of a 3-pyrrylethylene derivative.[85,86]

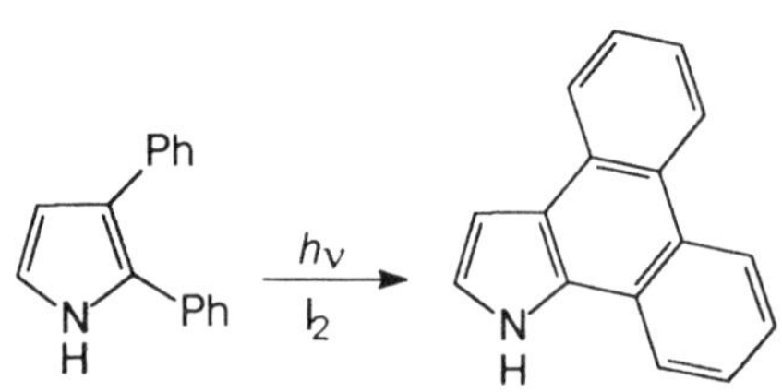

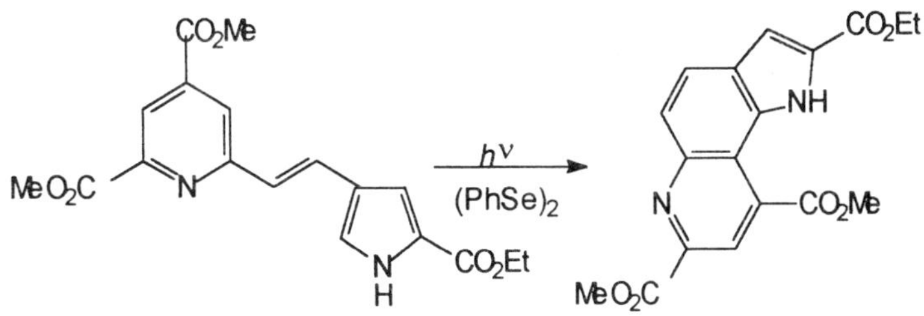

In this case, the cyclization required both the presence of diphenyldiselenide and very long irradiation time (3 - 4 weeks) to give the product. In the synthesis of units B and C of antitumor agent CC-1065 and of the related substances PDE-I and PDE-II that were able to inhibit cyclic adenosine-3',5'-monophosphate phosphodiesterase and showed a low toxicity, the photocyclization of pyrrylethylenes has been investigated with care. 1,2-Dipyrryl derivatives can be cyclized when irradiated in a Pyrex tube in the presence of air. The reaction can be performed also in the presence of a thiophene, a furan or a simple aromatic ring in the molecule.[87] However, this reaction is not useful for preparative purpose. In the presence of air the extensive oxidation of the indolic product was observed.

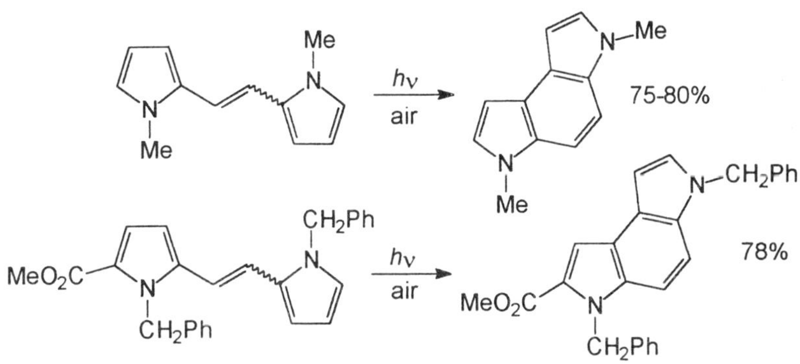

When the reaction was performed in the presence of 5% Pd/C and with a nitrocompound to avoid reduction of the carbon carbon double bond, the reaction seems to be a very useful and general synthetic procedure.[88]

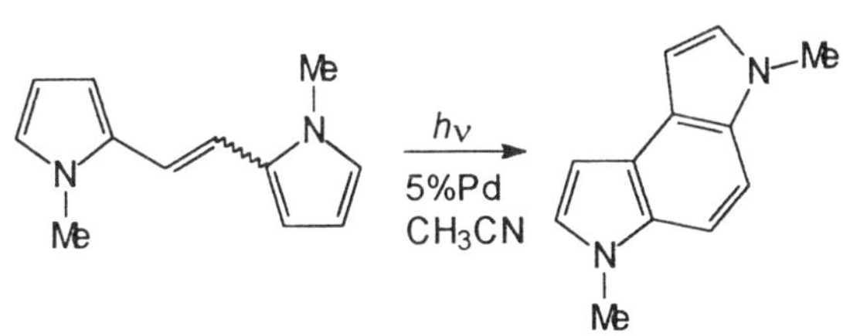

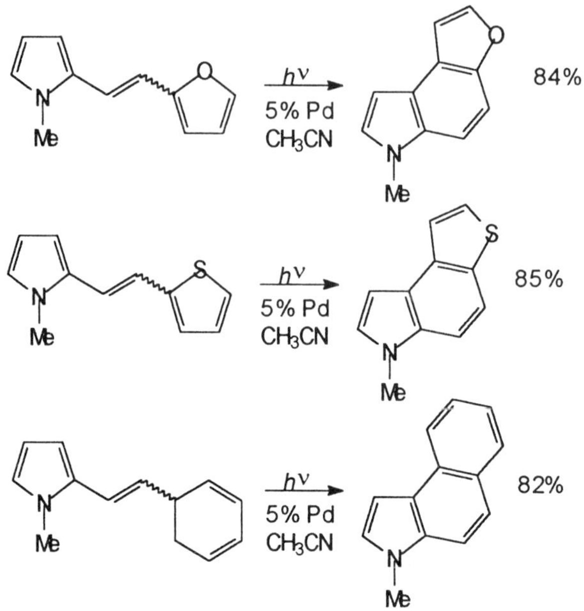

Unfortunately, this reaction did not work with *N*-BOC derivatives. This reaction worked in moderate yield in the presence of air and the cyclization product was used in the synthesis of dehydroxy PDE-I and PDE-II.[89] The synthesis of PDE-I and PDE-II using the described photochemical approach was also described. In this case the photochemical reaction was performed on a dimethoxy derivative. Using this substrate the above described procedure using 5% Pd/C seems to work very well.[90]

The same synthetic scheme was used to synthesise thiophene analogues of PDE-I and PDE-II.[91] In this case the photochemical reaction by using Pd on carbon methodology gave the product in 64% yield.[91]

The reaction of a haloquinone with a pyrryl derivative gave the ethylenic coupling product that, in the reaction mixture, underwent a cyclization reaction.[92-94] Nanosecond laser excitation showed that quinones and pyrroles can interact to give the corresponding pyrryl radical cation. This step could be the basis of the coupling reaction.[95]

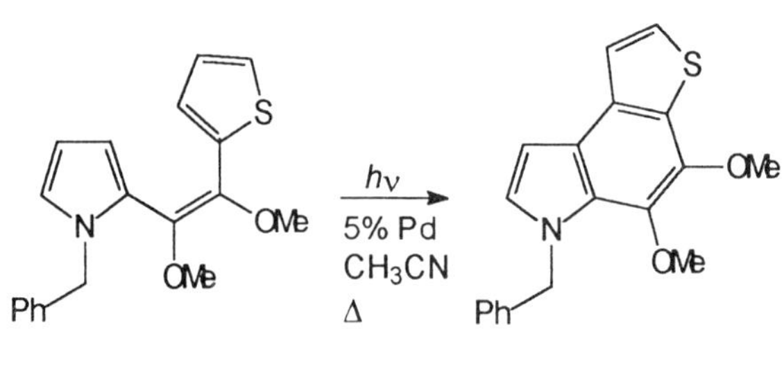

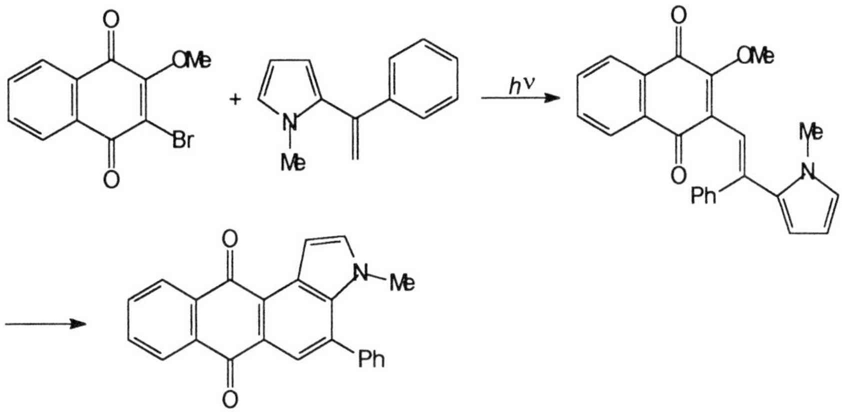

The same behaviour was observed using thiophene and furan derivatives. Thienyl-styrene gave the corresponding arylation product in quantitative yields.[96,97]

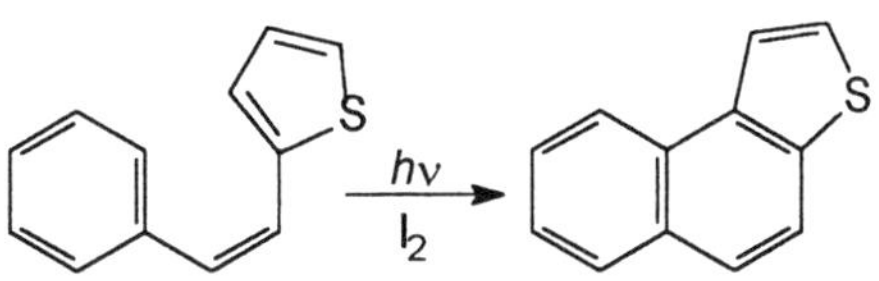

The irradiation of the dithienyl derivative gave the coupling product in 30% yield. The same reaction on the difuryl derivative gave only 5% of the coupling product, while furyl-styrene gave the condensed product in 9% yield.[98] Better results can be obtained in the presence of iodine.[99]

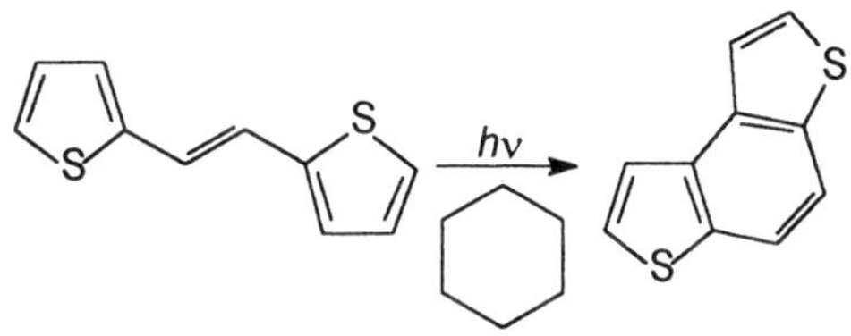

This type of reaction on furan derivatives seems to work better in the presence of electron-withdrawing substituents.[100]

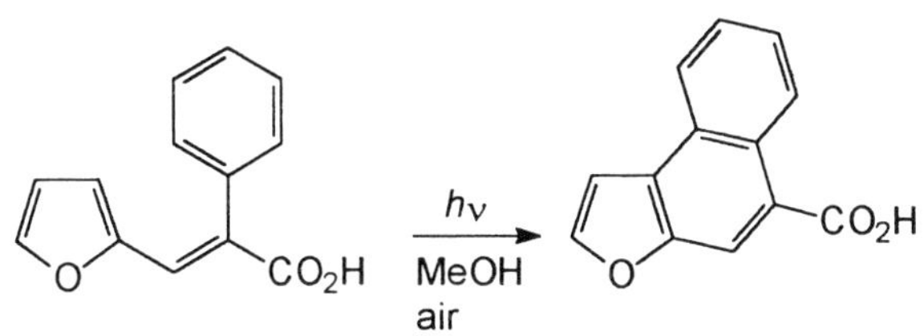

The same reactions can be performed with 3-substituted substrates. In this case, the reaction has been carried out in cyclohexane in the presence of iodine, or on iodine substituted derivatives.[101-102]

The most important use of this reaction was in the synthesis of heterohelicenes.[103-107]

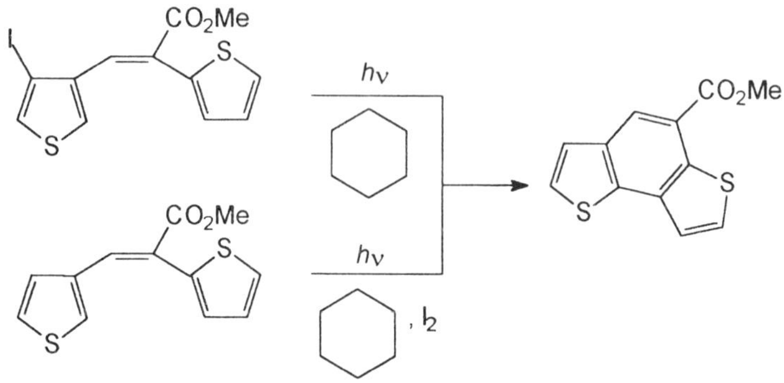

4. References

1. Wolf, W.; Kharasch, N. *J. Org. Chem.*, **1965**, *30*, 2493.
2. Birchall, J. M.; Hazard, R.; Haszeldine, R. N.; Wakalski, W. W. *J. Chem. Soc. (C)*, **1967**, 47.
3. Matsuura, T.; Omura, K. *Bull. Chem. Soc. Jpn*, **1966**, *39*, 944.
4. Robinson, G. E.; Vernon, J. M. *J. Chem. Soc. (C)*, **1971**, 3363.
5. Baum, E. J.; Pitts, J. N., Jr. *J. Phys. Chem.*, **1966**, *70*, 2066.
6. Kametani, T.; Fukumoto, K. *Acc. Chem. Res.*, **1972**, *5*, 212.
7. Hasabe, M.; Kogawa, K.; Tsuchiya, T. *Tetrahedron Lett.*, **1984**, *25*, 3887.
8. Benati, L.; Tiecco, M. *Boll. Sci. Fac. Chim. Ind. Bologna*, **1966**, *24*, 225.
9. Terashima, M.; Seki, K.; Yoshida, C. *Heterocycles*, **1981**, *15*, 1075.
10. Bratt, S.; Iddon, B.; Mack, A. G.; Suschitzky, H.; Taylor, J. A.; Wakefield, B. J. *J. Chem. Soc., Perkin Trans. I*, **1980**, 648.
11. Seki, K.; Ohkura, K.; Terashima, M.; Kanaoka, Y. *Chem. Pharm. Bull.*, **1988**, *36*, 940.
12. Seki, K.; Ohkura, K.; Terashima, M. *Heterocycles*, **1984**, *22*, 2347.
13. Allen, D. W.; Buckland, D. J.; Hutley, B. G.; Oades, A. C.; Turner, J. B. *J. Chem. Soc., Perkin Trans. I*, **1977**, 621.
14. Saito, I.; Ito, S.; Shinmura, T.; Matsuura, T. *Tetrahedron Lett.*, **1980**, *21*, 2813.
15. Seki, K.; Matsuda, K.; Ohkura. K. *Chem. Lett.*, **1987**, 175.
16. Nair, V.; Richardson, S. G.; Coffman, R. E. *J. Org. Chem.*, **1982**, *47*, 4520.

17. Nair, V.; Young, D. A. *J. Org. Chem.*, **1984**, *49*, 4340.

18. Benati, L.; Tiecco, M. *Boll. Sci. Fac. Chim. Ind. Bologna*, **1966**, *24*, 45.

19. D'Auria, M.; De Mico, A.; D'Onofrio, F.; Piancatelli, G. *Italian Pat. Appl.* 479799A90 (18 May 1990).

20. Martelli, G.; Spagnolo, P.; Tiecco, M. *J. Chem. Soc. (B)*, **1968**, 901.

21. Antonioletti, R.; D'Auria, M.; De Mico, A.; Piancatelli, G.; Scettri, A. *J. Chem. Soc., Perkin Trans. 1*, **1985**, 1285.

22. Antonioletti, R.; D'Auria, M.; De Mico, A.; Piancatelli, G.; Scettri, A. *Tetrahedron*, **1985**, *45*, 3441.

23. Fišer-Jakic, L.; Karminski-Zamola, G. *Croat. Chim. Acta*, 1986, 59, 891.

24. Antonioletti, R.; D'Auria, M.; D'Onofrio, F.; Piancatelli, G.; Scettri, A. *J. Chem. Soc., Perkin Trans. 1*, **1986**, 1755.

25. D'Auria, M.; De Mico, A.; D'Onofrio, F.; Mendola, D.; Piancatelli, G. *J. Photochem. Photobiol., A: Chem.*, **1989**, *47*, 191.

26. D'Auria, M.; Sleiter, G., unpublished results.

27. D'Auria, M.; Antonioletti, R.; De Mico, A.; Piancatelli, G. *Heterocycles*, **1986**, *24*, 1575.

28. D'Auria, M.; De Luca, E.; Mauriello, G.; Racioppi, R.; Sleiter, G. *J. Chem. Soc., Perkin Trans 1*, in press.

29. D'Auria, M. *Gazz. Chim. Ital.*, **1994**, *124*, 195.

30. D'Auria, M.; Mauriello, G. *Photochem. Photobiol.*, **1994**, *60*, 542.

31. D'Auria, M. *J. Photochem. Photobiol., A: Chem.*, **1995**, *91*, 187.

32. D'Auria, M.; De Mico, A.; D'Onofrio, F.; Piancatelli, G. *J. Chem. Soc., Perkin Trans. 1*, **1987**, 1777.

33. D'Agostini, A.; D'Auria, M. *J. Chem. Soc., Perkin Trans. 1*, **1994**, 1245.

34. D'Auria, M.; De Mico, A.; D'Onofrio, F.; Piancatelli, G. *Gazz. Chim. Ital.*, **1989**, *119*, 381.

35. Elisei, F.; Latterini, L.; Aloisi, G. G.; D'Auria, M. *J. Phys. Chem.*, **1995**, *99*, 5365.

36. D'Auria, M.; D'Onofrio, F. *Gazz. Chim. Ital.*, **1985**, *115*, 595.

37. Wynberg, H. *Acc. Chem. Res.*, **1971**, *4*, 65.

38. Kellogg R. M.; Dik, J. K.; van Driel, H.; Wynberg, H. *J. Org. Chem.*, **1970**, *35*, 2737.

39. Kagan, J. *Progr. Chem. Org. Nat. Prod.*, **1991**, *56*, 87.

40. D'Auria, M. *Tetrahedron Lett.*, **1994**, *35*, 3151.

41. Hietz, J. R. in *Light-Activated Pesticides*, Ed. by J. R. Heitz and K. R. Downum, American Chemical Society, Washington, DC, 1987, pp 1 - 21.

42. Towers, G. H. N.; Champagne, D. in *Chemistry and Biology of Naturally-Occurring Acetylenes and Related Compounds (NOARC)*, Ed. by J. Lam, H. Breteler, T. Arnason, and L. Hansen, Elsevier, Amsterdam, 1988, pp. 139-150.

43. D'Auria, M.; De Mico, A.; D'Onofrio, F.; Piancatelli, G. *J. Org. Chem.*, **1987**, *52*, 5243.

44. D'Auria, M.; De Mico, A.; D'Onofrio, F.; Piancatelli, G. *Synth. Commun.*, **1987**, *17*, 491.

45. D'Auria, M.; De Mico, A.; D'Onofrio, F. *Gazz. Chim. Ital.*, **1986**, *116*, 747.

46. Lozanova, A. V.; Moiseenkov, A. M.; Semenovskii, A. V *Izv. Akad. Nauk SSSR, Ser. Khim.*, **1980**, 958.

47. Lozanova, A. V.; Moiseenlov, A. M.; Semenovskii, A. V. *Izv. Akad. Nauk SSSR, Ser. Khim.*, **1982**, 2783.

48. D'Auria, M.; Volpe, F. *Gazz. Chim. Ital.*, **1993**, *123*, 527.

49. Kopecky, J. *Organic Photochemistry: a Visual Approach*, VCH Publishers, Inc., New York, 1991, pp. 125-128.

50. D'Auria, M.; Piancatelli, G.; Ferri, T. *J. Org. Chem.*, **1990**, *55*, 4019.

51. D'Auria, M.; De Mico, A.; D'Onofrio, F. *Heterocycles*, **1989**, *29*, 1331.

52. D'Auria, M.; D'Onofrio, F. *Gazz. Chim. Ital.*, **1993**, *123*, 129.

53. D'Auria, M.; Piancatelli, G.; Vantaggi, A. *J. Chem. Soc., Perkin Trans. I*, **1990**, 2999.

54. D'Auria, M.; D'Annibale, A.; Ferri, T. *Tetrahedron*, **1992**, *48*, 9323.

55. D'Auria, M. *Heterocycles*, **1996**, *43*, 959.

56. D'Auria, M.; Vantaggi, A. *Tetrahedron*, **1992**, *48*, 2523.

57. D'Auria, M.; D'Onofrio, F.; Vantaggi, A. *Tetrahedron*, **1990**, *46*, 7831.

58. D'Auria, M. *Tetrahedron*, **1991**, *47*, 9225.

59. Meng, J.; Fu, D.; Gao, Z.; Wang, R.; Wang, H.; Saito, I.; Kasatani, R.; Matsuura, T. *Tetrahedron*, **1990**, *46*, 2367.

60. D'Auria, M.; De Luca, E.; Esposito, V.; Mauriello, G.; Racioppi, R. *Tetrahedron*, **1997**, *53*, 1157.

61. D'Auria, M. *Heterocycles*, **1991**, *32*, 1059.

62. D'Auria, M. *J. Photochem. Photobiol., A: Chem.*, **1994**, *79*, 67.

63. Splitter, J. S.; Calvin, M. *J. Org. Chem.*, **1955**, *20*, 1086.

64. D'Auria, M.; Esposito, V.; Mauriello, G. *Tetrahedron*, **1996**, *52*, 14253..

65. D'Auria, M. *Tetrahedron Lett.*, **1994**, *35*, 633.

66. D'Auria, M. *Chem. Lett.*, **1995**, 109.

67. Marchant, Y. Y.; Ganders, F. R.; Wat, C. K.; Towers, G. H. N. *Biochem. Syst. Ecol.*, **1984**, *12*, 167.

68. Christensen, L. P.; Lam, J. *Phytochemistry*, **1991**, *30*, 11.

69. D'Auria, M. unpublished results.

70. D'Auria, M.; De Mico, A.; D'Onofrio, F.; Piancatelli, G. *Gazz. Chim. Ital.*, **1989**, *119*, 201.

71. D'Auria, M.; Tofani, D. *Tetrahedron*, **1992**, *48*, 9315.

72. Bastos, M. M. S. M.; Kijjoa, A.; Cardoso, J. M.; Gutierrez, A. B.; Hertz, W. *Planta Med.*, **1990**, *56*, 403.

73. D'Auria, M.; Ferri, T. *J. Org. Chem.*, **1995**, *60*, 8360.

74. D'Auria, M. *Tetrahedron Lett.*, **1995**, *36*, 6567.

75. Karminski-Zamola, G.; Pavlicic, D.; Bajic, M.; Blazevic, N. *Heterocycles*, **1991**, *32*, 2323.

76. Dogan, J.; Karminski-Zamola, G.; Boykin, D. W. *Heterocycles*, **1995**, *41*, 1659.

77. Karminski-Zamola, G.; Bajic, M. *Proceedings of the XII IUPAC Symposium on Photochemistry*, Bologna, **1988**, p. 617.

78. Crnjak, A.; Bajic, M.; Karminski-Zamola, G. *Proceedings of the XIV IUPAC Symposium on Photochemistry*, Leuven, **1992**, p. 255.

79. Dogan, J.; Karminski-Zamola, G.; Boykin, D. W. *Proceedings of the XVI IUPAC Symposium on Photochemistry*, Helsinki, **1996**, p. 190.

80. Malesevic, M.; Karminski-Zamola, G.; Bajic, M.; Boykin, D. W. *Proceedings of the XVI IUPAC Symposium on Photochemistry*, Helsinki, **1996**, p. 378.

81. Laarhoven, W. H. *Recl. Trav. Chim. Pays-Bas*, **1983**, *102*, 185.

82. Mallory, F. B.; Mallory, C. W. *Org. React.*, **1984**, **30**, 1.

83. Laarhoven, W. H. *Org. Photochem.*, **1989**, *10*, 163.

84. Padwa, A.; Ku, H.; Mazzu, A. *J. Org. Chem.*, **1978**, *43*, 381.

85. Hendrickson, J. B.; de Vries, J. G. *J. Org. Chem.*, **1982**, *47*, 1148.

86. Hendrickson, J. B.; de Vries, J. G. *J. Org. Chem.*, **1985**, *50*, 1688.

87. Rawal, V. H.; Cava, M. P. *J. Chem. Soc., Chem. Commun.*, **1984**, 1526.

88. Rawal, V. H.; Jones, R. J.; Cava, M. P. *Tetrahedron Lett.*, **1985**, *26*, 2423.

89. Rawal, V. H.; Jones, R. J.; Cava, M. P. *J. Org. Chem.*, **1987**, *52*, 19.

90. Rawal, V. H.; Cava, M. P., *J. Am. Chem. Soc.*, **1986**, *108*, 2110.

91. Jones, R. J.; Cava, M. P. *J. Chem. Soc., Chem. Commun.*, **1986**, 826.

92. Maruyama, K.; Otsuki, T.; Mitsui, K.; Tojo, T. *J. Heterocycl. Chem.*, **1980**, *17*, 695.

93. Maruyama, K.; Otsuku, T., Tai, S. *J. Org. Chem.*, **1985**, *50*, 52.

94. Maruyama, K.; Mitsui, K.; Otsuki, T. *Chem. Lett.*, **1978**, 323.

95. Petrushenko, K. B.; Vokin, A. I.; Turchaninov, V. K.; Korostova, S. E. *Izv. Akad. Nauk SSSR, Ser. Khim.*, **1988**, 41.

96. Carruthers, W.; Stewart, H. N. M. *Tetrahedron Lett.*, **1965**, 301.

97. Carruthers, W.; stewart, H. N. M. *J. Chem. Soc.*, **1965**, 6221.

98. Loader, C. E.; Timmons, C. J. *J. Chem. Soc. (C)*, **1967**, 1677.

99. Kellogg, R. M.; Groen, M. B.; Wynberg, H. *J. Org. Chem.*, **1967**, *32*, 3093.

100. Karminski-Zamola, G.; Bajic, M. *Synth. Commun.*, **1989**, *19*, 1325.

101. De Luca, G.; Martelli, G.; Spagnolo, P.; Tiecco, M. *J. Chem. Soc. (C)*, **1970**, 2504.

102. Croisy, A.; Jacquignon, P., Perin, F. *J. Chem. Soc., Chem. Commun.*, **1975**, 106.

103. Wynberg, H.; Groen, M. B. *J. Am. Chem. Soc.*, **1968**, *90*, 5339.

104. Wynberg. H.; Groen, M. B. *J. Am. Chem. Soc.*, **1970**, *92*, 6664.

105. Groen, M. B.; Schandenberg, H.; Wynberg, H. *J. Org. Chem.*, **1971**, *36*, 2797.

106. Dopper, J. H.; Oudman, D.; Wynberg, H. *J. Am. Chem. Soc.*, **1973**, *95*, 3692.

107. Lehman, P. G.; Wynberg, H. *Aust. J. Chem.*, **1974**, *27*, 315.

UV AND PHOTOELECTRON SPECTROSCOPY CONFORMATIONAL ANALYSIS OF RADICAL CATIONS AND PARENT MOLECULES OF HETEROCYCLIC RING ASSEMBLIES

Alexander I. Vokin and Valerii K. Turchaninov
Irkutsk Institute of Organic Chemistry, Siberian Division of the Russian Academy of Sciences, 1, Favorsky St., 664033 Irkutsk, Russia.
Fax: (7 3952)-35-60-46, E-mail: root@irioch.irkutsk.su

Abstract. *The results of investigations on the conformational analysis of heterocyclic ring assembly radical cations and parent molecules by means of ultraviolet and photoelectron spectroscopy have been discussed. They allow a series of conclusions common to all the molecules and π-radical cations of heterocyclic and aromatic ring assemblies to be drawn. The overwhelming majority of ring assemblies are non-planar in the gas phase, irrespectively of whether they are symmetrical like oligothiophenes and biphenyls, or contain various unsaturated rings like aryl(heteroaryl)pyrroles. The transfer of ring assemblies from gas to the condensed phase reduces the equilibrium torsional angle. The extent to which the angle varies depends on the structure of a given compound and the presence of substituents. One-electron ionization leads to an analogous, but a more pronounced effect.*

1. Introduction

A comprehensive and consecutive consideration of situations in which conformational problems are solved by means of ultraviolet photoelectron spectroscopy (UPS) has been given in a published recently review by Brown and Jorgensen[1]. The review provides data on equilibrium conformation of organic compounds of various types, collected from a great number of original works and several previous reviews. Among the compounds analyzed there are some ring assemblies, in particular, biphenyl derivatives, 2,2'- and 4,4'-bipyridines, some 2,2'- and 3,3'-bithiophenes and 1,1'-bipyrroles. It has been established that most of molecules with similar carbo- or heterocycles (X-X) are non-planar in the gas phase. Depending on the position of substituent and the chemical composition of fragments (X), the equilibrium dihedral angles (φ) of the above ring assemblies vary in a wide range ($\varphi = 0 - 90^0$). At present data on equilibrium conformation of a fairly large number of molecules having different unsaturated fragments (X-Y) have been obtained by means of UPS. This information has not been summarized so far, and it is one of the goals of the present review. Another goal is to disscuss the results of studies on equilibrium conformation of short-living π-radical cations (RC) of ring assemblies. These data were obtained on the basis of laser photolysis, pulse radiolysis and UPS or EPR. A number of data needed for these goals to be achieved, have been obtained in our laboratory just recently. Therefore, we find it possible to publish them together with the available data in the present review.

2. General consideration of ionic states

2.1. Analysis of radical cations UV spectra in terms of Koopmans' theorem

The correspondence of transition energies (caused by electron excitation from one of double occupied molecular orbitals (MO) to a single occupied one) observed in the UV spectra of radical cations of planar molecules and the difference in ionization potentials have provided a prerequisite for use of UPS data in conformational analysis of ring assembly π-radical cations. This approach requires the following conditions to be met: 1) allowed transitions in the RC UV spectrum; 2) measurement of the ionization potentials and characteristics of RC UV spectra carried out in the same phase state; 3) coincident geometries of the parent molecule and RC.

The first examples of correspondence of the transitions in the RC UV spectrum with the energy difference of the appropriate band maxima in UPS were revealed from gas-phase emission spectra of RC of rather simple molecules such as N_2S, N_2, CO_2 [2,3]. Somewhat later the emission spectra of electron-impact generated RC of more complex fluorobenzenes have been studied[4]. It has been found that the difference between the first and third adiabatic ionization potentials of the parent molecule is identical, within the error of determination, with the energy of $0 \leftarrow 0$ transition observed in the visible region of the RC fluorescence spectrum.

In most cases the UV spectra of RC are recorded in the liquid or in the solid phases. Here RC can more easily be generated and accumulated in high concentrations. However, a question arises as to how strong is the effect of medium on the transition energies in the RC UV spectra and whether the correspondence is retained.

Numerous empirical studies of the effect of gas → condensed phase transfer on the $\pi^* \leftarrow \pi$ transitions in the UV spectra of open shell systems, undertaken by means of UPS and UV spectroscopy in combination with gas-phase photodissociation, photoionization or γ-irradiation in rigid matrices have shown a red phase shift to be their feature[5-13]. Radical cations of benzenes, toluenes, naphthalenes, pyrenes and some other condensed aromatic hydrocarbons rank among this type of compounds. The red phase shift of absorption band in the RC UV spectrum is not constant. Thus, the shift of electron transition $^2B_{2g}(\pi^1{}_3) \leftarrow {}^2A_u(\pi^2{}_5)$ in the UV spectrum of naphthalene RC depends on the matrix material or on the solvent and decreases with an increase in the medium polarity[5]. This is explained by the fact that the naphthalene RC excitation involves positive charge redistribution between the periphery and the central part of the ion:

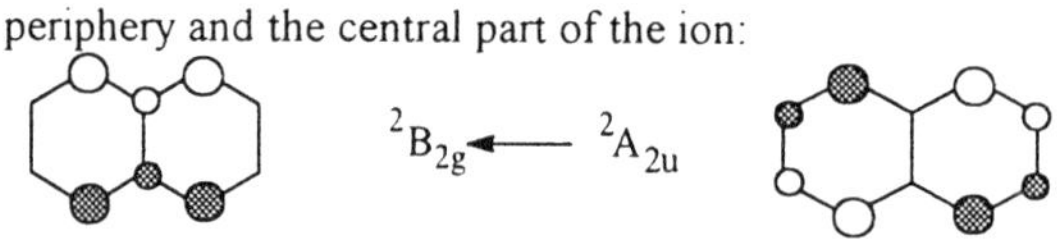

Alongside, in some cases solvatochromic effects do not manifest themselves. Thus, for RC of condensed aromatic hydrocarbons of various structure recorded in a boracic acid film[14] an average shift referred to the gas phase is 0. An analogous shift is 0.01 eV for benzanthracenes[15] and 0.02 eV for pyrenes[7].

A blue shift was observed only in few cases[16, 17]. This shift is typical, in particular, of RC of indene hetrocyclic analogs[17] and it is likely to be caused by much larger excitation-induced electron density redistribution compared with naphthalene RC:

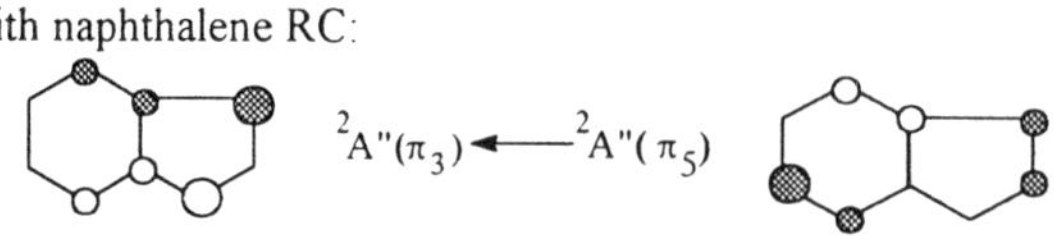

However, the blue shift value for planar heterocycle π-RC is rather small (approximately 0.04 eV).

The first scarce data concerning the correspondence of UPS of the parent molecules and UV spectra of RC of compounds different in electronic structure have been summarized and analyzed by Zahradnik *et al.*[18]. The correspondence has been found to exist both for simple di- and triatomic molecules and for rather complex species. In spite of the fact that the RC spectra were recorded in different media by different methods (electron impact, γ-irradiation in rigid matrix, electrochemical generation in solution), the agreement between spectral characteristics was fairly good in most cases. It should be noted that the authors[18] have demonstrated the fruitfulness of the approach based on comparison of spectral data found by UPS and UV spectroscopy, although sometimes they were not precise in comparing the results.

During the subsequent decades a large body of information on this problem has been accumulated. A great contribution has been made by Dunbar *et al.*[6, 10, 19-23], Andrews *et al.*[8, 11, 12, 24-31], Haselbach *et al.* [5, 32-38], Shida *et al.* [9, 37, 39-41], Khan[7, 14-16, 42, 43] and other researchers[4, 13, 44]. At present it is established that with constant geometry of the molecular system the absolute value of difference between the first adiabatic ionization potential and a subsequent one must be equal to the energy of low-energy transition in the UV spectrum of RC*. This is illustrated by following orbital diagrams:

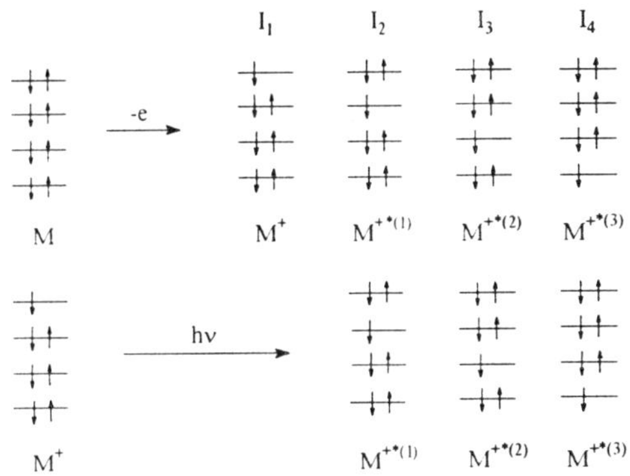

2.2. Factors responsible for inconsistency between the UV spectrum of radical cation and the photoelectron spectrum of the molecule

The correspondence between the two spectra is theoretically based on Koopmans' theorem which implies that the chemical structure and geometry of the molecule remain unchanged on ionization. Consequently, there is no correspondence when the above conditions are not fulfilled.

The problem of the effect of geometry on the UPS and UV spectra of RC has been given much attention in one of earlier work[45]. In the strict sense, correspondence must exist only when the bands in the UPS and UV spectra of RC have fine structure, which enables the adiabatic ionization potential and the position of 0 ← 0 transition to be determined. The absence of fine structure directly points to a change in geometry. The question is how sensitive the energy difference between the two ionic states is to the above change. When the difference in vertical ionization potentials (ΔE^0) is nearly equal to the transition energy in the RC UV spectrum

---

*Here, we deal with Koopmans' ionic states. Transitions to non-Koopmans' states have been discussed in detail in[5-7, 15, 16, 24, 32-36, 42, 43]

(ΔE^+), the changes in geometry are either small or correlated. Considerable ionization-induced changes in geometry can lead to inequality of ΔE^0 and ΔE^+. Thus, it has been found for iodine[45] that the bond length of RC in the first excited state is equal to that of the parent molecule in the ground state, but it differs much from the RC equilibrium bond length. The transition in the RC UV spectrum, calculated from the difference between the first and the second ionization potentials, must be 1.79 eV, whereas in the real spectrum there is a maximum at 1.94 eV. As for the difference between the first and the third ionization potentials, it corresponds to the transition energy in the RC UV spectrum (2.58 and 2.53 eV, respectively). Consequently, even for the simplest iodine molecule a change in geometry (bond length) causes non-correspondence between UPS and UV spectra of RC.

A considerable variation of valence angles presents another type of ionization-induced change in geometry. Changes of this kind were revealed on NH_3 ionization. The valence angles of the ammonia molecule are close to tetrahedral. In RC the valence angle increases to 120^0. This conclusion was based on analysis of RC EPR spectra[46], vibrational structure of Ridberg's transitions in the UV spectrum of the parent molecule[47] and results of *ab initio* STO-3G calculations of RC[48].

In another work[49] the UV spectra for RC of a series of aliphatic amines were obtained by photo-ionization in a polar medium. Correlation with UPS allowed qualitative assignment of transitions; quantitative correspondence is absent. According to[49], this is related to ionization-induced change in geometry of amines.

Analogous changes in geometry are observed on ionization of alkylhydrazines. Analysis of the EPR spectra of alkylhydrazine RC revealed an increase in the valence angles of the nitrogen atom[50]. If in the hydrazine molecule complete planarization occurs, the molecules of alkyl-substituted derivatives are partially flattened[50]. X-ray diffraction of the molecules and the radical ion salts of alkylhydrazines proved the increase in valence angles involving the nitrogen atom and revealed an ionization-induced shortening of the N-N bond[51-53]. The same is evidenced by AM1 quantum-chemical calculations[54].

Examination of the spectrum of 1,4-dithiane RC[40] (obtained by γ-irradiation in Freon matrix at 77K)

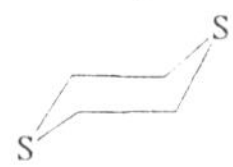

has revealed a decrease in band intensities in the 1.25-0.45 eV region and a simultaneous increase in the 2.50-1.25 eV region occurring in the dark as time goes on. Shida *et al.*[40] explained this in terms of "chair" → "boat" conformational transition. *Ab initio* STO-3G calculations show the parent molecule to be in the "chair" conformation. For RC the state of "boat" is more stable. Thus, a group of bands in the 1.25-0.45 region has been assigned to a metastable RC ("chair") taking into account the correspondence of the main band position (energy maximum 0.5 eV) with UPS data and calculations. Further changes in the absorption band intensities are stipulated by spontaneous RC transition to "boat", which is confirmed by agreement between the short wave-length transition and the value calculated for this conformation. Thus, the use of UPS data, RC UV spectra and calculations in combination makes it possible to investigate ionization-induced changes in the conformational structure of organic molecules.

Endo[4+2]-dicyclopentadiene is an example of a dramatic ionization-induced change in the molecular structure.

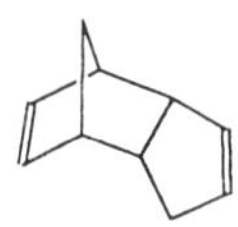

The RC of this compound was obtained by photo-induced electron transfer from the parent substrate to a chloranil in the triplet state, and studied by CIDNP[55]. Analysis of possible RC structures indicates that the experimental data are in agreement with a form with bond cleavage between cyclopentadienyl fragments. According to authors' hypothesis[56], the reduced ring strain and extended spin and charge delocalization are the driving force of this process. It is noted that rearrangement does not follow ionization. Recombination of geminal ion pairs leads to parent molecules rather than to species of RC-like structure. Similar ionization-induced structural changes are observed for anti[2+2]- and exo[4+2]dicyclopentadiene[56]:

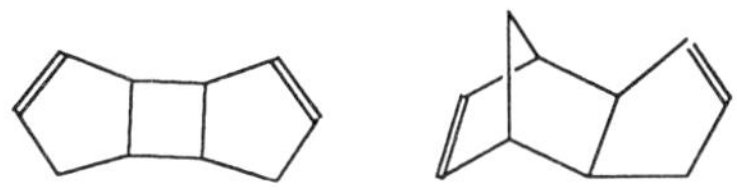

The UV spectra of toluene and cycloheptatriene RC (obtained by photoionization in an argon matrix at 20 K) have been studied[28]:

As expected[11], for toluene there is full agreement with UPS. From the UPS data the longest wave-length band of cycloheptatriene RC should be located in the 1.2 eV region. Experimentally it is found at 1.75 eV. On this basis Andrews *et al.*[28] believe that on ionization a rearrangement involving ring opening takes place and that the band recorded in the argon matrix corresponds to substituted hexatriene RC. This is supported by the fact that the band position is close to that of hexatriene RC:

The gas-phase photodissociation spectra of styrene-related molecules have been discussed in[57]. Comparison of these spectra with UPS of parent molecules has revealed correspondence to exist for styrene derivatives, however it is not available for cyclooctatetraene, allylbenzene and phenylbutenes. Dunbar *et al.* think this to be related to ionization-induced changes in the chemical structure of molecules. Thus, it has been established that the photodissociation spectrum of allylbenzene RC is practically identical to that of 1-phenylpropene RC. On this basis a conclusion has been drawn that the stable form of allylbenzene RC is structurally similar to that of phenylpropene and its RC:

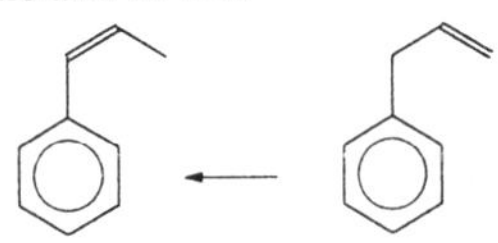

Other examples demonstrating the effect of structural changes of this kind are presented in[38, 41].

On ionization of ethylene there are observed geometry changes related to rotation about the double bond. Analysis of Ridberg's transitions in the UV spectrum[58] and vibrational band structure in UPS[59] allowed the determination of the equilibrium torsional angle of ethylene RC, $\varphi = 25^0$. Analogous, but more clearly defined changes were observed on ionization of octamethyl-substituted bicyclopentylidene. Whereas its molecule is planar, the radical cation shows nearly an orthogonal position of five-membered rings according to EPR spectroscopy[60]:

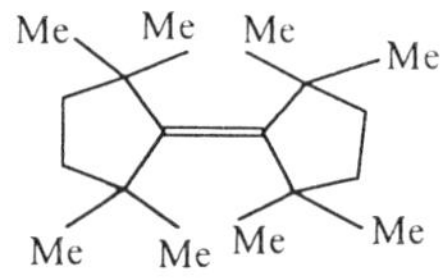

The results of detailed investigations of stilbene RC structural changes induced by internal rotation have been reported by Suzuki *et al.*[61-63].

Spectral shifts of the absorption bands of RC whose aryl fragments are bonded by σ-bonds were observed in defected matrices[64]. By analogy with substituent-induced shifts[65] (in the spectra of parent molecules of this type), the RC band shift has been attributed by Shida[64] to the rotation of RC fragments. For instance, on the basis of similarity of the spectrum of 1,1'-dinaphthylmono-sulfide RC to that of naphthalene RC it has been concluded that the naphthyl fragments of the former are orthogonal to each other. Heating or UV irradiation of the matrix leads to changes in the spectrum. Thus, in the near IR region there appears a highly intense band assigned to resonance absorption of RC with a planar arrangement of naphthyl fragments.

Biphenyl and its derivatives fall into a category of compounds in which conformational rearrangements are quite possible. Previously[64], biphenyl was regarded as a compound of invariable geometry, and it fitted into a general pattern. From this pattern it follows that formal correspondence between UPS and RC spectrum does exist in spite of deviations ranging from 0.25 to 0.6 eV (for various transitions). Later[66] the absence of exact correspondence between UPS and the long wave-length band of biphenyl RC has attracted attention. It has been suggested that the above phenomenon was due to a change in geometry (dihedral angle between benzene rings) in going from the parent molecule in the gas phase to RC in the argon matrix. A series of biphenyl methyl- and fluoro derivatives as well as fluorene, a biphenyl planar analog, have been considered[25]. Whereas correspondence is the case for planar compounds (naphthalene, phenanthrene, fluorene), for biphenyls there is a marked discrepancy between the results of the two methods. It is assumed[25] that the discrepancy is due to a considerable ionization-induced structural relaxation. A similar result was obtained by other procedures. An appreciable deviation of long wave-length transition of biphenyl RC (generated by photo-ionization in a boric acid film) from the relationship existing for condensed aromatic hydrocarbons, between energy transitions in the UV spectrum of RC and those calculated from UPS has been revealed[14]. By analogy with the shift observed in the UV spectrum of the biphenyl molecule in going from gas to crystal, caused by a change of the equilibrium torsional angle (φ)[67], the energy transition shift in the RC UV spectrum with respect to UPS has been explained by the same reason. An empirical linear correlation of energy gaps between π_3- and π_6-MOs on the one hand and cos(φ) on the other hand has been found[1, 68, 69] on the basis of UPS data for biphenyls, for which the equilibrium dihedral angles in the gas phase were measured. They are found to be of maximum value with $\varphi = 0^0$ and of minimum value with $\varphi = 90^0$. Consequently, if ionization leads to a decrease in the φ value, one should observe a blue (δE) shift of the transition in the RC UV spectrum (ΔE^+) compared to the position obtained from UPS ($\delta E = \Delta E^+ - \Delta E^0$). In this case the more φ, the more δE[25] (see scheme on next page). Andrews *et al.*[25] have drawn the analogy to the excited state of biphenyl[70] and suggested that, like the excited molecule, the biphenyl RC is planar. The UV spectra of substituted biphenyl RC have also been considered[25]. In contrast to biphenyl, RC of sterically hindered members of this series are not planar.

Theoretical conformational analysis of biphenyl parent molecules and RC has been undertaken[70]. With the use of configuration interaction formalism it has been shown that the biphenyl RC is structurally different from the parent molecule. This manifests itself in a smaller φ angle (26^0 against 40^0) and a higher barrier value

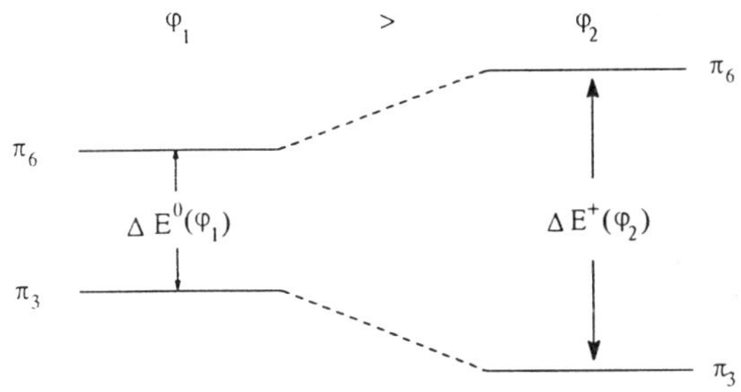

when $\varphi = 90^0$ (8.4 against 1.4 kcal/mol). The main contribution to the barrier increase is made by a significant drop of the π-electron component of RC full energy (relative to the planar form) compared to the parent molecule. Analogous results have been provided[72] by INDO calculation of biphenyl and 4,4'-dihydroxybiphenyl RC with partial optimization of geometry. For biphenyl RC $\varphi = 18^0$, rotation barrier at 90^0 is 30 kcal/mol. MNDO study of biphenyl[73] has also revealed ionization-induced changes of φ from 87^0 to 32^0. Calculation of the UV spectra of biphenyl RC[73] and 4,4'-bipyridyl RC[74] for different conformations has been carried out. An agreement with the experimental data was observed only with $\varphi \sim 0^0$. Thus, theoretical investigations of biphenyl and some biphenyl hetero analogs performed by various quantum chemical methods indicate stabilization of the RC planar conformation in all the cases.

A number of experimental studies of the RC structure of biphenyl derivatives have been carried out[75-81].

Raman spectra for short-living radical cations of biphenyl and biphenyl alkyl derivatives have been obtained[75]. Spectral analysis supports predictions about strengthening of the interannular bond and planarization of these molecules on ionization. The experimental data are in agreement with a planar quinoidal character of the RC structure.

Asmus *et al.* investigated RC of a series of biphenyl chloro derivatives in dichloroethane by pulse radiolysis[77]. The main feature attracted authors' attention is that the decay kinetics of the shortest wave-length absorption band of some RC differs from that of other bands of the same RC. However, when a reducing agent (prometazine) is introduced all the bands decay at the same rate. It follows that these bands correspond to substituted biphenyl RC. The kinetic behavior is explained in terms of a hypothesis on the effect of RC conformational structure on kinetic. Relatively short-living species giving rise to an absorption band in the 2.1-1.7 eV region are identified with planar RC[77]. Longer-living species bringing about an absorption band only in the short wave-length region are considered to be radical cations the geometrical structure of which remains intact on ionization and resembles conformation of the parent molecules. In other words, it is suggested that in the prebarrier region an ionization-induced change of the torsional potential from a one-trough to two-trough one takes place. It is pertinent to note that in previous investigations, such as[72, 76], for example, only one-trough potential minimum shift has been observed. Later, taking into consideration the work of Andrews *et al.*[25, 66] and the results of a more careful experiment, the authors[77] abandoned their concepts[79]. They have also investigated RC of biphenyl monohalo derivatives[78]. Data on the transition energies in the UV spectra of RC are correlated with the first ionization potential. The presence of uni-linear correlation is in agreement, at least in part, with ionization-induced planarization of the biphenyls studied.

In other words, from the literature cited it follows that the absence of correspondence between UPS and UV spectra of RC is often related to considerable ionization-induced changes of the chemical structure and geometry of molecules. Comparison of the results of the two methods provides information on these changes which can be used for the determination of the structure and geometry of radical cations.

In most cases, the investigation of energy characteristics of molecules and RC was carried out in different phase states and media. Due to investigations of this type, the overall transition: parent molecule in the gas phase - RC in the condensed phase is most frequently considered. Evidently, owing to this fact, it is necessary to consider in detail the effect of medium and phase state on the geometry and structure of the molecular system both before and after ionization.

2.3. Radical cations of planar ring assemblies

Fluorene is a planar reference compound of the biphenyl series. The MINDO/3-calculated electron charge distribution for fluorene in corresponding once-ionized configurations in the ground state shows that the $^2B_1(\pi_3) \leftarrow {}^2A_2(\pi_6)$ transition in the UV spectrum of fluorene RC does not induce electron charge redistribution:

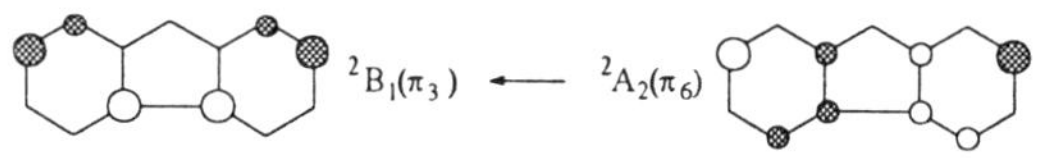

Consequently, the solvation effects in going from gas to solution should be small. In fact, the shift of the observed $\pi_6 \leftarrow \pi_3$ transition of fluorene RC in MeCN with respect to that calculated from UPS is 0.01 eV (Table 1, 2). A planar reference compound of the 2-arylpyrrole (2-AP) series in the gas phase[82] and in solution[83] is presented by 4,5-dihydrobenzo[g]indole (DHBI):

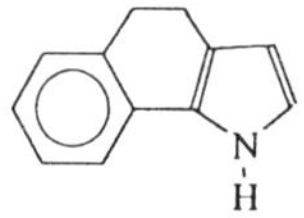

Analysis of electron charge distribution for the latter shows it to be similar in both the equilibrium ground state and in Franck-Condon's excited state of RC:

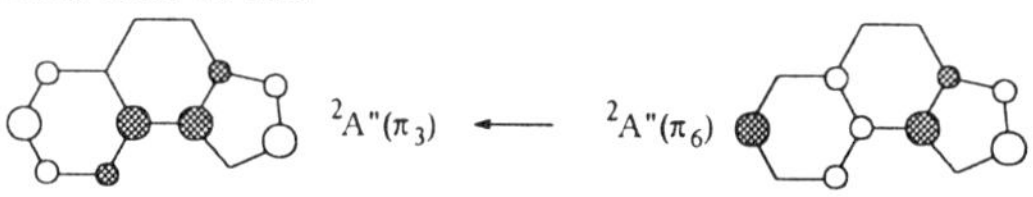

Moreover, it is qualitatively analogous to the charge distribution in the $^2B_1$ and $^2A_2$ states of fluorene (see the scheme for fluorene). In this connection, one can expect an agreement between the value experimentally found in MeCN and that calculated from UPS of energy of electron transition of type $^2A''(\pi_3) \leftarrow {}^2A''(\pi_6)$. This agreement is the case indeed (Table 1, 2). In 2,2'-furylpyrrole (2-FuP), a planar reference compound of 2-heteroarylpyrroles[84, 85], which shows an analogous charge distribution, the energy of transition calculated from UPS and that measured in MeCN are also in agreement (Table 1, 2):

According to the data reported in[34], a similar picture is observed for planar indeno[2,1-a]indene.

Thus, when the equilibrium ring assembly conformation is planar the energy of transition observed in the UV spectrum of RC in MeCN solution agrees with that calculated from UPS. Consequently, solvatochromic effects do not play any appreciable role , in contrast to indene hetero analogs.

3. N,C-Phenylpyrroles

3.1. 1-Phenylpyrroles

In the ultraviolet photoelectron spectra (UPS) of substituted 1-phenylpyrroles (1-PhP) in the energy region below 11 eV four bands are observed[86] (Table 1). According to MINDO/3 quantum chemical calculations and Koopmans' theorem these bands are caused by four ionic states arisen upon electron removal from four upper occupied π-MO's of the above compounds. For the π-MO's the orbital electron density of which is not equal to zero on the bonding atoms (R, S), the energy difference is presented by the following equation[86]:

$$\Delta E^0(\varphi) = [\Delta E^2(90^0) + 4\beta^2(\varphi)]^{1/2} \qquad (1)$$

Here $\Delta E(90^0)$ is the energy difference of levels in the conformation with a torsional angle $\varphi = 90^0$, $\beta(\varphi)$ corresponds to the energy of fragment π-orbital interactions. In order to describe the $\beta \sim \varphi$ dependence the authors[86] made use of the known relation $\beta(\varphi) = \beta(0)\cos(\varphi)$. Previously[1, 87] this relation was successfully used in estimating the dihedral angles in both symmetrical and asymmetrical molecules. The use of equation (1) for determining the 1-PhP torsional angles required a knowledge of π-MO energies of molecules in both planar and orthogonal forms. Rettig and Marschner[86] took 2,5,2',6'-tetramethyl-1-phenylpyrrole as a model of the orthogonal form. They determined $\Delta E(90^0)$ as the difference of corresponding ionization potentials of compounds modelling the 1-PhP fragments. Then the value obtained was corrected for the induction effect arising upon bonding the fragments. This effect was evaluated from the UPS of 2,5,2',6'-tetramethyl-1-phenylpyrrole, 1,3-dimethylbenzene, and 2,5-dimethylpyrrole. For the determination of $\beta(0) = C_R C_S \beta^{AI}$ a semi-empirical approach was employed. The AO values of the C_R and C_S coefficients were calculated by MINDO/3, whereas the β^{AI} values were determined from the UPS of model compounds containing a C-N bond with a length equal to that of the C-N bond in 1-phenylpyrrole. The values of evaluated equilibrium torsional angles (φ) are presented in Table 3. They provide evidence for a non-planar structure of methyl-substituted 1-PhP in the gas phase. When the benzene ring hydrogen atoms located near the interannular bond are substituted by methyl groups the φ value varies more significantly ($\Delta\varphi \sim 50^0$)[86], than when an analogous substitution occurs in the pyrrole fragment ($\Delta\varphi \sim 30^0$). In the authors' opinion, other compounds can be planar with a wide distribution function at the torsional angle. It is noted that MINDO/3 does not reproduce the φ values obtained from UPS data.

3.2. 2-Arylpyrroles

Four bands are also recorded in the low-energy region of the UPS of 2-arylpyrroles (2-AP) [82, 88-90] (Table 1). According to semi-empirical quantum chemical calculations these bands are due to four ionic states arising upon electron removal from $\pi_6(a_2\text{-}b_1)$-, $\pi_5(2b_1)$-, $\pi_4(1a_2)$- and $\pi_3(a_2+b_1)$-MO's of these compounds. From the orbital electron density distribution for the 2-phenylpyrrole (2-PhP) planar form it follows[91] that the π_4 and π_5 orbitals have nodes on bonding atoms (C_2 and C'_1). Therefore the rotation of the aryl and heteroaryl fragments around the interannular bond does not change the energy of these orbitals in 2-PhP. The π_6 orbital has a strong antibonding character between the rings. In going to a non-planar conformation the orbital considerably stabilizes. The π_3-MO is characterized by a bonding character on the C_2 - C'_1 bond. Upon rotation of the rings this orbital destabilizes (Fig. 1). The existence of this π-MO makes it possible to use the energy difference for getting information concerning the equilibrium conformation of both molecules and radical cations. In terms of Koopmans' theorem, for 2-AP the most appropriate for this purpose are energy

Table 1
UPS and calculated data for some ring assemblies

Compound	Method	Ionization energies (eV)					ΔE^0	Ref
		π_6	π_5	π_4	π_3			
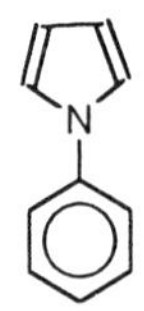	UPS	8.16	8.5sh	9.52	10.02			86
	MINDO/3 (φ=32°)	8.20	8.08	9.38	9.65			86
2,5-dimethyl	UPS	7.44	8.39	9.45	9.7sh			86
	MINDO/3 (φ=62°)	7.86	8.33	9.29	9.42			86
2',6'-dimethyl	UPS	8.09	8.87	9.0sh	9.2sh			86
	MINDO/3 (φ=81°)	8.15	8.38	9.06	9.01			86
2,5,2',6'-	UPS	7.40	8.38	8.84	9.03			86
tetramethyl	MINDO/3 (φ=90°)	7.83	8.51	9.00	9.09			86
4'-cyano	UPS	8.55	8.8sh	10.18	10.4sh			86
	MINDO/3 (φ=30°)	8.31	8.17	9.59	9.61			86
4'-nitro	UPS	8.60	8.9	10.25	10.6			86
	MINDO/3 (φ=30°)	8.77	8.82	10.24	10.49			86
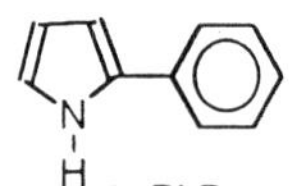	UPS	7.61	8.89	9.19	9.69		2.08	88,89
2-PhP	AM1 (φ=29°)[a]	7.60	8.88	9.07	9.44			82,90
4'-methyl	UPS	7.43	8.80	9.18	9.48		2.05	88
4'-ethyl	UPS	7.27	8.75	8.91	9.25sh		~2.0	91
4'-fluoro	UPS	7.72	9.14	9.56	9.75		2.03	90
	AM1(φ=29°)[a]	7.76	9.00	9.38	9.40			90
4'-chloro	UPS	7.70	9.18	9.55	9.55		1.85	88,95
	AM1(φ=29°)[a]	7.78	9.00	9.32	9.33			90
4'-bromo	UPS	7.71	9.00	9.30	9.46		1.75	90
4'-methoxy	UPS	7.18	8.52	8.96	9.30sh		~2.10	91
4'-phenoxy	UPS	7.47	9.22	9.22	9.67		2.20	88
5-methyl	UPS	7.36	8.88	9.22	9.50		2.14	82
1-ethyl-4'-ethyl-5-oxyethyl	UPS	7.56	8.70	9.02	9.18		1.62	91
3-n-heptyl	UPS	7.52	8.58	9.14	9.50		1.98	91
3-n-nonyl	UPS	7.50	8.60	9.19	9.55		2.05	82
3-ethyl	AM1(φ=51°)[a]	7.69	8.59	9.06	9.23			82
1-ethoxyethyl	UPS	7.54	8.50	9.18	9.32		1.78	91
3-methyl-2',5'-	UPS	7.50	8.45	8.60	9.04		1.54	91
dimethyl	AM1(φ=75°)[a]	7.79	8.50	8.56	8.97			91
3-ethyl-2',5'-dimethyl	UPS	7.51	8.41	8.53	9.03		1.52	82
1-ethoxyethyl-3-ethyl-2',5'-dimethyl	UPS	7.46	8.44	8.44	8.96		1.50	82
		π_7	π_6	π_5	π_4	π_3		
1-vinyl	UPS	7.76	8.37	9.18	9.58	10.47	1.82	88,89
	AM1(φ=52°)[a]	7.91	8.34	9.10	9.28	10.44		82
1-vinyl-3-methyl	UPS	7.56	8.13	9.07	9.4sh	10.19	~1.85	89
1-vinyl-3-ethyl	UPS	7.60	8.17	9.17	9.49	10.20	1.89	82

Table 1 (continued 1)

Compound	Method	Ionization energies (eV)					ΔE^0	Ref
1-vinyl-3-i-propyl	UPS	7.56	8.09	9.06	9.35sh	10.20	~1.80	89
1-vinyl-3-n-nonyl	UPS	7.53	8.16	9.13	9.45sh	10.3sh	~1.90	82
1-vinyl-3-ethyl-2',5'-dimethyl	UPS	7.50	8.0sh	8.40	8.98	10.06	1.48	82
		π_6	π_5	π_4	π_3			
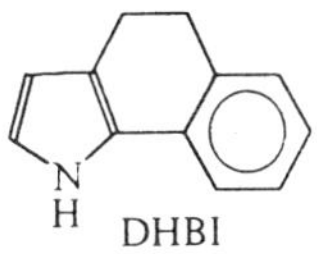 DHBI	UPS	7.14	8.53	8.79	9.49		2.35	91
	UPS				9.47		2.33	82
	AM1 (φ=0°)[a]	7.44	8.64	8.79	9.50			82
		π_7	π_6	π_5	π_4	π_3		
1-vinyl	UPS	7.26	8.26	8.78	9.47	10.29	2.21	82
		π_6	π_5	π_4	π_3			
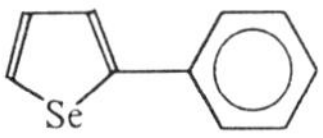	UPS	8.08	8.81	9.25	9.88		1.80	88,95
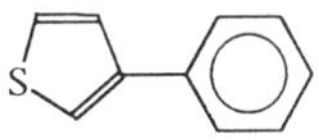	UPS	8.23	8.96	9.17	9.93		1.70	91
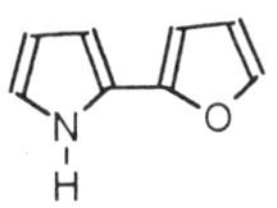	UPS	7.29	8.99	9.47	10.06		2.17	84
	AM1 (φ=0°)[a]	7.56	8.98	9.38	10.01			91
1-ethyl	UPS	7.36	8.53	9.20	9.99		1.84	91
	AM1 (φ=0°)[a]	7.37	8.63	9.28	9.81			91
		π_7	π_6	π_5	π_4	π_3		
1-vinyl	UPS	7.50	8.38	9.34	10.07	10.57	1.84	89
1-vinyl-3-methyl	UPS	7.51	8.16	9.15	9.95sh	10.31	1.64	89
		7.74sh[b]					1.41	
		π_6	π_5	π_4	π_3			
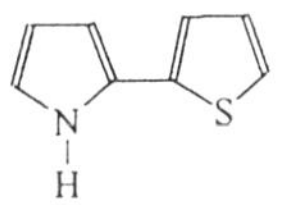	UPS	7.41	9.00	9.18	9.45		2.04	84
		π_7	π_6	π_5	π_4	π_3		
1-vinyl	UPS	7.55	8.47	9.10	9.3sh	10.56	~1.75	89
		7.83sh[b]					~1.45	
1-vinyl-3-methyl	UPS	7.72	8.23	9.03	9.23	10.31	1.51	91
		7.8sh[b]					1.43	
1-vinyl-3-n-propyl	UPS	7.64	8.24	8.98	9.24	10.26	1.60	91
		π_6	π_5	π_4	π_3			
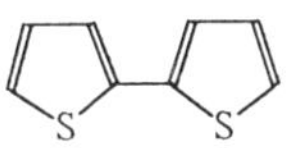	UPS	8.05	9.14	9.37	9.77		1.72	122
	UPS	7.94	9.11	9.22	9.82		1.88	88,95
	UPS	7.95	9.16	9.35	9.78		1.83	123
4,4'-dimethyl	UPS	7.60	8.67	8.90	9.40		1.80	119
3,4'-dimethyl	UPS	7.73	8.66	8.85	9.34		1.61	119
3,3'-dimethyl	UPS	8.18	8.69	8.91	8.91		0.73	119
		π_8	π_7	π_6	π_5			
5,5'-dichloro	UPS	8.05	9.70	9.70	11.60		1.65	177
5,5'-dibromo	UPS	7.96	9.4	9.5	9.6		~1.6	122
3,3'-dibromo	UPS	8.68	8.96	9.31	9.31		0.28	122

Table 1 (continued 2)

Compound	Method	Ionization energies (eV)					ΔE^0	Ref
5,5'-diiodo	UPS	7.80	9.00	9.40	9.4		~1.6	177
		π_9	π_8	π_7	π_6			
3,5,5'-trichloro	UPS	8.0	9.6	9.6	9.6		~1.6	178
3,5,5'-tribromo	UPS	7.95	9.40	9.55	9.55		1.60 1.45[c]	177
		π_{10}	π_9	π_8	π_7			
3,5,3',5'-tetrachloro	UPS	8.60	9.2	9.60	9.60		1.0 0.6[c]	178
3,5,3',5'-tetrabromo	UPS	8.4	8.87	9.27	9.27		0.87 0.47[c]	177
		π_6	π_5	π_4	π_3			
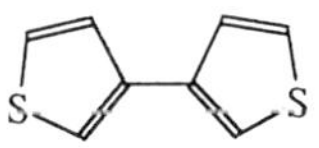	UPS	8.19	9.15	9.15	10.03		0.96	122
		π_8	π_7	π_6	π_5			
2,2'-dibromo	UPS	8.5	8.62	9.2	9.6		0.12	122
4,4'-dibromo	UPS	8.9	8.9	8.9	8.9		0	122
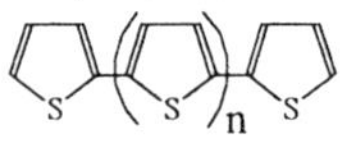		π_9	π_8	π_{7-5}	π_4			
n=1	UPS	7.43	8.77	9.05sh-9.23	9.88			123
		π_{12}	π_{11}	π_{10-7}	π_6	π_5		
n=2	UPS	7.28	8.25	9.1sh-9.20	9.3sh	9.92		123
		π_{15}	π_{14}	π_{13-7}	π_6			
n=3	UPS	7.11	7.85	8.7sh-9.18	9.97			123
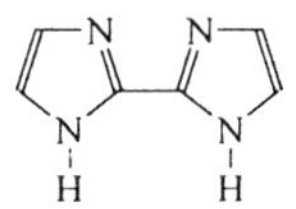		π_6	π_5	π_4	π_3			
1-ethyl	UPS	7.77	9.43	9.90	9.90		1.66	91
	AM1(φ=0°)[a]	7.70	9.10	9.48	9.97			91
1,1'-diethyl	UPS	7.65	9.22	9.64	9.64		1.57	91
		π_7	π_6	π_5	π_4			
1-vinyl	UPS	7.97	9.19	9.73	9.90		1.76	91
		π_8	$\pi_{7,6}$	π_5	n_N	$\pi_{4,3}$		
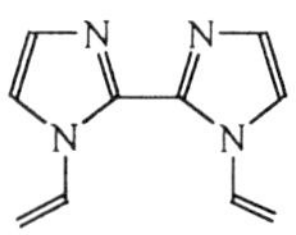	UPS	7.93	9.06	9.72	9.93	11.30	1.79	91
	OVGF AM1	8.30	9.07	9.91	10.49	11.48		this work
	AM1(φ=0°)[a]	8.59	9.65	10.20	11.46	12.06		
		π_6	π_5	π_4	π_3			
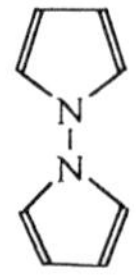	UPS	8.30	8.30	9.25	9.52		0.27	152
2,5-dimethyl	UPS	7.77	8.45	8.82	9.25		0.34	152
2,2',5,5'-tetramethyl	UPS	7.73	7.73	8.88	8.88		0	152

Table 1 (continued 3)

Compound	Method	Ionization energies (eV)				ΔE^0	Ref
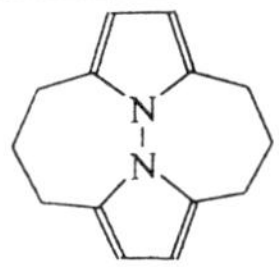	UPS	7.72	7.72	8.25	9.47	1.22	152
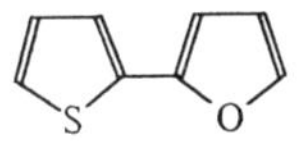	UPS	7.85	9.20	9.80	10.25	1.95	179
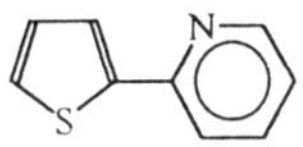	UPS	8.25	9.14 9.45(n_N)	10.00	10.46	1.75	180
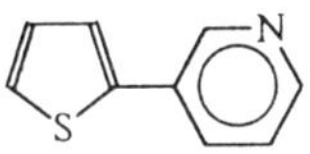	UPS	8.40	9.38 9.38(n_N)	9.90	10.55	1.50	180
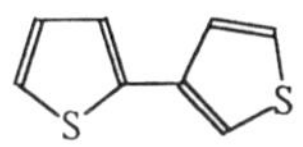	UPS	7.95	8.82	9.13	9.87	1.18	177
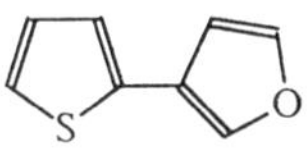	UPS	8.10	9.25	9.25	10.65	1.15	179
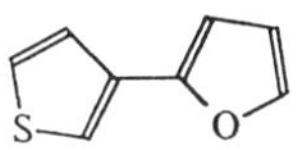	UPS	7.82	8.95	9.80	10.2	1.13	179
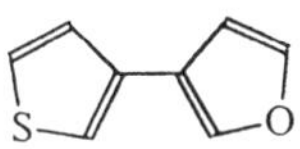	UPS	8.10	9.0	9.33	10.68	0.9	179
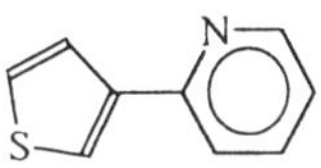	UPS	8.50	9.1 9.5(n_N)	10.15	10.55	1.65	180
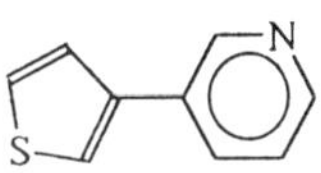	UPS	8.65	9.4 9.4(n_N)	10.0	10.7	1.35	180
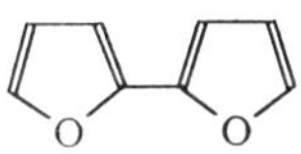	UPS	7.90	9.73	10.30	10.30	1.83	179
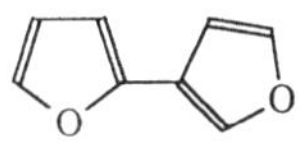	UPS	8.10	9.30	10.19	10.85	1.20	179
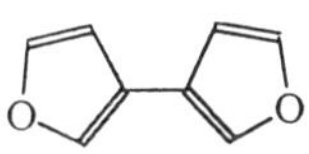	UPS	8.35	9.0	9.74	11.33	0.65	181
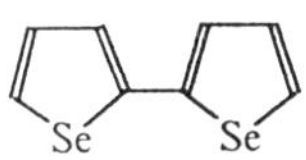	UPS	8.06	9.00	9.25	9.86	1.8	182

Table 1 (continued 4)

Compound	Method	Ionization energies (eV)				ΔE^0	Ref
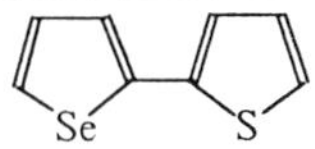	UPS	8.13	9.08	9.50	9.85	1.72	182
		π_8	π_7	π_6	π_5		
5,5'-dichloro	UPS	8.15	9.38	9.69	9.69	1.54	182
5,5'-dibromo	UPS	7.75	8.98	9.34	9.34	1.59	182
		π_6	π_5	π_4	π_3		
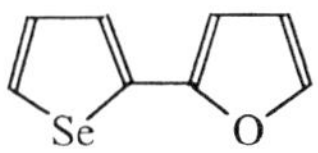	UPS	7.75	8.84	9.62	10.15	1.87	182
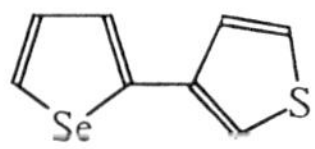	UPS	8.17	8.97	8.97	9.98	0.80	182
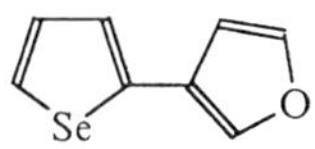	UPS	8.20	9.0	9.2	10.6	1.0	182
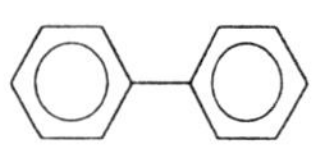	UPS	8.32	9.05	9.15	9.80	1.48	68
	UPS	8.35	9.15	9.15	9.85	1.50	160
	UPS	8.34	9.04	9.20	9.82	1.48	117
	UPS	8.35	9.1	9.25	9.80	1.45	69
	AM1(φ=40°)[a]	8.32	9.02	9.04	9.56		90
4-methyl	UPS	8.15	9.00	9.15	9.60	1.45	68
	UPS	8.15	9.1	9.1	9.65	1.50	69
4-fluoro	UPS	8.35	9.35	9.45	9.80	1.45	68
2-fluoro	UPS	8.55	9.18	9.35	9.95	1.40	68
4,4'-difluoro	UPS	8.38	9.50	9.50	9.80	1.42	68
2,2'-difluoro	UPS	8.85	9.20	9.38	10.00	1.15	68
perfluoro	UPS	9.80	9.96	10.10	10.70	0.90	68
	UPS	9.8	9.95	10.10	10.75	0.95	69
2-methyl	UPS	8.60	8.85	9.10	9.50	0.90	68
	UPS	8.55	9.0	9.0	9.45	0.90	69
2,2'-dimethyl	UPS	8.55	8.80	9.00	9.25	0.70	68
	UPS	8.7	9.0	9.0	9.4	0.7	69
4-chloro	UPS	8.46	9.43	9.56	9.75	1.29	68
	UPS	8.45	9.50	9.50	9.80	1.35	69
4,4'-dichloro	UPS	8.40	9.50	9.50	9.75	1.35	69
	UPS	8.28	9.49	9.49	9.71	1.43	95
	AM1(φ=40°)[a]	8.43	9.37	9.40	9.44		95
4-bromo	UPS	8.40	9.40	9.40	9.65	1.25	68
4,4'-dibromo	UPS	8.43	9.46	9.46	9.7sh	1.30	90
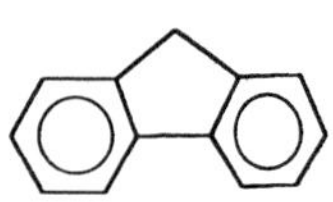	UPS	7.93	8.83	9.13	9.91	1.98	68
	UPS	7.91	8.77	9.09	9.84	1.93	117
	UPS	7.95	8.85	9.15	9.90	1.95	69
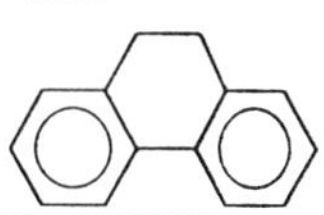	UPS	7.96	8.80	8.95	9.8	1.84	68

[a]Ionization energies were derived by Koopmans' theorem from calculated orbital energies: $I_i = -0.93\varepsilon_i$; [b]Secon conformer is detected; [c]The assignment is made by analogy with[122].

gaps ΔE^0, ΔE^+ between the π_3- and π_6-MO's the energies of which change in opposite directions with varying the φ angle (Fig. 1). Analytical form of the $\Delta E^0(\varphi) = f(\varphi)$ dependence for the molecules of 2-aryl(heteroaryl)pyrroles and some other ring assemblies was studied by semi-empirical quantum chemical

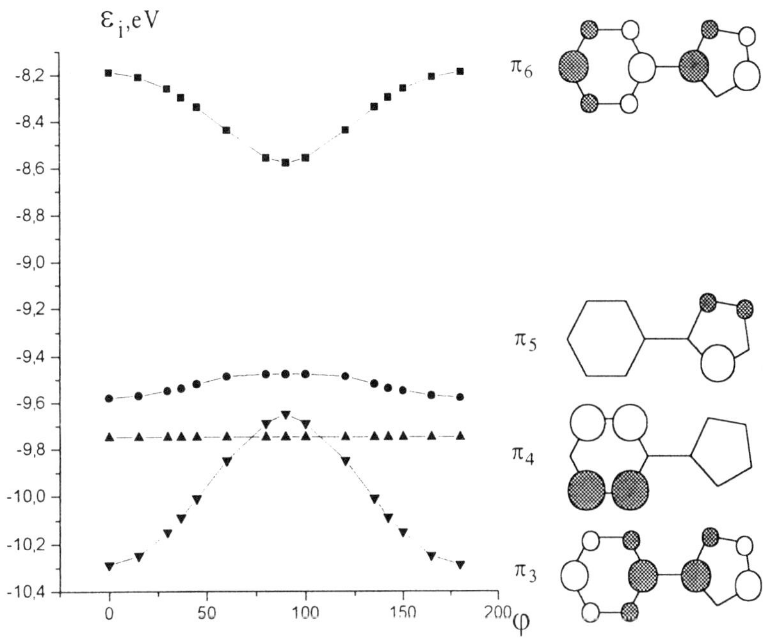

Fig. 1. Orbital energies as function of dihedral angle. Data of AM1 calculations for 2-phenylpyrrole.

methods[90]. It is found to be best approximated by the following semi-empirical relationship:

$$\Delta E^0(\varphi) = \Delta E^0(0^0) - 1.03 \sin^2(\varphi) \qquad (2)$$

Here $\Delta E^0(0^0)$ is the energy gap between conformationally dependent ionic states of the planar conformer. It should be noted that equation (2) is an approximated form of equation (1). Equation (2), when applied to a set of experimentally determined ΔE^0 values and dihedral φ angles of biphenyls, has shown[90] to reproduce an empirically established[90] relationship $\Delta E^0 \sim f(\varphi)$ in the range $0 \leq \varphi \leq 70^0$. Besides, it has been found that the $\Delta E^0(0^0)$ value is determined by the structure of molecules and depends on valence approximation. Therefore, in evaluating the φ angle of molecules the use should be made of the $\Delta E^0(0^0)$ values experimentally determined from the UPS of a planar reference compound. In the series of alkyl(vinyl)-substituted 2-arylpyrroles this compound is presented by 4,5-dihydrobenzo[g]-indole (DHBI)[82]. The 2-AP φ angles thus determined are presented in Table 3. They display the same tendency that the dihedral angles of 1-PhP and biphenyls do, i.e., their values rise with increasing the number of substituents in the interannular bond surrounding; the φ values for 2',3-substituted 2-arylpyrroles are comparable with those for 2,2'-difluorobiphenyl and 2,5-dimethyl-1-

phenylpyrrole. However, in 2-phenylpyrrole and 4',5-substituted 2-PhP derivatives the equilibrium torsional angles are smaller than in unsubstituted biphenyl and 4,4'-difluorobiphenyls. Rings bonded with an ethane bridge leads to planarization of molecules (42^0, 20^0, 30^0, and 0^0 for biphenyl, 9,10-dihydrophenanthrene, 2-phenylpyrrole, and DHBI, respectively). Thus, 2-AP which have no substituents near the C_2 - C'_1 bond or connected by an ethane bridge, are characterized in the gas phase by a more planar structure than analogous biphenyls. Two alkyl groups or alkyl and vinyl groups in the 1 and 3 position of the 2-PhP molecule produce a lower effect on the equilibrium torsional angles than two methyl groups in the 2 and 5-position of the 1-PhP molecule:

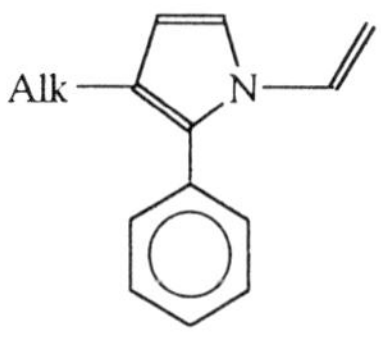

$\varphi = 39 \div 47^{\circ}$

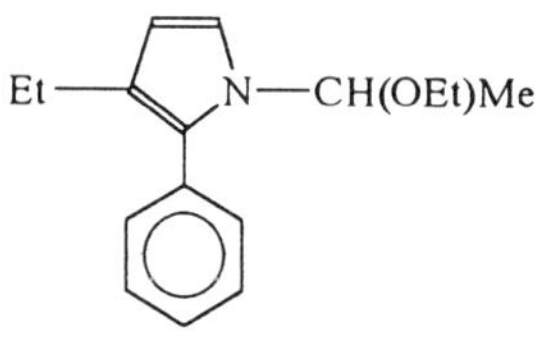

$\varphi = 47 \pm 5^{\circ}$

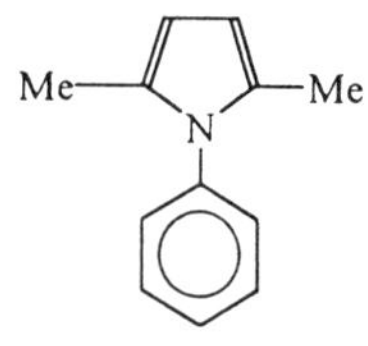

$\varphi = 62^{+8}_{-14}{}^{\circ}$

Nevertheless, as was the case with 1-PhP, the alkyl substituent in the 2-PhP pyrrole ring affects the φ value to a less extent that in the benzene ring:

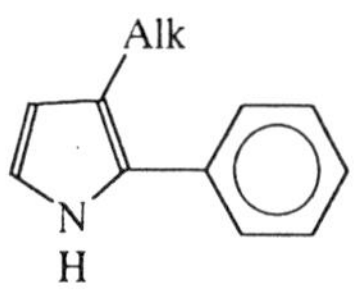

$\varphi = 32{-}36^{\circ}$

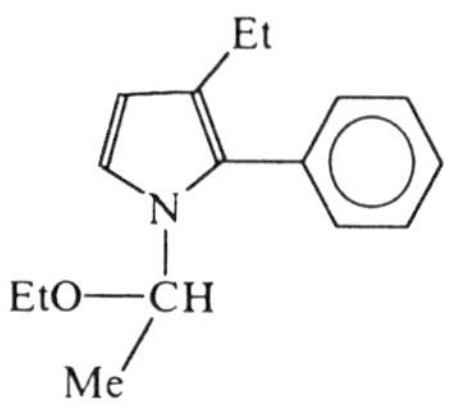

$\varphi = 47^{\circ}$

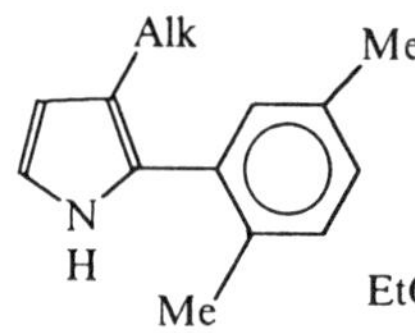

$\varphi = 62{-}63^{\circ}$

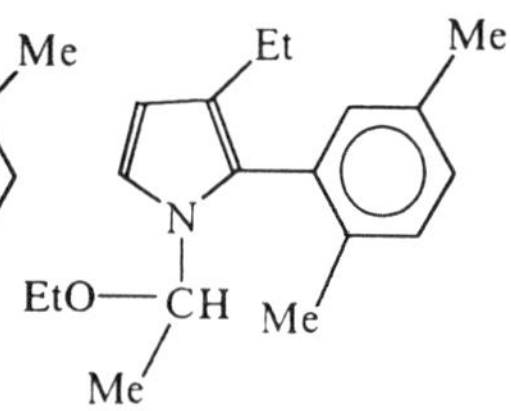

$\varphi = 64^{\circ}$

However, in the presence of two alkyl groups near the 2-PhP interannular bond further substitution of hydrogen at the pyrrolic nitrogen atom by an alkyl group practically does not lead to any increase in the equilibrium torsional angle. Similar tendencies were also observed on vinylic substitution in the 2-PhP 1-position:

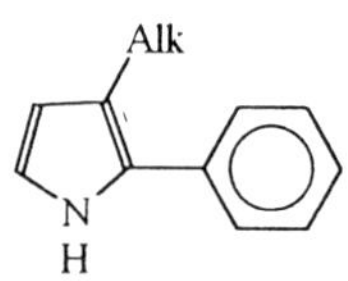

$\varphi = 32{-}36^{\circ}$

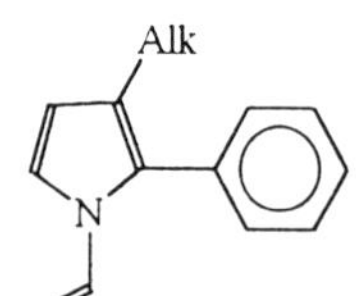

$\varphi = 39 \div 47^{\circ}$

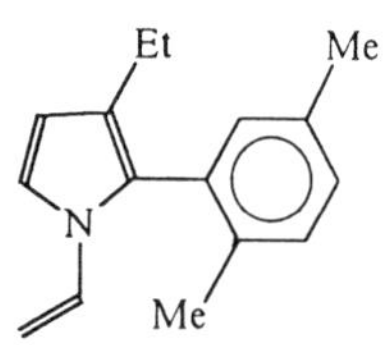

$\varphi = 66^{\circ}$

The equilibrium torsional angles of 2-AP free molecules were theoretically evaluated as well[82, 90, 92-94]. Of semi-empirical methods AM1 was used for this purpose as the most adequate approach among semiempirical methods for the examination of ring assembly conformation[94]. The theoretical 2-AP φ values are in good agreement with the angles determined from experimental data (Table 3).

Table 2
UV spectroscopy data for radical cations of some ring assemblies

Radical cation	Medium	Transition energies (eV) ΔE^{+}	$\varepsilon \cdot 10^{-3}$ l mol^{-1}cm^{-1}	δE	Ref.
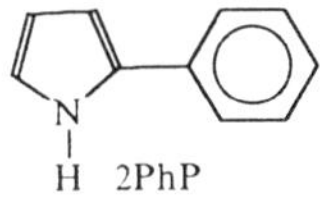 2PhP	MeCN	2.34	2.8	0.26	88, 95
4'-methyl	MeCN	2.21	5.1	0.16	91
4'-fluoro	MeCN	2.32	5.2	0.29	90
4'-chloro	MeCN	2.18	4.3	0.32	90
4'-bromo	MeCN	2.18	5.6	0.32	90
5-methyl	MeCN	2.36	3.3	0.22	82
3-n-nonyl	MeCN	2.32	3.8	0.27	82
1-vinyl	MeCN	2.32	2.8	0.50	82
1-vinyl-3-methyl	MeCN	2.26	3.4	0.41	91
1-vinyl-3-ethyl	MeCN	2.30	3.4	0.41	82
1-vinyl-3-n-nonyl	MeCN	2.30	3.2	0.40	82
1-vinyl-3-i-propyl	MeCN	2.30	3.5	0.50	91
3-ethyl-2',5'-dimethyl	MeCN	2.32	3.5	0.80	82
1-ethoxyethyl-3-ethyl-2',5'-dimethyl	MeCN	2.34	1.6	0.84	82
1-vinyl-3-ethyl-2',5'-dimethyl	MeCN	2.30	1.4	0.82	82
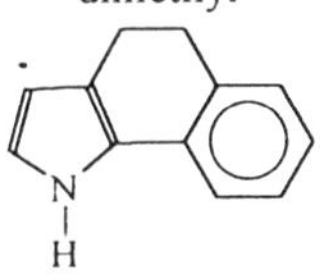	MeCN	2.35	2.6	0.0 0.02	91 82
1-vinyl	MeCN	2.36	4.1	0.15	82
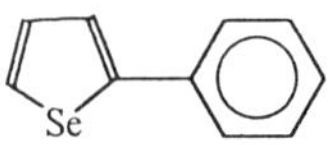	MeCN	2.03		0.23	88, 95
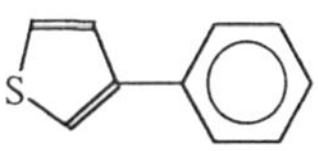	MeCN	2.08		0.38	91
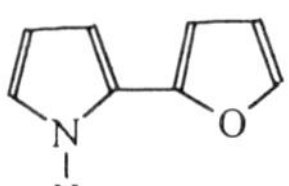	MeCN	2.18	4.3	0.01	91
1-ethyl	MeCN	2.19	4.4	0.35	91
1-vinyl	MeCN	2.16	4.9	0.32	91
1-vinyl-3-methyl	MeCN	2.17	4.7	0.53	88
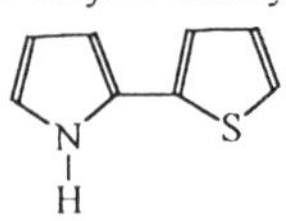	MeCN MeCN	2.32 2.12	2.5 2.3	0.28	88, 95 88, 95
1-vinyl	MeCN MeCN	2.29 2.14	3.0 3.0	0.54 0.69	91 91
1-vinyl-3-methyl	MeCN	2.16	3.4	0.65	91

Table 2 (continued 1)

Radical cation	Medium	Transition energies (eV) ΔE^+	$\varepsilon \cdot 10^{-3}$ $l\,mol^{-1}cm^{-1}$	δE	Ref.
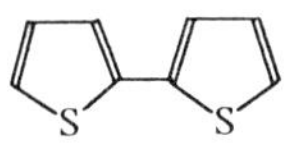	MeCN	2.11		0.23	88, 95
	MeCN	2.92			88
	MeCN	2.14			124
	MeCN	2.95			124
	zeolites	3.05			158
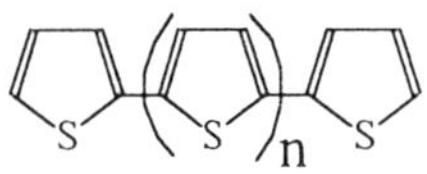					
n=1	MeCN	2.34			124
	CH_2Cl_2	1.46			135
	CH_2Cl_2	2.25			135
	CH_2Cl_2	2.5-2.6sh			135
	zeolites	2.38			158
n=2	CH_2Cl_2	1.91			135
	CH_2Cl_2	2.2-2.6sh			135
	CH_2Cl_2	1.16			136
	CH_2Cl_2	1.92			136
	CH_2Cl_2	2.0sh			136
	zeolites	2.02			158
n=3	CH_2Cl_2	0.98			136
	CH_2Cl_2	1.1sh			136
	CH_2Cl_2	1.72			136
	CH_2Cl_2	1.86			136
n=4	CH_2Cl_2	1.59			135
	CH_2Cl_2	1.8sh			135
	CH_2Cl_2	2.25			135
	CH_2Cl_2	0.84			136
	CH_2Cl_2	1.0sh			136
	CH_2Cl_2	1.59			136
	CH_2Cl_2	1.82			136
	zeolites	1.60			158
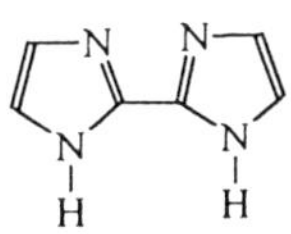					
1-ethyl	MeCN	1.98	4.2	0.32	91
1,1-diethyl	MeCN	1.86	6.3	0.29	91
1-vinyl	MeCN	1.89	3.6	0.13	91
1,1'-divinyl	MeCN	1.89	6.2	0.10	91
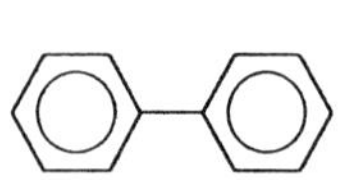	solid argon	1.85		0.37	25
	boric acid	1.84		0.36	14
	$C_2H_4Cl_2$	1.80		0.32	78
	MeCN	1.87		0.39	95
4-methyl	solid argon	1.83		0.35	25
4-fluoro	$C_2H_4Cl_2$	1.81		0.36	78
2-fluoro	solid argon	1.83		0.43	25
	$C_2H_4Cl_2$	1.76		0.36	78
4,4'-difluoro	$C_2H_4Cl_2$	1.82		0.40	161
2,2'-difluoro	solid argon	1.76		0.61	25
perfluoro	solid argon	1.38		0.48	25

Table 2 (continued 2)

Radical cation	Medium	Transition energies (eV) ΔE^{+}	$\varepsilon \cdot 10^{-3}$ $l\ mol^{-1}cm^{-1}$	δE	Ref.
2-methyl	solid argon	1.66		0.76	25
2,2'-dimethyl	solid argon	1.65		0.95	25
4-chloro	$C_2H_4Cl_2$	1.72		0.38	78
4,4'-dichloro	$C_2H_4Cl_2$	1.54		0.15	79
	MeCN	1.84		0.41	95
4-bromo	$C_2H_4Cl_2$	1.63		0.38	78
4,4'-dibromo	$C_2H_4Cl_2$	1.51		0.21	79
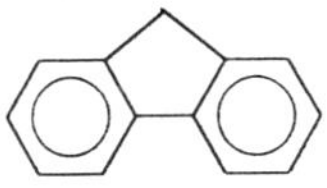	solid argon	1.93		0.0	25
	MeCN	1.94		0.01	95
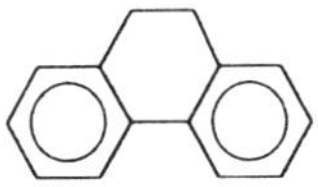	solid argon	1.93		0.09	25
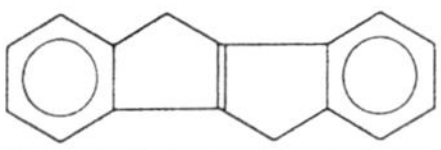	freon matrix	1.95		-0.01	34

Moreover, the orbital energies calculated in terms of Koopmans' theorem under the conditions of equilibrium theoretical geometry exactly reflect the sequence and absolute values of ionization potentials found from UPS (Table 1).

Reliable data on the equilibrium conformation of substituted 2-AP short-lived radical cations (RC) were obtained by comparison of ΔE^{0} values calculated from photoelectron spectra and ΔE^{+} values found from the RC electronic absorption spectra recorded by means of nanosecond laser photolysis in a MeCN solution[82]. The UV spectra of 2-AP RC are characterized by absorption bands in the 500-700 nm region. These bands are fairly intense ($\varepsilon > 10^{3}$) and their position depends greatly on the type and number of rings in an assembly. Thus, in the UV spectrum of pyrrole RC there are no absorption bands in the above region. In the UV spectrum of 2-PhP RC there is a band with λ_{max} = 530 nm, whereas in that of 2,5-diphenylpyrrole RC a band with λ_{max} = 620 nm and a shoulder on 580 nm are observed[91]. The high intensity and a batochromic shift as the π-system extends make it possible to assign these bands to $\pi^{*} \leftarrow \pi$ transitions. Comparison with UPS data makes the assignment more precise. The transition energy in the visible region of the spectrum of DHBI (planar reference compound for the series of 2-AP) RC is similar to the energy difference of π_3- and π_6-MO's of the parent molecule (δE = 0)[82] (Table 2). This should have been expected owing to the known correspondence between the energy gaps found from the UPS of planar aromatic and heterocyclic compounds, and the energy transitions in the UV spectra of relevant RC[6-9, 14-18, 20-22, 24-31, 37, 42]. A great discrepancy between the ΔE^{0} and ΔE^{+} values has been found for other 2-arylpyrroles ($\delta E > 0^{0}$)[88, 95] (Table 2). The $\delta E = \Delta E^{+} - \Delta E^{0}$ increases with increasing the number of substituents near the interannular bond. In compounds having one substituent the δE value is ~ 0.25 eV, in those with two and three substituents this value is equal to ~ 0.4 and ~ 0.8 eV, respectively (Table 2). These rather high δE values may arise for several reasons: (a) the fact that the above assignment is incorrect and the observed transition is a transition to "non-Koopmans'" state, i.e., that with an electron on the vacant π-orbitals of radical cation; (b) unequivalency of π- and σ-MO interactions in equilibrium states of the molecule and radical cation, and (c) great changes in the geometry of molecules on ionization.

Table 3
Equilibrium torsional angles of ring assemblies and their radical cations

Compound	Method	Medium	Reference compound	φ°	Ref.
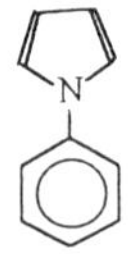	UPS	gas	2,5,2',6'-tetramethyl-1-PhP	32^{+17}_{-32}	86
1-PhP	dipole moment	benzene		42	187
2,5-dimethyl	UPS	gas	2,5,2',6'-tetramethyl-1-PhP	62^{+8}_{-14}	86
	dipole moment	benzene		58	187
2',6'-dimethyl	UPS	gas	2,5,2',6'-tetramethyl-1-PhP	81^{+6}_{-9}	86
	dipole moment	benzene		59	187
4'-cyano	UPS	gas	2,5,2',6'-tetramethyl-1-PhP	30^{+19}_{-30}	86
	dipole moment	benzene		32	187
4-nitro	UPS	gas	2,5,2',6'-tetramethyl-1-PhP	38^{+16}_{-38}	86
	dipole moment	benzene		40	187
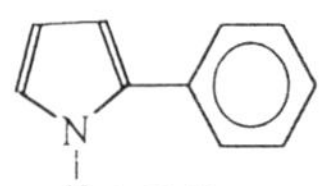	UPS	gas	DHBI	30	82,89
	AM1			29	82
2-PhP	UV	solution	comp. in polym. matr.	23±6	83
radical cation	UV	MeCN	DHBI	3	90
	AM1			10	90
4'-methyl	UPS	gas	DHBI	32	91
	AM1			28	91
	UV	solution	comp. in polym. matr.	23±6	83
4'-fluoro	UPS	gas	DHBI	33	90
	AM1			29	90
	UV	solution	comp. in polym. matr.	23±9	83
radical cation	UV	MeCN	DHBI	10	90
	AM1			2	90
4'-ethyl	UPS	gas	DHBI	36±13	91
4'-methoxy	UPS	gas	DHBI	27±16	91
5-methyl	UPS	gas	DHBI	26	82
	AM1			34	91
	UV	solution	comp. in polym. matr.	22±9	83
radical cation	UV	MeCN	DHBI	0	82

Table 3 (continued 1)

Compound	Method	Medium	Reference compound	φ°	Ref.
4'-chloro	UPS	gas	DHBI	34	90
	AM1			29	90
	UV	solution	comp. in polym. matr.	25±5	83
radical cation	UV	MeCN	-	9	90
	AM1			8	90
4'-bromo	UPS	gas	DHBI	40	90
	UV	solution	comp. in polym. matr.	26	90
radical cation	UV	MeCN	-	10	90
3-n-heptyl	UPS	gas	DHBI	36	91
	UV	MeCN	DHBI	26±6	83
3-n-nonyl	UPS	gas	DHBI	32	82
	UV	solution	comp. in polym. matr.	29±4	83
radical cation	UV	MeCN	DHBI	10	82
1-ethyl-4'-ethyl-5-oxyethyl	UPS	gas	DHBI	57	91
3-ethyl	AM1			51	82
radical cation	AM1			23	91
3-i-propyl	UV	MeCN	comp. in polym. matr. +DHBI	29±7	83
1-ethoxyethyl	UPS	gas	DHBI	47	91
3-methyl-2',5'-dimethyl	UPS	gas	DHBI	62	91
	AM1			57	91
	UV	MeCN	DHBI	55±3	83
3-ethyl-2',5'-dimethyl	UPS	gas	DHBI	63	82
	UV	MeCN	DHBI	53±3	83
radical cation	UV	MeCN	DHBI	10	82
1-ethoxyethyl-3-ethyl-2',5'-dimethyl	UPS	gas	DHBI	64	82
	UV	MeCN	DHBI	64±3	83
1-vinyl	UPS	gas	DHBI	45	82,89
	AM1			52	91
	UV	solution	1-vinyl-DHBI	32±9	83
	NMR	$CDCl_3$		30	105,106
radical cation	UV	MeCN	DHBI	11	82
	AM1			15-25	82
1-vinyl-3-methyl	UPS	gas	DHBI	44±8	91
	UV	solution	1-vinyl-DHBI	43±9	83
	NMR	$CDCl_3$		35-40	105,106
radical cation	UV	MeCN	DHBI	17	91
1-vinyl-3-ethyl	UPS	gas	DHBI	41	82
	UV	solution	1-vinyl-DHBI	53±3	83
	NMR	$CDCl_3$		40	105,106
radical cation	UV	MeCN	DHBI	14	82

Table 3 (continued 2)

Compound	Method	Medium	Reference compound	φ°	Ref.
1-vinyl-3-i-propyl	UPS	gas	DHBI	47±7	91
	UV	solution	1-vinyl-DHBI	58±3	83
	NMR	$CDCl_3$		40	105,106
radical cation	UV	MeCN	DHBI	13	91
1-vinyl-3-n-nonyl	UPS	gas	DHBI	39±10	91
	UV	solution	1-vinyl-DHBI	53±3	83
radical cation	UV	MeCN	DHBI	14	82
1-vinyl-3-ethyl-2',5'-dimethyl	UPS	gas	DHBI	66	82
	UV	solution	1-vinyl-DHBI	65±3	83
	NMR	$CDCl_3$		50	106
radical cation	UV	MeCN	DHBI	16	82
1-ethyl-4'-methyl	AM1			43	91
1-ethyl-5-methyl	AM1			44	91
1-ethyl-3-methyl	AM1			47	91
1-ethyl	UV	ethanol	2-PhP	41	104
1-ethyl-3-methyl	UV	ethanol	2-PhP	46	104
1-ethyl-3-n-propyl	UV	ethanol	2-PhP	48	104
1-ethyl-4'-methoxy	UV	ethanol	4'-methoxy-2-PhP	41	104
1-thioethyl-dimethyl-4'-ethyl	UV	ethanol	4'-methyl-2-PhP	40	104
1-vinyl-4'-methyl	NMR	$CDCl_3$		30	106
1-vinyl-4'-t-butyl	NMR	$CDCl_3$		30	106
1-vinyl-4'-methoxy	NMR	$CDCl_3$		30	106
1-vinyl-4'-amino	NMR	$CDCl_3$		30	106
1-vinyl-4'-phenyl	NMR	$CDCl_3$		30	106
1-vinyl-4'-chloro	NMR	$CDCl_3$		30	106
1-vinyl-4'-bromo	NMR	$CDCl_3$		30	106
1-vinyl-4'-thiomethyl	NMR	$CDCl_3$		30	106
1-vinyl-4'-thioethyl	NMR	$CDCl_3$		30	106
1-vinyl-3'-amino	NMR	$CDCl_3$		30	106
1-vinyl-3-methyl-4'-ethyl	NMR	$CDCl_3$		30-35	106
1-vinyl-3-methyl-4'-t-butyl	NMR	$CDCl_3$		30-35	106
1-vinyl-3-methyl-4'-thiomethyl	NMR	$CDCl_3$		30-35	106
1-vinyl-3-methyl-4'-thioethyl	NMR	$CDCl_3$		30-35	106
1-vinyl-3-n-propyl	NMR	$CDCl_3$		40	105,106
1-vinyl-3-n-buthyl	NMR	$CDCl_3$		40	105,106
1-vinyl-3-n-amyl	NMR	$CDCl_3$		40	105,106

Table 3 (continued 3)

Compound	Method	Medium	Reference compound	φ°	Ref.
1-vinyl-3-methyl -2',5'-dimethyl	NMR	$CDCl_3$		50	105,106
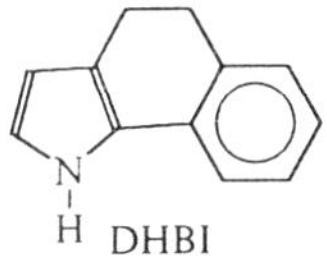 DHBI	UPS	gas		0	82,90
	AM1			0	82,90
	UV	solution		0	83
radical cation	UV	MeCN		0	82,90
	AM1			0	82,90
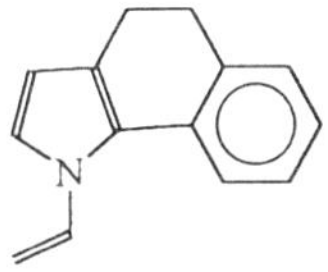	UPS	gas	DHBI	20	82
	UV	solution	-	0	83
	NMR	$CDCl_3$		15	106
radical cation	UV	MeCN	DHBI	0	91
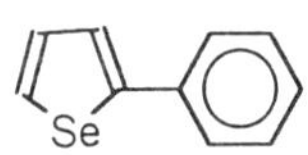	UPS	gas	radic. cation	~30	a
	chromato-struct. anal.	gas		31	112
radical cation	UV	MeCN		~0	a
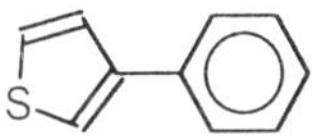	UV and dipole moment	solution		0	113
	NMR	nemat. phase		24	114
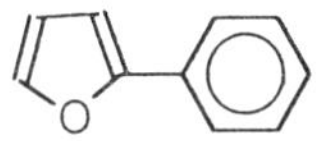	chromato-struct. anal.	gas		24	112
	UV	solution		0	115
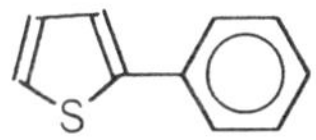	chromato-struct. anal.	gas		29	112
	UV	solution		0	115
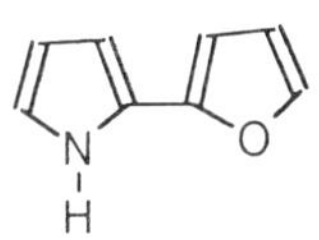	UPS	gas		0	84
	ab initio (4-31G)			0(*syn*)	84
	UV	solution		0	91
	dipole moment	benzene		0	84
radical cation	UV	MeCN		0	91
1-ethyl	UPS	gas	2-FuP	35	91
	AM1			11	91
radical cation	UV	MeCN	2-FuP RC	0	91
	AM1			1	91
1-vinyl	UPS	gas	2-FuP	35	89
radical cation	UV	MeCN	2-FuP RC	8	91
1-vinyl-3-methyl	UPS	gas	2-FuP	46	89
radical cation	UV	MeCN	2-FuP RC	6	91
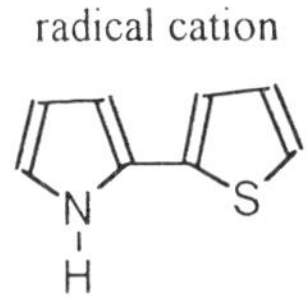	UPS	gas	pyr., thioph.	0	84
	ab initio (STO-3G)			0	84
	UPS	gas	radic. cation	~31	a
	UV	solution		~25	a
	UV	polym. matr.		~0	a
radical cation	UV	MeCN		~0	a
	UV	MeCN	2-TP RC	~26	a

Table 3 (continued 4)

Compound	Method	Medium	Reference compound	φ°	Ref.
1-vinyl	UPS	gas	radic. cation	~48	a
			of 2-TP	~67	a
radical cation	UV	MeCN	radic. cation	~10	a
			of 2-TP	~25	a
1-vinyl-3-methyl	UPS	gas	radic. cation	~62	a
			of 2-TP	~68	a
radical cation	UV	MeCN	2-TP RC	~23	a
1-vinyl-3-n-propyl	UPS	gas	radic. cation of 2-TP	~57	a
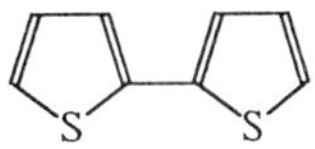	electron diffraction	gas		34(*anti*)+85(*syn*)	154
	electron diffraction	gas		32(*anti*)+36(*syn*)	118
α2T	UPS	gas		~35(*anti*)	122
	ab initio (3-21G*)			33-34(*anti*)+ 43-44(*syn*)	118-120
	ab initio (6-31G*)			32-33(*anti*)+44(*syn*)	118, 120
	AM1			30(*anti*)+~30(*syn*)	121
	AM1			2	177
	NMR	nemat. phase		0(*anti*)+0(*syn*)	155,156
	NMR	nemat. phase		0(*anti*)+24(*syn*)	157
	Kerr constant	solution		23(*anti*)	188
	dipole moment	solution		78(*anti*)	189
	X-ray diffraction	crystal		0(*anti*)	101,137,138
radical cation	UV	MeCN	parent mol.	<10	a
4,4'-dimethyl	UPS	gas	α2T, its RC	37	a
	ab initio (3-21G*)			28(*anti*)+41(*syn*)	119
	UV+UPS	solution		31	a
	MMP2	solution		27(*anti*)+0(*syn*)	127
3,4'-dimethyl	UPS	gas		~40(*anti*)+~50(*syn*)	119
	UPS	gas	α2T, its RC	47	a
	ab initio (3-21G*)			71(*anti*)	120
	ab initio (6-31G*)			64(*anti*)	120
	UV+UPS	solution		39	a
	MMP2	solution		38(*anti*)+0(*syn*)	127
3,3'-dimethyl	UPS	gas	α2T, its RC	82	a
	ab initio (3-21G*)			86-87	119,120
	ab initio (6-31G*)			86	120
	UV+UPS	solution		69	a
	MMP2	solution		46(*anti*)+35(*syn*)	127
5,5'-dimethyl	X-ray diffraction	crystal		0(*anti*)	125
radical cation	EPR	CH_2Cl_2+ $AlCl_3$		0(*anti*)+0(*syn*)	125
	EPR	CF_3COOH+ H_2SO_4		0(*anti*)+0(*syn*)	110
3,3'-diethyl	UPS+UV	gas		72	a
	UV+UPS	solution		60	a
3-methyl	UPS+UV	gas		38	a
	UV+UPS	solution		32	a
3-ethyl	UPS+UV	gas		43	a
	UV+UPS	solution		36	a

Table 3 (continued 5)

Compound	Method	Medium	Reference compound	φ°	Ref.
5,5'-dichloro	UPS	gas		0	178
	AM1			7	177
	NMR	nemat. phase		0	183
3,5,5'-trichloro	UPS	gas		0	178
	X-ray diffraction	crystal		2(*anti*)	178
3,5,3',5'-	UPS	gas		twisted	178
tetrachloro	X-ray diffraction	crystal		0(*anti*)	178
5,5'-dibromo	UPS	gas		twisted	122
	AM1			10	177
	NMR	nemat. phase		0	183
	X-ray diffraction	crystal		0(*anti*)	184
3,3'-dibromo	UPS	gas		twisted	122
3,5,5'-tribromo	UPS	gas		0	178
	AM1			17	177
	X-ray diffraction.	crystal		5	185
3,5,3',5'-	UPS	gas		twisted	178
tetrabromo	AM1			44	177
5,5'-diiodo	UPS	gas		0	178
	AM1			7	177
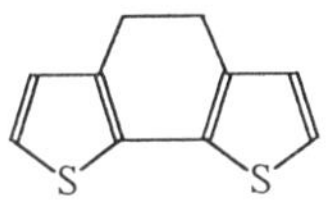	UPS+UV	gas		15	a
	UV+UPS	solution		13	a
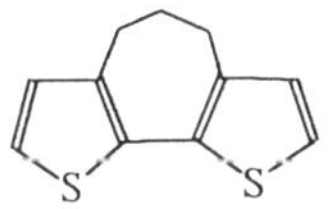	UPS+UV	gas		27	a
	UV+UPS	solution		23	a
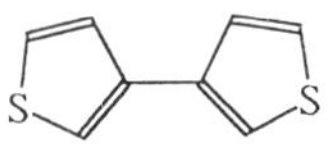	UPS	gas		~35	122
	AM1			9	177
	NMR	nemat. phase		22(*anti*)+22(*syn*)	101
	X-ray diffraction	crystal		0	102
2'2'-dibromo	UPS	gas		twisted	122
4,4'-dibromo	UPS	gas		twisted	122
	linear dichroism	polym. matr.		90	186
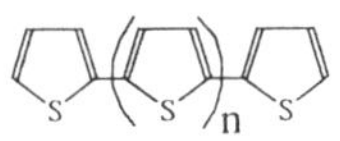					
n=1	X-ray diffraction	crystal		6-9	140
5,5''-dimethyl radical cation	EPR	CF_3COOH+ H_2SO_4		0	110
n=2					
4,4',3'',4'''-tetramethyl	X-ray diffraction	crystal		~0	139
5,5'''-dimethyl radical cation	EPR	CF_3COOH+ H_2SO_4		0	110

Table 3 (continued 6)

Compound	Method	Medium	Reference compound	φ°	Ref.
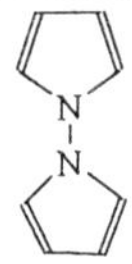	UPS	gas		77	152
2,5,2',5'-tetramethyl	UPS	gas		90	152
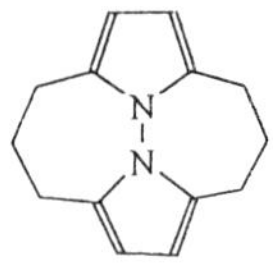	UPS	gas		0	152
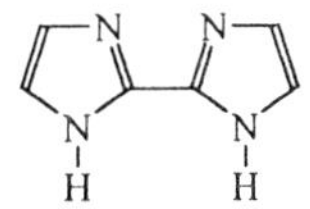	UV	DMSO		0	91
	X-ray diffraction	crystal		4-6(***anti***)	149
1-ethyl	UPS	gas	1-ethyl-α2Im RC	34	a
	UV	MeCN		0	91
radical cation	UV	MeCN		0	91
1-vinyl	UPS	gas	1-ethyl-α2Im RC	21	a
radical cation	UV	MeCN	1-ethyl-α2Im RC	17	91
1,1'-diethyl	UPS	gas	1-ethyl-α2Im RC	32	a
radical cation	UV	MeCN	1-ethyl-α2Im RC	20	91
1,1'-divinyl	UPS	gas	1-ethyl-α2Im RC	18	a
	dipole moment	solution		30(***anti***)	147
radical cation	UV	MeCN	1-ethyl-α2Im RC	17	91
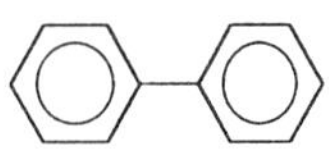	electron diffraction	gas		44	97
	UPS	gas		45	68
	UPS	gas	fluorene	41	90
	UPS	gas	fluorene	44	91
	ab initio (6-31G*)			45	162
	AM1			40	82,90
	UV	solution		20	65
	UV	solution		23	91
	electro-optic	cyclohexane		30	98
	Raman intensity	CCl_4		29	98
	Kerr constant	cyclohexane		~35	163
	IR	solution		32	164
	IR	solution		37	168
	NMR	nemat. phase		34•	164
	NMR	solution		~35	168
	Rayleigh scattering	solution		31	166

Table 3 (continued 7)

Compound	Method	Medium	Reference compound	φ°	Ref.
	X-ray diffraction	crystal		0	99
	Raman spectra	crystal		0	167
radical cation	UV	solution and rigid matrix	radic. cation fluorene	17	90
	AM1			18	90
4-methyl	UPS	gas		45	68
	UPS	gas		42	69
	UPS	gas	fluorene	43	a
	Rayleigh scattering	solution		34	166
radical cation	UV	solid argon	fluorene RC	18	a
4-fluoro	electron diffraction	gas		45	97
	UPS	gas		46	68
	UPS	gas	fluorene	43	90
	electro-optic	cyclohexane		30	98
	Raman intensity	CCl_4		44	98
radical cation	UV	solution	fluorene RC	16	90
2-fluoro	electron diffraction	gas		49	97
	UPS	gas	fluorene	47	a
radical cation	UV	solution	radic. cation	25	a
	UV	solid argon	fluorene	18	a
4,4'-difluoro	electron diffraction	gas		45	97
	UPS	gas		47	68
	UPS	gas	fluorene	43	90
	IR and Raman	gas		45	169
	AM1			41	90
	electro-optic	cyclohexane		30	98
	Raman intensity	CCl_4		56	98
	IR and Raman	solution		45	169
	X-ray diffraction	crystal		0	170
	IR and Raman	crystal		0	169
	NMR	crystal		0	171
radical cation	UV	solution	fluorene RC	14	90
	AM1			18	90
	UV	$C_2H_4Cl_2$		21	91
2,2'-difluoro	electron diffraction	gas		60	97
	UPS	gas	fluorene	62	91
radical cation	UV	solid argon	fluorene RC	24	a
perfluoro	electron diffraction	gas		64	172
	UPS	gas		70	68
	Kerr constant	cyclohexane		63	98
	neutron diffraction	crystal		59	173
	X-ray diffraction	crystal		60	174
radical cation	UV	solid argon	fluorene RC	47	a
2-methyl	UPS	gas		70	68
	UPS	gas		69	69
	AM1			70	82
	UV	solution		58	65
	Rayleigh scattering	solution		46	166
radical cation	UV	solid argon	fluorene RC	32	82
	AM1			34	82

Table 3 (continued 8)

Compound	Method	Medium	Reference compound	φ°	Ref.
2,2'-dimethyl	UPS	gas		78	68
	UPS	gas		78	69
	UV	solution		70	65
	Rayleigh scattering	solution		48	166
radical cation	UV	solid argon	fluorene RC	31	a
4-chloro	electron diffraction	gas		44	97
	UPS	gas		53	68
	UPS	gas		50	69
	UPS	gas	fluorene	49	90
	AM1			40	90
radical cation	UV	$C_2H_4Cl_2$		20	90
	AM1			17	90
4,4'-dichloro	electron diffraction	gas		45	97
	UPS	gas		50	69
	UPS	gas	fluorene	47	90
	AM1			40	90
	NMR	nemat. phase		38	175
	NMR	nemat. phase		35	176
	X-ray diffraction	crystal		40	170
radical cation	UV	$C_2H_4Cl_2$		13	91
	AM1			18	90
4-bromo	UPS	gas		55	68
	UPS	gas	fluorene	54	90
	electro-optic	cyclohexane		30	98
	Raman intensity	CCl_4		22	98
	Rayleigh scattering	solution		22	166
	X-ray diffraction	crystal		19	170
radical cation	UV	$C_2H_4Cl_2$		20	90
4,4'-dibromo	UPS	gas	fluorene	51	90
	UV	solution		28	91
	electro-optic	cyclohexane		≤30	98
	Raman intensity	CCl_4		14	98
	X-ray diffraction	crystal		40	170
radical cation	UV	$C_2H_4Cl_2$		15	91
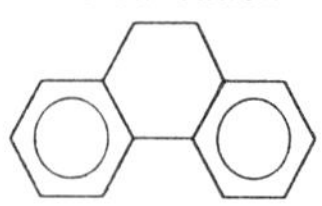	UPS	gas		21	68
	UPS	gas	fluorene	17	90
	AM1			14	90
	UV	ethanol	fluorene	18	165
	UV	solution		20	65
	Rayleigh scattering	solution		24	166
radical cation	UV	solid argon	fluorene RC	0	90
	AM1			0	90

[a] The equilibrium torsional angles were evaluated in this work using experimental data presented in Tables 1, 2, 4

The electron transitions to non-Koopmans' states have been found in the UV spectra of radical cations of many unsaturated compounds[5-7, 15, 16, 24, 32-36, 42, 43]. In particular, Haselbach *et al.*[5, 32-36] examined characteristics of electron transition to non-Koopmans' states in the UV spectra of radical cations of aromatic compounds and acetylenes. Transitions of this kind from single- occupied or double-occupied π-MO to vacant π-orbitals occur upon electron excitation. A number of characteristic features of these transitions have been found. Thus, the lowest-energy transitions are characterized by correspondence of Franck-Condon's profile of the absorption bands in the UV spectra for radical cation and the parent molecule[32, 34]. They have been established to obey the SDT-equation relating the energies of the singlet (S) and triplet (T) states of the molecule with the energy of the doublet (D) state of radical cation[32, 34]. The absorption bands in the UV spectra of 2-AP RC do not have features of this kind. They neither follow the STD equation, nor correlate with Franck-Condon's profile of long wave-length absorption bands of the parent molecules[91]. Moreover, between ΔE^0 and ΔE^+ there is a linear correlation, though with a low angle coefficient[91]. On this basis, it has been concluded[91] that the 2-AP RC absorption bands in the visible spectral region are caused by electron transitions to Koopmans' states perturbed either by σπ-interaction or by an ionization-induced change in equilibrium conformation.

Manifestations of orbital σπ(nπ) interactions for planar molecules can be detected by comparison of ΔE^0 and ΔE^+ of halogen-substituted benzenes and toluenes[19, 96]. The δE value increases from ~0.1 to 0.6 eV in the series Cl < Br < I. Examination of 4,(4')-halogen-substituted 2-phenylpyrroles and biphenyls has shown[90] that the UPS of the molecule and the UV spectrum of RC are also influenced by halogen (with the exception of fluoro derivatives) to a different extent. Consequently, δE may arise from non-equivalency of σπ interactions. However, the ΔE^0 variations in the series of compounds and the appearance of δE in other 2-AP or biphenyls can be explained in terms of unequivalent σπ interactions only on condition that the ionization-induced change in their equilibrium conformation is taken into account.

Therefore, the most general approach is that based on analysis of equilibrium torsional angle values, since it involves both direct (π-electronic) and possible indirect (σπ) mechanisms for the formation of ΔE^0, ΔE^+, and δE.

The equilibrium torsional angles of RC in solution were evaluated by an equation analogous to (2):

$$\Delta E^+(\varphi) = \Delta E^+(0^0) - 1.03 \sin^2(\varphi) \quad (3)$$

where $\Delta E^+(0^0)$ corresponds to RC of a planar reference compound[82, 90, 91]. The φ values obtained show (Table 3) that the 2-AP RC equilibrium conformation is close to periplanar. Unlike that of 2-AP molecules, the equilibrium torsional angle of RC does not depend much on the number of substituents near the interannular bond. And this explains the effect of the number of substituents on δE. The angles determined for MeCN solutions were close to those calculated for free radical cations in semi-empirical AM1 approximation[82, 90] (Table 3).

In studying the ionization effect it was of certain interest to compare equilibrium dihedral angles of radical cations and parent molecules. However, it does not seem possible to draw any conclusions on the basis of comparison of the φ angle values of 2-AP in the gas phase and of their RC in solution. This is caused by a high sensitivity of equilibrium geometry in the ring assemblies to the phase state. For example, the equilibrium torsional angle of biphenyl changes from 44^0 in the gas phase[97] to 30 degrees[98] (or 20 degrees[65]) in the liquid state and to 0^0 in crystal degrees[99]. Among heterocyclic ring assemblies 3,3'-bithienyl having the equilibrium torsional angle equal to 30^0 in the gas phase degrees[100], 22^0 in the nematic state[101] and 0^0 in crystal[102] can be cited as an example. The 2-AP φ values in solution were obtained by UV and NMR spectroscopy[83, 103-106]. According to CNDO/S calculation of the UV spectra of 2-PhP and 1-vinyl-2-PhP[83], the shift of long wave-length absorption maximum (band 1) when φ varies from 0^0 to 75^0 is described by a simple relation:

$$\lambda_{\hat{E}} = \lambda_{Ð} + \alpha\cos(\varphi) \quad (4)$$

where $\lambda_{Ð}$ is the wave length of the absorption band maximum in a conformation with $\varphi = 90^0$ (as shown in[83],

Table 4
UV spectroscopy data of some ring assemblies

Compound	Medium	Band 1		Band 2		Reference	α		φ		Ref.
		λ_{00}	λ_{max}	λ_{00}	λ_{max}	compound	00	max	00	max	
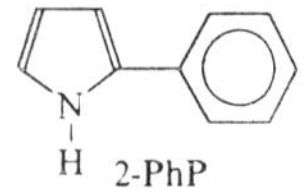 2-PhP	hexane	302	288	231	224	cpm[a]	76	71.5	21	26	91
	MeCN	303	286.5	231	223	cpm[a]	76	71.5	19	27	91
4'-methyl	hexane	302	283	231	224.5	cpm[a]	76	67	21	29	91
	MeCN	301	285.5	230	223	cpm[a]	76	67	21	21	91
4'-fluoro	hexane	297.5	279	229.5	222	cpm[a]	72	62.5	19	24	91
	MeCN	294		229.5		cpm[a]	72		26		
4'-chloro	hexane	309	294	233	225.5	cpm[a]	85.5	75	27	24	91
	MeCN	309.5	293	234	223	cpm[a]	85.5	75	28	21	91
4'-bromo	hexane	310.5	296	234	227.5	cpm[a]	84.5	77	25	27	91
	MeCN	308.5	296	234.5	224.5	cpm[a]	84.5	77	29	22	91
5-methyl	hexane		298.5		227.5	cpm[a]		78.5		25	91
	MeCN	309.5	297.5	234.5	225.5	cpm[a]	79	78.5	18	24	91
3-n-heptyl	MeCN	300	284	236.5	224.5	cpm[a]	74	64	31	22	91
	MeCN	300	284	236.5	224.5	DHBI	70.5	64.5	26	23	91
3-n-nonyl	hexane	294	282.5	234.5	228	cpm[a]	72	62	34	28	91
	MeCN	296.5	281.5	232.5	227	cpm[a]	72	62	27	28	91
	MeCN	296.5	281.5	232.5	227	DHBI	70.5	64.5	25	32	91
3-i-propyl	MeCN	301	283	235.5	227	cpm[a]	76.5	65.5	31	31	91
	MeCN	301	283	235.5	227	DHBI	70.5	64.5	22	30	91
3-methyl-2',5'-dimethyl	MeCN	281.5	271.5	241	234	DHBI	70.5	64.5	55	54	91
3-ethyl-2',5'-dimethyl	MeCN	285.5	271.5	241	233.5	DHBI	70.5	64.5	51	54	91
1-ethoxyethyl-2',5'-dimethyl	MeCN	277	268	247	237	DHBI	70.5	64.5	65	63	91
1-vinyl	cyclo-hexane		270		248	1-vinyl DHBI		25		28	91
	MeCN		266.5		247	1-vinyl DHBI		24		36	91
1-vinyl-3-methyl	cyclo-hexane		270		254	1-vinyl DHBI		25		50	91
	MeCN		271.5		252	1-vinyl DHBI		24		36	91
1-vinyl-3-ethyl	cyclo-hexane		269		254	1-vinyl DHBI		25		53	91
	MeCN		266		252	1-vinyl DHBI		24		54	91
1-vinyl-3-n-nonyl	hexane		267.5		252.5	1-vinyl DHBI		25		53	91
	MeCN		265		251	1-vinyl DHBI		24		54	91
1-vinyl-3-i-propyl	cyclo-hexane		265		252	1-vinyl DHBI		25		59	91
	MeCN		264		250.5	1-vinyl DHBI		24		56	91
1-vcnyl-3-ethyl-2',5'-dimethyl	hexane		261		250.5	1-vinyl DHBI		25		65	91
	MeCN		259.5		250	1-vinyl DHBI		24		66	91

Table 4 (continued)

Compound	Medium	Band 1		Band 2		Reference	α		φ		Ref.
		λ_{00}	λ_{max}	λ_{00}	λ_{max}	compound	00	max	00	max	
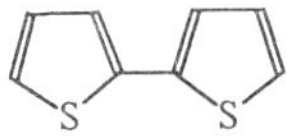	$CHCl_3$		302								126
	cyclo-hexane		302								130
	hexane		301								131
	benzene		305								131
	hexane		301								132
	solution		300								159
4,4'-dimethyl	$CHCl_3$		312								126
	MeCN		311								127
3,4'-dimethyl	$CHCl_3$		302								126
	MeCN		299								127
3,3'-dimethyl	$CHCl_3$		270								126
	MeCN		268								127
3,3'-diethyl	$CHCl_3$		279								128
3-methyl	$CHCl_3$		299								128
3-ethyl	$CHCl_3$		295								128
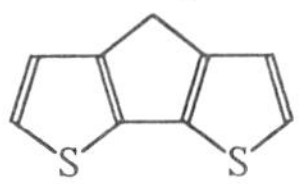	$CHCl_3$		311								128
	cyclo-hexane		312								130
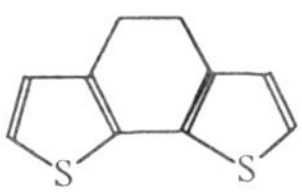	ethanol		320								129
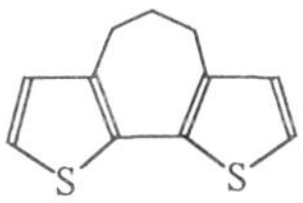	ethanol		315								129
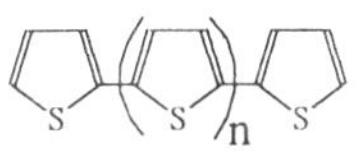											
n=1	$CHCl_3$		355								126
	hexane		350								131
	benzene		355								131
	solution		354								159
n=2	$CHCl_3$		390								126
	$CHCl_3$		390								142
	hexane		385								131
	benzene		391								131
	CH_2Cl_2		387								136
	solution		390								159
n=3	benzene		418								131
	CH_2Cl_2		403								136
	$CHCl_3$		416								142
n=4	$CHCl_3$		432								126
	$CHCl_3$		438								142
	CH_2Cl_2		432								136
	solution		434								159

[a] cpm - compound in polymer matrix.

this value corresponds to that of a short wave-length absorption band in the 2-AP UV spectrum (band 2), and α denotes the angle coefficient (nm). This equation can be employed for evaluation of 2-AP φ angles in solution[83] (Table 4) when the following conditions are fulfilled: 1) for determining the angle coefficient α it is necessary to employ the experimental values $\Delta\lambda = \lambda_{\hat{E}} - \lambda_D$ of planar forms; 2) in 3-alkyl-substituted 2-PhP the use should be made of the analytical value from the UV spectrum of planar DHBI in a given solution; 3) with 4'- or 5-substituted and unsubstituted 2-PhP, parameterization should be performed from spectra in a polymeric matrix (where these compounds are planar) with an account for a red phase shift found for planar systems; 4) for N-vinyl derivatives the needed values should be obtained from the UV spectra of 1-vinyl-DHBI in a given solution.

The angle values thus calculated indicate a more planar structure of 2-PhP and 4',5-substituted 2-PhP in solution compared with the gas phase (Table 3). In these compounds the change in the φ angle value in the phase transition is 6^0 on average. Analogous changes are also observed in compounds having alkyl substituents in the positions 3 and 3, 2', 5'. Steric hindrances produced by substituents in 1-ethoxyethyl-3-ethyl-2',5'-dimethyl-2-PhP and 1-vinyl-3-ethyl-2',5'-dimethyl-2-PhP prevent their planarization in going to solution. Thus, for 2-AP both in solution and gas a clearly defined dependence of the torsional angle on the number of substituents near the interannular bond is observed (Table 3). 2-AP in solution without substituents of this kind show an average angle of 24^0, for those with one, two and three substituents these values are 29^0, 53^0, and 65^0, respectively.

Similar equilibrium torsional angles thus obtained for the series of 2-AP (2-PhP, 4'-methyl-2-PhP, 5-methyl-2-PhP, 4'-fluoro-2-PhP, 4'-chloro-2-PhP, 4'-bromo-2-PhP) and biphenyl in solution (φ_S) allowed an independent check of the approach to be undertaken[83]. If in these compounds the φ_S values are close in fact, there should exist a linear relationship with a slope angle other than 1 between the wave length of the long wave-length band maximum in a given solvent ($\lambda_{\hat{E},S}$) and the corresponding value in the rigid matrix ($\lambda_{\hat{E},R}$) where $\varphi = 0^0$. This dependence is fulfilled in fact:

$$\lambda_{\hat{E},R} = (14 \pm 11) + (0.93 \pm 0.04)\lambda_{\hat{E},S}, \quad n = 7, r = 0.996. \qquad (5)$$

Its angular coefficient is close to a value calculated from estimates $\varphi_S \sim 24^0$ ($\cos(24^0) = 0.91$). Planar molecules obey linear relationship with a single slope angle ($\cos(0^0) = 1.0$). 3-Alkyl-substituted 2-PhP whose φ_S values markedly differ from 24^0, are not satisfied with the relationships found.

The φ angle values for 1-ethyl-2 AP in solution were evaluated from the extinction coefficient of their long wave-length band[104]. However, the authors imply the 2-PhP taken as a reference compound to be planar in solution, which does not agree with the data presented above[83]. For 1-vinyl-2-AP in solution the dihedral φ angles were independently determined from ^{1}H NMR spectra[105, 106]. For 1-vinyl-2-PhP and 1-vinyl-3-methyl-2-PhP the φ values found by NMR spectroscopy, are consistent with those determined from UV spectra (Table 3). According to NMR spectroscopy the φ values of 3-ethyl-, 3-isopropyl- and 3-ethyl-2',5'-dimethyl-1-vinyl-2-PhP are slightly lower, although one can observe the same tendency to increasing these values as the number of alkyl substituents near the C_2 - C'_1 bond enhances (Table 3). In 1-vinyl-DHBI the φ value obtained from ^{1}H NMR spectra is, on the contrary, slightly higher, than it follows from the data of UV spectroscopy.

Comparison of equilibrium torsional angles φ in 2-AP neutral molecules and ions in solution shows (Table 3) that in all cases the decrease in the φ values in going to RC exceeds 15^0. It should be noted that the more the molecular non-planarity, the more the molecule flattens on ionization (for instance, 1-vinyl-2-PhP and 1-vinyl-3-ethyl-2',5'-dimethyl-2-PhP, Table 3).

The results of AM1 calculations[91] of internal rotation potentials of 2-PhP and 3-ethyl-2-PhP molecules and RC allows an analogous conclusion to be drawn: on ionization a decrease in the equilibrium torsional angles must take place (Table 3). Moreover, the theoretical data suggest a considerable increase in the barriers of internal rotation when $\varphi = 90^0$ (from 1.22 to 9.65 kcal/mol for 2-PhP and from 0.86 to 4.75 kcal/mol for 3-ethyl-2-PhP) (Fig. 2).

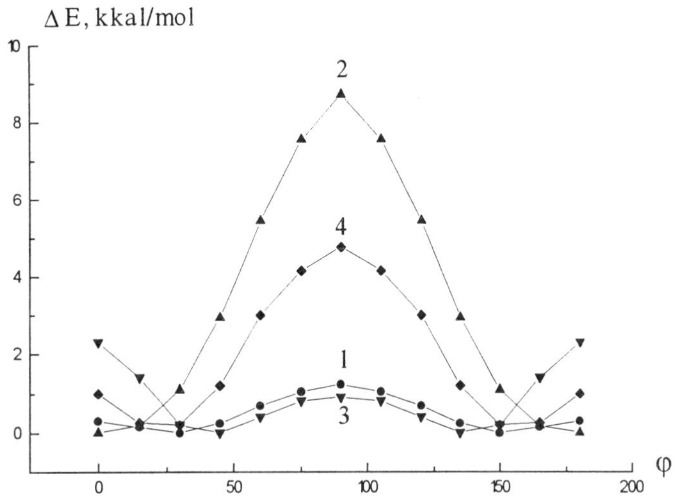

Fig. 2. The potential of internal rotation of 1) 2-phenylpyrrole molecule, 2) 2-phenylpyrrole radical cation, 3) 3-ethyl-2-phenylpyrrole molecule, 4) 3-ethyl-2-phenylpyrrole radical cation. Data of AM1 calculation.

The fact that the φ angle decreases as a result of ionization is treated[25, 72, 107-110, 125] as follows. When RC are formed in the ground state the electron is removed from the highest occupied π-MO (π-HOMO) of the molecule, which has a strong antibonding character between the rings, i.e., the double-electron destabilization of the molecular system is changed for a single-electron one. Theoretically it should lead to shortening the interannular bond and stabilization of the planar conformation. These predictions were supported by analysis of the Raman spectra of short-living RC of biphenyl, its alkyl derivatives[75] and fluorene[109]. The planar quinoidal character of the RC structures is in agreement with experimental data. Planarization of the biphenyl molecules on the electron transition to the excited singlet state[70] or to the triplet state[75, 111] is explained analogously. In the latter case this effect should be more pronounced since in the first excited state of biphenyls the orbital having a strong bonding character on the C_1 - C'_1 fragment is occupied.

3.3. Electronic analogs of phenylpyrroles.

The UV spectra of short-living RC of compounds of this type have been obtained only for 2-phenylselenophene (2-PhS)[88, 95] and 3-phenylthiophene (3-PhT)[91]. Like with 2-PhP, the transition energies in the visible region of the UV spectrum for 2-PhS and 3-PhT RC are characterized by blue shifts relative to those evaluated from UPS (Table 1, 2). For 2-PhS and 3-PhT the δE values are 0.23 and 0.38 eV, respectively. The high δE values (Table 2) indicate a non-planar equilibrium conformation of these compounds in the gas phase. The absence of information on 2-PhS and 3-PhT planar analogs does not allow the UPS data to be directly used in conformational analysis of these compounds. Nevertheless, it is possible to calculate the 2-PhS φ value. Since 2-PhS and 2-PhP are of similar structure, the equilibrium conformation of 2-PhS RC seems to be close to periplanar ($\varphi \sim 0^0$) likewise with 2-PhP RC. As the $\sin^2(\varphi)$ function does not change much around $\varphi \sim 0^0$, the $\Delta E^0(0^0)$ in equation (2) can be replaced by ΔE^+ in good approximation. This approach gives $\varphi \approx 30^0$ for gaseous 2-PhS (Table 3). This value is in agreement with that estimated ($\varphi = 31^0$) by chromato-structural analysis performed by comparison of the measured and calculated Henry constants of adsorption equilibrium[112]. For 3-PhT an analogous approach is not suitable as the 3-PhT RC φ value is likely to considerably differ from 0^0. It is only possible to use equation $\delta E \approx 1.03\sin^2(\Delta\varphi)$[95] for a rough estimation of the change in the torsional angle in going from a neutral molecule in the gas phase to radical cations in solution. For 3-PhT the $\Delta\varphi$ value is approximately 35^0. Data on the equilibrium conformation of 3-PhT in solution are variable. Thus, according to the UV spectra and dipole moments, the 3-PhT molecule is planar[113]. NMR spectroscopy, on the contrary, shows[114] the equilibrium torsional angle of 3-PhT in the nematic phase to be 24^0 (Table 3). From the data of chromato-structural analysis, free molecules of 2-phenylfuran and 3-phenylthiophene have a non-planar equilibrium conformation ($\varphi = 24^0$ and 29^0, respectively)[112]. The presence of fine structure of long wave-length absorption bands in the UV spectra points to a planar structure of these compounds in solution[115] (Table 3).

4. Heteroarylpyrroles

According to the data of laser photolysis and UPS, radical cations of substituted 2,2'-furyl(thienyl)pyrroles are characterized by a fairly high δE value (Table 2). This is a result of non-planar structure of molecules in the gas phase.

4.1. 2,2'-Furylpyrroles

AM1 quantum-chemical calculations have shown that the π_4- and π_6-MO energies in these compounds are conformationally dependent[91]. The same conclusion has been drawn in an analysis of orbital correlation diagrams[84]. Consequently, the difference between the first and third ionization potentials can be used as the ΔE^0 value of 2,2'-furylpyrrole derivatives. For unsubstituted 2,2'-furylpyrrole (2-FuP) the ΔE^0 and ΔE^+ values are nearly equal (Table 1, 2) which indicates a planar equilibrium conformation of both the molecules and RC (Table 3). The same is indicated by the presence of the first band vibrational structure in UPS[84] and from the results of *ab initio* 4-31G calculations[85]. These calculations show the planar conformation of molecules having *syn*-orientation of heteroatoms to be most stable. The planar *syn*-form does not contradict the experimentally determined dipole moment of 2-FuP in solution (Table 3). From these data it follows that an assembly of two planar rotational isomers (*cis*- and *trans*-) in the 92:8 ratio exists in the liquid phase[84]. The planar structure of 2-FuP in the gas phase enables its ΔE^0 value to be employed as $\Delta E^0(0^0)$ in equation (2), thus providing information on the equilibrium torsional angles of substituted 2,2'-furylpyrroles. The calculated φ values are presented in Table 3. They show the ethyl and vinyl groups to lead to the equal rotation of the heterocycles relative to each other. As with 2-AP, non-coplanarity of molecules of 2-FuP derivatives increases with increasing the number of substituents in positions near the interannular bond (Table 3). The molecules of 1-vinyl-2-FuP are more planar than those of 1-vinyl-2-PhP, analogously to the molecules of 2-FuP compared with 2-PhP (Table 3). But in contrast to 1-vinyl-2-PhP, the methyl group in the position 3 of 1-vinyl-2-FuP induces further increase in the φ angle. As shown by the total energy partitioning technique in AM1 approximation, this behavior is explained by the fact that the resonance component of the interannular bond energy of 2-FuP and its derivatives is 0.6-0.8 kcal/mol greater compared with that of 2-AP[91].

The equilibrium torsional angles of radical cations of substituted 2,2'-furylpyrroles in solution were evaluated by the equation (3), where $\Delta E^+(0^0)$ corresponds to 2-FuP RC[91]. As seen from Table 3, all RC of

2,2'-furylpyrroles have a conformation close to periplanar. Contrary to the parent molecules, the RC torsional angle does not depend much on the number of substituents near the interannular bond (Table 3).

The UV spectra of unsubstituted, 3- and 5'-methyl-2-FuP have been studied in[116] and those of 1-vinyl- and -1-vinyl-3-methyl-2-FuP in[91]. According to CNDO/S semi-empirical calculations, the energy of long wave-length electron transition in the UV spectrum of unsubstituted 2-FuP depends on the torsional angle in the same manner as with 2-PhP[91]. Besides, from the above calculations it follows that the 2-FuP *syn-* and *anti-*forms having the same φ angle are indistinguishable by UV spectra. Therefore the *syn-* and *anti-*isomers of these compounds can be revealed and examined only when their equilibrium conformations are quite different. This situation is realized with 1-vinyl-2,2'-furylpyrroles. Vinyl substitution at the pyrrole nitrogen atom considerably changes the shape of long wave-length absorption. Instead of a single intense transition there are three at once (Fig 3). Among them the shortest wave-length transition is assigned to that localized on the pyrrole ring[91]. The other two correspond to the *syn-* and *anti-*isomers of 1-vinyl-2-FuP. This assignment is supported by a significant change in the relative intensities of these bands on methyl substitution in the position 3 of 1-vinyl-2-FuP (Fig. 3).

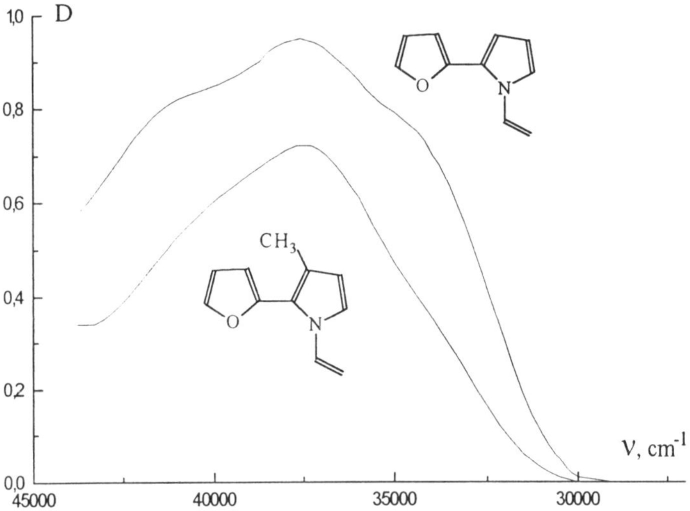

Fig. 3. UV spectra of 1-vinyl-2,2'-furylpyrroles in hexane.

In the UV spectrum of 2-FuP, in accord with the planar structure of its molecules, a fine structure of the long-wave absorption band is observed (Table 5). The energy of the corresponding transition is minimal among 2,2'-furylpyrroles. The introduction of alkyl or vinyl substituents in the positions 1 and 3 of 2-FuP leads to loss of the absorption band fine structure and to a blue shift of the band maximum[91]. These data suggest that the alkyl- and vinyl-substituted 2-FuP derivatives remain non-planar in the liquid phase. Consequently, ionization affects their equilibrium conformation to a much greater extent than the transition from gas to solution.

Table 5
UV absorption bands of 2,2'-heteroarylpyrroles and 2,2'-biimidazoles

compound	solvent	band 1	band 2	band 3
		ν, cm^{-1}		
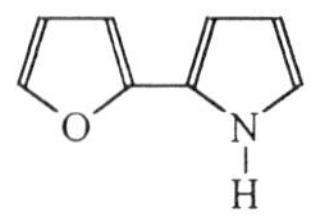	hexane	33300; 34950; 35800	-	~43500w
	MeCN	33500sh; 35050; 35700sh	-	w
	propylene carbonate	33400sh; 34900; 35000	-	w
	polymer film	32800sh; 34350; 35000sh	-	~43500w
1-ethyl	MeCN	35800	-	-
1-vinyl	hexane	34600sh	37600	40600sh
	MeCN	34800sh	37500	41000sh
	propylene carbonate	34700sh	37600	40800sh
1-vinyl-3-methyl	hexane	34500sh	37500	40400sh
	MeCN	34800sh	37800	40600sh
	propylene carbonate	34700sh	38100	40200sh
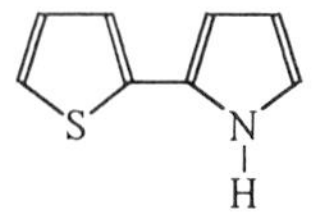	hexane	33300	-	-
	MeCN	32900	-	-
	polymer film	32000	-	-
1-vinyl	MeCN	34900	38000sh	40300
1-vinyl-3-methyl	MeCN	34500sh	38900	40200sh
1-vinyl-	hexane	34000sh	38900	40100sh
3-i-propyl	MeCN	34300sh	38900	39900sh
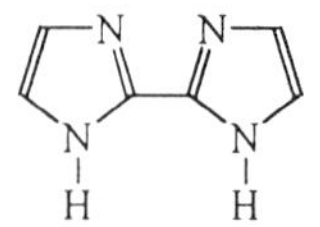	DMSO	33800; 35500; 36350; 37000sh	-	-
1-ethyl	MeCN	34000sh; 35600; 36450; 37000sh	-	-
1,1'-diethyl	MeCN	37000	-	-
1-vinyl	MeCN	34600	-	44500
1,1'-divinyl	hexane	33350	-	41300
	MeCN	34300	-	42100

4.2. 2,2'-Thienylpyrroles

The UPS for 2,2'-thienylpyrrole (2-TP) has been discussed in[84]. Band assignment in the low-energy region of the spectrum was based on qualitative analysis of molecular orbitals. As a result of bonding five-membered heterocycles to form a 2-TP molecule the parent $\pi_3(a_2)$-MO's are most perturbed. According to the data presented in[84], their interaction gives rise to π_6- and π_4-MO's with a low electron density on the pyrrole nitrogen atom. The unsaturated substituents in the 2-TP position 1 can influence the energy of π_6- and π_4-orbitals only at the expense of the inductive effect and the change in the dihedral angle between its heteroaryl fragments. Derivatives of the compound under consideration are presented by N-vinyl-2,2'-thienylpyrroles the UPS of which were examined in[89, 91]. The vinyl group has a low inductive effect, therefore its influence on the 2-TP π_6- and π_4-MO's is mainly determined by a change in the molecular conformation. Having a fairly high orbital electron density in the position 1, the 2-TP π_5-MO will be strongly perturbed by orbital interactions on N-vinyl substitution. The vinyl group effect on the 2-TP π-MO established by Koopmans' theorem is shown in Fig. 4. From its character it follows that predictions based on assignments[84] are fulfilled only in part. It is true for the 2-TP π_6- and π_5-MO's but not for the 2-TP π_4-MO. The 2-TP π_3-MO is destabilized on vinyl substitution. Consequently, it is this molecular orbital that presents bonding combination $\pi_3(a_2)$-MO of the parent heterocycles. A direct support for the new assignment is gained from UPS of 1-vinyl-3-methyl-2-TP in which further decrease in the energy gap between the first and forth ionic states is observed (Fig. 4).

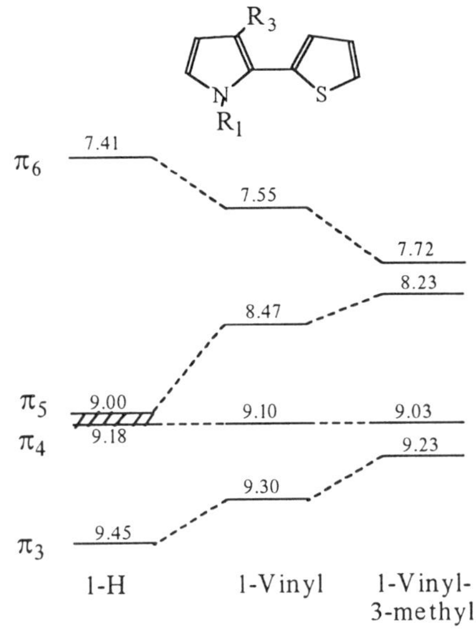

Fig. 4. Orbital correlation diagram of 2,2'-thienylpyrroles.

In the UPS of 1-vinyl-, 1-vinyl-3-methyl-2-TP and 1-vinyl-3-methyl-2-FuP the first band shows a shoulder on its high-energy slope (Table 1). This indicates that in the gas phase there is a mixture of two conformers which differ much from each other by equilibrium torsional angle.

It has been emphasized[84] that the conformationally dependent energy gap between the 2-TP ionic states (1.8-2.0 eV) is much greater than the difference between the first ionization potentials of pyrrole and thiophene (~ 0.8 eV). Being an evidence for a strong π-interaction between the two rings, this fact has led Galasso *et al.* [84] to a conclusion that in the gas phase 2-TP is planar. This conclusion does not contradict the results of *ab initio* STO-3G calculations of 2-TP potential of internal rotation[84]. However, analysis of the UV spectrum of short-living 2-TP π-RC[88, 95] has found a considerably increased ($\delta E \sim 0.3$ eV) transition energy of RC in the visible region (ΔE^+) compared with that calculated from UPS (ΔE^0) (Table 1, 2). As shown in previous sections, this strongly supports the non-planar structure of the parent molecule in the gas phase. Consequently, effective mesomeric interactions between unsaturated 2-heteroarylpyrrole fragments are also possible if the molecule has the non-planar equilibrium conformation.

It should be noted that in the visible region of the UV spectra of 2-TP and 1-vinyl-2-TP RC there are two close to each other absorption bands, and only one band is observed in the spectrum of 1-vinyl-3-methyl-2-TP RC (Table 2) [91]. Such a strong substituent effect on the UV spectrum of RC of 2-TP derivatives and the absence of effect in those of 2-furyl- and 2-phenylpyrrole RC point to conformational inhomogeneity of 2-TP radical cations. The 2-TP π-RC rotational isomer, characterized by maximum ΔE^+ value (2.32 eV) seems to be close to planar. This originates from the following: 1) ΔE^+ value of this 2-TP π-RC conformer is close to that of planar DHBI (2.35 eV), which is the highest value among ionic forms of the bicyclic ring assemblies studied (Table 2); 2) 2-TP δE is nearly equal to 2-PhP δE (Table 2); 3) The ΔE^+ value under consideration is highest in the series of π-RC of 2-TP derivatives. Consequently, under assumptions analogous to those made for 2-phenylselenophene, the 2-TP ΔE^+ value can be used as $\Delta E^0(0^0)$ and $\Delta E^+(0^0)$ in conformational analysis of both the molecules and π-RC of substituted 2-TP. The equilibrium torsional angles thus evaluated are presented in Table 3.

Analogously to 2-AP and 2-FuP, the conformers of free molecules characterized by a smaller φ values show an enhancing non-coplanarity with an increase in the number of substituents near the interannular bond. The same tendency is observed with RC of 2,2'-thienylpyrroles, but this is not valid for 2-AP and 2-FuP RC (Table 3). Probably, this is related to the difference in non-valent interactions in the *syn*- and *anti*-forms of substituted 2-TP π-RC. It should be emphasized that due to the high symmetry of $\Delta E^0(\varphi)$ and $\Delta E^+(\varphi)$ relationships it is impossible to distinguish *syn*- and *anti*-conformers of 2,2'-thienylpyrroles and their π-RC basing on only UPS and UV spectra.

Data on the 2-TP equilibrium comformation in solution were obtained on the basis of dipole moment and UV spectrum characteristics.

Considering only the 2-TP dipole moment in benzene the authors[84] give preference to neither of the two models: (a) equilibrium mixture of planar *syn*- and *anti*-conformers at a ratio of 64:36, and (b) one non-planar *syn*-conformer with $\varphi = 73^0$. The UV spectra of 2-TP and 2-TP 1-vinyl-(3-alkyl) derivatives have been studied in[116] and[91], respectively. Unlike the long wave-length band in 2-FuP, that of 2-TP in MeCN is structureless. Its red shift in transfer to polymeric matrix (1100 cm^{-1}, Table 5) exceeds that of planar 2-FuP (600 cm^{-1}, Table 5). These data indicate a non-planar structure of 2-TP in solution. The characteristics of 2-TP long wave-length bands found for the liquid and solid phase obey correlation equation (5) obtained for ring assemblies with $\varphi \sim 24^0$. Consequently, 2-TP is characterized by $\varphi \sim 25^0$ in solution and becomes planar in a polymeric matrix. The introduction of a vinyl group into the 2-TP position 1 leads to the appearance of three intense bands instead of one band in the low-energy UV region (Table 5). Of these, the shortest wave-length band corresponds to the transition localized on the pyrrole fragment. Other two bands result from its *syn*- and *anti*-isomerism. This assignment is justified by the dependence of relative intensities on the presence or the absence of pyrrole ring substituent. Thus, on the introduction into the position 3 of methyl group instead of the hydrogen atom the band observed as a maximum in the spectrum of 1-vinyl-2-TP, becomes a shoulder and *vice versa* (Fig. 5).

5. 2,2'-Bithiophenes and some analogs

Recently the equilibrium conformation of 2,2'-bithiophene (α2T) has been reinvestigated by means of gas phase electron diffraction[118]. According to the data of this direct experimental method, α2T exists in the gas phase as a mixture of *syn*- and *anti*-like non-planar conformers with $\varphi = 36\pm5^0$, 148 ± 3^0 and conformational weights of 44±4 and 56±4%, respectively[118]. The experimental data are in agreement with the results of *ab initio* 3-21G* and 6-31G* calculations[118-120] and AM1 semi-empirical method[121]. The little difference between the angles of deviation from planar *syn*- and *anti*-forms (32^0 and 36^0) inhibits the use of UPS in a study of α2T *syn-anti*-isomerism. However, these results can be employed in conformational analysis of α2T π-RC. Taking an average dihedral angle φ as 34^0 and making use of the energy gap ΔE^0 it is possible to calculate $\Delta E^0(0^0)$ for the unstable planar conformer. Then the value obtained can be compared with the conformationally dependent energy of transition to π-RC Koopmans' state. According to well-consistent data reported in[88, 95, 123] the average ΔE^0 value is 1.85 eV. From equation (2), applicable to α2T in the $0 \leq \varphi \leq$

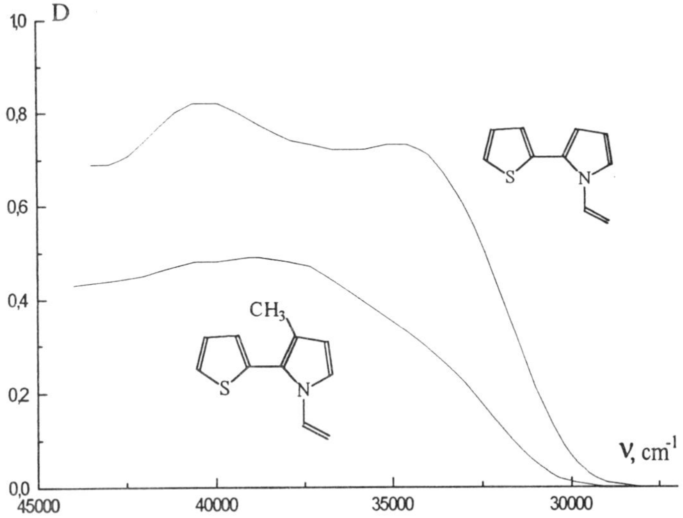

Fig. 5. UV spectra of 1-vinyl-2.2'-thienylpyrroles in MeCN.

70^0 range, $\Delta E^0(0^0)$ is equal to 2.17 eV with ΔE^0 = 1.85 eV and $\varphi = 34^0$. The energy of the long wave-length absorption band maximum recorded by laser photolysis in the visible region of the spectrum of short-living α2T π-RC is 2.13 eV on average[88, 95, 124]. Consequently, the α2T π-RC is practically planar, and the transition occurs between the ionic states having vacancies on the π_6- and π_3-MO's. An independent conclusion that dimerization-resistant 5,5'-substituted α2T π-RC in solution are planar has been made[110, 125] on the basis of EPR spectroscopy data. These data provide evidence for the existence in the liquid an equilibrium mixture of planar *syn*- and *anti*-isomers. The same conclusion follows from the data of EPR and ENDOR spectroscopy for π-RC of sterically unhindered oligothiophenes[110].

The data available for α2T $\Delta E^0(34^0)$[88, 95, 123] and $\Delta E^+(0^0)$ [88, 95, 124] allow conformational analysis of 2,2'-bithiophene alkyl derivatives whose UPS have been discussed in[119]. It will suffice for the parameterization of equation $\Delta E^0(\varphi) = a\cos(\varphi) + b$, used in UPS-based evaluation of dihedral angles of ring assemblies consisting of two uniform rings X-X[69, 87]. For α2T derivatives this equation can be written as follows:

$$\Delta E^0(\varphi) = 0.49(\text{eV}) + 1.64(\text{eV})\cos(\varphi) \qquad (6)$$

The angular coefficient in this equation is close to that in an analogous relationship for biphenyl[69]. The φ values for dimethyl-α2T derived from equation (6) are presented in Table 3. The presence of methyl groups in the α2T positions 4 and 4' does not influence much the gas phase equilibrium conformation. This is in agreement with the results of *ab initio* 3-21G* calculations[119] (Table 3). Basing on the dependence of 4,4'-dimethyl-α2T orbital energies calculated in 3-21G* basis on the dihedral angle and UPS data, Distefano *et al.*[119] estimated the φ value for the molecule of 3,4'-dimethyl-α2T (Table 3). The result is in good agreement with the φ value estimated by equation (6). However this contradicts the data of *ab initio* 3-21G* calculations[120]. From the latter it follows that the free molecule of 3,4'-dimethyl-α2T has a single stable conformer with $\varphi = 71^0$. This clearly disagrees with the UPS data[119]. The *ab initio* calculation[119, 120] predicts the existence of a single conformer with nearly orthogonal rings for a sterically hindered molecule of 3,3'-dimethyl-α2T (Table 3). Qualitatively this is compatible with the experimental data[119].

If the equilibrium torsional angle in the ring assembly in solution (φ_s) is assumed to be directly proportional to that in the gas phase (φ_g): $\varphi_s = r\ \varphi_g$, it is clearly seen that $\cos(\varphi_s) \approx a + b\cos(\varphi_g)$, $a = \sin(r\pi)/r\pi$, $b = 2r^2a/(1-r^2)$. Under this condition it follows from equation (4) that the wave length of the low-energy band maximum of ring assembly in solution (λ_{max}) will be linearly related to $\cos(\varphi_g)$. The λ_{max} values for 4,4'-, 3,4'- and 3,3'-dimethyl-α2T in $CHCl_3$ and MeCN reported in[126, 127] are linearly related to $\cos(\varphi_g)$ derived from equation (6):

$$\lambda_{max} = 260(\text{nm}) + 62(\text{nm})\cos(\varphi_g), \quad r = 0.995, \quad n = 6 \qquad (7)$$

This relationship enables evaluation of the equilibrium torsional angles of some other α2T disubstituted derivatives in the gas phase for which only λ_{max} values are known[128, 129] (Table 3, 4). For monoalkyl-substituted derivatives α2T the parameters of equation (7) can be calculated from the known λ_{max} values of unsubstituted ($\varphi_g = 34^0$) and β,β'-CH_2-bridged 2,2'-bithiophene ($\varphi = 0^0$) [126, 128, 130-132]:

$$\lambda_{max} = 250(\text{nm}) + 62(\text{nm})\cos(\varphi_g) \qquad (8)$$

Knowing the physical sense of equation (8) coefficients and taking 3,3'-dithienylmethane (λ_{max} = 237 nm[130]) as the orthogonal form in solution, it is possible to determine r to be 0.84 and to calculate the equilibrium dihedral angles for the liquid phase $\varphi_s = r\varphi_g$. The φ_s values (Table 3) obtained for dimethyl-α2T are in good agreement with those calculated by the MMP2 method with modeling the solvent effect for *anti*-conformers[127]. However, in the UV spectra of these compounds there are no indications for the existence of stable *syn*-conformers predicted by calculation.

At the present time a great attention is attracted to conductive polymers, in particular, those based on thiophene derivatives[133]. Due to investigations of conductivity and its dependence on the structure of polymers a large body of information on thiophene oligomers and their radical cations has been accumulated. The data available allow assignment of transitions in the UV spectra of radical cations of oligothiophenes, which is ambiguous so far even for α2T. In[88] two transitions in the visible region of the spectrum of α2T π-RC were assigned to the transitions from deep lying occupied π-MO's onto a singly occupied orbital, whereas in[134] the same transitions were assigned to highest occupied MO (HOMO) → singly occupied MO (SOMO) and SOMO → lowest unoccupied MO (LUMO) (see scheme on next page). The assignments of the two types suffer from certain drawbacks. On the one hand, as follows from the first part of this section, the consideration of the singly occupied orbital and next double-occupied π-MO's leads to an incomplete result. On the other hand, a model not including unoccupied π-MO requires a very strong interaction in the equilibrium ionic state of fragmentary π-MO's of the thiophene (T) molecule and radical cation. Therefore, an approach combining the above two would be more adequate. According to this approach the UV spectrum of π-RC of α2T and other oligomers may contain transitions of several types: 1) to Koopmans' state with a vacancy on deep lying π-MO's, 2) to non-Koopmans' state, in particular, SOMO → LUMO, 3) between close lying occupied π-MO's and a singly occupied π-MO.

The whole body of available data on the UV spectra of oligothiophene RC's can be described by the scheme depicted in Figure 6.

From this Figure it follows that the intense band with energy in maximum at ~ 2.9 eV in the transient spectrum of α2T π-RC[88, 124] corresponds to excitation to the first non-Koopmans' state, as interpreted in[134]. This assignment is favored by results reported in[43] according to which the transitions of this kind are most intense and sensitive to topological changes among all long wave-length transitions of π-RC. The energy of

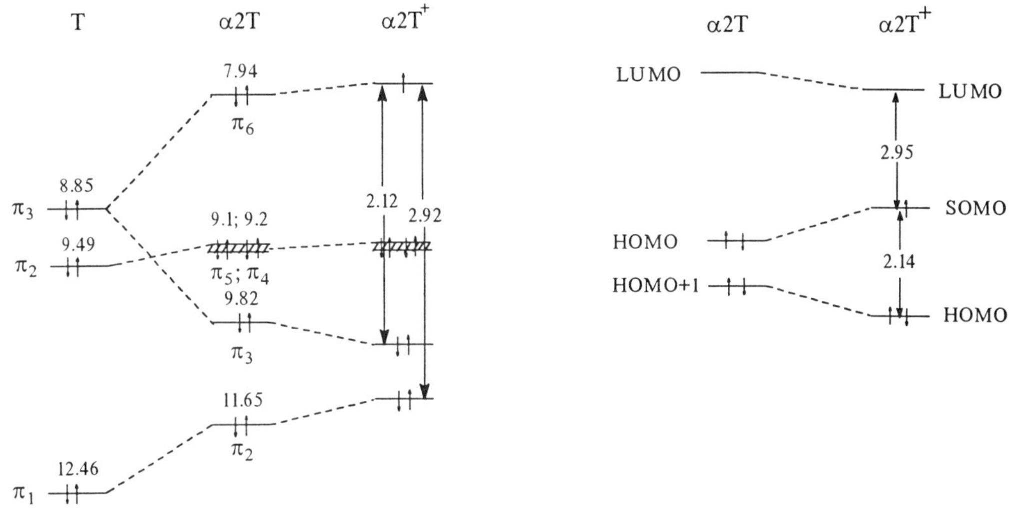

α2T π-RC band in maximum ~ 2.1 eV corresponds to transition to Koopmans' state with a vacancy on the π_3-MO as follows from comparison with the UPS data[87, 94]. In the UV spectrum of 2,2',5',2''-terthiophene (α3T) π-RC the transition between singly occupied and unoccupied π-MO is shifted towards the low-energy region. The corresponding band is overlapped with a less intense one caused by Koopmans' transition (Fig. 6). For this reason in the visible region of the spectrum of α3T π-RC there is a broad absorption band with a maximum of 2.25 eV and a less distinct shoulder in the 2.5-2.6 eV range[135].

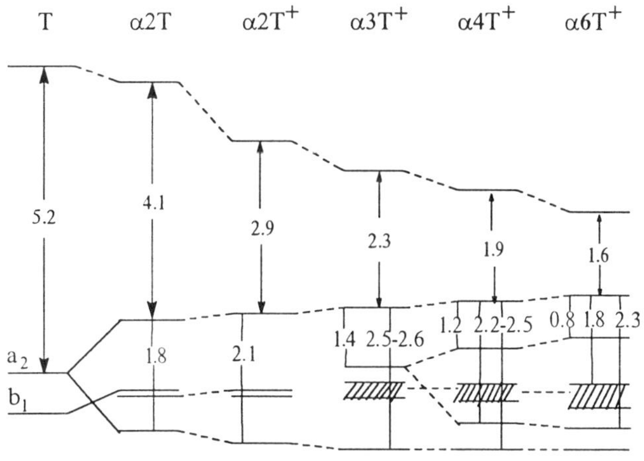

Fig. 6. The assignment of low energy transitions in the oligothiophene radical cations.

Simultaneously, a transition to Koopmans' state with a vacancy on the α3T π_8-MO appears in the near IR region of the α3T π-RC spectrum[135]. This orbital arises from the interaction of α2T π-MO with $\pi_3(a_2)$-MO of thiophene. Further increase in the number of RC rings (up to 4 in α4T and 6 in α6T) leads to a shift of the transition on the unoccupied π-MO towards the low-energy region (Fig. 6). At the same time, on the short wave-length slope of the corresponding absorption band one can see a low-intense shoulder (α4T RC, 2.2-2.6 eV) or clearly defined shoulder and a maximum (α6T RC 1.8 and 2.25 eV, respectively). The authors[135] do not discuss these features of UV spectra obtained by laser photolysis. However, these shoulders are of importance in transition assignments, all the more that some additional bands in the UV spectra of α4T and α6T π-RC are recorded in the steady-state conditions[136]. These are transitions to Koopmans' state (Fig. 6). Assignment of the intense band in the visible region of the spectra of oligothiophene RC to the transition from a singly occupied to unoccupied orbital is confirmed by linear correlation of the energy of its maximum (E^+ Table 2) and the energy of low-energy transition in the UV spectrum of the parent molecule (E^0, Table 4):

$$E^+(eV) = 1.12E^0(eV) - 1.63 \quad r = 0.999 \quad n = 5 \tag{9}$$

By this relationship it is possible to approximately determine the energy of transition to non-Koopmans' state of π-RC of other oligothiophenes.

In order to empirically verify adequacy of the assignment of transitions to Koopmans' state of oligothiophene π-RC, data of UPS are needed. The energy gaps recorded in the solid phase are most suitable for a direct comparison, since the α-oligothiophene molecules are nearly planar in crystal[125, 137-140], as well as π-RC in solution[110, 125]. The UPS of α2T - α8T and α3βα3T solid films have been reported in[141-143]. The most informative spectrum has been recorded for only one film, that of α3βα3T. In the low-energy region of this spectrum there are four bands with energies of maxima 1.5, 3.0, 3.9; 4.0 eV and 4.8; 4.5 eV corresponding to the ionic states[142, 143]. Fujimoto *et al.* have developed an interesting approach to analysis of UPS of oligothiophenes. The α3βα3T molecular geometry was optimized by the semi-empirical MNDO method. The intermediate thiophene ring bonded in the β,β'-positions to the α3T terminal fragments has been found to be orthogonal with respect to these fragments. With this geometry UPS of α3βα3T thin film is superposition of the spectra of two planar α3T and thiophene[142-144]. From α3βα3T UPS the energies of allowed α3T π-RC transitions to Koopmans' states must be 1.5 and 3.0 eV. Comparison with the flash photolysis data[135] shows that among the predicted energies only that of low-energy transition is close to the observed values. As next transition to Koopmans' state is conformationally dependent, the absence of correlation should be related to the fact that the α3βα3T geometry in the solid state differs from that found by MNDO for the free molecule. Thus, sterically hindered 4,4',3'',4'''-tetramethyl-2,2':5',2'':5'',2'''-tetrathiophene which, by analogy with 3,3'-dimethyl-α2T, must show a nearly orthogonal position of the central rings in the gas phase, remains non-planar in solution, but acquires a planar structure in crystal[139]. The presence of interaction in α3βα3T in solution is evidenced by data of the UV spectra of oligothiophenes. The maximum of α3βα3T long wave-length absorption band (370 nm[143]) is intermediate between the corresponding α3T and α4T maxima (Table 4). Thus, the UPS data for α3βα3T in the solid phase cannot directly be compared with the spectrum of α3T π-RC.

The photoelectron spectra of α-oligothiophenes (α3T, α4T and α5T) in the gas phase have been investigated[123]. The energy gaps between π-MO's of these molecules in the gas phase (Fig. 7) are slightly smaller than the energies of transitions to Koopmans' state of their π-RC in solution (Fig. 6). As follows from the above discussion, this is connected with an ionization-induced planarization of molecules. Thus, the UPS data for the gas phase support the assignment of transitions in π-RC of α-oligothiophenes.

Data on the UV spectra of molecules and π-RC of α-oligothiophenes containing terminal methyl groups have been considered in[145]. Analysis of the UV spectra of α-oligothiophenes in the solid phase has been carried out in[145, 146]. In accord with the planar structure and symmetry (C_{2h} or C_{2v}) of these compounds, the long wave-length absorption band exhibits a fine structure caused by vibrational splitting of the $S_0 \rightarrow S_1$ electron transition.

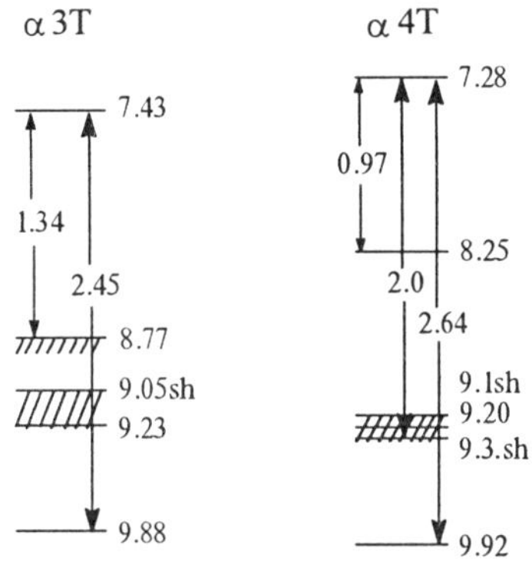

Fig. 7. Data of photoelectron spectroscopy for some oligothiophenes[123].

6. 2,2'-Biimidazoles and some biphenyls

6.1. 2,2'-Biimidazoles

Some features of the orbital structure of these compounds will be considered with 1,1'-divinyl-2,2-biimidazole (1,1'-divinyl-α2Im) as an example:

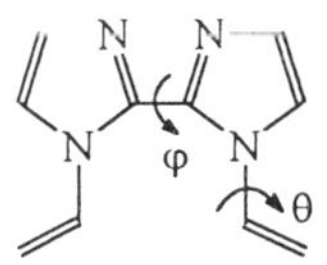

In the 7-12 eV region of the photoelectron spectrum of this compound there are four main bands arisen from ionization from six upper occupied π-MOs and two quasi-degenerated orbitals of non-bonding electron pairs. The sequence of 1,1'-divinyl-α2Im molecular orbitals is demonstrated by correlation diagram constructed on the basis of UPS data and semi-empirical quantum chemical calculations (AM1, OVGF AM1) (Fig. 8). The interaction of two π_4-MO's of 1-vinyl-imidazole (1-vinyl-Im) leads to the formation of bonding π_5-MO and anti-bonding π_8-MO of 1,1'-divinyl-α2Im. The energy gap between the ionic states corresponding to these π-orbitals is mainly determined by conformation of the compound under consideration. The two π_3-MO's of 1-vinyl-Im form a quazi-degenerated set of π_6- and π_7-MO's of 1,1'-divinyl-α2Im. In UPS the set corresponds to an intense unresolved band with maximum at 9.06 eV. The second quazi-degenerated set of π_3- and π_4-MO's of 1,1'-divinyl-α2Im arises from π_2-MO's of the parent heterocycle and leads to the appearance of band overlapping in the 10.3 eV region. The degeneracy of $\pi_{6,7}$- and $\pi_{3,4}$-MO's is indicated by two facts: 1) two-fold intensity drop in the above bands in UPS of 1-vinyl-α2Im where the band maxima are 9.19 and 11.26 eV, 2) complete disappearance of high-energy band in UPS of 1,1'-diethyl-α2Im and 1-ethyl-α2Im (Table 1). To the ionic states characterized by conformationally dependent energies, there correspond bands with energy in maximum at 7.97(π_7) and 9.73(π_5) eV in UPS of 1-vinyl-α2Im; 7.65(π_6) and 9.22(π_5) eV in UPS of 1,1'-diethyl-α2Im; 7.77(π_6) and 9.43(π_5) eV in that of 1-ethyl-α2Im (Table 1).

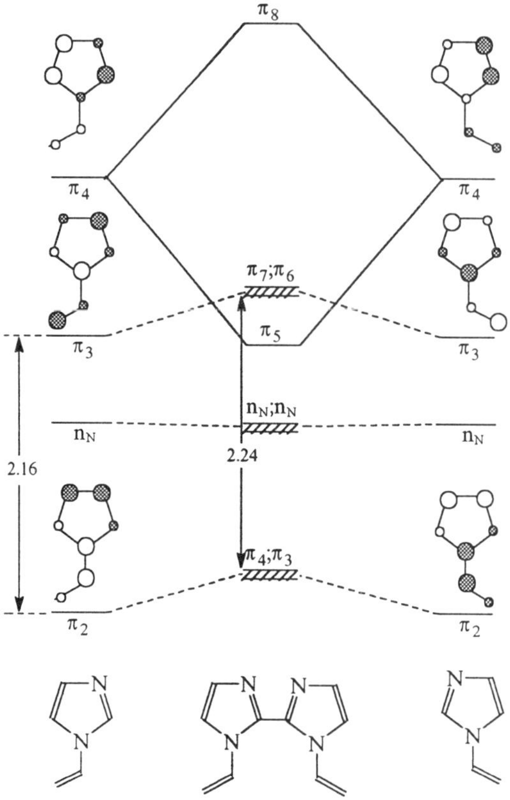

Fig. 8. Orbital correlation diagram of 2,2'-biimidazole vinyl derivatives.

The equilibrium conformation of 1,1'divinyl-α2Im in solution has been studied in[147]. According to the data obtained these molecules exist as non-planar *anti*-conformers with a rotation angle (φ) of heterocycles relative to each other of approximately 150^0. The vinyl groups of 1,1'-divinyl-α2Im are oriented in opposite directions, but equilibrium angle of their plane rotation relative to the heterocycles (θ) is not reported. An analogous 1-vinyl-Im orientation of vinyl groups is likely to be anticipated, since the dipole moment of 1-vinyl-Im was used in the analysis of equilibrium conformation. The NMR data obtained later[148], corroborate the *anti*-conformation of 1,1'-divinyl-α2Im. Some information concerning the θ angle value can be derived from UPS data. According to Fig. 8, a change of the θ angle affects the $\pi_{6,7}$- $\pi_{3,4}$-MO energies due to π-interactions. Therefore, if the orientation of 1,1'-divinyl-α2Im vinyl groups relative to the heterocycle planes is markedly different from that in 1-vinyl-Im, the energy gaps between the corresponding ionic states of these compounds will differ much too. From UPS data the differences are equal to 0.08 eV, i.e., they are within the range of ΔE^0 determination error (Fig. 8). Consequently, the vinyl group orientation remains unchanged when 1-vinylimidazole fragments are united to make a molecule of 1,1'-divinyl-α2Im. It should be noted that due to the specific electron density distribution in π_8- and π_5-MO of 1,1'-divinyl-α2Im (Fig. 8) the vinyl groups will affect the energy of these MO's only at the expense of changing the torsional angle φ.

In the UV spectra of α2Im, 1-ethyl-α2Im and 1,1'-diethyl-α2Im in solution only one long wave-length band has been fixed (Table 5). In the spectra of α2Im these bands display fine structure and do not differ much in energy. As mentioned previously, the presence of distinct structure of the long wave-length absorption band in the UV spectra of conformationally labile ring assemblies in the condensed phase is indicative of planar equilibrium conformation. Consequently, α2Im and 1-ethyl-α2Im are planar in solution analogously to α2Im in the crystal state (X-ray diffraction) [149]. The UV spectrum of 1,1'-diethyl-α2Im is characterized by a blue shift of major maximum of the band under consideration compared to α2Im and 1-ethyl-α2Im, and by complete loss of fine structure. This points to a non-planar equilibrium conformation of 1,1'-diethyll-α2Im in solution. Unlike 2-FuP and 2-TP, in the UV spectra of α2Im vinyl derivatives no signs of *syn*- and *anti*-isomerism are observed. It is likely to be due to difference in the potentials of internal rotation in compounds of these classes.

A conclusion on equilibrium conformation of 1-ethyl-α2Im in the gas phase can be drawn judging by the fact that its molecule in solution is planar, and basing on the data of UPS and π-RC absorption spectra. In the UV spectrum of 1-ethyl-α2Im short-living π-RC recorded by laser photolysis the energy of transition to Koopmans' state with a vacancy on the π_5-MO is 1.98 eV. The energy calculated from UPS is 1.66 eV. It follows that the 1-ethyl-α2Im molecule is non-planar in the gas phase. As 1-ethyl-α2Im and 1-vinyl-α2Im are ring assemblies of type X-Y, evaluation of equilibrium torsional angles of their molecules and π-RC of 1-vinyl-α2Im was carried out by equation (2) parameterized by the spectrum of 1-ethyl-α2Im planar RC. The values obtained are presented in Table 3. Equation (2) can also be applied to the spectra of symmetrically substituted α2Im (X-X) since their δE values do not exceed 0.4 eV and the equlibibrium torsional angles are within the $0 \div 70^0$ range (Table 3).

6.2. Biphenyl radical cations

A review of the research on the electronic structure and conformation of biphenyl RC has been performed in the section 2. Evaluations of equilibrium torsional angles reported in[89] are presented in Table 3. These values were determined by an equation found for ring assemblies of type X-Y. Owing to small φ values this equation is also applicable to systems X-X (Table 3). RC of geometrically labile biphenyl derivatives remain non-planar in solution, though their equilibrium torsional angle is smaller than that of the parent molecule in the gas phase.

7. Conclusions

The results of researches discussed in the present review allow a series of conclusions common to all the molecules and π-radical cations of heterocyclic and aromatic ring assemblies to be drawn. The overwhelming majority of ring assemblies are non-planar in the gas phase, irrespectively of whether they are symmetrical like oligothiophenes and biphenyls, or contain various unsaturated rings like aryl(heteroaryl)pyrroles. The transfer of ring assemblies from gas to the condensed phase reduces the equilibrium torsional angle. The extent to which the angle varies depends on the structure of a given compound and the presence of substituents. One-electron ionization leads to an analogous, but a more pronounced effect.

Due to the difference in equilibrium torsional angles the parent molecule of ring assembly and its π-radical cation show a slightly different position of energy levels: the energy of transition to Koopmans' state with a vacancy on a deep lying π-MO of π-RC of non-planar ring assemblies is higher than that calculated from UPS.

Since the π-radical cations of 2,2'-furylpyrrole, 2,2'-bithiophene, 4,5-dihydrobenzo[g]indole, 9,10-dihydrophenantrene and, probably, 2,2'-thienylpyrrole are planar, their characteristics can be employed in solving the problems of spectroscopy and stereochemistry of ionic forms. For example, the energy transition to certain Koopmans' state of planar π-RC is necessary in evaluating the equilibrium torsional angles of other short-living π-RC of the same series. Direct measurement of these angles cannot be made by steady-state physics and chemical methods. The transition energy under consideration is also of help in conformational analysis of the parent molecules by means of UPS. The role of this characteristics is of even greater importance in the absence of planar member of a set of compounds.

The accumulated information on planar neutral and ionic forms of ring assemblies consisting of either various number of similar rings or of various carbo- and heterocycles will make it possible to adequately evaluate mesomeric orbital effects and to reveal their structural dependence. Although limited the literature available on symmetrical ring assemblies and their π-RC[25, 88, 95, 124, 135, 150-152] shows that this goal can be attained. (With 2,2'-bipyridine and 2,2'-bipyrrole the use is made of the energy gap between the corresponding π-MO's found from the UPS of these planar molecules; for *p*-terphenyl a value found from UPS in the solid state where this compound is planar at T ~ 300 K, was employed[153]):

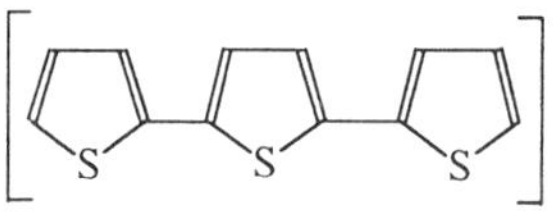
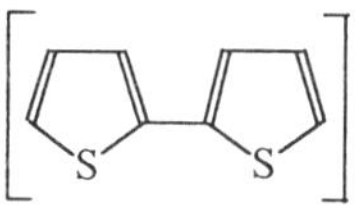
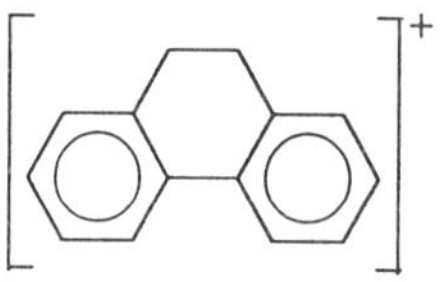

$\Delta E^{a}_{\pi_i,\pi_j}$, эВ 2.5 - 2.6 2.13 1.93

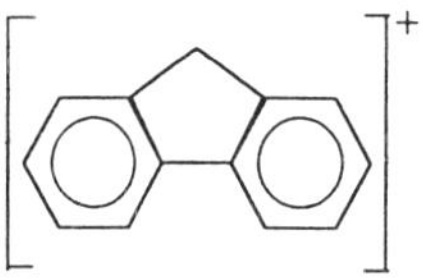
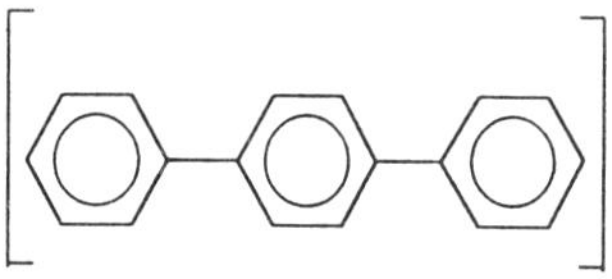
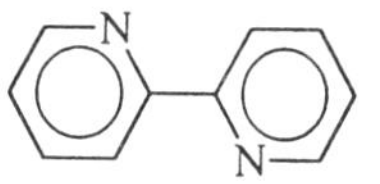
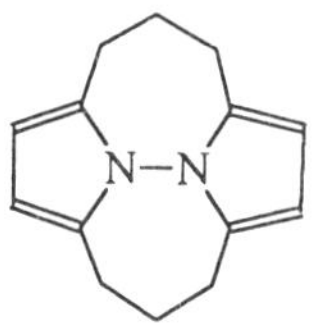

1.94 2.1 1.82 1.22

From the above data it is seen that orbital interactions are most effective in planar 2,2'-bithiophene derivatives. This appears to be responsible for the interesting electrophysical properties of oligothiophenes.

References

1. Brown, R. S.; Jorgensen, F. S. In: *Electron Spectroscopy;*. Academic Press: London, 1984; Vol 5, p. 1.
2. Dixon, R. N.; Duxbury, G.; Horani, M.; Rostas, J. *J. Molec. Phys.* **1971,** *22,* 977.
3. Herzberg, G. *Quart. Rev.* **1971**, *25*, 201.
4. Allan, M.; Maier, J. P. *Chem. Phys. Lett.* **1975**, *34*, 442.
5. Haselbach, E.; Bally, T.; Gschwind, R.; Klemm, U.; Lanyiova, Z. *Chimia.* **1979**, *33*, 405.
6. Dunbar, R. C.; Klein, R. *J. Am. Chem. Soc.* **1976**, *98*, 7994.
7. Khan, Z. H. *Spectrochim. Acta (A)* **1989**, *45A*, 253.

8. Lurito, J. T.; Andrews, L. *Chem. Phys.* **1985**, *97*, 121.
9. Shida, T.; Nosaka, Y.; Kato, T. *J. Phys. Chem.* **1978**, *82*, 695.
10. Dunbar, R. C. *Chem. Phys. Lett.* **1975**, *32*, 508.
11. Andrews, L.; Keelan, B. W. *J. Am. Chem. Soc.* **1980**, *102*, 5732.
12. Keelan, B. W.; Andrews L. J. *Am. Chem. Soc.* **1981**, *103*, 822.
13. Bondybey, V. E.; Miller, T. A.; English, J. H. *J. Am. Chem. Soc.* **1979**, *101*, 1248.
14. Khan, Z. H. *Can. J. Spectrosc.* **1984**, *29*, 63.
15. Khan, Z. H. *Z. Naturforsch. (A).* **1984**, *39A*, 668.
16. Khan, Z. H. *Z. Naturforsch. (A).* **1987**, *42A*, 91.
17. Petrushenko, K. B.; Turchaninov, V. K.; Vokin, A. I.; Ermikov, A. F.; Frolov, Yu. L. *Khim. Geterotsikl. Soedin.* **1985**, 750. *Chem. Abstr.* **1986**, *104*, 5421p.
18. Zahradnik, R.; Carsky, P.; Slanina, Z. *Collect. Czech. Chem. Commun.* **1973**, *38*, 1886.
19. Fu, E. W.; Dymerski, P. P.; Dunbar, R. C. *J. Am. Chem. Soc.* **1976**, *98*, 337.
20. Dymerski, P. P.; Fu, E.; Dunbar, R. C. *J. Am. Chem. Soc.* **1974**, *96*, 4109.
21. Dunbar, R. C. *J. Phys. Chem.* **1979**, *83*, 2376.
22. Dunbar, R. C. *J. Am. Chem. Soc.* **1973**, *95*, 6191.
23. Asamoto, B.; Dunbar, R. C. *Int. J. Mass Spectr.* **1988**, *86*, 387.
24. Andrews, L.; Kelsall, B. J.; Blankenship, T. A. *J. Phys. Chem.* **1982**, *86*, 2916.
25. Andrews, L.; Arlinghaus, R. T.; Payne, C. K. *J. Chem. Soc. Faraday Trans. 2.* **1983**, *79*, 885.
26. Andrews, L.; Friedman, R. S.; Kelsall, B. J. *J. Phys. Chem.* **1985**, *89*, 4550.
27. Andrews, L.; Friedman, R. S.; Kelsall, B. J. *J. Phys. Chem.* **1985**, *89*, 4016.
28. Kelsall, B. J.; Andrews, L. *J. Am. Chem. Soc.* **1983**, *105*, 1413.
29. Keelan, B. W.; Andrews, L. *J. Am. Chem. Soc.* **1981**, *103*, 829.
30. Andrews, L.; Harvey, J. A.; Kelsall, B. J.; Duffey, D. C. *J. Am. Chem. Soc.* **1981**, *103*, 6415.
31. Andrews, L.; Kelsall, B. J.; Harvey, J. A. *J. Chem. Phys.* **1982**, *77*, 2235.
32. Forster, P.; Gschwind, R.; Haselbach, E.; Klemm, U.; Witz J. *Noiv. J. Chem.* **1980**, *4*, 365.
33. Haselbach, E.; Klemm, U.; Buser, U.; Gschwind, R.; Jungen, J.; Kloster-Jensen, E.; Maier, J. P.; Marthaler, O.; Christen, H.; Baertschi P. *Helv. Chim. Acta* **1981**, *64*, 823.
34. Haselbach, E.; Klemm, U.; Gschwind, R.; Bally, T.; Chassot, T.; Nitsche, S. *Helv. Chim. Acta.* **1982**, *65*, 2464.
35. Bally, T.; Neuhaus, L.; Nitsche, S.; Haselbach, E.; Janssen, J.; Luttke, W. *Helv. Chim. Acta.* **1983**, *66*, 1288.
36. Bally, T.; Nitsche, S.; Roth, K.; Haselbach, E. *J. Am. Chem. Soc.* **1984**, *106*, 3927.
37. Shida, T.; Haselbach, E.; Bally, T. *Acc. Chem. Res.* **1984**, *17*, 180.
38. Bally, T.; Haselbach, E.; Lanyiova, Z.; Baertschi, P. *Helv. Chim. Acta* **1978**, *61*, 2488.
39. Kato, T.; Shida, T. J. *Am. Chem. Soc.* **1979**, *101*, 6869.
40. Momose, T.; Suzuki, T.; Shida, T. *Chem. Phys. Lett.* **1984**, *107*, 568.
41. Momose, T.; Shida, T.; Kobayashi, T. *Tetrahedron* **1986**, *42*, 6337.
42. Khan, Z.H. *Spectrochim. Acta (A)* **1988**, *44A*, 313.
43. Khan, Z.H. *Spectrochim. Acta (A)* **1988**, *44A*, 1125.
44. Freiser, B. S.; Beauchamp, J. L. *J. Am. Chem. Soc.* **1976**, *98*, 3136.
45. Herring, F. G.; McLean, R. A. N. *Inorg. Chem.* **1972**, *11*, 1667.

46. Cole, T. *J. Chem. Phys.* **1961**, *35*, 1169.
47. Walsh, A. D.; Warsop, P. A. *Trans. Faraday Soc.* **1961**, *57*, 345.
48. Claxton, T. A.; Smith, N. A. *Trans. Faraday Soc.* **1970**, *66*, 1825.
49. Halpern, A. M.; Forsyth, D. A.; Nosowitz, M. *J. Phys. Chem.* **1986**, *90*, 2677.
50. Nelsen, S. F.; Weistman, G. R.; Hintz, P. J.; Olp, D.; Fahey, M. R. *J. Am. Chem. Soc.* **1974**, *96*, 2916.
51. Nelsen, S. F.; Hollinsed, W. C.; Kessel, C.R.; Calabrese, J. C. *J. Am. Chem. Soc.* **1978**, *100*, 7876.
52. Nelsen, S. F.; Cunkle, G. T.; Evans, D. H.; Haller, K. J.; Kaftory, M.; Kirste, B.; Kurreck, H.; Clark, T. *J. Am. Chem. Soc.* **1985**, *107*, 3829.
53. Nelsen, S. F.; Blackstock, S. C.; Haller, K. J. *Tetrahedron* **1986**, *42*, 6101.
54. Nelsen, S. F.; Frigo, T. B.; Thompson-Colon, J. A. *J. Am. Chem. Soc.* **1986**, *108*, 7926.
55. Roth, H. D.; Schilling, M. L. M. *J. Am. Chem. Soc.* **1985**, *107*, 716.
56. Roth, H. D.; Schilling, M. L. M.; Abelt, C. J. *Tetrahedron* **1986**, *42*, 6157.
57. Fu, E. W.; Dunbar, R. C. *J. Am. Chem. Soc.* **1978**, *100*, 2283.
58. Merer, F. J.; Schoonveld, L. *J. Chem. Phys.* **1968**, *48*, 522.
59. Koppel, H.; Domcke, W.; Cederbaum, L. S.; Von Niessen, W. *J. Chem. Phys.* **1978**, *69*, 4252.
60. Todres, Z. V. *Ion-radicals in Organic Synthesis;* Khimiya: Moskow, 1986. *Chem. Abstr.* **1987**, *106*, 195522u.
61. Suzuki, H.; Koyano, K.; Shida, T.; Kira, A. *Bull. Chem. Soc. Jpn.* **1979**, *52*, 2794.
62. Suzuki, H.; Koyano, K.; Shida, T.; Kira, A. *Bull. Chem. Soc. Jpn.* **1982**, *55*, 3690.
63. Suzuki, H.; Ogava, K.; Shida, T.; Kira, A. *Bull. Chem. Soc. Jpn.* **1983**, *56*, 66.
64. Shida, T. *J. Phys. Chem.* **1978**, *82*, 991.
65. Suzuki, H. *Electronic Absorption Spectra and Geometry of Organic Molecules;* Academic Press: New York. 1967.
66. Puiu, A. C.; Andrews, L.; Chupka, W. A.; Colson, S. D. *J. Chem. Phys.* **1982**, *76*, 3854.
67. Suzuki, H. *Bull. Chem. Soc. Jpn.* **1959**, *32*, 1340.
68. Maier, J. P.; Turner, D. W. *Discuss. Faraday Soc.* **1972**, *54*, 149.
69. Dynes, J. J.; Baudais, F. L.; Boyd, R. K. *Can. J. Chem.* **1985**, *63*, 1292.
70. Baca, A.; Rossetti, R.; Brus, L. E. *J. Chem. Phys.* **1979**, *70*, 5575.
71. Golebiewski, A.; Parczewski, A. *Z. Naturforsch(A).* **1970**, *25A*, 1710.
72. Sullivan, P. D.; Fong, J. P. *Chem. Phys. Lett.* **1976**, *38*, 555.
73. Heidenreich, A.; Munzel, N.; Schweig, A. *Z. Naturforsch(A).* **1986**, *41A*, 1415.
74. Krainov, I. P.; Tsyguleva, O. M.; Kramarenko, S. F. *Teor. i eksp. khim.* **1987**, *23*, 752. *Chem. Abstr.* **1988**, *109*, 128254x.
75. Buntix, G.; Poizat, O. *J. Chem. Phys.* **1989**, *91*, 2153.
76. Sullivan, P. D.; Fong, J. J. *J. Phys. Chem.* **1977**, *81*, 71.
77. Monig, J.; Asmus, K-D.; Robertson, L. W.; Oesch F. *J. Chem. Soc. Perkin Trans. 2.* **1986**, 891.
78. Anklam, E.; Asmus, K-D.; Robertson, L. W. *J. Chem. Soc. Perkin Trans. 2.* **1989**, 1569.
79. Anklam, E.; Asmus, K-D.; Robertson, L. W. *J. Chem. Soc. Perkin Trans. 2.* **1989**, 1573.
80. Forbes, W. F.; Sullivan P. D. *Can. J. Chem.* **1968**, *46*, 325.
81. Sullivan, P. D.; Norman, L. J. *Can. J. Chem.* **1972**, *50*, 2141.

82. Turchaninov, V. K.; Vokin, A. I.; Toryashinova, D. S. D.; Ermikov, A. F.; Petrushenko, K. B.; Korostova, S. E. *Izv. Akad. Nauk, Ser. Khim.* **1993**, 319 [*Russ. Chem. Bull.* **1993**, *42*, 275 (Engl. Transl.)].
83. Turchaninov, V. K.; Vokin, A. I.; Korostova, S. E. To be published in *Izv. Akad. Nauk, Ser. Khim.* **1997**.
84. Galasso, V.; Klasinc, L.; Sabljic, A.; Trinajstic, N.; Pappalardo, G. C.; Steglich, W. *J. Chem. Soc. Perkin Trans. 2.* **1981**, 127.
85. Orti, E.; Sanchez-Marin, J.; Merchan, M.; Tomas, F. *J. Phys. Chem.* **1987**, *91*, 545.
86. Rettig, W.; Marschner, F. *Noiv. J. Chem.* **1983**, *7*, 425.
87. Klessinger, M.; Rademacher, P. *Angew. Chem.* **1979**, *91*, 885.
88. Ermikov, A. F.; Turchaninov, V. K.; Petrushenko, K. B.; Vokin, A. I.; Korostova, S. E.; Ostroukhova, L. A.; Frolov, Yu. L. *Zh. Obshch. Khim.* **1988**, *58*, 450 [*J. Gen. Chem.* **1988**, *58* (Engl. Transl.)]. *Chem. Abstr.* **1989**, *110*, 114125q.
89. Turchaninov, V. K.; Ermikov, A. F.; Korostova, S. E.; Shagun, V. A. *Zh. Obshch. Khim.* **1989**, *59*, 791 [*J. Gen. Chem.* **1989**, *59* (Engl. Transl.)]. *Chem. Abstr.* **1989**, *111*, 231565q.
90. Turchaninov, V. K.; Vokin, A. I.; Toryashinova, D. S. D.; Korostova, S. E. *Izv. Akad. nauk, Ser. Khim.* **1993**, 869 [*Russ. Chem. Bull.* **1993**, *42*, 822 (Engl. Transl.)]. *Chem. Abstr.* **1995**, *122*, 159984h.
91. Vokin, A. I. *Thesis.* Irkutsk. 1994.
92. Russo, N.; Toscano, M.; Barone, V.; Minichino, C. *J. Chim. Phys. Phys-Chim. Biol.* **1987**, *84*, 735.
93. Fabian, W. M. F. *Z. Naturforsch (A).* **1987**, *42A*, 641.
94. Fabian, W. M. F. *J. Comput. Chem.* **1988**, *9*, 369.
95. Turchaninov, V. K.; Petrushenko, K. B.; Ermikov, A. F.; Vokin, A. I. *Khim. Geterotsikl. Soedin.* **1989**, 1213 [*Chem. Heterocycl. Comp.* **1989** (Engl. Transl.)] *Chem. Abstr.* **1990**, *113*, 171350v.
96. Dunbar, R. C.; Teng, H. H. I.; Fu, E. W. *J. Am. Chem. Soc.* **1979**, *101*, 6506.
97. Almenningen, A.; Bastiansen, O.; Fernholt, L.; Cyvin, B. N.; Cyvin, S. J.; Samdal, S. *J. Mol. Struct.* **1985**, *128*, 59.
98. Proutiere, A.; Legoff, D.; Chabanel, M.; Megnassan, E. *J. Mol. Struct.* **1988**, *178*, 49.
99. Trotter, J. *Acta Crystallogr.* **1961**, *14*, 1135.
100. Skancke, A. *Acta Chem. Scand.* **1970**, *24*, 1389.
101. Lunazzi, L.; Salvetti, F.; Veracini, C.A. *J. Chem. Soc. Perkin Trans. 2.* **1976**, 1796.
102. Visser, G. J.; Heeres, G. J.; Wolters, J.; Vos, A. *Acta Crystallogr. Sect. B.* **1968**, *24B*, 467.
103. Trofimov, B. A.; Sigalov, M. V.; Bzhezovskii, V.M.; Kalabin, G. N.; Korostova, S. E.; Mikhaleva, A. I.; Balabanova, L. N. *Khim. Geterotsikl. Soedin.* **1978**, 768. *Chem. Abstr.* **1978**, *89*, 107165x.
104. Golovanova, N. I.; Korostova, S. E.; Mikhaleva, A. I.; Nesterenko, R. N. *Izv. Sib. Otd. Akad. Nauk SSSR, Ser. Khim. Nauk* **1986**, 122. *Chem. Abstr.* **1987**, *107*, 216811r.
105. Afonin, A. V.; Sigalov, M. V.; Korostova, S. E.; Voronov, V. K.; Aliev, I. A. Izv. *Akad. Nauk SSSR, Ser. Khim.* **1988**, 2765. *Chem. Abstr.* **1989**, *111*, 56696j.
106. Afonin, A. V.; Sigalov, M. V.; Korostova, S. E.; Aliev, I. A.; Vashchenko, A. V.; Trofimov, B. A. *Magn. Res. Chem.* **1990**, *28*, 580.
107. Hofmann, H. J.; Cimiraglia, R.; Tomasi, J. *J. Chem. Res (S).* **1987**, 48.
108. Davies, A. G.; Schiesser, C. H. *J. Organomet. Chem.* **1990**, *389*, 301.
109. Buntix, G.; Poizat, O. *Laser Chem.* **1990**, *10*, 333.

110. Kirste, B.; Tian, P.; Kossmehl, G.; Engelmann, G.; Jugelt, W. *Magn. Res. Chem.* **1995**, *33*, 70.

111. Wagner, P. J.; Scheve, B. J. *J. Am. Chem. Soc.* **1977**, *99*, 2888.

112. Dallakyan, P. B.; Petrova, R. S. *Vestn. Mosk. Univ. Ser. 2: Khim.* **1987**, *28*, 46. *Chem. Abstr.* **1987**, *107*, 58426n.

113. Hilal, R.; Abu-Eittah, R. *Tetrahedron.* **1978**, *34*, 785.

114. Chidichimo, G.; Liguori, A.; Longeri, M.; Veracini C. A. *J. Magn. Reson.* **1983**, *51*, 438.

115. Abu-Eittah, R.; Hilal, R.; Hamed, M. M. *Int. J. Quant. Chem.* **1981**, *19*, 383.

116. Strashnikova, N. V.; Sigalov, M. V.; Korostova, S. E.; Mikhaleva, A.I.; Trofimov, B. A. *Izv. Akad. Nauk, Ser. Khim.* **1993**, 1060. *Chem. Abstr.* **1995**, *122*, 105142p.

117. Ruscic, B.; Kovac, B.; Klasinc, L.; Gusten, H. *Z. Naturforsch.(A).* **1978**, *33A*, 1006.

118. Samdal, S.; Samuelsen, E. J.; Volden, H. V. *Synth. Met.* **1993**, *59*, 259.

119. Distefano, G.; Dal Colle, M.; Jones, D.; Zambianchi, M.; Favaretto, L.; Modelli A. *J. Phys. Chem.* **1993**, *97*, 3504.

120. Hernandez, V.; Lopez Navarrete, J. T. *J. Chem. Phys.* **1994**, *101*, 1369.

121. Hernandez, V.; Lopez Navarrete, J. T. *Synth. Met.* **1992**, *51*, 211.

122. Meunier, P.; Coustale, M.; Guimon, C.; Pfister-Guillouzo, G. *J. Mol. Struct.* **1977**, *36*, 233.

123. Jones, D.; Guerra, M.; Favaretto, L.; Modelli, A.; Fabrizio, M.; Distefano G. *J. Phys. Chem.* **1990**, *94*, 5761.

124. Evans, C. H.; Scaiano, J. C. *J. Am. Chem. Soc.* **1990**, *112*, 2694.

125. Aleman, C.; Brillas, E.; Davies, A. G.; Fajari, L.; Giro, D.; Julia, L.; Perez, J. J.; Rius, J. *J. Org. Chem.* **1993**, *58*, 3091.

126. Cunningham, D. D.; Laguren-Davidson, L.; Mark, H. B.; Pham, C. V.; Zimmer H. *J. Chem. Soc. Chem. Commun.* **1987**, 1021.

127. Arbizzani, C.; Barbarella, G.; Bongini, A.; Mastragostino, M.; Zambianchi M. *Synth. Met.* **1992**, *52*, 329.

128. Amer, A.; Burkhardt, A.; Nkansah, A.; Shabana, R.; Galal, A.; Mark, H. B.; Zimmer H. *Phosphorus, Sulfur and Silicon* **1989**, *42*, 63.

129. Zerbi, G.; Radaelli, R.; Veronelli, M.; Brenna, E.; Sannicolo, F.; Zotti G. *J. Chem. Phys.* **1993**, *98*, 4531.

130. Kraak, A.; Wynberg, H. *Tetrahedron* **1968**, *24*, 3881.

131. Sease, J. W.; Zechmeister, L. *J. Am. Chem. Soc.* **1947**, *69*, 270.

132. Birnbaum, D.; Kohler, B. E. *J. Chem. Phys.* **1991**, *95*, 4783.

133. Roncali, J. *Chem. Rev.* **1992**, *92*, 711.

134. Horowitz, G.; Yassar, A.; von Bardeleben, H. J. *Synth. Met.* **1994**, *62*, 245.

135. Wintgens, V.; Valat, P.; Garnier, F. *J. Phys. Chem.* **1994**, *98*, 228.

136. Fichou, D.; Horowitz, G.; Xu, B.; Garnier, F. *Synth. Met.* **1990**, *39*, 243.

137. Chaloner, P. A.; Gunatunga, S. R.; Hitchcock, B. B. *Acta Crystallogr. Sect. C.* **1994**, *50C*, 1941.

138. Pelletier, M.; Brisse, F. *Acta Crystallogr. Sect. C.* **1994**, *50C*, 1942.

139. Barbarella, G.; Zamdianchi, M.; Bongini, A.; Antolini L. *Adv. Mater.* **1992**, *4*, 282.

140. van Bolhuis, F.; Wynberg, H.; Havinga, E. E.; Meijer, E. W.; Staruing, E. G. J. *Synth. Met.* **1989**, *30*, 381.

141. Tourillon, G.; Jugnet, Y. *J. Chem. Phys.* **1988**, *89*, 1905.

142. Fujimoto, H.; Nagashima, U.; Inokuchi, H.; Seki, K.; Cao, Y.; Nakahara, H.; Nakayama, J.; Hoshino, M.; Fukuda, K. *J. Chem. Phys.* **1990**, *92*, 4077.
143. Fujimoto, H.; Nagashima, U.; Inokuchi, H.; Seki, K.; Nakahara, H.; Nakayama, J.; Hoshino, M.; Fukuda, K. *J. Chem. Phys.* **1988**, *89*, 1198.
144. Nagashima, U.; Fujimoto, H.; Inokuchi, H.; Seki, K. *J. Mol. Struct.* **1989**, *197*, 265.
145. Hotta, S.; Waragai, K. *J. Phys. Chem.* **1993**, *97*, 7427.
146. Fichou, D.; Horowitz, G.; Xu, B.; Garnier F. *Synth. Met.* **1992**, *48*, 167.
147. Keiko, V. V.; Baikalova, L. V.; Domnina, E. S.; Skvortsova, G. G.; Toryashinova, D. C. D.; Chipanina, N. N.; Voronov, V. K.; Nikolenko, L. N. *Zh. Obshch. Khim.* **1981**, *51*, 892. *Chem. Abstr.* **1981**, *95*, 79967p.
148. Afonin, A. V.; Voronov, V. K.; Baikalova, L. V.; Domnina, E. S.; Chipanina, N. N. *Zh. Obshch. Khim.* **1987**, *57*, 1594. *Chem. Abstr.* **1987**, *107*, 189520m.
149. Cromer, D. T.; Ryan, R. R.; Storm, C. B. *Acta Crystallogr. Sect. C.* **1987**, *43C*, 1435.
150. Hino, S.; Seki, K.; Inokuchi, H. *Chem. Phys. Lett.* **1975**, *36*, 335.
151. Danovich, D. K.; Turchaninov, V. K.; Zakrzewski V. G. *J. Mol. Struct. (Theochem).* **1990**, *209*, 77.
152. Flitsch, W.; Peeters, H.; Schulten, W.; Rademacher P. *Tetrahedron* **1978**, *34*, 2301.
153. Rietveld, H. M.; Maslen, E. N.; Clews, C. J. B. *Acta Crystallogr. Sect. B.* **1970**, *26B*, 693.
154. Almenningen, A.; Bastiansen, O.; Svendsas, P. *Acta Chem. Scand.* **1958**, *12*, 1671.
155. Bucci, P.; Longeri, M.; Veracini, C. A.; Lunazzi, L. *J. Am. Chem. Soc.* **1974**, *96*, 1305.
156. Khetrapal, C. L.; Kunwar, A. C. *Mol. Phys.* **1974**, *28*, 441.
157. ter Beek, L. C.; Zimmerman, D. S.; Burnell, E. E. *Mol. Phys.* **1991**, *74*, 1027.
158. Caspar, J. V.; Ramamurthy, V.; Corbin, D. R. *J. Am. Chem. Soc.* **1991**, *113*, 600.
159. Martinez, F.; Voelkel, R.; Naegele, D.; Haarmann, H. *Mol. Cryst. Liq. Cryst.* **1989**, *167*, 227.
160. Koenig, T.; Tuttle, M. *J. Org. Chem.* **1974**, *39*, 1308.
161. Anklam, E.; Asmus, K-D. *J. Fluor. Chem.* **1988**, *38*, 209.
162. Hafelinger, G.; Regelmann, C. *J. Comput. Chem.* **1987**, *8*, 1057.
163. Aroney, M. J.; Calderbank, K. E.; Pierens, R. K.; Chia, L. H. L. *J. Mol. Struct.* **1985**, *127*, 175.
164. Eaton, V. J.; Steele, D. *J. Chem. Soc., Faraday Trans. 2.* **1973**, *18*, 1601.
165. Braude, E. A.; Forbes, W. F. *J. Chem. Soc.* **1955**, 3776.
166. Unanue, A.; Bothorel, P. *Bull. Soc. Chim. Fr.* **1966**, 1640.
167. Bree, A.; Edelson, M. *Chem. Phys. Lett.* **1977**, *46*, 500.
168. Akiyama, M.; Watanabe, T.; Kakihana, M. *J. Phys. Chem.* **1986**, *90*, 1752.
169. Barrett, R. M.; Steele, D. *J. Mol. Struct.* **1972**, *11*, 105.
170. Brock, C. P.; Minton, R. P. *J. Am. Chem. Soc.* **1989**, *111*, 4586.
171. Halstead, T. K.; Spiess, H. W.; Haeberlen, U. *Mol. Phys.* **1976**, *31*, 1569.
172. Bastiansen, O.; Gundersen, S.; Samdal, S. *Acta Chem. Scand.* **1989**, *43*, 6.
173. Mackenzie, G. A.; Pawley, G. S.; Dietrich, O. W. *Acta Crystallogr. Sect. A.* **1975**, *31A*, 851.
174. Gleason, W. B.; Britton, D. *Cryst. Struct. Commun.* **1976**, *5*, 483.
175. Niederberger, W.; Diehl, P.; Lunazzi L. *Mol. Phys.* **1973**, *26*, 571.
176. Field, L. D.; Sternhell, S.; Tracey A.S. *J. Am. Chem. Soc.* **1977**, *99*, 5249.
177. Novak, I.; Ng, S. C.; Huang, H. H.; Mok, C. Y., Khoi, E., Kovac B. *J. Phys. Org. Chem.* **1991**, *4*, 675.

178. Novak, I.; Ng, S. C.; Mok, K. F.; Mok, C. Y.; Huang, H. H. *J. Chem. Res. (S).* **1992**, 350.

179. Novak, I.; Ng, S. C.; Chua, Y. T.; Mok, C. Y.; Huang, H. H. *J. Electron Spectrosc. Relat. Phenom.* **1992**, *61*, 143.

180. Novak, I.; Ng, S. C.; Mok, C. Y.; Huang, H. H.; Fang, J.; Wang, K. K. T. *J. Chem. Soc. Perkin Trans. 2.* **1994**, 1771.

181. Novak, I.; Ng, S. C.; Chua, Y. T.; Mok, C. Y.; Huang, H. H. *J. Electron Spectrosc. Relat. Phenom.* **1994**, *69*, 177.

182. Novak, I.; Ng, S. C.; Chua, Y. T.; Mok, C. Y.; Huang, H. H. *J. Phys. Chem.* **1994**, *98*, 5240.

183. Veracini, A.; Macciantelli, D.; Lunazzi, L. *J. Chem. Soc. Perkin Trans. 2.* **1973**, 751.

184. Pyrka, G. J.; Fernando, Q.; Inoue, M. B.; Inoue, M.; Velazquez, E. F. *Acta Crystallogr. Sect. C.* **1988**, *44C*, 562.

185. Pyrka, G. J.; Fernando, Q.; Inoue, M. B.; Inoue M. *Acta Crystallogr. Sect. C.* **1988**, *44C*, 1800.

186. Norden B. *Chem. Scrip.* **1975**, *7*, 226.

187. Lumbroso, H.; Bertin, D. M.; Marschner, Ḟ. *J. Mol. Struct.* **1988**, *178*, 187.

188. Aroney, M. J.; Lee, H. K.; LeFevre, R. J. W. *Aust. J. Chem.* **1972**, *25*, 1561.

189. Lumbroso, H.; Carpanelli, C. *Bull. Soc. Chim. Fr.* **1964**, 3198.

RECENT APPROACHES TO γ-LACTAMS *via* C-C BOND FORMATION

Roberta Galeazzi, Giovanna Mobbili, Mario Orena *

Dipartimento di Scienze dei Materiali e della Terra - Università di Ancona

Via Brecce Bianche - I-60131 Ancona, Italy

Abstract: *During the last few years γ-lactams (pyrrolidin-2-ones) have attracted the interest of synthetic organic chemists for the preparation of a lot of nitrogen-containing products with biological interest, such as alkaloids and non-proteinogenic amino acids. The different ways of generating these compounds via C-C bond formation are presented here, with a particular attention to the stereoselective construction of mono- and polysubstituted systems.*

1. Introduction

A number of compounds containing the γ-lactam (pyrrolidin-2-one) moiety show interesting biological and pharmacological activities and are key intermediates for the synthesis of antibiotics and naturally occurring products.

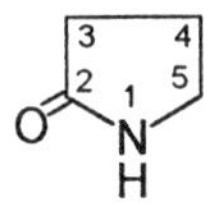

Lactacystine **1**, [1] cotinine **2** [2] and Rolipram **3** [3] are reported here as examples.

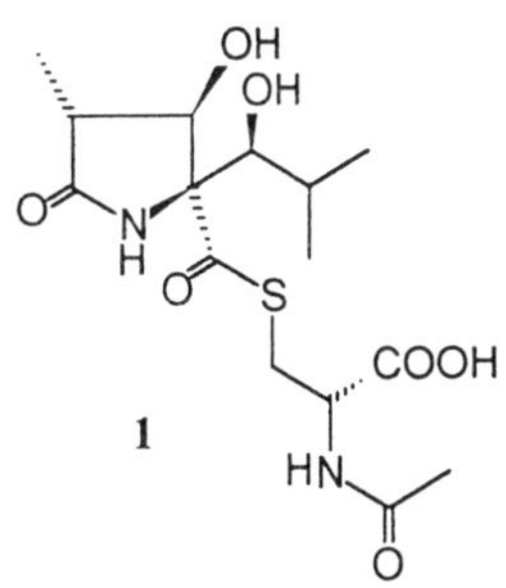

1

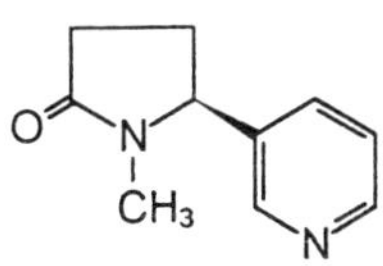

2

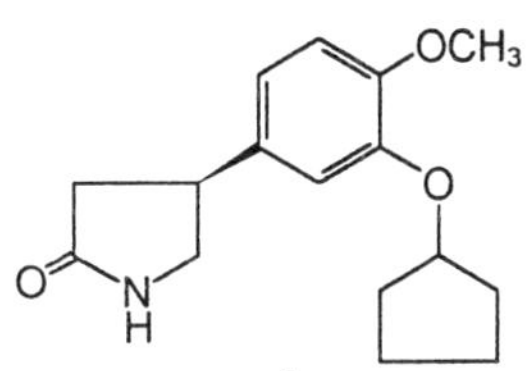

3

On the other hand, pyrrolizidinic [4] and indolizidinic [5] alkaloids formally derive from bicyclic lactams pyrrolizidinone **4** and indolizidinone **5**, respectively.

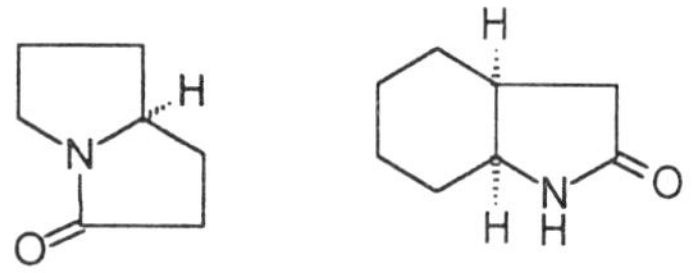

4 **5**

In addition, by removal of the carbonyl group, γ-lactams can be converted into pyrrolidines, so that every synthesis of γ-lactams allows a convenient approach to pyrrolidines. [6,7]

Therefore, a number of useful syntheses of γ-lactams have been reported in the literature. Most approaches exploit cyclisation involving bond formation between nitrogen and an acyl group [8] or between an amido group and an activated double bond. [9] However, this review deals with recent synthetic methods aimed to build the γ-lactam ring starting from acyclic precursors, through the formation of new carbon-carbon bonds. In fact, this approach constitutes an alternative route for this class of compounds, particularly suitable when stereocentres with a definite configuration must be created on the heterocyclic ring.

2. Formation of the C_3-C_4 Bond

2.1. Cyclisation involving free radical intermediates

The efficiency of cyclisation *via* free radical intermediates relies on both structural and temperature effects. Amides generally exist as a *syn-anti* rotameric mixture owing to the rotational barrier of the amidic bond which is about 16-22 Kcal/mol. This barrier is higher than the activation energy for radical cyclisation, so that only the *anti* rotamer undergoes cyclisation, whereas the *syn* one leads to chain termination. Moreover, the rotameric population strongly depends on the bulkiness of the substituents of the amidic nitrogen. [10,11]

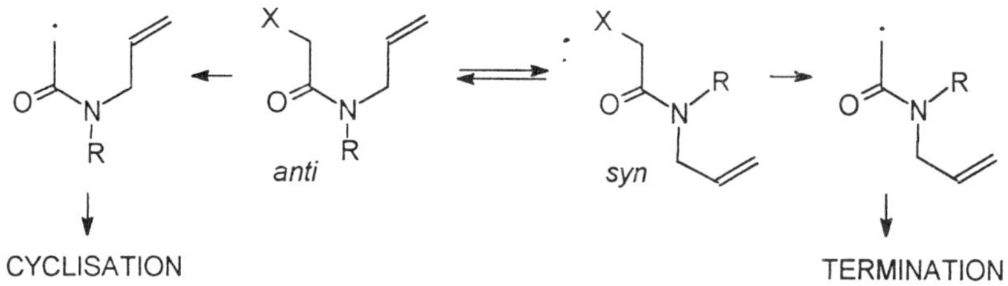

Thus, when *N,N*-diallyl iodoacetamide **6a** is treated under atom transfer conditions, one allyl group is always available for the cyclisation and high yields of γ-lactam **7a** are observed, independent of the temperature.

On the contrary, *N*-allyl-*N*-methyl iodoacetamide **6b** at 25 °C gives the cyclisation product **7b** in low yield. On heating, however, the rotameric interconversion rate increases and at 80 °C the cyclisation yield raises to about 80%.[12]

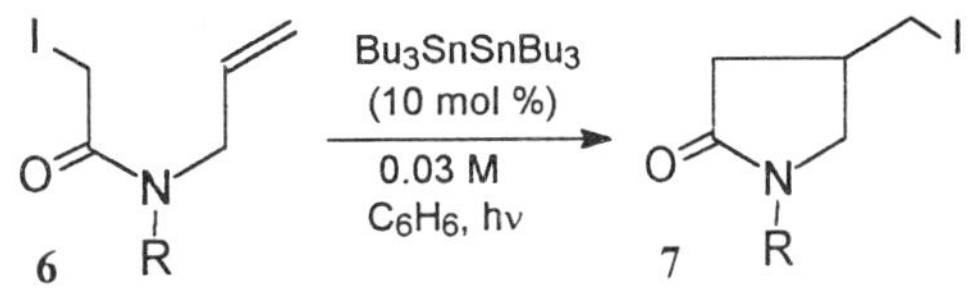

a. R = Allyl, 99% at 25 °C, 99% at 80 °C **b.** R = CH_3, 39% at 25 °C, 87% at 80 °C

The concentration of the reagents is the third factor which strongly affects both cyclisation yields and products distribution. In fact, when the reaction is performed under high dilution conditions, the probability for the termination step to occur is low, allowing the cyclisation product to be recovered in high yield.[10,11]

From a direct comparison between hydrogen transfer and atom or group transfer, it results that the hydrogen transfer is effective when a rapid cyclisation can occur, whereas the atom transfer method is useful when relatively slow intermediate cyclisations are involved.[10,11]

A net reduction occurs in the hydrogen transfer. On the contrary, the atom or group transfer method always leads to isomerisation. Moreover, the atom transfer cyclisation of ω-halo olefins is valuable as a synthetic tool for the construction of heterocyclic molecules, since, in contrast to the hydride-mediated radical cyclisation, in which the last step consists of a simple hydrogen transfer by the hydride, the atom transfer method can introduce a versatile halogen atom in the cyclisation product which is susceptible to useful transformations.

2.1.1. Reactions terminating with hydrogen transfer

A number of useful studies on the construction of γ-lactams has been carried out by hydrogen transfer reactions using mainly tin hydride as hydrogen donor. Depending on the starting material, these cyclisations can proceed via either a 5-*exo* mode or a 5-*endo* mode, which are both allowed.[13,14]

2.1.1.1. Cyclisations proceeding *via* a 5-*exo* mode

The radical mediated cyclisations proceeding *via* a 5-*exo* mode in which a hydrogen transfer is involved, are generally carried out starting from *N*-allyl haloamides and the rate of the initial homolytic cleavage of the C-halogen bond strongly affects the reaction yield. In fact the cyclisation of chloroacetamides **8a,b** gives lactams **9a,b** in poor yield.[15]

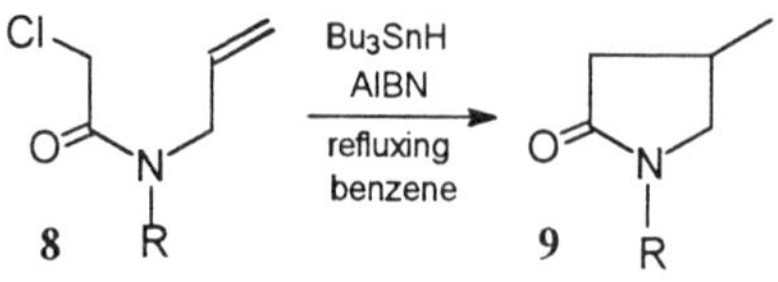

a. R = CH_3, **24%** **b.** R = Ph, 12%

On the other hand, starting from (*S*)-*N*-allyl-*N*-(phenylethyl) iodoacetamides **10**, the corresponding γ-lactams **11** and **12** can be easily obtained in moderate to good yield as equimolar diastereomeric mixtures. The diastereomers can be separated by silica gel chromatography and their configuration is assigned by inspection of the ^{1}H NMR spectra. This synthetic approach allows a convenient route to enantiomerically pure pyrrolidines, which can be prepared after removal of both the carbonyl and the chiral auxiliary group. [16]

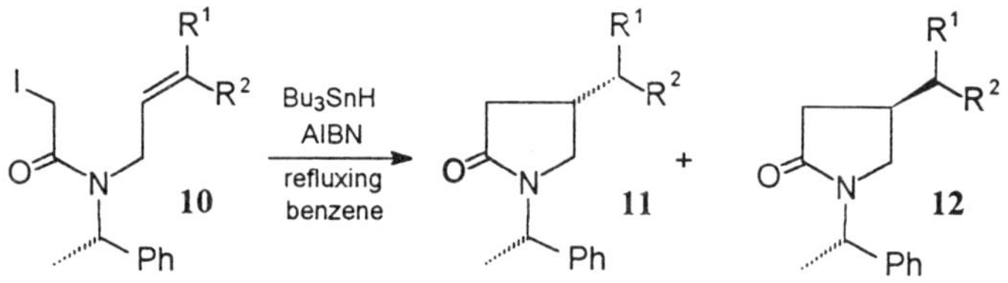

a. $R^1 = R^2$ = H, **54%**. **b.** $R^1 = CH_3$, R^2 = H, 65%.

c. $R^1 = R^2 = CH_3$, 81**%**. **d.** $R^1 = COOCH_3$, R^2 = H, 48%.

Nickel powder, in combination with acetic acid, allows a practical synthesis of γ-lactams starting from *N*-allyl trichloroacetamides. [17]

By homolytic cleavage of a C-halogen bond of the trichloroacetamide **13**, a carbon centered radical is formed which cyclises to give a five-membered ring, thus generating another radical species. In the presence of *t*-dodecanethiol which acts as hydrogen donor, the corresponding 4-methylpyrrolidin-2-one **14** is obtained in good yield. [17]

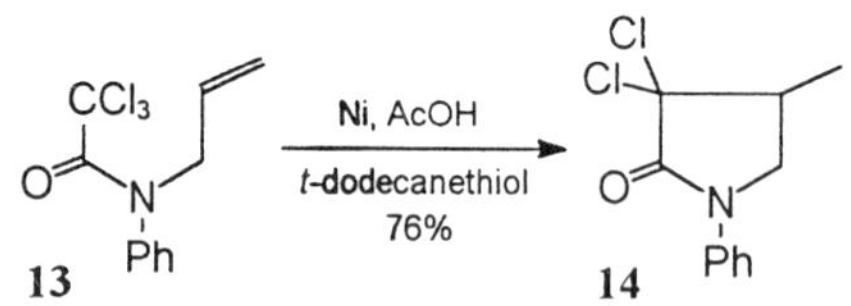

4-Substituted γ-lactams can be prepared also by electroreductive cyclisation of *N*-allyl bromoamides, performed in the presence of the Ni(II) catalyst **15**. [16]

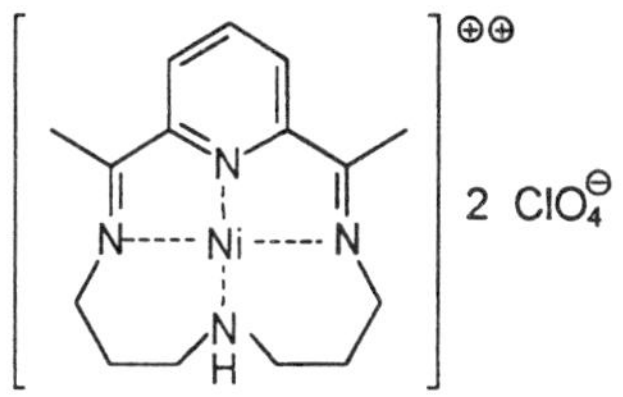

15

This approach to γ-lactams differs from the tin hydride method in that there is no active hydrogen atom donor like tin hydride in the reaction system since the hydrogen is transfered from Ph_2PH. This means that hydrogen transfer to the initial radical and to the cyclised one is much slower than those in the tin hydride method.[18]

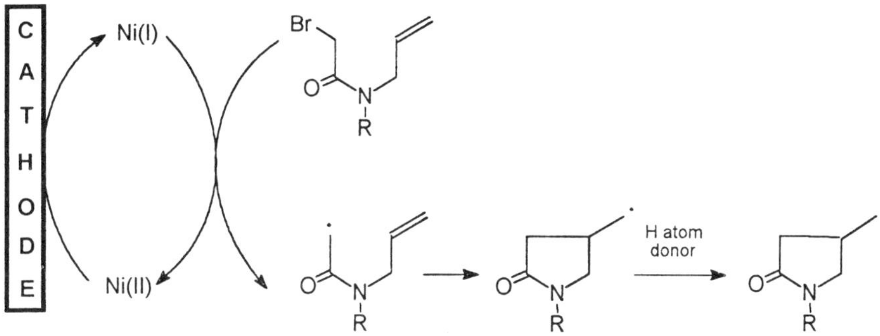

The γ-lactams **17** are obtained in moderate yield starting from the corresponding bromoacetamides **16**, but no example in which asymmetric induction is involved has been reported.[18]

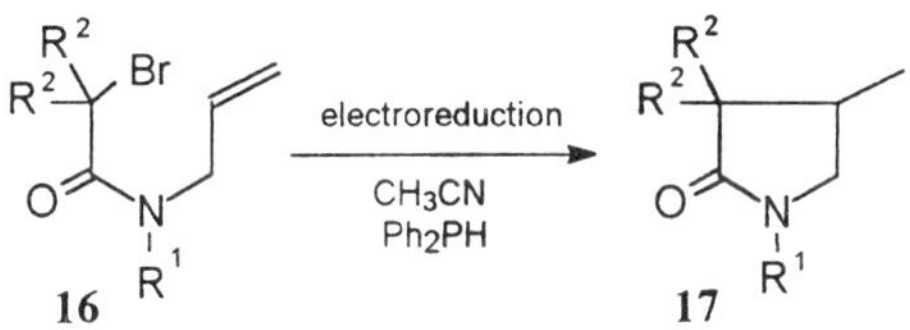

16 **17**

a. R^1 = Ts, R^2 = H, 58% **b.** R^1 = Allyl, R^2 = H, 50% **c.** R^1 = Ts, R^2 = CH_3, 67%

In addition, even 4-methylene γ-lactams **18** can be obtained, although in low yield.[18]

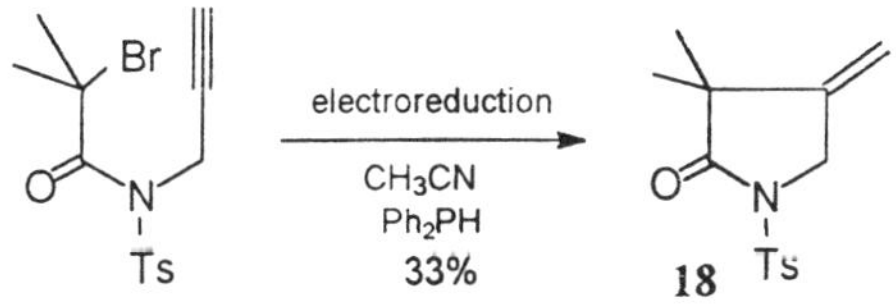

18

Also Ni in acetic acid, in the presence of *t*-dodecanethiol, is effective in producing methylene lactams. In fact, the *N*-propargyl amide **19** gives the corresponding 4-methylene lactam **20** in moderate yield. [17]

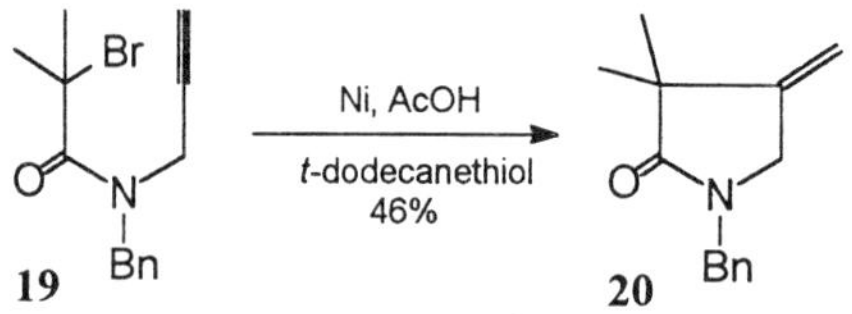

Methylene lactams however can be prepared with better yields by hydrogen transfer reaction mediated by Bu_3SnH. [19]

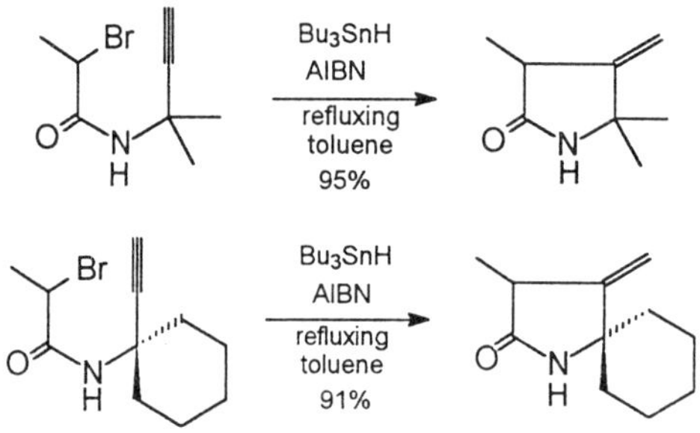

The cyclisation of dichloroacetamides **21a,b** or α-chloroamide **21c** proceeds with moderate asymmetric induction, leading to diastereomeric mixtures of **22** and **23** in which the 3,4-*trans* isomer is the major one. [15]

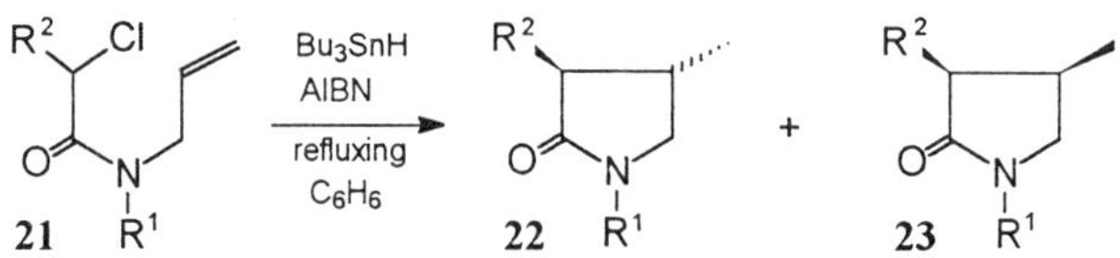

a. $R^1 = CH_3$, $R^2 = Cl$, 50%, d.r. 73:27. **b.** $R^1 = Ph$, $R^2 = Cl$, 85%, d.r. 79:21.
c. $R^1 = Ph$, $R^2 = CH_3$, 74%, d.r. 74:26.

In analogy, the cyclisation of *N*-allyl α-chloro-α-methylthioacetamides **24**, leads to 3,4-disubstituted γ-lactams **25** and **26** with moderate stereoselection. [15]

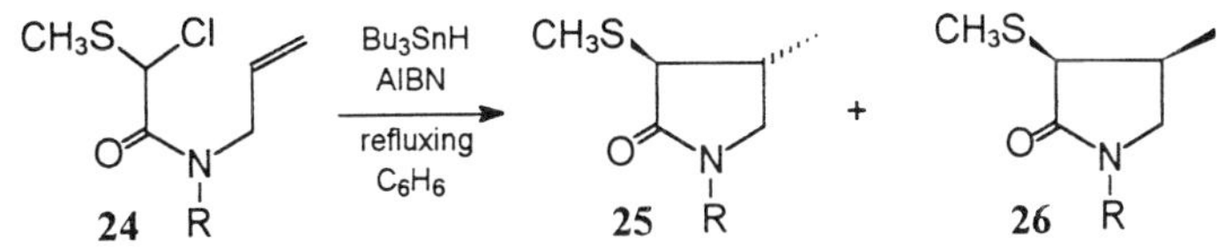

a. R = CH_3, 68%, d.r. 71:29 **b.** R = Bn, 80%, d.r. 72:28 **c.** R = Ph, 90%, d.r. 83:17

By changing the methylthio group with a chiral phenyl sulphoxide, the cyclisation proceeds with very high diastereoselection. In fact, by treating the *N,N*-diallylamide **27** with either Bu_3SnH or $(Me_3Si)_3SiH$ at low temperature, the 3,4-*trans* isomer **28** is the major component of the reaction mixture and this result has been explained by inspection of the transition states leading to the cyclised products. [20]

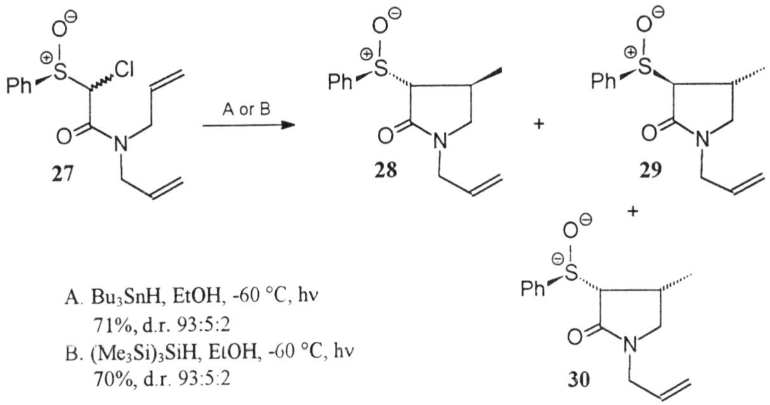

A. Bu_3SnH, EtOH, -60 °C, hν
71%, d.r. 93:5:2
B. $(Me_3Si)_3SiH$, EtOH, -60 °C, hν
70%, d.r. 93:5:2

N-Allyl acrylamides **31** react with Me_3SnCl and $NaBH_3CN$ to give γ-lactams in moderate to good yield. The addition of a tributylstannyl radical to the conjugate double bond constitutes the key step of the reaction, which is followed by cyclisation and hydrogen extraction. Compounds **32** and **33** are obtained with complete regioselection, and good diastereoselection, the 3,4-*trans*-disubstituted product **32** being the major component of the reaction mixture. [21]

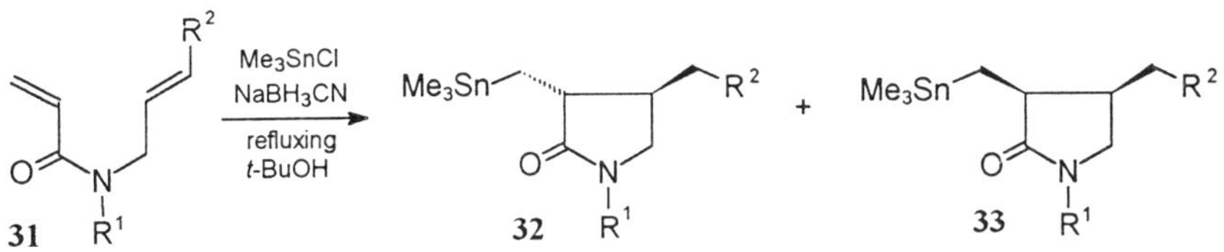

a. R^1 = Bn, R^2 = H, 58%, d.r. 81:19. **b.** R^1 = *t*-Boc, R^2 = H, 32%. d.r. 87:13.
c. R^1 = Allyl, R^2 = H, 64%, d.r. 83:17. **d.** R^1 = Bn, R^2 = $COOCH_3$, 80%, d.r. 80:20
e. R^1 = H, R^2 = Ph, 57%, d.r. 82:18

Moreover, when a stereogenic centre lies α to the amidic nitrogen in the starting material, as in **34**, the *anti,anti* isomer **35** is prevalently formed. [21]

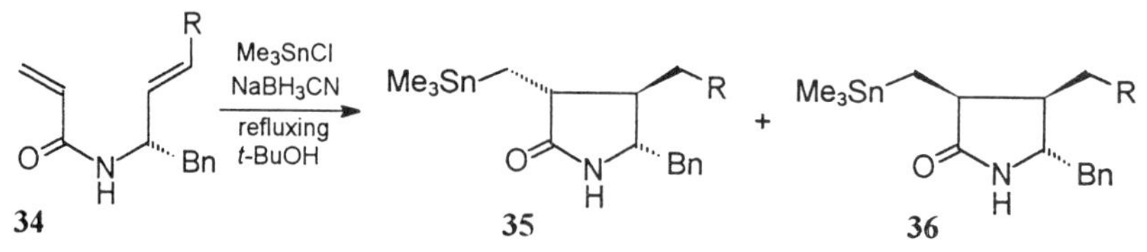

R = $COOCH_3$ 64%, d.r. 85:15

Following this synthetic approach, enantiopure α-methano amino acid analogues such as **37** have been prepared, which have a potential activity on the central nervous system. [21]

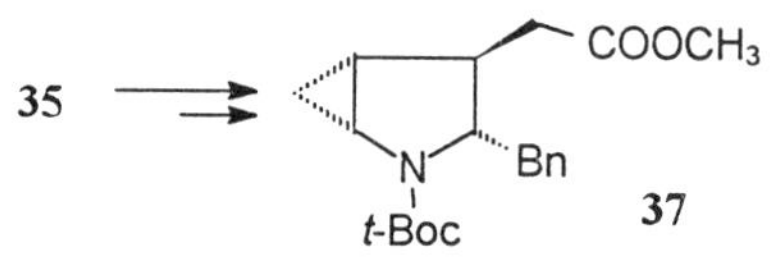

When the *N*-alkyl-*N*-allyl haloacetamide **38**, in which a substituent lies α to the amidic nitrogen, undergoes cyclisation, a diastereomeric mixture is obtained in moderate yield, the 4,5-*trans* isomer **39** being the prevalent component of the reaction mixture. [22]

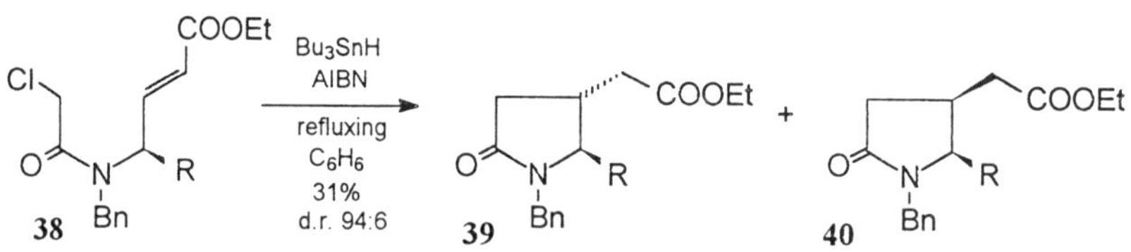

R = $CH_2OTBDMS$

On the other hand, starting from the trichloroacetamide **41**, the 3,3'-dichloro derivatives **42** and **43** are recovered in better yield, although the diastereoselection results slightly decreased. The fully dehalogenated products can be obtained by subsequent treatment with an excess of Bu_3SnH. [22]

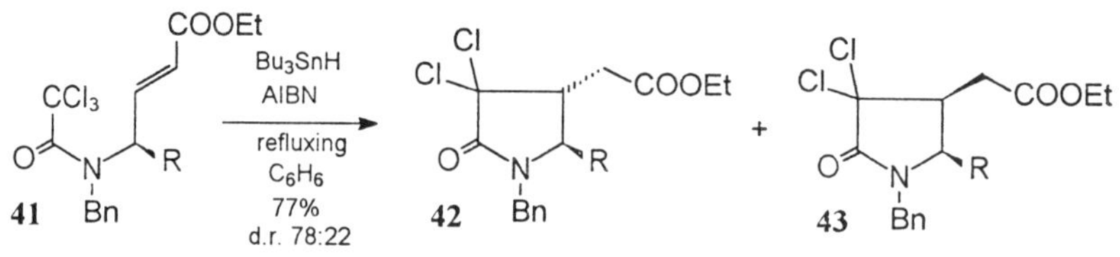

R = $CH_2OTBDMS$

The same behaviour is observed in the cyclisation of the secondary trichloroacetamide **44**, which affords with good yield and moderate stereoselection a diastereomeric mixture of 4,5-disubstituted pyrrolidin-2-ones **45** and **46**.[23]

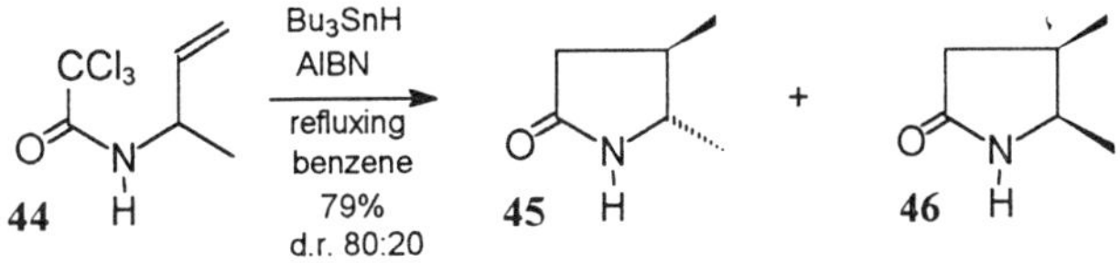

On the contrary, when the corresponding imide **47** is cyclised under the same conditions, a reversal of the stereoselection is observed, and a diastereomeric mixture is recovered in which the 4,5-*cis* isomer **48** is the prevalent one.[23]

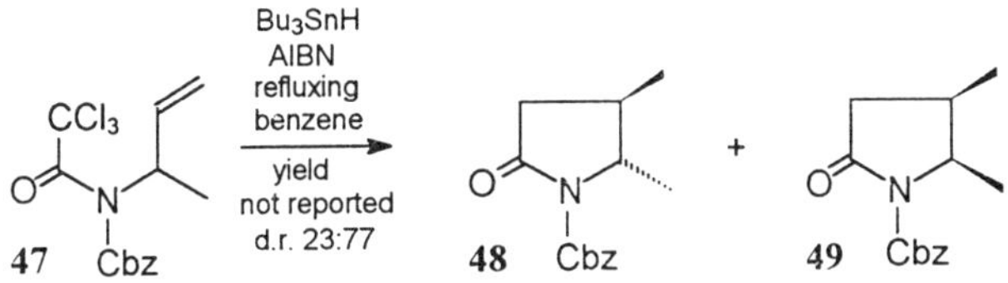

Thus, the stereochemical outcome of the cyclisation leading to 4,5-disubstituted γ-lactams changes simply by changing the substituent at the amidic nitrogen, although this result remains unexplained. Eventually, when the haloacetamides **50** are converted into 3,4,5-trisubstituted γ-lactams, only complex product mixtures of compounds **51**, **52** and **53** are recovered from the reaction.[22]

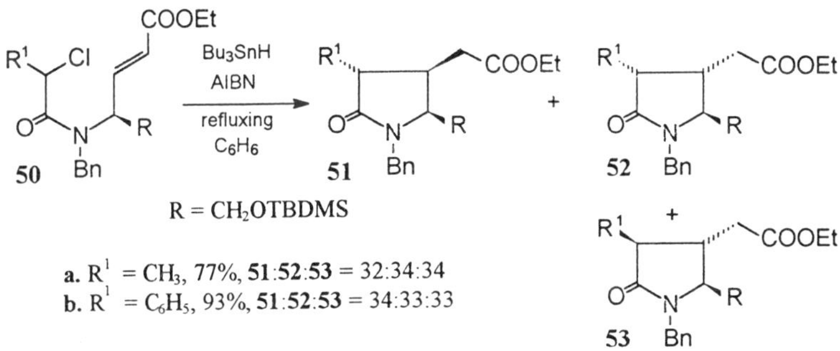

It is worth mentioning, however, that a 3,4,5-trisubstituted γ-lactam is formed with complete diastereoselection when a bicyclic compound can be obtained. In fact, the α-chloro α-phenylthioacetamide **54** affords **55** exclusively, probably due to steric constraints in the bicyclic system formation. This latter product is the key intermediate in the synthesis of enantiopure **56**, an amino acid which acts on the central nervous system.[24]

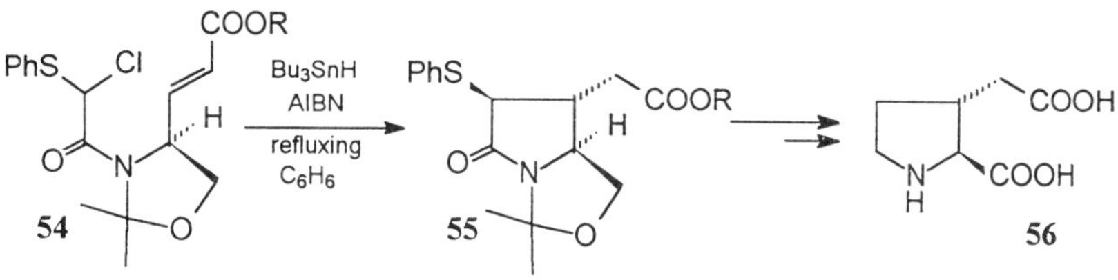

a. R = C_2H_5, 57% **b.** R = t-C_4H_9, 51%

Even indolizinones can result from a 5-*exo* cyclisation terminated by hydrogen transfer. The reaction, starting from either **57** or **59**, invariably leads to a single product, with complete regio and diastereoselection.[25,26]

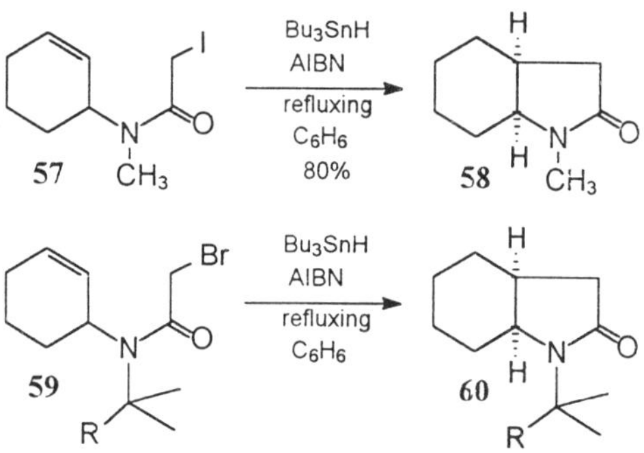

a. R = CH_3, 77% **b.** R = Ph, 73%

Indolizidinones **62** can be prepared also starting from dichloroacetamides **61**. In this case, however, an excess of the hydrogen donor is required, in order to obtain the fully dehalogenated products.[27]

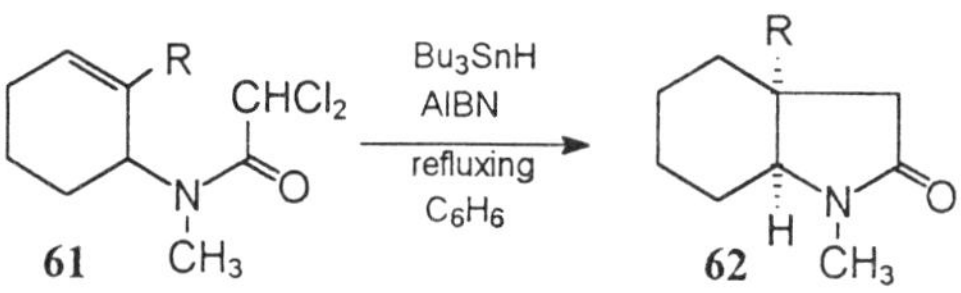

a. R = H, 93% **b.** R = C_6H_5, 43%

Together with amides, imides **63** constitute a convenient source for indolizidinones **64** which can be obtained in good yield by using as hydrogen donor either Bu_3SnH or Ph_3GeH.[26]

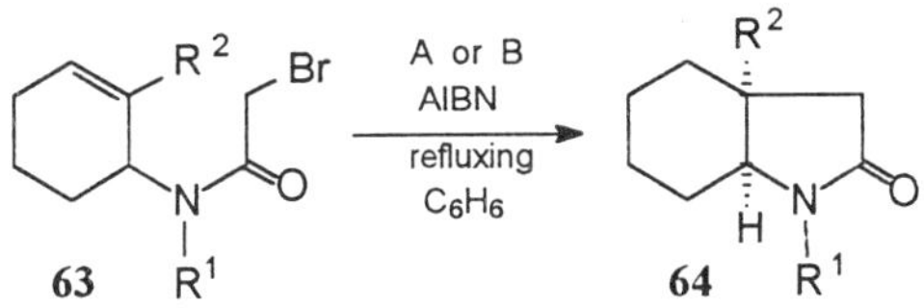

A = Bu_3SnH, **B** = Ph_3GeH

a. $R^1 = CF_3CO$, $R^2 = H$, 90% **b.** $R^1 = Ts$, $R^2 = H$, 85% **c.** $R^1 = CF_3CO$, $R^2 = CH_3$, 75%(**A**), 82%(**B**)

d. $R^1 = CF_3CO$, $R^2 = Ph$, 69%(**A**), 69%(**B**)

Radical mediated cyclisation of *nor*-ephedrine derived α-iodoamides **55** proceeds with very good diastereoselection to give the bicyclic products **56** and **57** with good asymmetric induction. The major isomers have been employed to prepare enantiomerically pure pyrrolidines.[28]

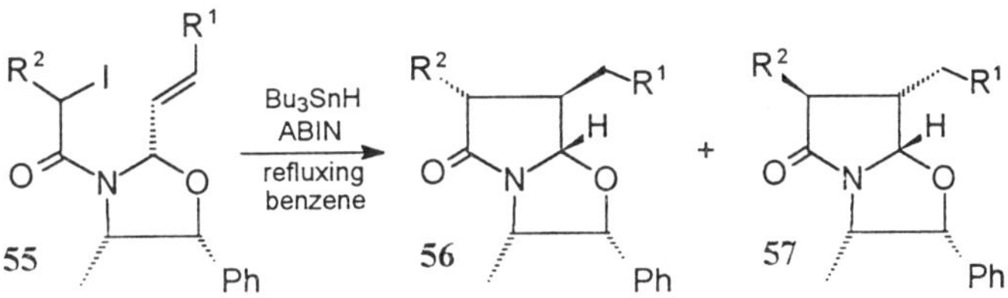

a. $R^1 = R^2 = H$, 75%, d.r. 97:3 **b.** $R^1 = CH_3$, $R^2 = H$, 60%, d.r. 97:3 **c.** $R^1 = H$, $R^2 = CH_3$, 85%, d.r. 97:3

Bicyclic compounds **59** and **60** have been prepared in a somewhat different manner which involves attack of a radical species to a conjugate double bond. The cyclisation of **58** proceeds with complete regioselection in a 5-*exo* mode, and diastereoselection is due to the stereogenic centres present in the chiral auxiliary bonded to amidic nitrogen.[29]

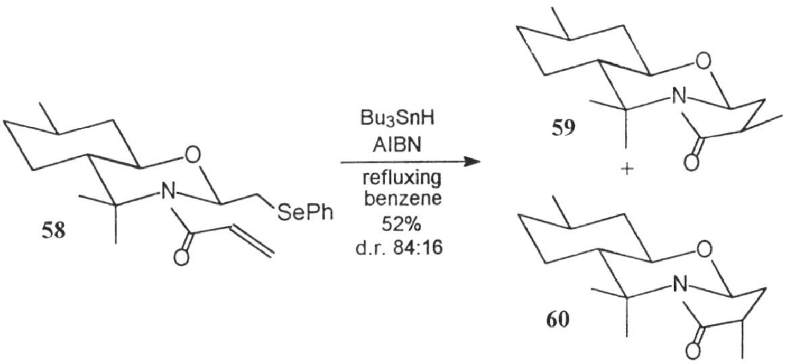

Starting from the pure isolated diastereomers **59** and **60**, 4-methylpyrrolidines in the enantiomerically pure form are prepared.

A radical attack to an α,β-unsaturated system, proceeding via a 5-*exo* mode, has stressed the dramatic effect of a Lewis acid on the stereochemical outcome of the cyclisation. In fact, when the chiral imide **61** is treated with Bu_3SnH in refluxing benzene, the cyclisation occurs with good diastereoselection. This result is ascribed to the chelating effect of the tin halide which is formed and constrains the two carbonyl groups in a *syn* conformation. On the contrary, $(Me_3Si)_3SiH$ alone affords mainly the reduction product, owing to the preferred conformation in which the carbonyl groups lie *anti* each other. If Bu_3SnCl is added to $(Me_3Si)_3SiH$, however, very high diastereoselection takes place, due to the chelating effect of the tin halide which constrains the carbonyl groups of **61** to lie *syn* each other.[30]

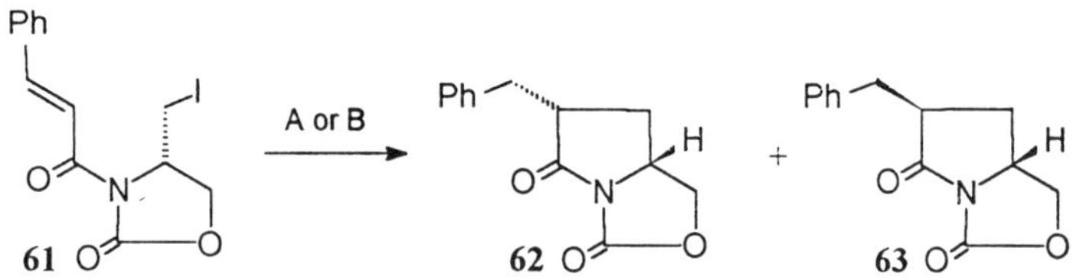

A. Bu_3SnH, AIBN, refluxing benzene, 82%, d.r. 99:1
B. $(Me_3Si)_3SiH$, Bu_3SnCl, AIBN, refluxing benzene, 87%, d.r. 99:1

Finally, the reduction of organomercury derivatives, performed with $NaBH_4$, produces a transient organomercury hydride which, by homolytic cleavage of the C-Hg bond, affords a radical which is able to attack a double bond. Thus, the reductive cleavage of the organomercury derivative **64** leads to the formation of tricyclic lactams **65** and **66** in moderate yield and good diastereoselection.[31]

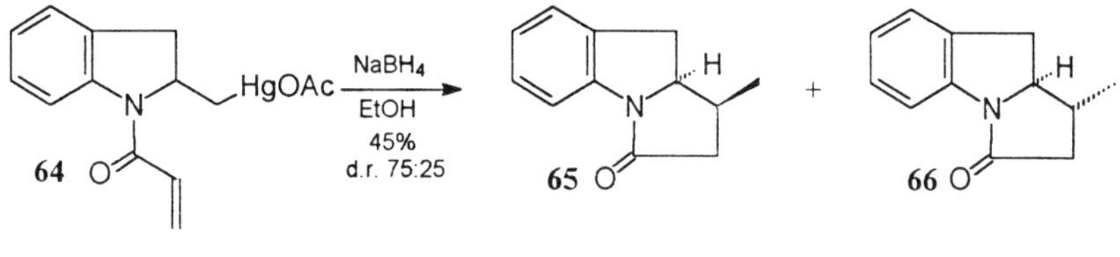

2.1.1.2. Cyclisations proceeding *via* a 5-*endo* mode

The radical approach to γ-lactams generally proceeds through a 5-*exo* cyclisation of *N*-allyl carbamoyl radicals. On the other hand, when *N*-vinyl carbamoyl radicals undergo cyclisation, the reaction affords γ-lactams through a "disfavoured" 5-*endo* mode. Thus, a number of α-halo or α-phenylthio enamides **67**, by reaction with Bu_3SnH in refluxing benzene, afford the corresponding γ-lactams **68** in moderate yield.[32,33]

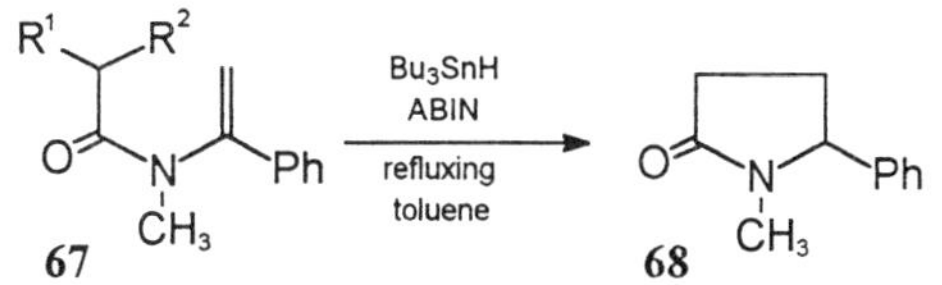

67 68

a. R^1 = Cl, R^2 = H, 17% **b.** R^1 = I, R^2 = H, 47% **c.** R^1 = R^2 = SPh, 75%

This reaction allows a convenient route to the alkaloid (±)-cotinine **2**. [2, 32]

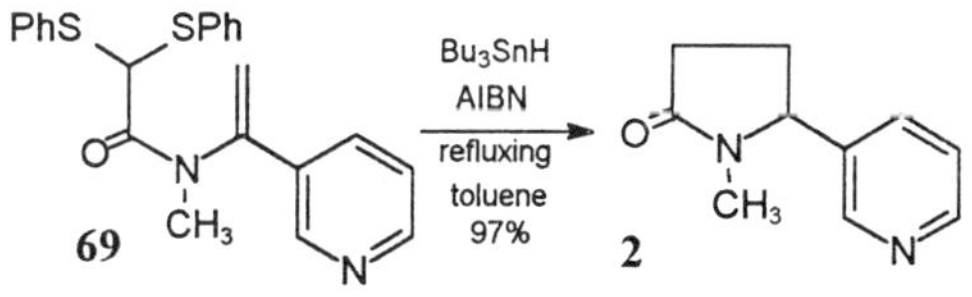

69 2

The formation of the 4,5-*cis*-isomer **71** as the major product of the cyclisation of enamide **70** is in agreement with a kinetically controlled tin hydride attack at the less hindered side of the intermediate radical. [32]

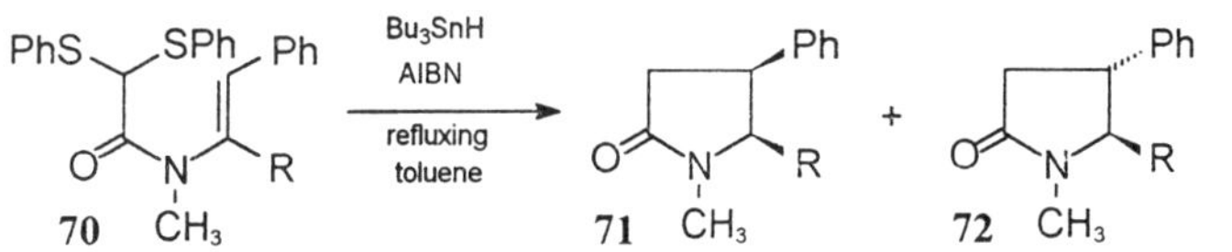

70 71 72

a. R = Ph, 92%, d.r. 73:27 **b.** R = Bn, 46%, d.r. 72:28 **c.** R = CH_3, 53%, 81:19

On the contrary, when 3,5-disubstituted γ-lactams **74** and **75** can arise by 5-*endo* cyclisation of the enamide **73**, only low diastereoselection is observed. [34]

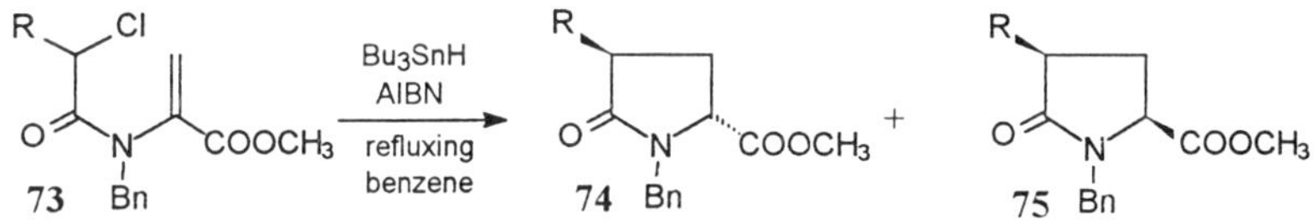

73 74 75

a. R = H, 52% **b.** R = CH_3, 47%, d.r. 63:37 **c.** R = Ph, 56%, d.r. 68:32
d. R = Cl, 33%, d.r. 75:25 **e.** R = PhS, 46%, d.r. 60:40

Ni powder in acetic acid is also effective to promote the cyclisation of an enamide via a 5-*endo* mode. In fact, by reacting the enamide **76** in the presence of 2-propanol, 2,2-dichloropyrrolidin-2-one **77** is first obtained, which then undergoes cleavage of a C-Cl bond to give the lactam **78** in moderate yield and good diastereoselection, although the configuration of the reaction products is not assigned. [35]

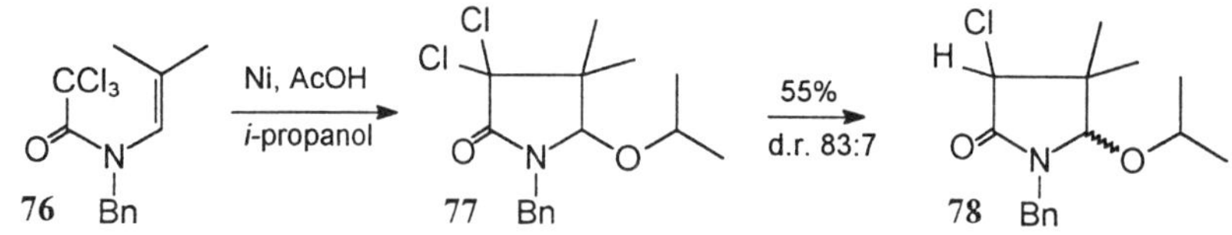

76 → 77 → 78

By cyclisation of α-halo enamides **79** and **80** indolizinones are exclusively formed with complete regioselection. When a substituent is present α to the carbonyl group, diastereomeric mixtures of **81** and **82** are obtained with moderate stereoselection.[36]

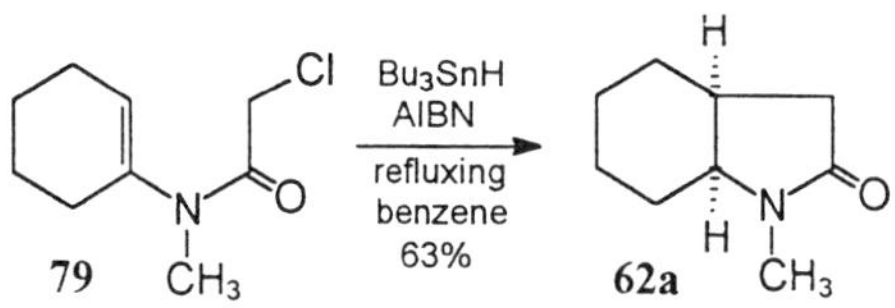

79 → **62a**

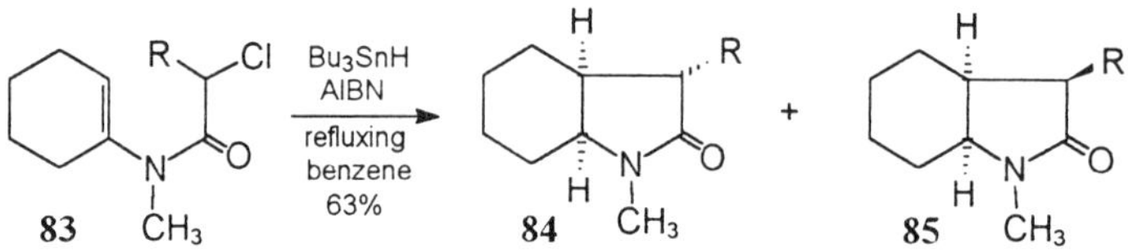

83 → **84** + **85**

a. R = CH_3, 73%, d.r. 84:16 **b.** R = Ph, 75%, d.r. 40:60

On the contrary, when the enamide **83**, derived from diphenylthioacetic acid, undergoes cyclisation, no asymmetric induction is observed and the reaction gives an equimolar diastereomeric mixture of indolizinones **84** and **85**.[36,37]

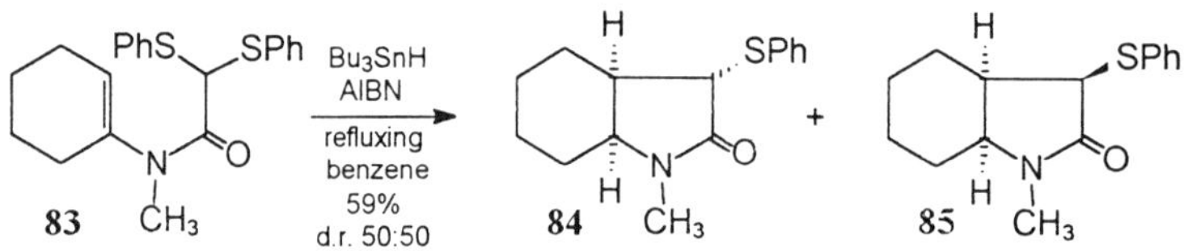

83 → **84** + **85**

Eventually, the 5-*endo* ring closure can give rise to a tandem process, useful for building an alkaloid framework. Thus, starting from the enamide **86**, the erithrinane derivative **87** is obtained in low yield but with complete diastereoselection.[38]

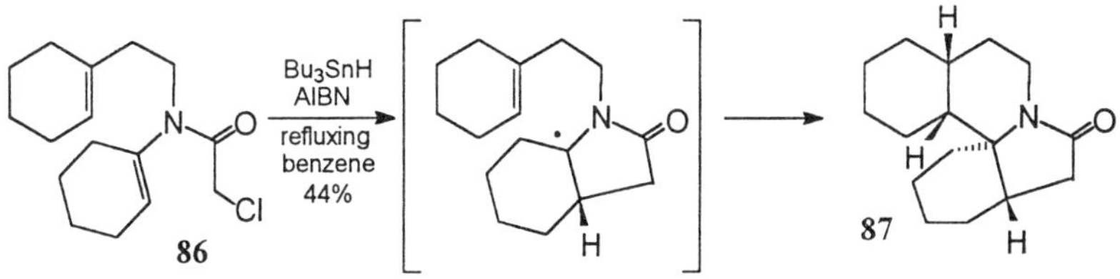

2.1.2. Reactions terminating *via* atom transfer

2.1.2.1. Cyclisations mediated by hexabutyl ditin

The radical cyclisation proceeding via halogen transfer, besides simple γ-lactams (see section **1.1.**) allows to prepare indolizinones, pyrrolizidinones and spiro lactams. This reaction is carried out by treatment of *N*-allyl iodoacetamides in the presence of an alkyl iodide. Iodoethane, 2-iodopropane and iodomethane lead to comparable yields of cyclised products.

The formation of both pyrrolizidinones **89,90** and indolizidinones **92,93** proceeds with complete regioselection and moderate to good diastereoselection. [25]

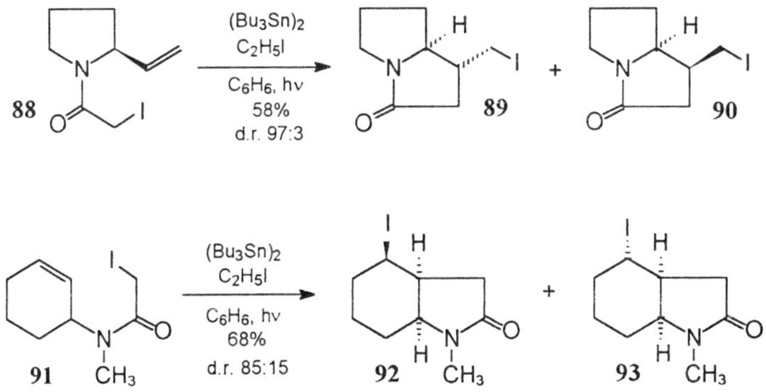

This approach has been utilised in a straightforward synthesis of the pyrrolizidine alkaloid trachelantamidine **94** in the enantiomerically pure form. [25]

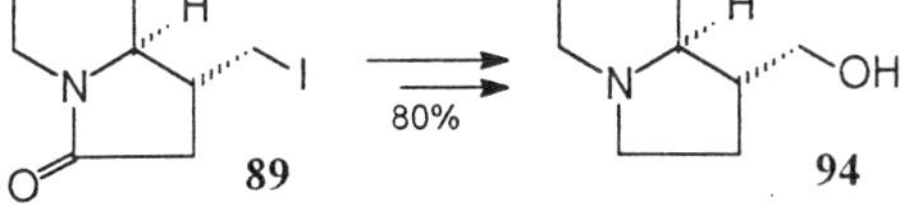

On the other hand, whereas the cyclisation of **95** proceeds with complete regioselection, leading to the spiro lactams **96** and **97**, very low diastereoselection is observed and a practically equimolar diastereomeric mixture is recovered from the reaction nixture. [28]

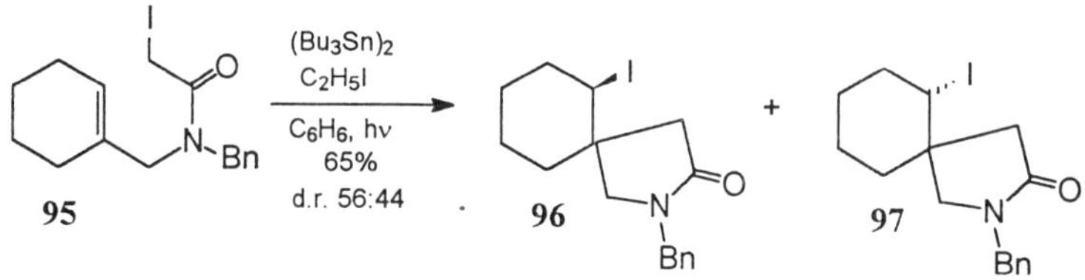

95 96 97

Finally, by cyclisation of compound **98**, containing *nor*-ephedrine as the chiral auxiliary, lactam **99** is exclusively obtained, probably due to steric constriction in the bicyclic system.[28]

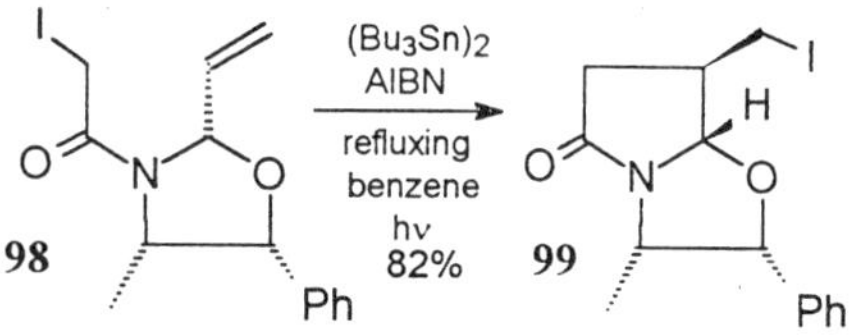

98 99

2.1.2.2. Cyclisations mediated by transition metals

2.1.2.2.1. Reactions mediated by Ru(II)

By treatment with $RuCl_2(PPh_3)_3$ in benzene, *N*-allyl trichloroacetamides **100** undergo cyclisation to give γ-lactams **101** in good yield. The cyclisation proceeds with complete regioselection, the products being formed through a 5-*exo*-closure, exclusively.[39]

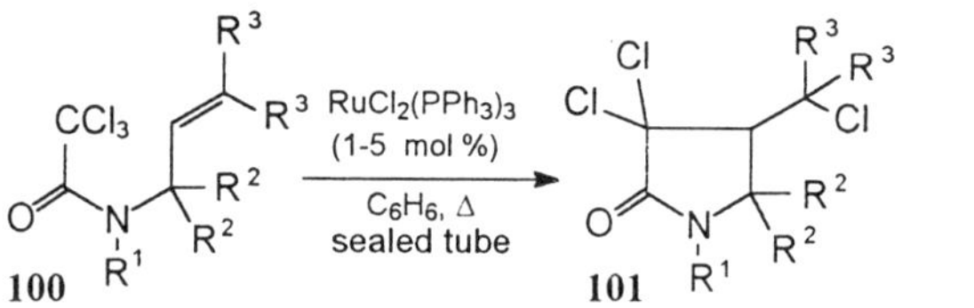

100 101

a. $R^1 = R^2 = R^3 = H$, 68% **b.** $R^1 =$ Allyl, $R^2 = R^3 = H$, 84%

b. **c.** $R^1 =$ Bn, $R^2 = R^3 = H$, 68% **d.** $R^1 = R^3 = H$, $R^2 = CH_3$, 77%

In order to obtain 4,5-disubstituted γ-lactams, **103** and **104**, secondary amides **102**, bearing a substituent α to the nitrogen are cyclised with high diastereoselection, due to 1,2-asymmetric induction, leading mainly to the 4,5-*trans* diastereomer **103**.[40]

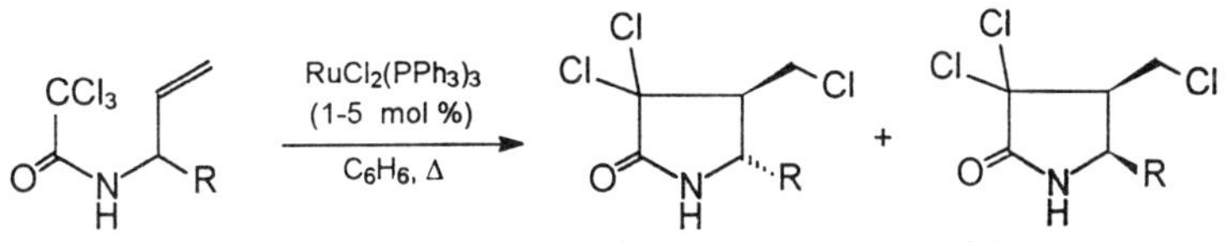

102 103 104

a. R = CH_3, 64%, d.r. 84:16 **b.** R = n-C_3H_7, 83%, d.r. 89:11
c. R = Ph, 60%, d.r. 100:0 **d.** R = CH_2OAc, 85%, d.r. 85:15

The same behaviour is observed also for the tertiary amide **105a**. On the other hand, when a tosyl group lies on the nitrogen, a reversal of the stereoselection is observed, and the 4,5-*cis* diastereomer becomes the major component of the reaction mixture. Thus, by properly choosing the amidic nitrogen substituent, both 4,5-*trans*- and 4,5-*cis*-dialkyl pyrrolidin-2-ones can be prepared. These results are in agreement with the cyclisation of analogous systems which proceeds *via* hydrogen transfer (see section **2.1.1.**).[23]

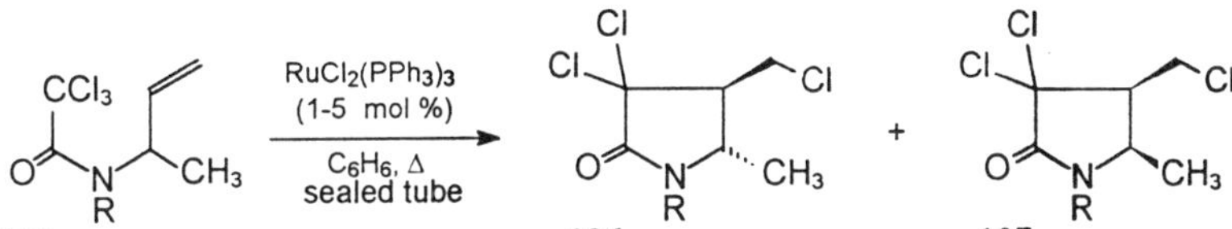

105 106 107

a. R = $CH_2C_6H_5$, 86%, d.r. 71:29 **b.** R = Ts, 74%, d.r. 21:79

Even 3,3',4-trisubstituted γ-lactams are prepared starting from *N*-allyl-*N*-tosyl dichloroacetamides **108**. In fact, by treating these compounds in benzene in the presence of a catalytic amount of $RuCl_2(PPh_3)_3$, 2-alkyl-2-chloro-3-chloromethyl-1-tosyl pyrrolidin-2-ones **109** and **110** are obtained in moderate yield. The *cis*-isomer is the major product of the reaction when R = H, CH_3, owing to a closure proceeding under kinetic control. On the contrary, when R = H or CH_3, the *trans*-isomer is prevalent since in this case an equilibrium takes place in the presence of the catalyst. In fact, when the cyclisation of the amide **108a** is carried out in refluxing isobutyronitrile in the presence of the CuCl(bpy) complex, (see section 1.1.2.2.2.) the pyrrolidin-2-ones **109a** and **110a** are obtained in 80% yield and 75:25 d.r., since the reaction is not reversible.[41,42]

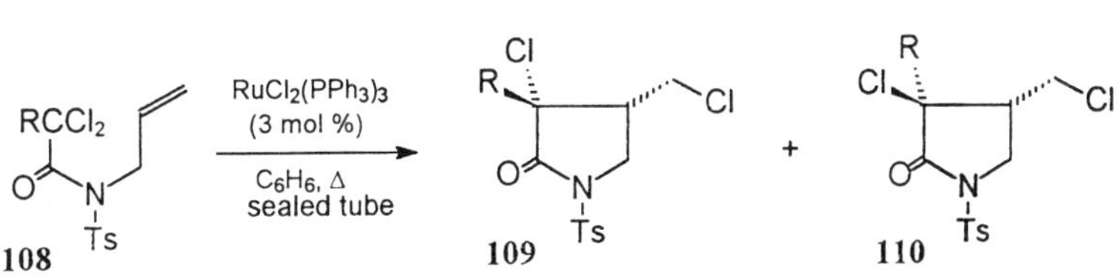

108 109 110

a. R = H, 65%, d.r. 22:78 **b.** R = CH_3, 69%, d.r. 27:73
c. R = *i*-C_3H_7, 54%, d.r. 77:23 **d.** R = $CH_2C_6H_5$, 57%, d.r. 95:5
e. R = *t*-C_5H_{11}, 70%, d.r. > 99:1

Eventually, by cyclisation of the *N*-allyl-*N*-tosyl-2-chloropropanamide **111** in the presence of $RuCl_2(PPh_3)_3$, a diastereomeric mixture of 3,4-disubstituted pyrrolidin-2-ones **112** and **113** is obtained in moderate yield with good diastereoselection, the 3,4-*trans*-isomer being the major component of the reaction mixture.[41]

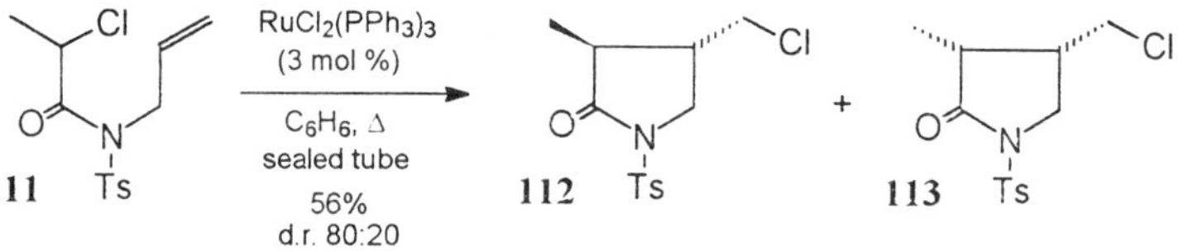

111 **112** **113**

N-Allyl α-chloro-α-thioacetamides **114** undergo cyclisation to pyrrolidin-2-ones by Ru(II) catalysed chlorine atom transfer. The reaction is carried out in hot benzene, in the presence of a catalytic amount of $RuCl_2(PPh_3)_3$ and proceeds with complete regioselection through a 5-*exo* mode, leading to 3,4-disubstituted pyrrolidin-2-ones **115** and **116** in moderate yield. The reaction proceeds with 1,2-asymmetric induction and the 3,4-*trans*-disubstituted isomer **115** is the major product.[43]

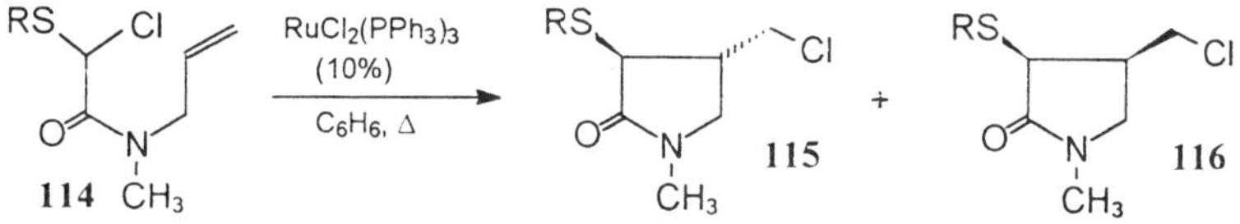

114 **115** **116**

a. R = C_6H_5, 62%, d.r. 70:30 **b.** R = CH_3, 57%, d.r. 60:40

With the aim to prepare enantiomerically pure pyrrolizidinic alkaloid trachelantamidine **94**, two similar approaches have been devised, both exploiting the Ru(II)-mediated atom transfer. In the former, the trichloroacetamide **117** is heated in the presence of the catalyst, to give the α-chloromethyl derivative **118**, as the only product, which is then converted into trachelantamidine **94**.[19,44]

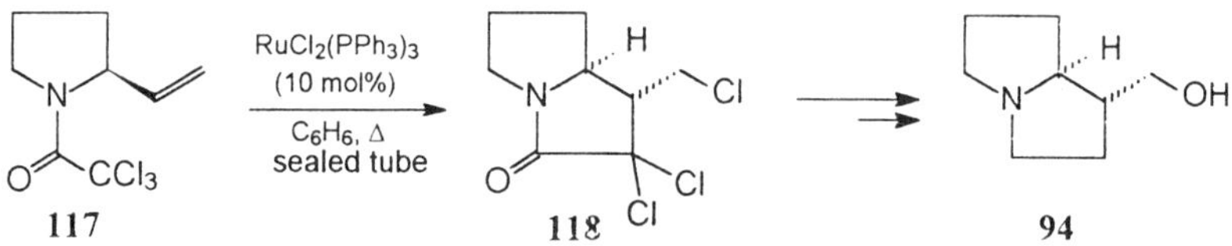

117 **118** **94**

Instead, in the latter approach, the chiral amide **119** is cyclised to give the bicyclic lactams **120** and **121** in moderate to good yield. A diastereomeric mixture is formed with moderate stereoselection, but in both diastereomers the chloromethyl group results to be exocyclic.[43]

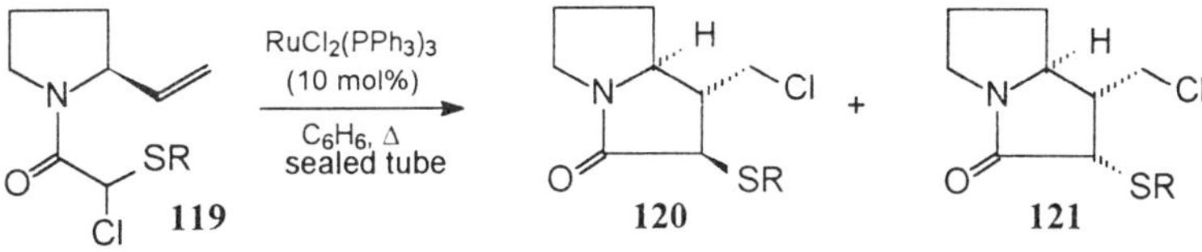

119 120 121

a. R = C_6H_5, 59%, d.r. 84:11 **b.** R = CH_3, 67%, d.r. 70:30

Thus, after cleavage of the C-S bond with Ni-Raney, nucleophilic displacement of the chlorine and final removal of the carbonyl group, trachelantamidine **94** is obtained in the enantiomerically pure form.[43]

120 + **121** 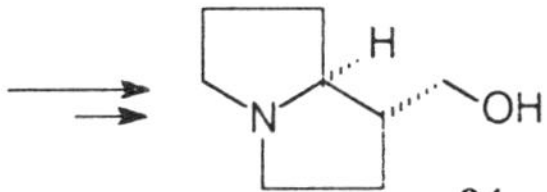

94

Pyrrolizidinones **123** and spiro pyrrolidin-2-ones **125** are prepared in good yield and complete diastereoselection, although the configuration of the spiro derivative remains unassigned.[45]

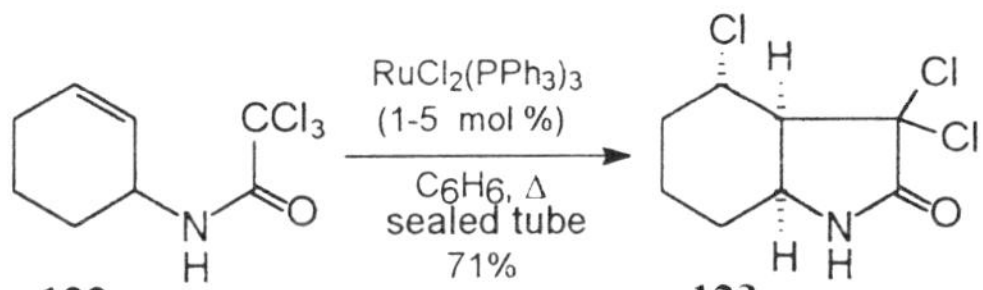

122 **123**

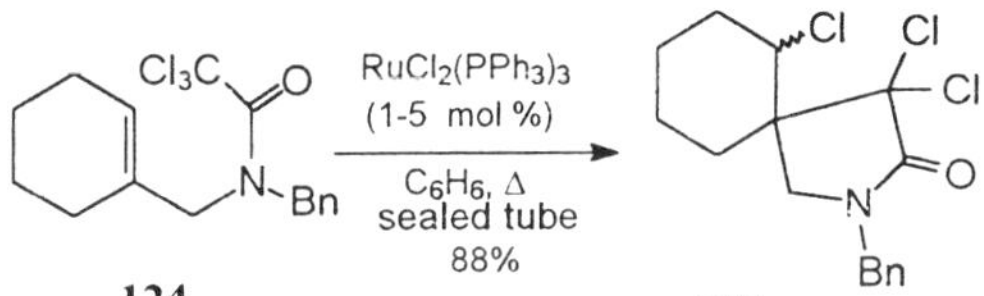

124 **125**

A rather surprising result is indeed observed when *N*-cyclohexenyl amides **126** and **129** undergo atom transfer cyclisation. In fact, by treating **126** with a catalytic amount of $RuCl_2(PPh_3)_3$ in hot benzene, the bicyclic lactams **127** and **128** are obtained in good yield and high diastereoselection, the chlorine atom resulting to be exocyclic.[43]

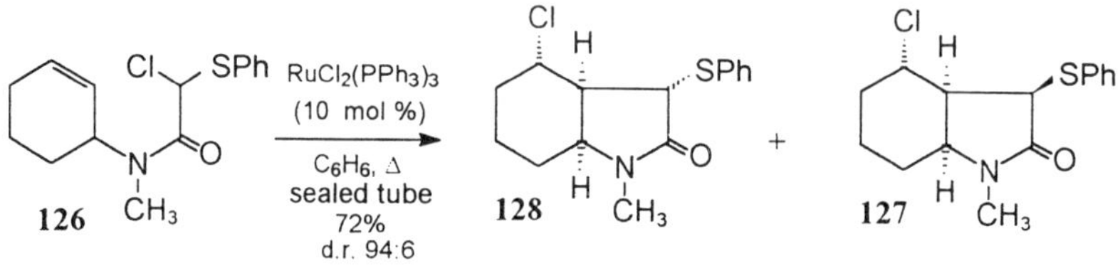

On the contrary, when **129a** or **129b** are treated under the same conditions, a single product, **130a** or **130b**, respectively, is obtained in moderate yield and complete diastereoselection. In this case, however, the stereochemical outcome of the reaction is reversed, the chlorine atom being endocyclic.[43,46]

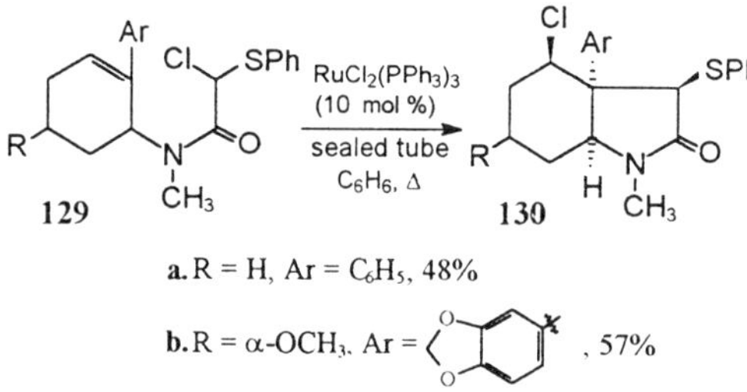

This result indicates that in the intermediate (**A**), leading to **127**, the cyclisation occurs *via* an *anti*-addition to the double bond, whereas in the intermediate (**B**), leading to **130**, the cyclisation proceeds via a *syn*-addition. This latter behaviour can be ascribed to the steric bulk of the angular phenyl group, which is large enough to direct the chlorine atom to the concave face of the radical intermediate.[43]

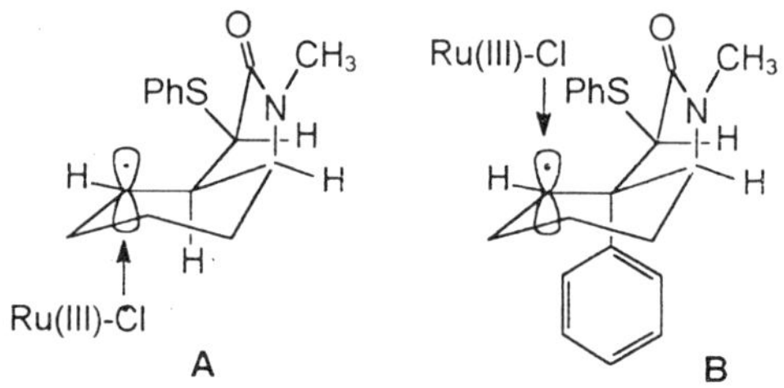

2.1.2.2.2. Reactions mediated by Cu(I)

N-Allyl trichloroacetamides **131** can be cyclised to γ-lactams by treatment with a catalytic amount of either CuCl alone or CuCl and a bidentate amine, such as 2,2'-bipyridine (bpy). The reaction gives 2,2-dichloro-3-chloroalkylpyrrolidin-2-ones **132**, exclusively, in very good yield *via* a 5-*exo* closure.[47]

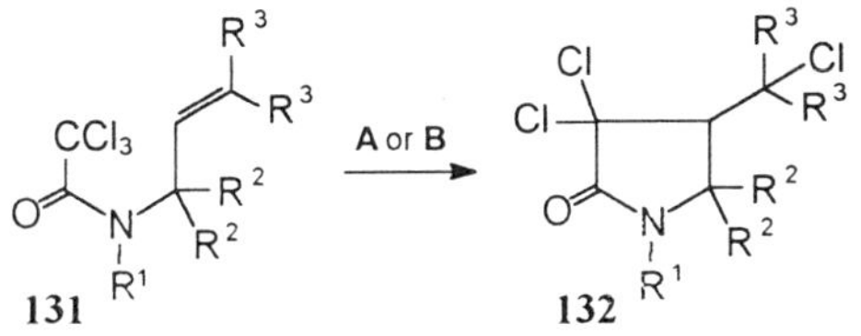

A: CuCl, C_6H_6, Δ(sealed tube) **B:** CuCl(Bpy), CH_2Cl_2, -70 °C

a. $R^1 = R^2 = R^3$ = H, Method **A**. 57% **b.** R^1 = Allyl, $R^2 = R^3$ = H, Method **A**, 90%

c. R^1 = Bn, R^2 = H, R^3 = CH_3, Method **B**, 98% **d.** R^1 = Bn, R^2 = CH_3, R^3 H, Method **B**, 96%

It is noteworthy that the formation of 4,5-disubstituted pyrrolidin-2-ones proceeds with opposite stereoselection, depending on the substituent on the nitrogen. In fact, *N*-alkyl trichloroacetamides **133a-c** cyclise to give predominantly the 4,5-*trans*-disubstituted product **134**, whereas *N*-tosyl, *N*-Cbz and *N*-*t*-Boc trichloroacetamides **133d-g** lead mainly to the 4,5-*cis*-isomer **135** (see sections 1.1.1. and 1.1.2.2.1.).[47]

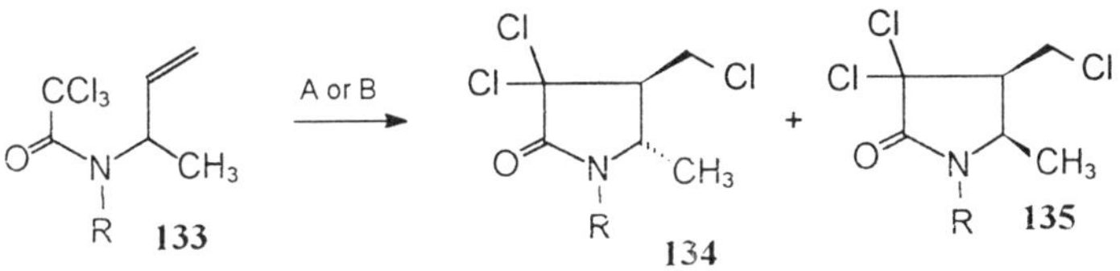

A: CuCl, C_6H_6, Δ, sealed tube
B: CuCl(bpy), CH_2Cl_2, -70°C

a. R = $CH_2C_6H_5$, method **A**, 78%, d.r. 70:30
b. R = $CH_2C_6H_5$, method **B**, 98%, d.r. 88:12
c. R = CH_3, method **B**, 93%, d.r. 95:5
d. R = Ts, method **A**, 90%, d.r. 27:73
e. R = Ts, method **B**, 99%, d.r. 22:78
f. R = Cbz, method **A**, 99%, d.r. 25:75
g. R = Cbz, method **B**, 83%, d.r. 14:86
h. R = *t*-Boc, method **A**, 74%, d.r. 29:71
i. R = *t*-Boc, method **B**, 74%, d.r. 15:85

Thus, the Cu(I) or Ru(II) catalysed cyclisations of *N*-allyl trichloroacetamides and the tin-mediated reductive cyclisations of analogous systems showed that either 4,5-*trans*-or 4,5-*cis*-disubstituted pyrrolidin-2-ones can be prepared with good diastereoselection by simply changing the substituents on the amide nitrogen, although the origin of this change in diastereoselection remains unclear. However, since the diastereomeric ratios obtained from these cyclisations are very similar for the same products, it is likely that these reactions proceed via similar transition states.

N-Allyl α,α-dichloromalonamide **136**, when treated with a catalytic amount of CuCl(bpy) in refluxing isobutyronitrile, affords pyrrolidin-2-ones **137** and **138** with moderate yield and low diastereoselection. However, when the substrate **136** reacts in the presence of an equimolar amount of catalyst, the γ-lactams **137** and **138** are obtained with reversed stereoselection, together with a minor amount of 3-aza-2-oxobicyclo[3.1.0]hexane derivative **139a**. [48]

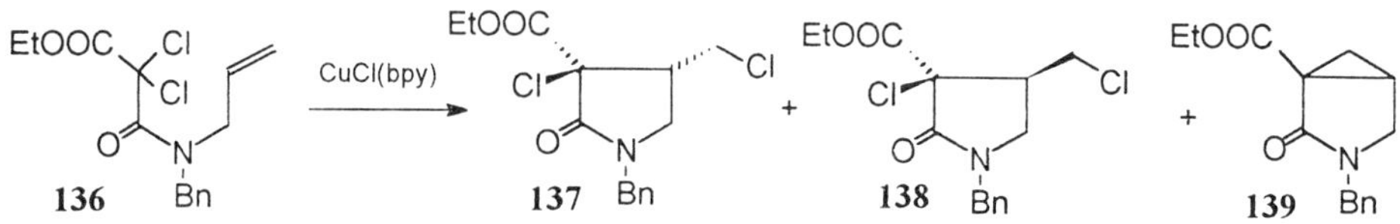

0.1 equiv catalyst:substrate, 58% overall yield, **137**:**138**:**139** ratio 60:40:0
1.0 equiv catalyst:substrate, 86% overall yield, **137**:**138**:**139** ratio 28:42:30

Finally, when the reaction is carried out by using a 2.1 Cu(I):substrate ratio, the bicyclic compound **139** is exclusively obtained in good yield. [48]

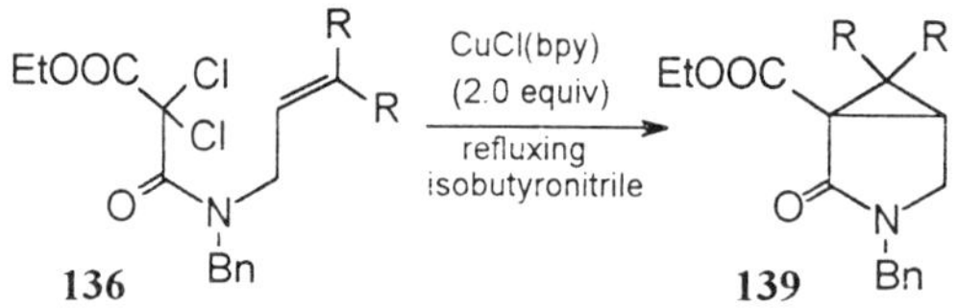

a. R = H, 77%; **b.** R = CH_3, 70%

This result is probably due to an intramolecular S_N2 reaction, following the reduction of the intermediate radical by Cu(I). [48]

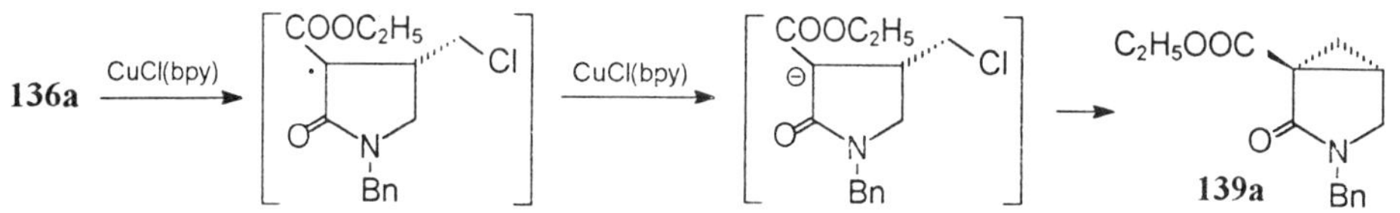

2.1.3. Reactions involving fragmentation

The fragmentation of a C-Se bond, followed by cyclisation, allows to prepare a γ-lactam. In fact, by irradiation of the phenylselenoamide **141** in benzene, the lactam **142** is obtained, in good yield and with complete diastereoselection. [49]

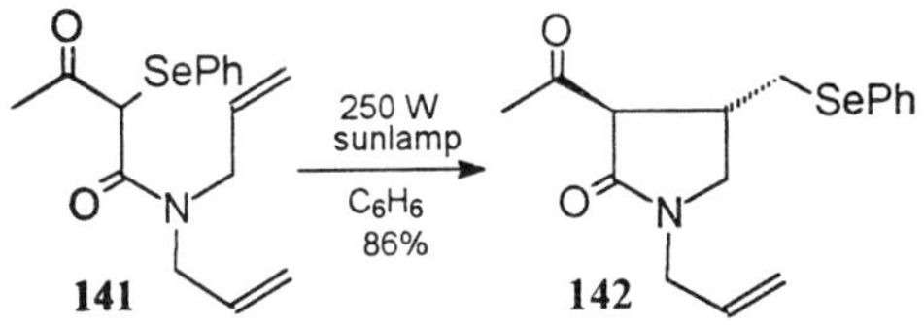

When α-haloacetamides are treated with Ni in acetic acid, the C-halogen bond is cleaved to give a radical which, by a reaction proceeding at higher speed than hydrogen transfer, undergoes termination depending on the reagent which is present in the reaction medium. A variety of functional groups can be introduced in the termination step. In fact, when the reaction is carried out in the presence of PhSeSePh, the corresponding phenylselenomethyl derivatives **144** and **146** are obtained. Following this approach bicyclic compounds are prepared as single products too.[17]

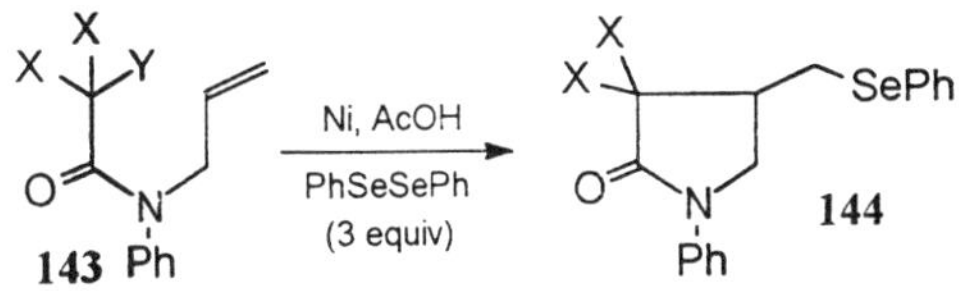

a. X = Cl, Y = Cl, 55% **b.** X = CH_3, Y = Br, 73%

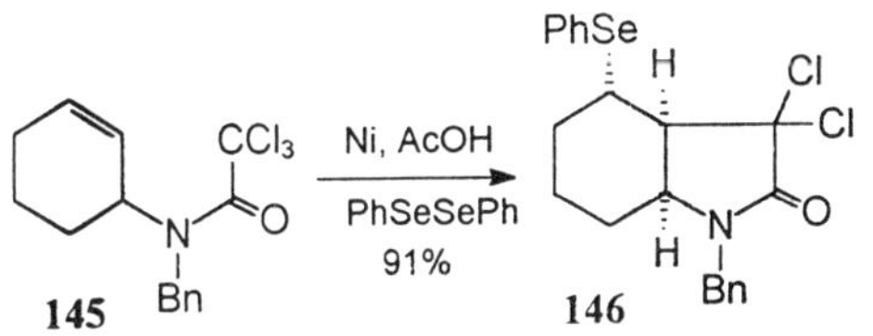

On the other hand, if the cyclisation is conducted in the presence of TEMPO, compound **148** may be obtained, which can lead to the corresponding alcohol by cleavage of the N-O bond.[50]

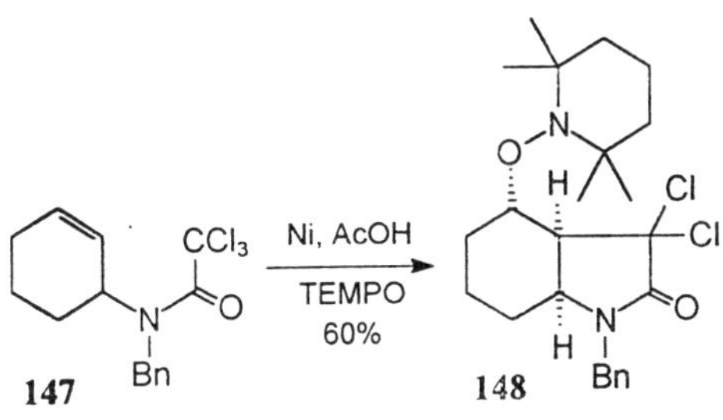

An interesting approach to γ-lactams based on fragmentation methodology exploits xantates. In fact, upon heating *N*-allyl α-xantyl acetamides **149** in chlorobenzene, in the presence of di-*t*-butyl peroxide, a cyclisation occurs via a 5-*exo* mode, to give the corresponding γ-lactams **150**.[50]

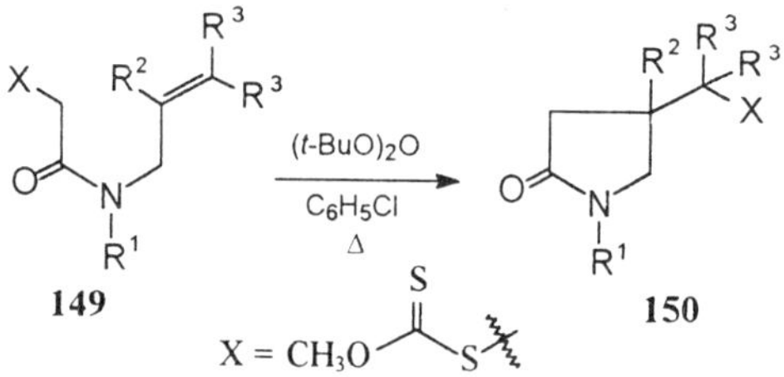

149 **150**

a. R^1 = Ph, R^2 = R^3 – H, 60% **b.** R^1 = Allyl, R^2 = R^3 = H, 55% **c.** R^1 = Bn, R^2 = CH_3, R^3 = H, 72%
d. R^1 = Bn, R^2 = H, R^3 = CH_3, 74%

The bicyclic lactam **152** is obtained in moderate yield as single diastereomer. On the other hand, the spiro lactam **154** is isolated as an equimolar diastereomeric mixture.[50]

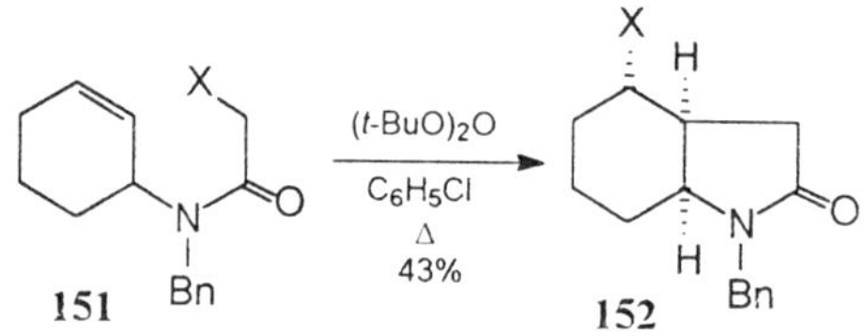

151 **152**

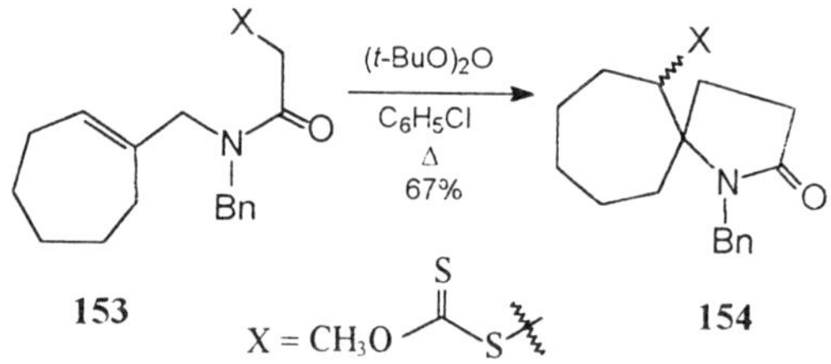

153 **154**

This approach results useful since the reaction products can be further elaborated through the radical or ionic chemistry of the organo sulphur derivatives. In addition, the xantate group can be removed with Bu_3SnH to give the corresponding desulphurised products.

The transfer of the xantate group is also obtained starting from *N*,*N*-diallyl bromoacetamides by reaction with triphenyltin neopentyl xantate **155**.[50]

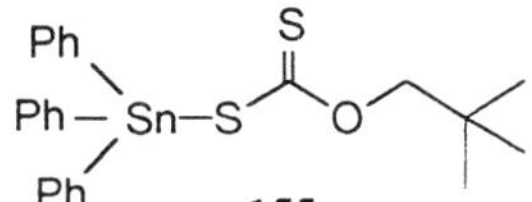

155

By using this reagent, the reaction proceeds with complete regioselection and good diastereoselection, the 3,4-*trans*-isomer **159** being the major component of the reaction mixture. Moreover there is no need for high dilution or slow addition of the tin reagent is necessary, in contrast to similar stannane-mediated cyclisations.[51]

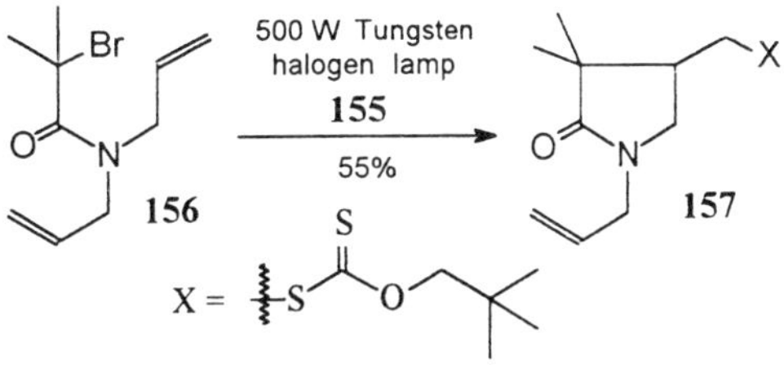

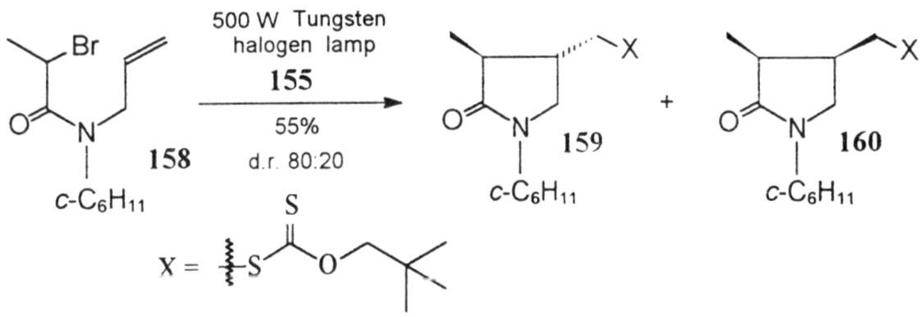

Finally, by irradiation of a mixture containing vinylcyclopropanes **161** and diphenyl disulphide, bicyclic γ-lactams **162** are formed in moderate yield. The reaction involves cyclopropane ring opening, followed by tandem ring closures, both proceeding via a 5-*exo* mode. The final step consists in the elimination of a phenylthio radical, leading to the formation of a vinyl group. The high stereoselection of the 5-*exo* closure, which gives *cis*-fused γ-lactams **162**, exclusively, is probably due to steric constraints in the cyclisation step.[52]

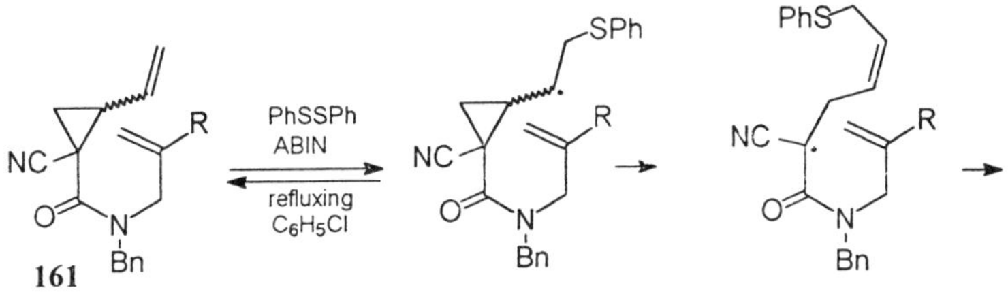

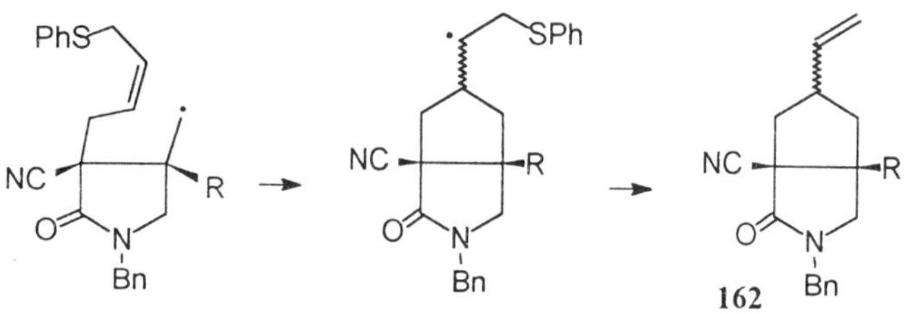

162

a. R = COOEt, 63%, **b.** R = CH_3, 50%

2.1.4. Reactions performed under oxidising conditions

The Mn(III) oxidation of C-H bonds activated by electron withdrawing groups does not follow a chain mechanism, and stoichiometric amounts of Mn(III) are required for generating a radical via a Mn(III) enolate.

When a double or a triple bond is present in the molecular framework, an intramolecular cyclisation occurs with good regio and stereoselection. Thus, when the *N*-allyl amides **163** containing the (*S*)-phenylethylamine as chiral auxiliary, are treated with $Mn(OAc)_3$ in the presence of $Cu(OAc)_2$, 3,4-*trans* disubstituted γ-lactams **164** and **165** are exclusively obtained in moderate to good yield. Moreover, cyclisation proceeds with moderate diastereoselection and the products can be easily separated by silica gel chromatography.[53,54]

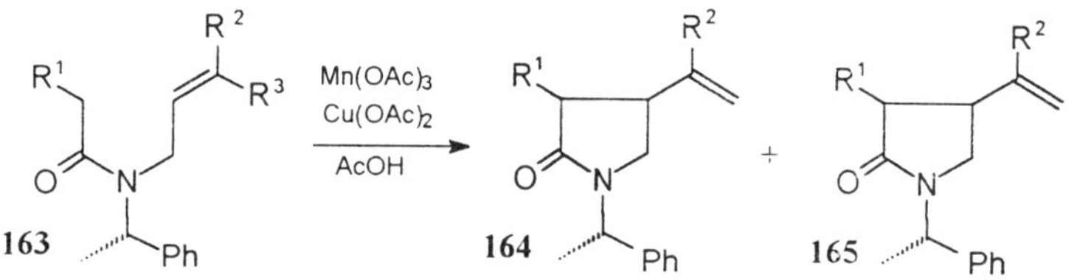

163 164 165

a. *Substrate*: R^1 = CH_3OOC, R^2 = CH_3, R^3 = H. *Product*: R^1 = CH_3OOC, R^2= H, 53%, d.r. 70:30
b. *Substrate*: R^1 = CH_3CO, R^2 = CH_3, R^3 = H. *Product*: R^1= CH_3CO, R^2= H, 53%, d.r. 68:32
c. *Substrate*: R^1 = CH_3CO, R^2 = R^3 = CH_3. *Product*: R^1 = CH_3CO, R^2 = CH_3, 68%, d.r. 67:33

The pure isolated diastereomer **164a** is employed in a synthesis of (*R*)-pyrrolidineacetic acid **166**, an inhibitor of GABA uptake in synapses.[54]

164a 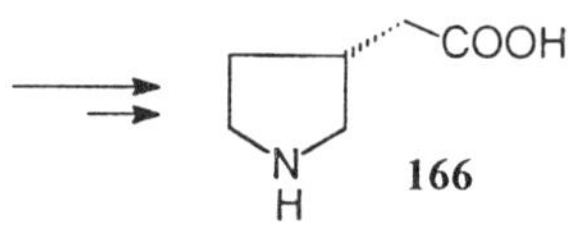 **166**

On the other hand, when *N*-allyl β-oxoamides **167** are cyclised under the same conditions, diastereomeric mixtures of 3-aza-2-oxobicyclo[3.1.0]hexanes, **168** and **169**, are obtained in moderate yield and stereoselection. The bicyclic products obtained from the cyclisation can be intermediates in the preparation of conformationally restricted amino acids.[55]

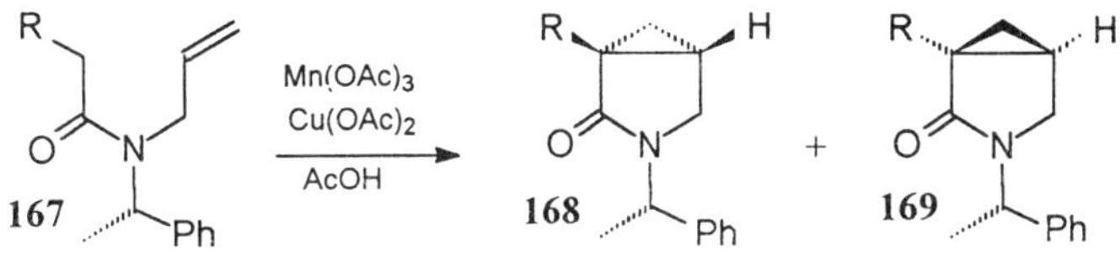

a. R = CH_3OOC, 42%, d.r. 67:33 **b.** R = CH_3CO, 47%, d.r. 68:32

The cyclisation reaction can be performed by using Mn(III) in ethanol to give spiro derivatives **171** and **172** in moderate yield and good diastereoselection starting from amides **170** containing carbocyclic rings.[56]

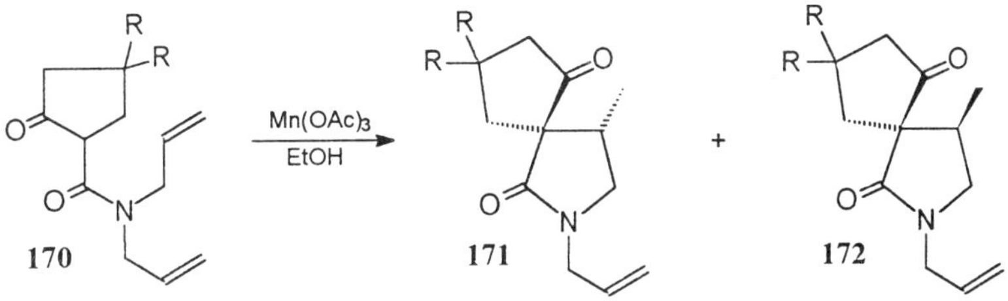

a. R = H, 50%, d.r. 94:6 **b.** R = CH_3, 50%, d.r. 94:6

Following this approach, even β-methylene spirolactams **174** and **175** have been prepared in moderate yield.[56]

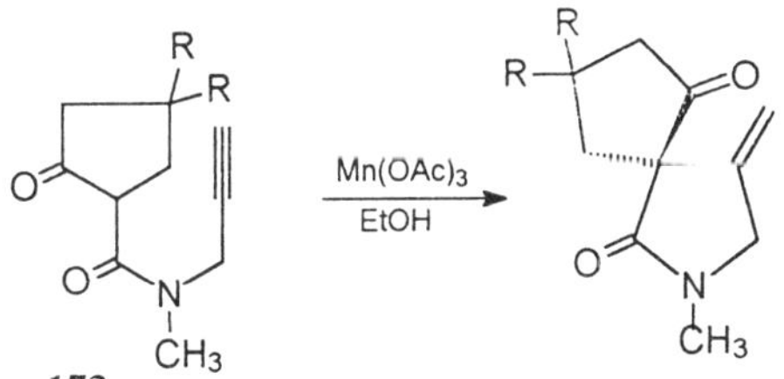

174. R = H, 60%
175. R = CH_3, 55%

The spiro derivatives **177** and **178** can be also prepared in low yield but with very high stereoselection by cyclisation of the enamines **176**.[57]

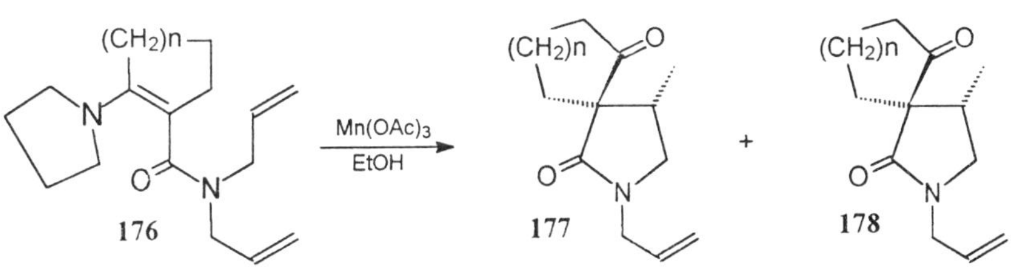

a. n = 1, 38%, d.r. 95:5 **b.** n = 2, 34%, d.r. = 95:5

2.2. Cyclisations involving carbocation intermediates

α-Thio carbocations **179** which can initiate a cationic alkene cyclisation are easily obtained under the Pummerer reaction conditions, starting from α-acyl sulphoxides and sulphides.[58]

179

These species are key intermediates for the synthesis of γ-lactams when *N*-allyl amides bearing a sulphoxide or an alkylthio group in the α position are the starting material. In fact the sulphoxides **180**, upon treatment with trifluoroacetic anhydride, give in good yield and with complete regioselection and moderate stereoselection the corresponding γ-lactams **181** and **182** as diastereomeric mixtures.[58]

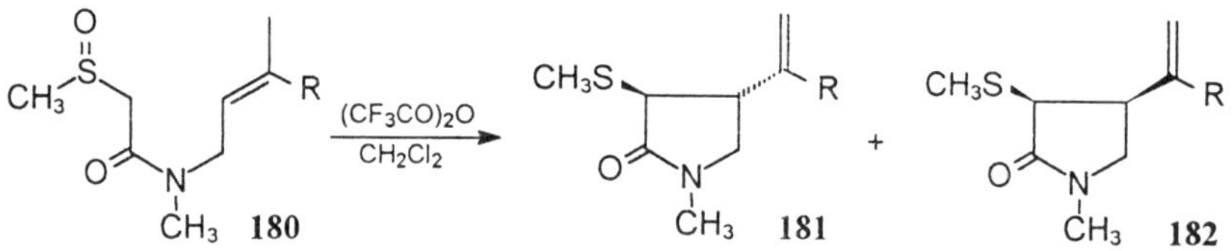

a. R = H, 92%, d.r. 69:31 **b.** R = CH_3, 50%, d.r. 50:50

As an alternative route, by treating the unsaturated methylthioacetamide **183** with phenyl iodosyl *bis*-trifluoroacetate (PIFA), the 3-methylthio-4-ethenyl pyrrolidin-2-one **184** is obtained in moderate yield as an equimolar diastereomeric mixture.[59]

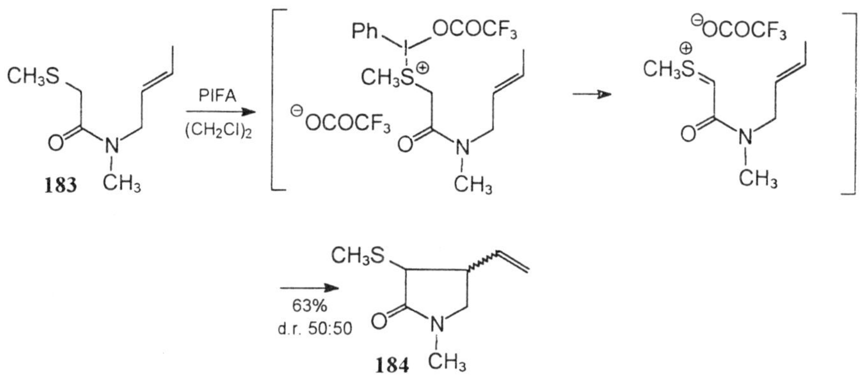

An intermediate cation can arise also when *N*-allyl α-chloro-α-methylthio acetamides react in the presence of a Lewis acid. Thus, by treating either **185** or **188** with $SnCl_4$, the corresponding γ-lactams **186**,**187** or **189**, respectively, are obtained in moderate yield and good diastereoselection.[58]

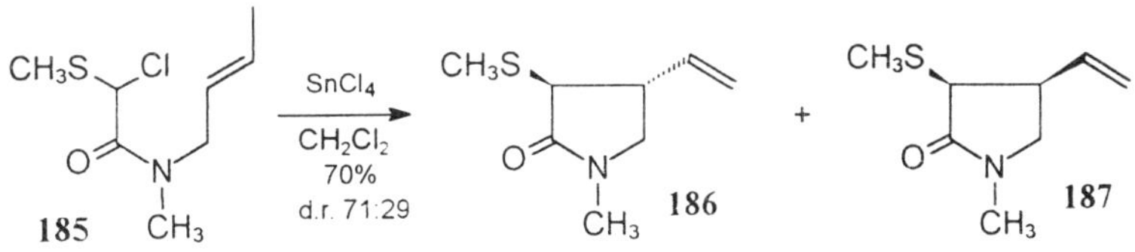

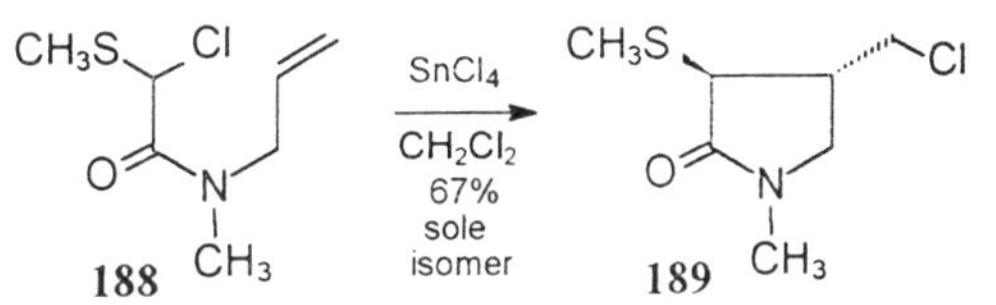

It is also worth mentioning that the chloro derivatives **190** give, upon heating at 100 °C and after silica gel chromatography, the corresponding 5-hydroxy derivatives **191** via a 5-*endo* mode as diastereomeric mixtures. This result is probably due to the reaction of an intermediate acyliminium salt with the water adsorbed on silica gel.[60]

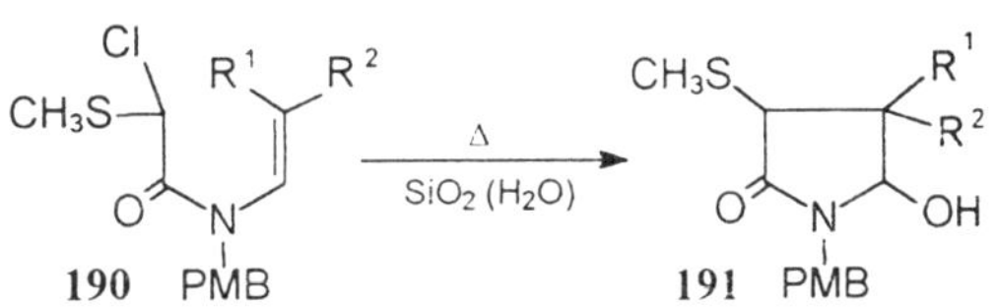

a. $R^1 = R^2 =$ Ph, 58%   **b.** $R^1 = CH_3$, $R^2 = C_2H_5$, 64%   PMB = *p*-methoxybenzyl

2.3. Cyclisations involving carbanion intermediates

2.3.1. Reactions proceeding *via* conjugate addition

A convenient approach to prepare the γ-lactam ring consists in the intramolecular conjugate addition of amide enolates to an α,β-unsaturated ester moiety. The reaction is carried out starting from chiral β-oxo or β-phenylsulphonyl amides, **192**, containing (S)-phenylethylamine as the chiral auxiliary. The cyclisation proceeds with complete 1,2-asymmetric induction, since 3,4-*trans*-disubstituted γ-lactams **193** and **194** are exclusively obtained. Owing to the presence of the chiral auxiliary, a good diastereoselection is observed, which depends on the reaction conditions. In fact, when the cyclisation proceeds under thermodynamic control (using NaH in

THF), the diastereomer **193** is the major component of the reaction mixture. On the contrary, if the cyclisation of **192** is carried out under kinetic control (using Na ethoxide in ethanol), the diastereomer **194** prevails. [61,62]

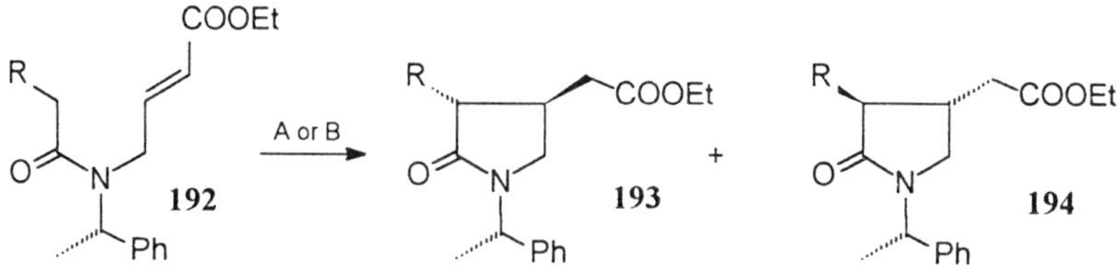

A: NaH, THF, -78 °C
a. R = CH_3CO, 80%, d.r. 70:30
b. R = $PhSO_2$, 75%, d.r. 80:20
c. R = $COOCH_3$, 75%, d.r. 80:20

B: EtONa, EtOH, -78 °C
a. R = CH_3CO, 80%, d.r. 15:85
b. R = $COOCH_3$, 76%, d.r. 17:83

The pure diastereomer **193c** has been employed in the synthesis of (*S*)-pyrrolidineacetic acid **195**. [61]

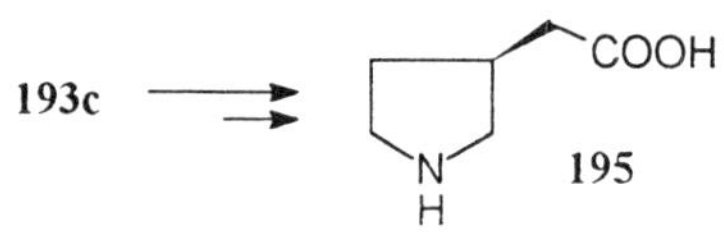

A tandem conjugate addition-S_N2 reaction has been used for the construction of the 3-aza-2-oxobicyclo[3.1.0] ring system. In fact, by treatment with a base, the dichloroacetamide **196** affords first the γ-lactam ring and subsequently the fused cyclopropane ring, to give in moderate yield the bicyclic product **197** with complete diastereoselection. [63]

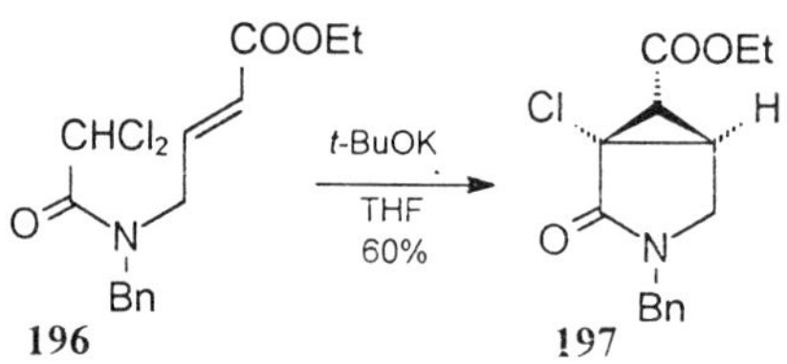

196 **197**

2.3.2. Reactions proceeding *via* intramolecular alkylation or acylation

The intramolecular alkylation of an acylated oxazolidine derived from L-serine constitutes a synthetic route to the chiral 4-methoxycarbonyl-4-hydroxymethyl pyrrolidin-2-one **200**. The bromoderivative **198** on treatment with LDA gives the bicyclic compound **199** with complete stereoselection, albeit in low yield. Further hydrolysis of the amidal affords the pyrrolidin-2-one **200** in good yield. [64]

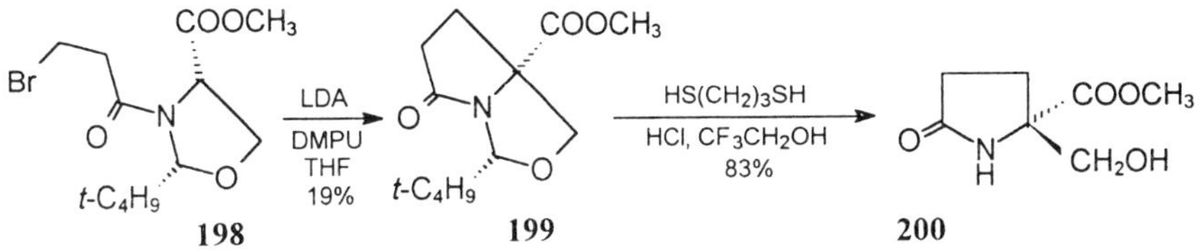

The "iodocarbocyclisation reaction", which involves the intramolecular attack of a carbon nucleophile on a double bond activated by an electrophilic iodinating agent, is useful in order to obtain pyrrolidin-2-ones. In fact the reaction of the *N,N*-diallyl 3-oxobutanamide **201** with NaH in THF, followed by addition of *N*-iodosuccinimide (NIS), allows to prepare the cyclic derivative **202** in moderate yield as an equimolar diastereomeric mixture.[65]

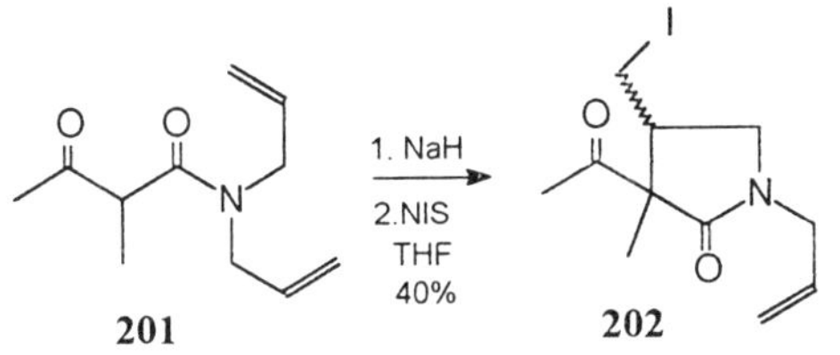

Also spirolactams **204** can be obtained in moderate yield, when cyclic amides such as **203** undergo cyclisation, but the configuration of the products remains unassigned.[65]

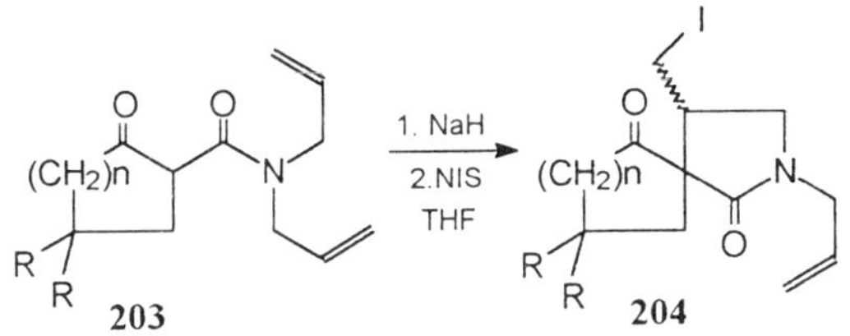

a. n = 1, R = H, 72% **b.** n = 1, R = CH_3, 81% **c.** n = 2, R = H, 48%

4-Oxopyrrolidin-2-one **207**, an intermediate for the synthesis of the pyrrolizidine ring system, can be easily obtained by an intramolecular acylation reaction. Thus, starting from *N*-acetyl proline **205**, the corresponding *N*-methyl-*N*-methoxy amide **206** is prepared, *via* a mixed anhydride. Then, by using LHMDS as the base, cyclisation occurs in good yield to give the product **207** as a single enantiomer, which is stereoselectively converted into the hydroxypyrrolizidine **208** by reduction with $NaBH_4$.[66]

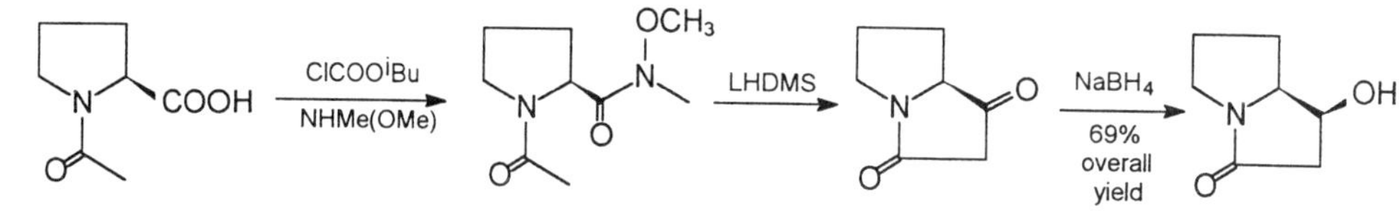

205 **206** **207** **208**

2.4. Cyclisations involving carbene intermediates

2.4.1. Carbene insertion into a C-H bond

The insertion reaction involves a carbene which interposes itself in a preexisting bond and this generally occurs at the C-H bond which is closest to the generated carbene. When an intramolecular insertion takes place, a cyclic derivative is formed.

In a few cases, the carbene is generated by irradiation. Thus the photochemical decomposition of *N,N*-diethyldiazoacetamide **209** affords the two possible intramolecular insertion products and the γ-lactam **211** is the minor component of the reaction mixture. [67]

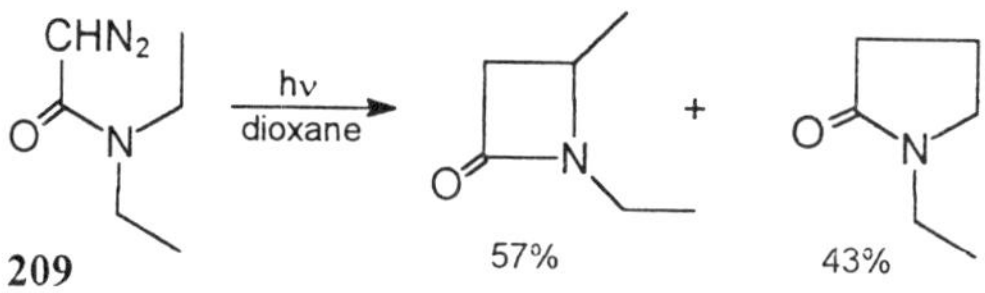

209 **210** **211**

On the other hand, dirhodium(II) carboxylates are the catalysts of choice for the catalytic C-H insertion reaction. When α-diazoamides are the starting products, γ-lactams are mainly formed, although β-lactams too can be obtained because of heteroatom activation for insertion into the C-H adjacent to the nitrogen. In fact, by treating the diazoacetamides **212** with a catalytic amount of $Rh_2(OAc)_4$ in refluxing dichloromethane, the pyrrolidin-2-ones **213** are obtained in good yield. [68]

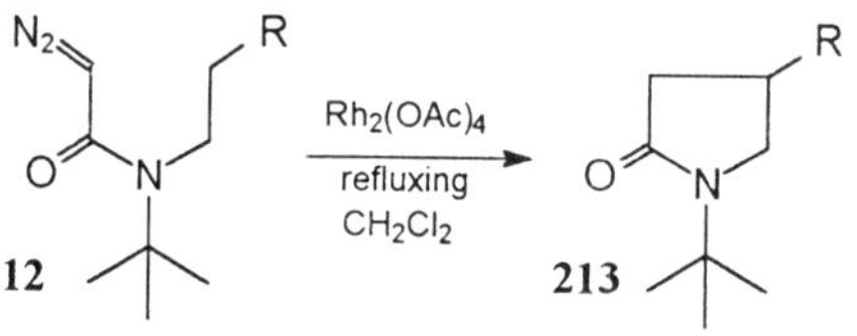

212 **213**

a. R = C_2H_5, 95%; **b.** R = C_6H_5, 68%

In addition, it is worth mentioning that product distribution is strongly affected by choosing the dirhodium(II) ligand, and this behaviour depends on the electron demand from ligands on the Rh(II) carbene intermediate. [69]

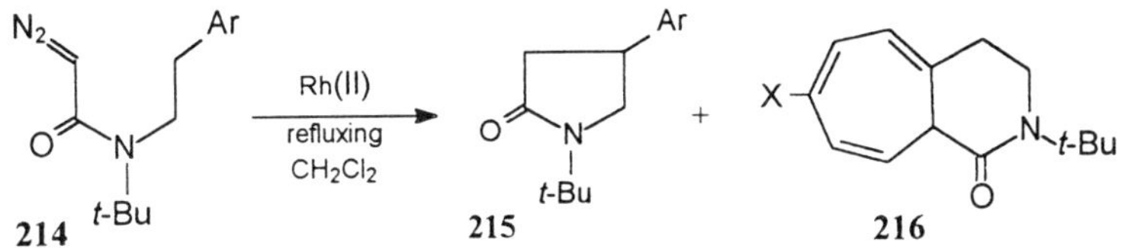

a. Ar = C_6H_5, X = H $Rh_2(OAc)_4$, 85%, 32:68
b. Ar = p-$CH_3OC_6H_4$, X = CH_3O Rh_2(perfluorobutyrate)$_4$, 80%, 5:95
c. Ar = p-$NO_2C_6H_4$, X = NO_2 Rh_2(caprolactam)$_4$, 82%, 97:3

The C-H insertion of a carbene in the amide **217** gives rise to the 4,5-*trans*-disubstituted product **218**, exclusively. The complete regio and stereoselection observed are in agreement with the rotamer population of **217**, since the amide conformation is fixed in order to minimize the $A^{1,3}$-strain between the allylic group and the methyl on the amidic nitrogen.[70]

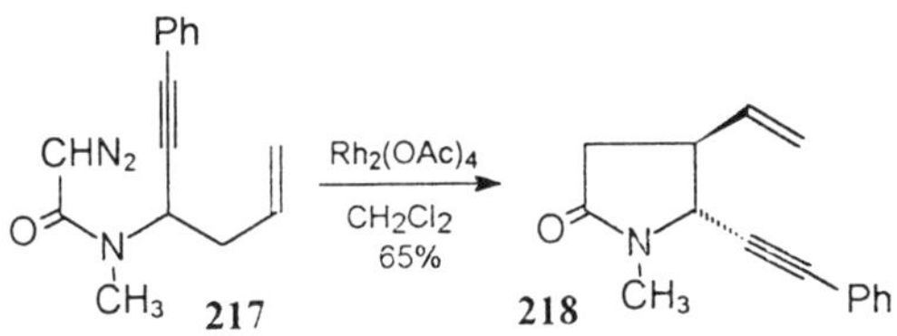

Chiral pyrrolidin-2-ones can be obtained by enantioselective C-H insertion reaction when dirhodium(II) catalysts which possess chiral ligands are employed, such as Rh_2(5S-MEPY)$_4$ **219** and Rh_2(4S-MEOX)$_4$ **220**.[71]

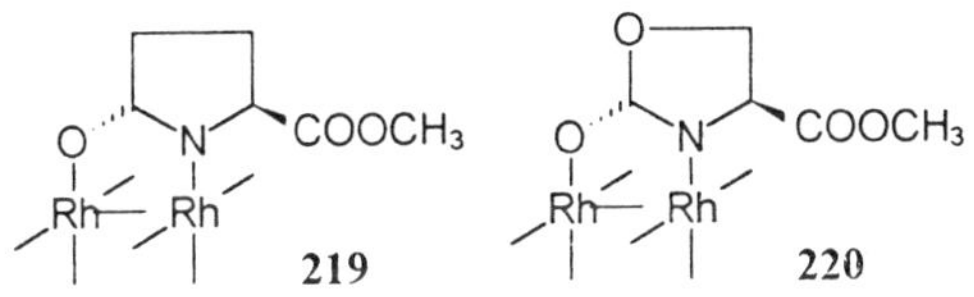

Rh_2(5*S*-MEPY)$_4$ Rh_2(4*S*-MEOX)$_4$

In fact, when a diazoacetamide reacts in the presence of these catalysts, the C-H insertion reaction proceeds with moderate to high enantioselection, leading mainly to 4-substituted pyrrolidin-2-ones **222**. It is worth mentioning, however, that 3,4-disubstituted pyrrolidin-2-ones cannot be prepared, since these chiral catalysts are unable to cause dinitrogen loss from the corresponding diazoacetacetamides.[71]

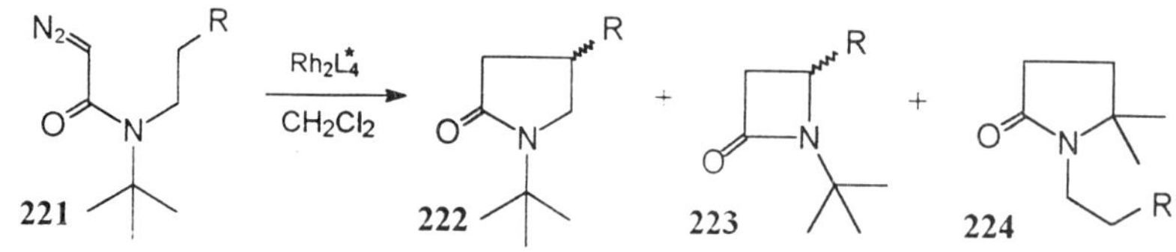

a. R = C_2H_5 **b.** R = *i*-C_3H_7 **c.** R = OC_2H_5

	Rh_2L_4		**222** : **223** : **224** (e.e.)	Rh_2L_4		**222** : **223** : **224** (e.e.)
a	Rh_2(5*S*-MEPY)$_4$	74%	88(63):12(73):0	Rh_2(4*S*-MEOX)$_4$	82%	91(71):9(80):0
b.	Rh_2(5*S*-MEPY)$_4$	91%	80(58):20(72):0	Rh_2(4*S*-MEOX)$_4$	93%	82(69):18(65):0
c.	Rh_2(5*S*-MEPY)$_4$	91%	100(58):0:0	Rh_2(4*S*-MEOX)$_4$	97%	100(78):0:0

Also, chiral bicyclic systems containing the pyrrolidin-2-one moiety can be prepared by C-H insertion reaction. Better stereoselection however is observed by using the chiral catalysts Rh_2(4*S*-MACIM)$_4$ **225** and Rh_2(4*S*-MPPIM)$_4$ **226** [72]

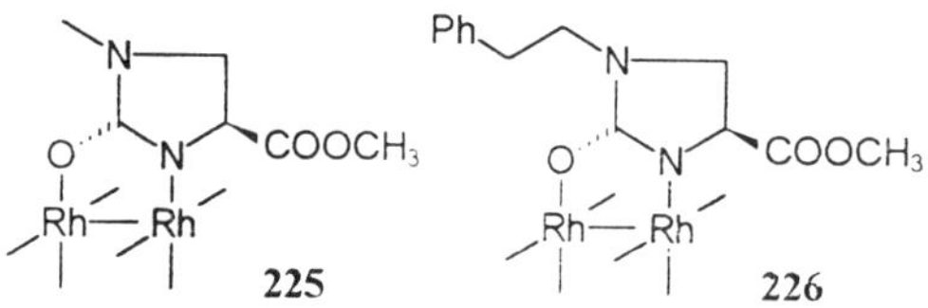

225 Rh_2(4*S*-MACIM)$_4$ **226** Rh_2(4*S*-MPPIM)$_4$

In fact the cyclisation of the diazoacetamide **227** proceeds to give the corresponding bicyclic products **228** and **229** in good yield and with very high d.e. [72]

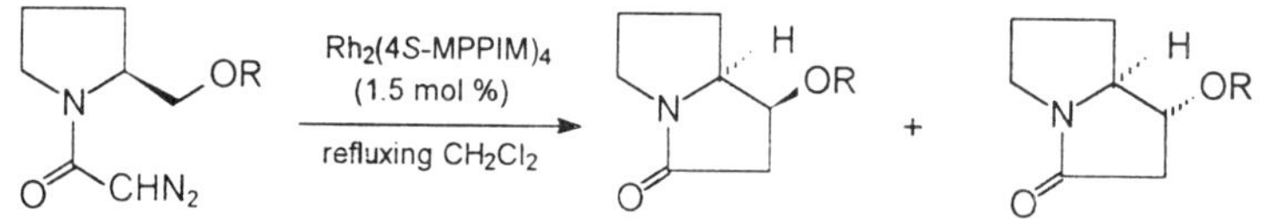

227 **228** **229**

a. R = CH_3, 97%, d.r. 97:3 **b.** R = $CH_2C_6H_5$, 93%, d.r. 97:3

The C-H insertion mediated by Rh(II) constitutes also the key step of a straightforward synthesis of (-)-heliotridane **232**, a pyrrolizidine alkaloid. Starting from **230**, the bicyclic compound **231** is obtained in good yield and high stereoselection. Eventual removal of the carbonyl group leads to the natural product in the enantiomerically pure form. [72]

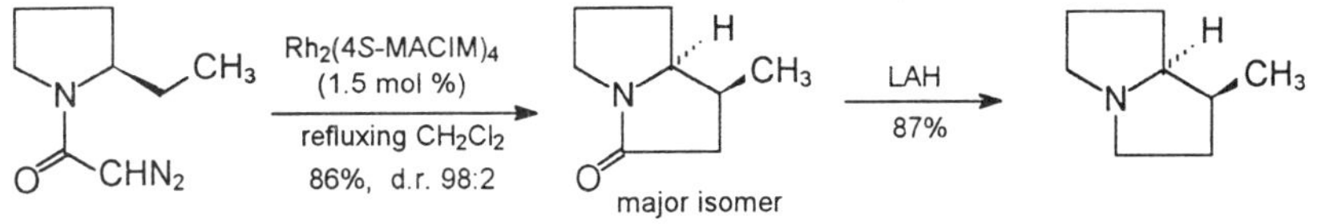

230 **231** **232**

2.4.2. Carbene insertion into a C=C bond

The addition of a carbene to an alkene allows to form cyclopropanes and when an intramolecular reaction takes place, a bicyclic compound is generally formed. Thus, the anion of *N,N*-diallyl phenylthioacetamide **233**, upon treatment with *p*-tosylazide in THF, gives rise directly to a 3-aza-2-oxo[3.1.0]bicyclohexane derivative **234**, and this result can be easily rationalised in terms of intramolecular ring closure of a carbenic species.[73]

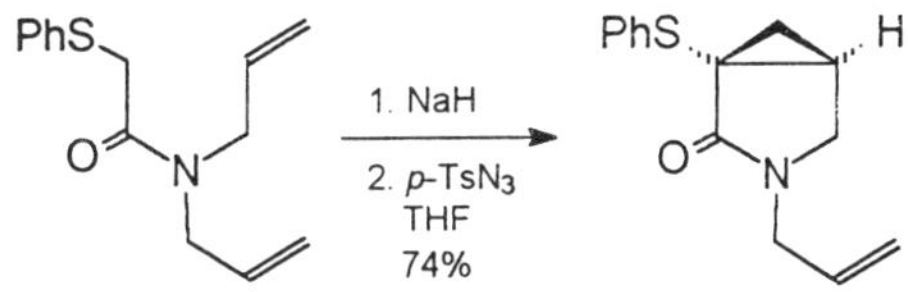

233 **234**

The formation of a carbene which then adds to a double bond can be catalysed by $Rh_2(OAc)_4$. In fact, starting from the allylic imide **235**, the bicyclic compound **236** is obtained in moderate yield.[74,75]

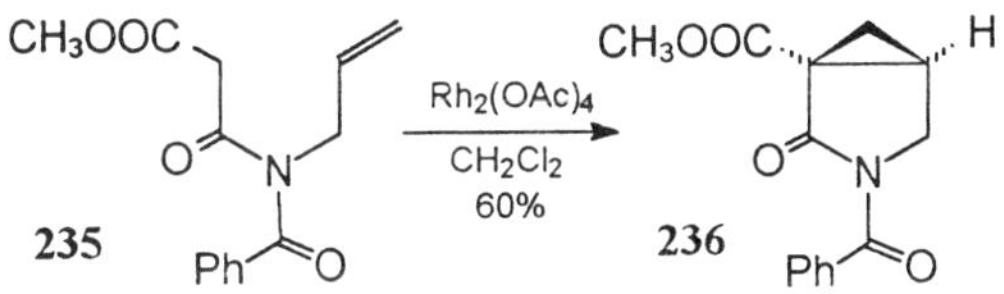

Also $Pd(OAc)_2$ is effective in promoting addition of a carbene species to a double bond. Thus, the chiral amide **237** is converted into **238** and **239** in moderate yield and good diastereoselection.[76]

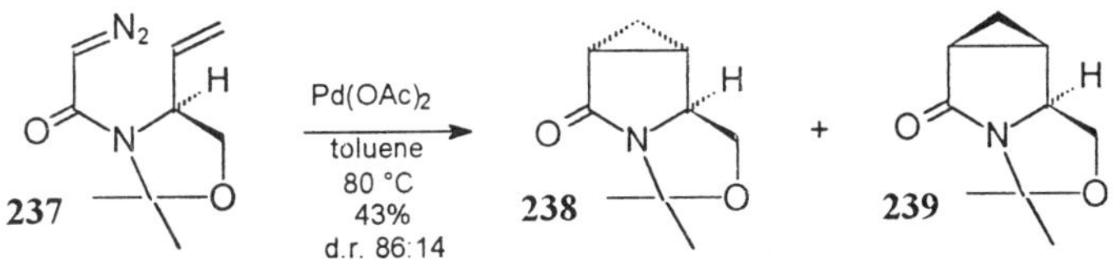

A sole isomer **241**, on the contrary, is obtained starting from **240**, probably owing to steric interactions in the transition state.[77,78]

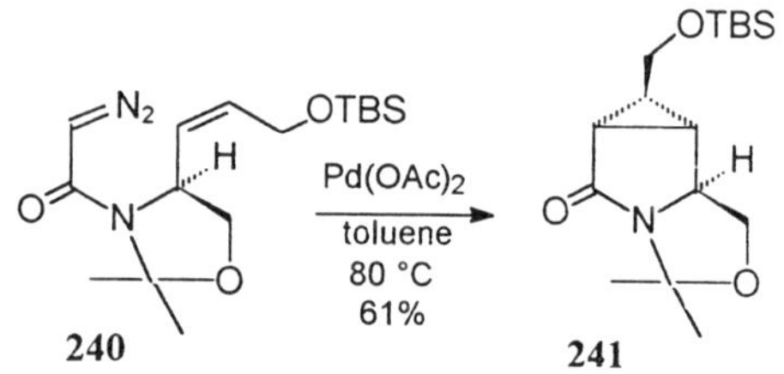

240 **241**

2.5. Cyclisation *via* 1,3-dipolar cycloaddition

Dipolar cycloadditions constitute an important route to prepare a number of heterocyclic molecules. Intramolecular 1,3-dipolar cycloadditions are particularly useful in synthesis and have been devised both to form a carbon-carbon bond and to introduce new functional groups. For example, the cycloaddition nitrone-alkene affords isoxazolines, and the N-O bond can be easily cleaved to give a hydroxy and an amino group. Even nitrile oxides can be employed as 1,3-dipoles. By addition to a double bond, an isoxazole is formed, which is precursor of both a keto and a hydroxy group.

2.5.2. 1,3-Cycloaddition of nitrones and nitrile oxides

N-Allyl diazoderivatives **242** give on standing the corresponding γ-lactams **243** with complete regio and stereoselection, the *cis*-isomer being the sole product recovered from the reaction mixture.[79]

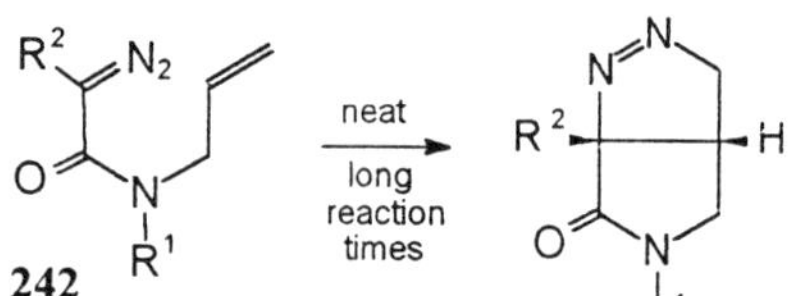

242 **243**

a. R^1 = Allyl, R^2 = H, 65%. **b.** R^1 = Ph, R^2 = H, 80%.

c. R^1 = Allyl, R^2 = COOEt, 90%. **d.** R^1 = Ph, R^2 = COOEt, 72%.

The same result is observed starting from α-diethoxyphosphinyl α-diazoacetamides **244**.[73]

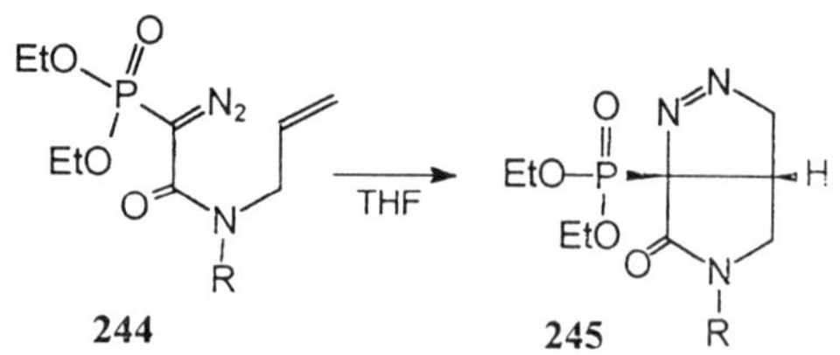

244 **245**

a. R = Allyl, 78%, **b.** R = Ph, 74%, **c.** R = Bn, 74%

2.5.2. 1,3-Cycloaddition of nitrones and nitrile oxides

A novel approach to functionalised γ-lactams relies on intramolecular cycloaddition to a double bond by nitrones obtained starting from unsaturated α-aminoaldehydes. Thus, by treating compounds **246** with hydroxylamine or *N*-methyl hydroxylamine, the corresponding nitrones are obtained, which directly cyclise to give the bicyclic products **247**. The reaction proceeds with complete regio and stereoselection, leading exclusively to *cis*-fused products.[80,81]

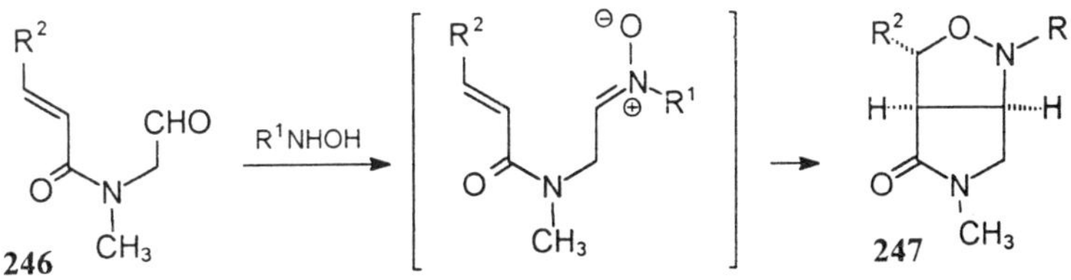

246 **247**

a. R^1 = H, R^2 = CH_3, 68%. **b.** R^1 = H, R^2 = C_2H_5, 72%. **c.** R^1 = H, R^2 = Ph, 87%.
d. R^1 = CH_3, R^2 = Ph, 78%. **e.** R^1 = R^2 = CH_3, 72%. **f.** R^1 = CH_3, R^2 = C_2H_5, 68%.

When a chiral amino aldehyde **248** is the starting material for the nitrone preparation, the cycloaddition proceeds with very high asymmetric induction. In fact 3-oxa-2,7-diazabicyclo[3.3.0]octan-6-ones **249** are obtained with very high diastereoselection, since the stereogenic centre α to the nitrone group directs the formation of the new adiacent centres, so that only one diastereomer is formed.[82]

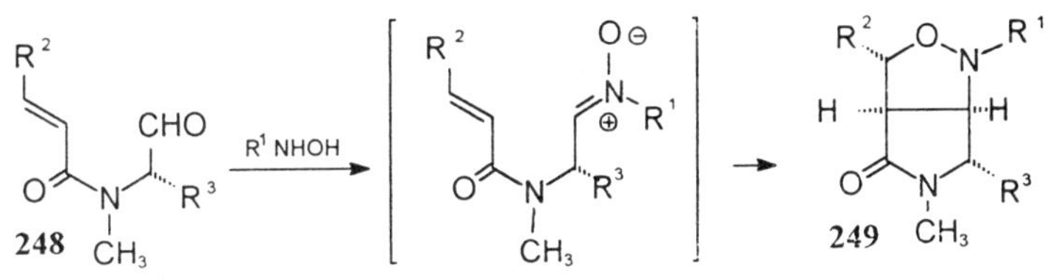

248 **249**

a. R^1 = R^2 = CH_3, R^3 = Ph, 80% **b.** R^1 = CH_3, R^2 = Bn, R^3 = Ph, 65% **c.** R^1 = R^2 = Bn, R^3 = Ph, 65%
d. R^1 = H, R^2 = Ph, R^3 = CH_3, 80% **e.** R^1 = H, R^2 = Ph, R^3 = Bn, 75%

In addition, when the nitrone derived from L-proline undergoes cycloaddition, the tricyclic compound **251** is recovered as single isomer.

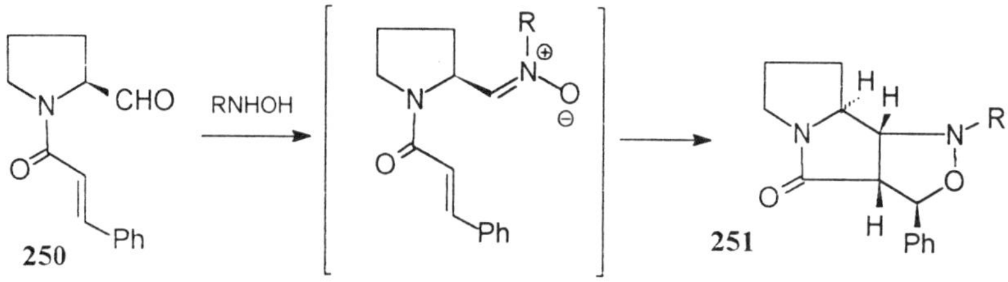

a. R = H, 90%
b. R = CH_3, 75%

The 1,3-dipolar cycloaddition can be carried out starting a also from nitrile oxides. When α-unsubstituted oximes **252** react giving the corresponding nitrile oxide, the unsaturated bicyclic lactams **253** are obtained in low to moderate yield.[83]

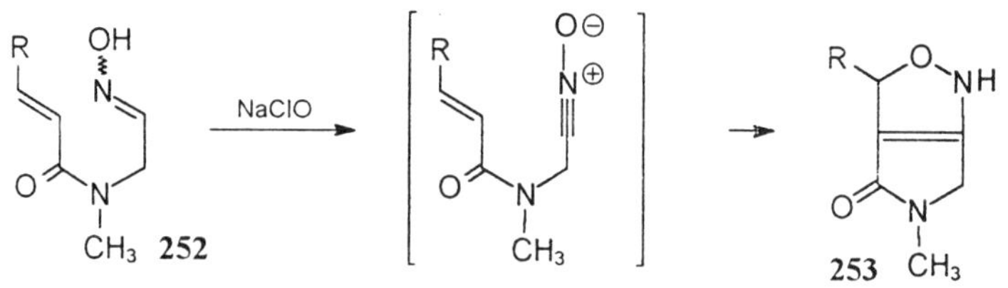

a. R = CH_3, 25% **b.** R = C_2H_5, 30% **c.** R = Ph, 75%

On the other hand, starting from α-substituted nitrile oxides, diastereomeric mixtures of bicyclic products are obtained.[83]

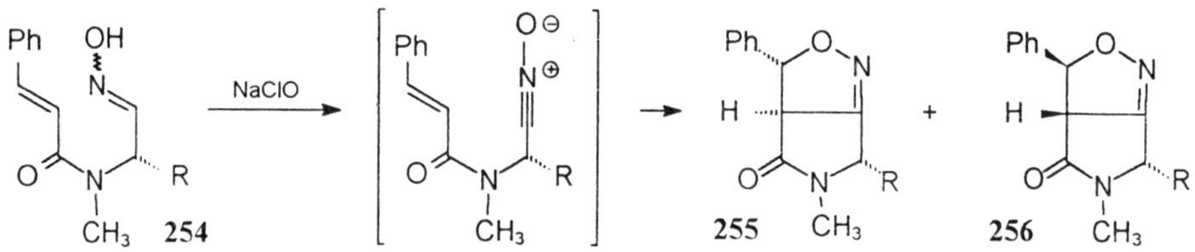

a. R = $(CH_3)_2CH$, 80%, d.r. 80:20 **b.** R = $(CH_3)_2CHCH_2$, 90%, d.r. 50:50 **c.** R = $C_6H_5CH_2$, 60%, d.r. 50:50

The oxime **257** derived from L-proline, by oxidation to the corresponding nitrile oxide, undergoes cycloaddition to give the tricyclic compounds **258** and **259** with moderate asymmetric induction.[83]

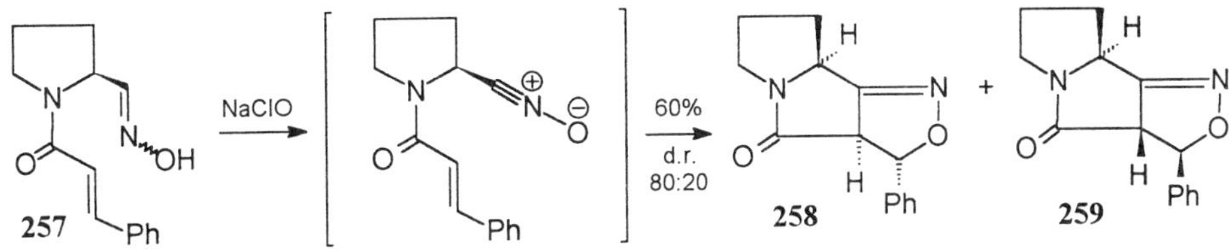

It is worth noting that the 1,3-dipolar cycloaddition results useful since the N-O bond of all γ-lactam derivatives can be cleaved under mild conditions, to afford products which can be employed in the synthesis of natural products, such as alkaloids.

2.6. Pd(II) mediated cyclisations

Pyrrolidin-2-ones can be obtained in low yield by reaction of *N*-allyl iodo and bromoacetamides in the presence of $Pd(PPh_3)_4$ and the reaction probably proceeds by means of a σ-alkylpalladium complex. Thus, by treating *N*-allyl-*N*-benzyliodoacetamides **260** with a catalytic amount of $Pd(PPh_3)_4$ in DMF, the corresponding 4-iodoalkyl pyrrolidin-2-ones **261** are recovered in low yield. [84]

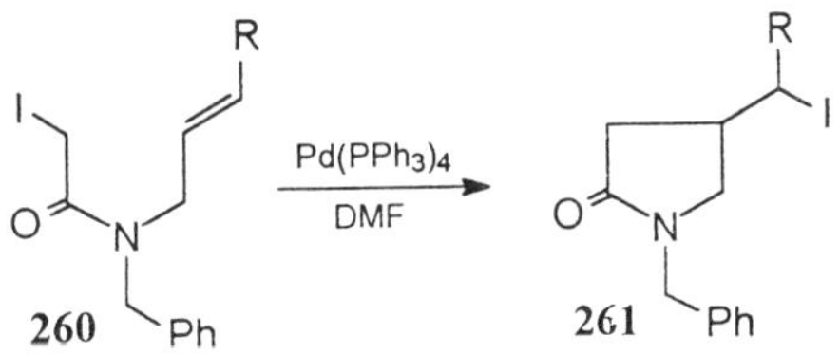

a. R = H, 25%; **b.** R = CH_3, 45%

On the other hand, by treating bromoacetamides **262** and **264** with $PdCl_2$-PPh_3 in CH_3CN in the presence of Et_3N, elimination products are exclusively obtained, although in low to moderate yields. [84,85]

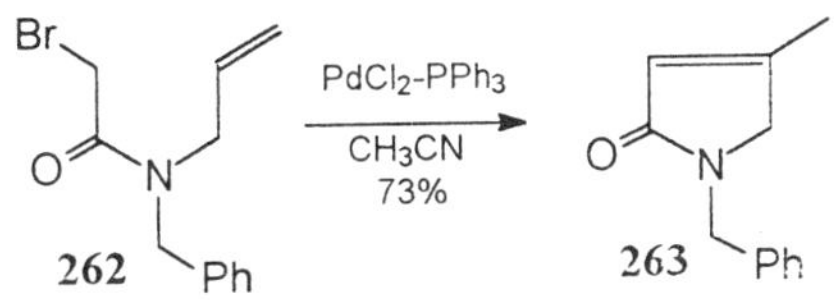

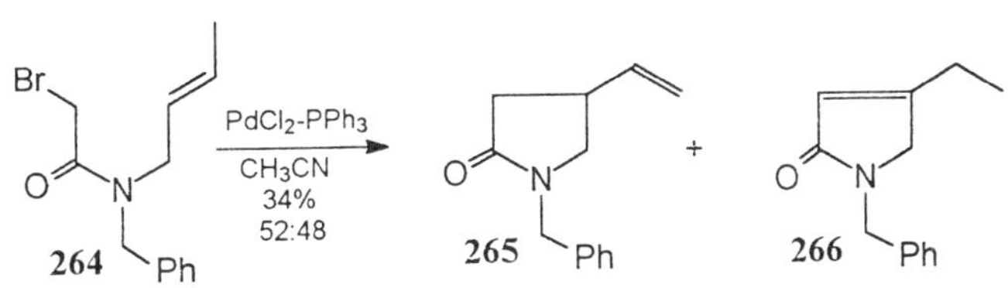

3. Formation of the C_2-C_3 bond

α-Methylene γ-lactams deserve interest since they resemble naturally occurring antitumor α-methylene butyrolactones. An useful access to these compounds has been devised by intramolecular cyclisation of a derivative of cobalt(salophen) **267**, which proceeds through transfer of the Co(III) and final elimination of a Co-H species. [86,87]

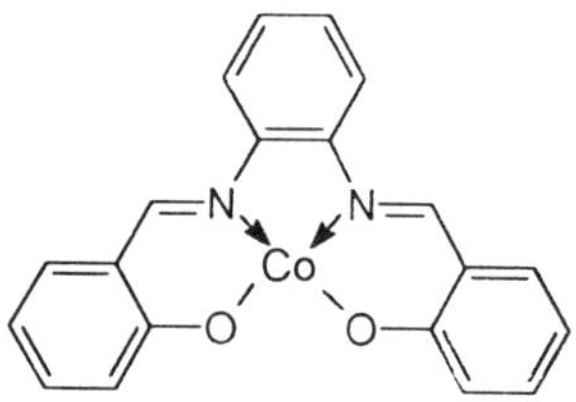

267

Thus, compound **268** in refluxing toluene, affords the corresponding α-methylene γ-lactam **269** in moderate yield. [86,87]

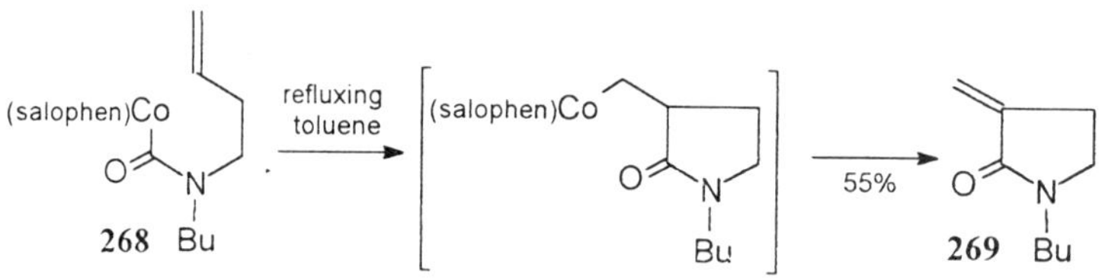

In addition, by irradiation of **268** in dichloromethane in the presence of an equimolar amount of TEMPO, derivative **270** can be obtained in moderate yield. [86,87]

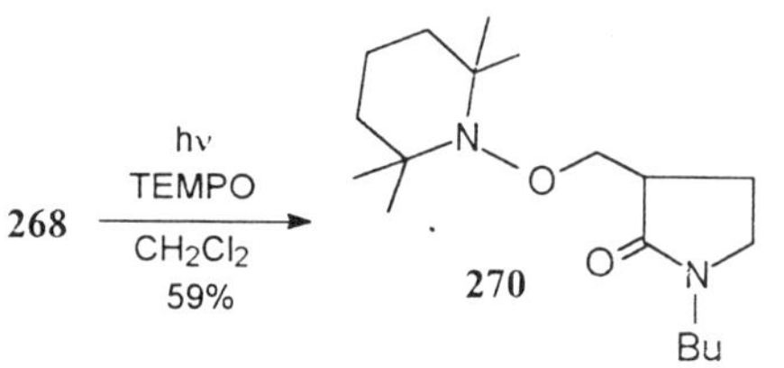

4. Formation of the C_4-C_5 bond

The cyclisation involving nucleophilic attack of the double bond of **271** to a carbon atom bearing a phosphate ester allows to prepare 4,5-disubstituted pyrrolidin-2-ones **272** and **273** in moderate yield. [88-90]

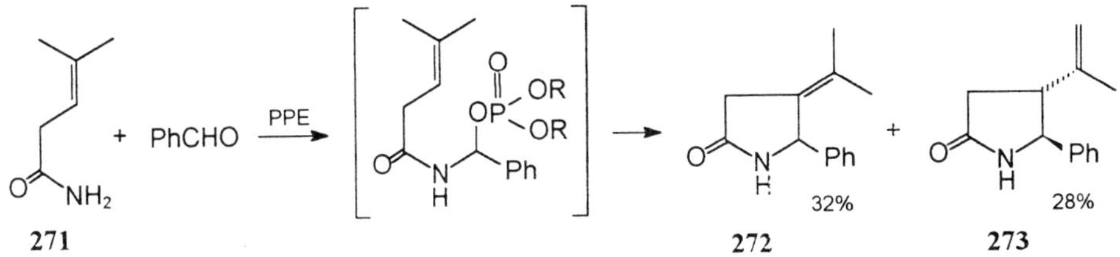

Following this approach, a number of polycyclic pyrrolidin-2-ones **275** have been prepared in moderate to good yield with complete *cis* diastereoselection, by using polyphosphoric acid as condensating agent.[88-90]

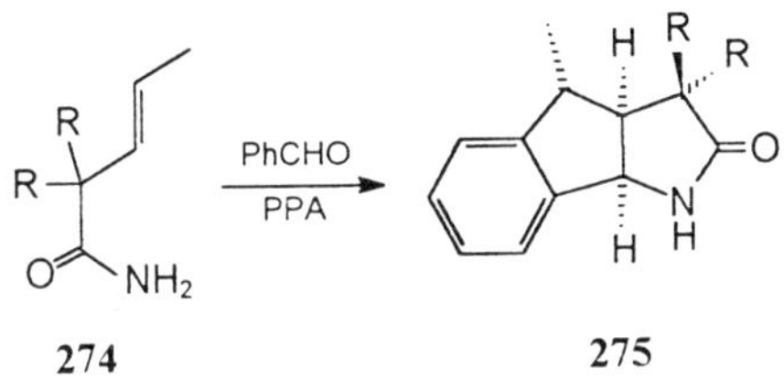

a. R = H, 74%; **b.** R = CH_3, 32%

The C_4-C_5 bond of a γ-lactam can arise also from an intramolecular alkylation reaction. Thus, by treating the acyclic amide **276** with tetrabutylammonium fluoride (TBAF), the corresponding γ-lactam **277** is obtained in moderate yield. It is worth mentioning that simply by changing the base (LDA instead of TBAF), the regioselection is dramatically changed and the β-lactam **278** is exclusively formed.[91]

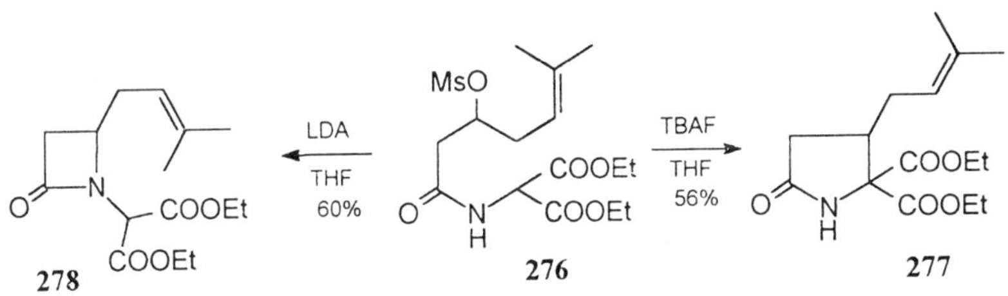

5. Simultaneous formation of the C_2-C_3 and C_4-C_5 bonds

Intramolecular [3+2] cycloadditions of azomethine ylides have been widely exploited for the construction of the pyrrolidine ring, in order to obtain valuable synthetic intermediates for the preparation of alkaloids. Thus, when the aminophosphonate **280** reacts with the aldehyde **279**, the tricyclic compound **281** is obtained in moderate yield, with complete regioselection and stereoselection , since only *cis*-fused diastereomers are recovered from the reaction mixture.[92]

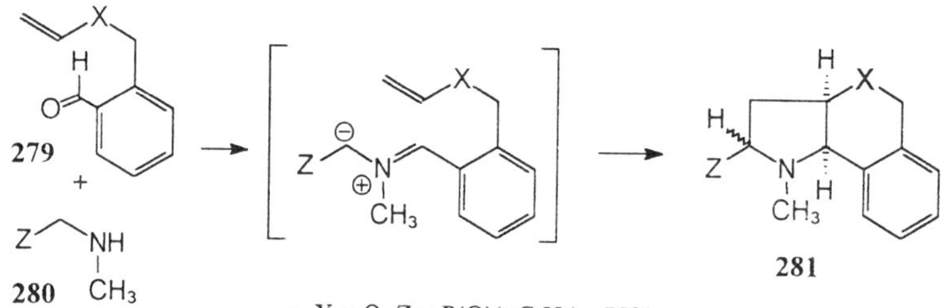

279 + 280 → 281

a, X = O, Z = $P(O)(i\text{-}C_3H_7)_2$, 75%
b, X = CH_2, Z = $P(O)(i\text{-}C_3H_7)_2$, 50%
c, X = $NHSO_2CH_3$, Z = $P(O)(i\text{-}C_3H_7)_2$, 55%

By simple oxygenation and hydrolysis of the P,O acetal, a γ-lactam is obtained, although the yields of the reaction are not reported.[92]

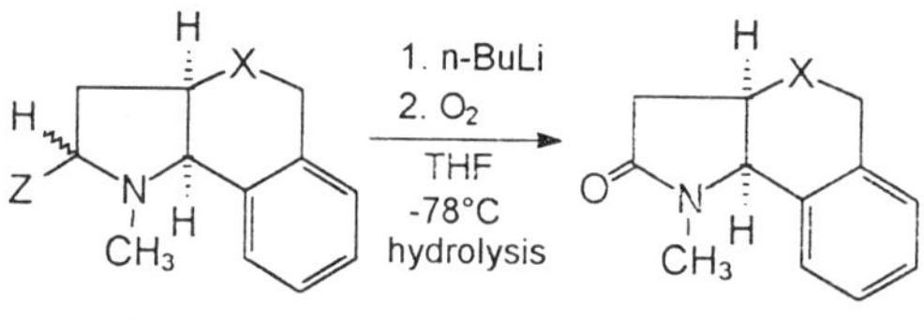

282 **283**

a, X = O, Z = $P(O)(i\text{-}C_3H_7)_2$ **b**, X = CH_2, Z = $P(O)(i\text{-}C_3H_7)_2$

γ-Lactams are also obtained by slight modification of a framework prepared by means of a 1,3-dipolar addition strategy. In fact, the dithiolane-isocyanate imminium methylide, generated by desilylation of the salt **284**, undergoes an efficient cycloaddition to electron deficient dipolarophiles to yield pyrrolidin-2-ones **286** after hydrolysis of the intermediate dithiolane **285**.[93]

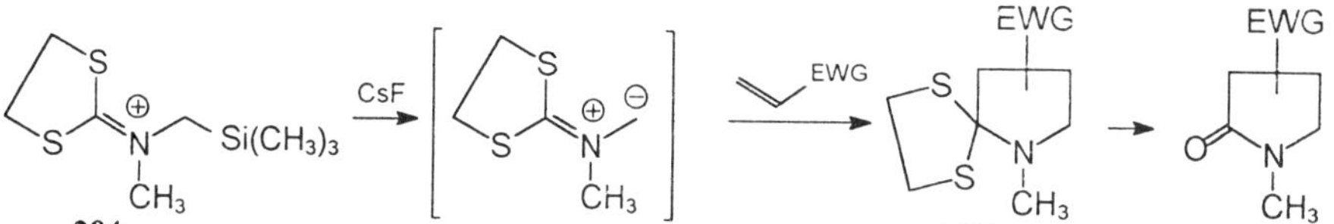

284 **285** **286**

Following this approach, a number of pyrrolidin-2-ones have been prepared in good yield and this methodology is employed in a short stereoselective synthesis of cynometrine **290**, an alkaloid obtained from *Cynometria Hankei* which has important analgesic properties.[94]

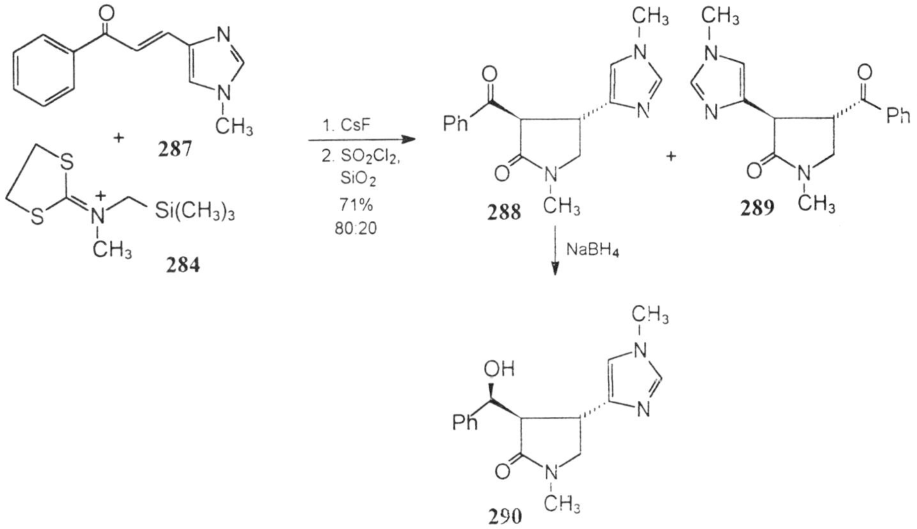

References

1. Corey, E.J.; Choi, S. *Tetrahedron Lett.*, **1993**, *34*, 6969-6972.
2. Sato, T.; Machigashira, N.; Ishibashi, H.; Ikeda, M. *Heterocycles*, **1992**, *33*, 139-142.
3. Schwabe, U.; Miyake, M.; Ohga, Y.; Daly, J. *Mol. Pharmacol.*, **1976**, 900-912.
4. Ikeda, M.; Sato, T.; Ishibashi, H. *Heterocycles*, **1988**, *27*, 1465-1487.
5. Daly, J.W.; Spande, T.F. In *Alkaloids: Chemical and Biological Perspectives*; Pelletier, S.W., Ed.; Wiley: New York, **1986**; Vol. 4, Chapter 1.
6. Kornet, M.J.; Thio, P.A.; Tan, S.I. *J. Org. Chem.*, **1968**, *33*, 3637-3639.
7. Russ, P.L., Caress, E.A. *J. Org. Chem.*, **1976**, *41*, 149-151.
8. Beckwith, A.L.J. In *The Chemistry of Amides*; Zabicky, J, E., Ed.; Interscience: London, 1970, pagg. 75-185.
9. Orena, M. In *Houben-Weyl. Methods of Organic Chemistry: Stereoselective Synthesis*; Helmchen, G.; Hoffmann, R.W.; Mulzer, J.; Schaumann, E., Eds, Thieme Verlag, Stuttgart, **1995**, Vol. E21e, pagg. 5291-5355.
10. Curran, D.P. *Synthesis*, **1988**, 417-439.
11. Curran, D.P. *Synthesis*, **1988**, 489-513.
12. Curran, D.P.; Tamine, J. *J. Org. Chem.*, **1991**, *56*, 2746-2750.
13. Baldwin, J.E. *J. Chem. Soc., Chem. Commun.*, **1976**, 734-736.
14. Baldwin, J.E.; Cutting, J.; Dupont, W.; Kruse, L.; Silberman, L.; Thomas, R.C. *J. Chem. Soc., Chem. Commun.*, **1976**, 736-738.

15. Sato, T.; Wada, T.; Nishimoto, M.; Ishibashi, H.; Ikeda, M. *J. Chem. Soc., Perkin Trans. I*, **1989**, 879-886.

16. Cardillo, B.; Galeazzi, R.; Mobbili, G.; Orena, M.; Rossetti, M. *Heterocycles*, **1994**, *38*, 2663-2676.

17. Boivin, J.; Yousfi, M.; Zard, S.Z. *Tetrahedron Lett.*, **1994**, *35*, 5629-5632.

18. Ozaki, S.; Matsushita, H.; Ohmori, H. *J. Chem. Soc., Perkin Trans. I*, **1993**, 2339-2344.

19. Clough, J.M.; Pattenden, G.; Wight, P.G. *Tetrahedron Lett.*, **1989**, *30*, 7469-7472.

20. De Mesmaeker, A.; Waldner, A.; Hoffmann, P.; Mindt, T. *Synlett*, **1993**, 871-874.

21. Hanessian, S.; Reinhold, U.; Ninkovic, S. *Tetrahedron Lett.*, **1996**, *37*, 8967-8970.

22. Parsons, A.F.; Taylor, R.J.K. *J. Chem. Soc., Perkin Trans. I*, **1994**, 1945-1951.

23. Nagashima, H.; Ozaki, N.; Ishii, T.; Seiki, K.; Washiyama, M.; Itoh, K. *J. Org. Chem.*, **1993**, *58*, 464-470.

24. Sato, T.; Matsubayashi, K.; Yamamoto, K.; Ishikawa, H.; Ishibashi, H.; Ikeda, M. *Heterocycles*, **1995**, *40*, 261-270.

25. Jolly, R.S.; Livinghouse, T. *J. Am. Chem. Soc.*, **1988**, 110, 7536-7537.

26. Stork, G.; Mah, R. *Heterocycles*, **1989**, *28*, 723-727.

27. Ishibashi, H.; So, T.S.; Sato, T.; Kuroda, K.; Ikeda, M. *J. Chem. Soc., Chem. Commun.*, **1989**, 762-764.

28. Belvisi, L.; Gennari, C.; Poli, G.; Scolastico, C.; Salom, B.; Vassallo, M. *Tetrahedron*, **1992**, *48*, 3945-3960.

29. Andrès, C.; Duque-Soladana, J.P.; Iglesias, J.M.; Pedrosa, R. *Tetrahedron Lett.*, **1996**, *37*, 9085-9086.

30. Sibi, M.P.; Ji, J. *J. Am. Chem. Soc.*, **1996**, *118*, 3063-3064.

31. Danishefsky, S.; Taniyama, E. *Tetrahedron Lett.*, **1983**, *24*, 15-18.

32. Sato, T.; Chono, N.; Ishibashi, H.; Ikeda, M. *J. Chem. Soc., Perkin Trans. I*, **1995**, 1115-1120.

33. Goodall, K.; Parsons, A.F. *J. Chem. Soc., Perkin Trans. I*, **1994**, 3257-3258.

34. Goodall, K.; Parsons, A.F. *Tetrahedron*, **1996**, *32*, 6739-6758.

35. Quiclet-Sire, B.; Saunier, J.-B.; Zard, S.Z. *Tetrahedron Lett.*, **1996**, *37*, 1397-1400.

36. Ishibashi, H.; Nakamura, N.; Sato, T.; Takeuchi, M.; Ikeda, M. *Tetrahedron Lett.*, **1991**, *32*, 1725-1728.

37. Sato, T.; Machigashira, N.; Ishibashi, H.; Ikeda, M. *Heterocycles*, **1992**, *33*, 139-142.

38. Sato, T.; Nakamura, N.; Ikeda, K.; Okada, M.; Ishibashi, H.; Ikeda, M.; *J. Chem. Soc., Perkin Trans. I*, **1992**, 2399-2407.

39. Nagashima, H.; Wakamatsu, H.; Itoh, K. *J. Chem. Soc., Chem. Commun.*, **1984**, 652-653.

40. Nagashima, H.; Wakamatsu, H.; Ozaki, N.; Ishii, T.; Watanabe, M.; Tajima, T.; Itoh, K. *J. Org. Chem.*, **1992**, *57*, 1682-1689.

41. Rachita, M.A.; Slough, G.A. *Tetrahedron Lett.*, **1993**, *34*, 6821-6824.

42. Slough, G.A. *Tetrahedron Lett.*, **1993**, *34*, 6825-6828.

43. Ishibashi, H.; Uemura, N.; Nakatani, H.; Okazaki, M.; Sato, T.; Nakamura, N.; Ikeda, M. *J. Org. Chem.*, **1993**, *58*, 2360-2368.

44. Seijas, J.A.; Vazquez-Tato, M.P.; Castedo, L.; Estevez, R.J.; Onega, M.G.; Ruiz, M. *Tetrahedron*, **1992**, *48*, 1637-1642.

45. Nagashima, H.; Ara, K.; Wakamatsu, H.; Itoh, K. *J. Chem. Soc., Chem. Commun.*, **1985**, 518-519.

46. Ishibashi, H.; So, T.S.; Okochi, K.; Sato, T.; Nakamura, N.; Nakatani, H.; Ikeda, M. *J. Org. Chem.*, **1991**, *56*, 95-102.

47. Nagashima, H.; Ozaki, N.; Seki, K.; Ishii, M.; Itoh, K. *J. Org. Chem.*, **1989**, *54*, 4497-4499.

48. Baldovini, N.; Bertrand, M.P.; Carriere, A.; Nouguier, R.; Plancher, J.-M. *J. Org. Chem.*, **1996**, *61*, 3205-3208.

49. Byers, J.H.; Gleason, T.G.; Knight, K.S. *J. Chem. Soc., Chem. Commun.*, **1991**, 354-356.

50. Boivin, J.; Camara, J.; Zard, S.Z. *J. Am. Chem. Soc.*, **1992**, *114*, 7909-7910.

51. Axon, J.; Boiteau, L.; Boivin, J.; Forbes, J.E.; Zard, S.Z. *Tetrahedron Lett.*, **1994**, *35*, 1719-1722.

52. Bertrand, M.P.; Nouguier, R.; Archavlis, A.; Carrière, A. *Synlett*, **1994**, 736-738.

53. Cardillo, B.; Galeazzi, R.; Mobbili, G.; Orena, M. *Synlett*, **1995**, 1159-1160.

54. Galeazzi, R.; Mobbili, G.; Orena, M., *Tetrahedron*, **1996**, *52*, 1069-1084.

55. Galeazzi, R.; Geremia, S.; Mobbili, G.; Orena, M. *Tetrahedron: Asymmetry*, **1996**, *7*, 3573-3584.

56. Cossy, J.; Leblanc, C. *Tetrahedron Lett.*, **1989**, *30*, 4531-4532.

57. Cossy, J.; Bouzide, A.; Leblanc, C. *Synlett*, **1993**, 202-204.

58. Ishibashi, H.; Ikeda, M.; Maeda, H.; Ishiyama, K.; Yoshida, M.; Akai, S.; Tamura, Y. *J. Chem. Soc., Perkin Trans. I*, **1987**, 1099-1105.

59. Tamura, Y.; Yakura, T.; Shirouchi, Y.; Haruta, *Chem. Pharm. Bull.*, **1986**, *34*, 1061-1066.

60. Ishibashi,H.; Nakaharu, T.; Nishimura, M.; Nishikawa, A.; Kameoka, C.; Ikeda, M. *Tetrahedron*, **1995**, *51*, 2929-2938.

61. Galeazzi, R.; Geremia, S.; Mobbili, G.; Orena, M. *Tetrahedron: Asymmetry*, **1996**, *7*, 79-88.

62. Orena M., unpublished results.

63. Chan, S.; Braish, T.F. *Tetrahedron*, **1994**, *50*, 9943-9950.

64. Andrews, M.D.; Brewster, A.G.; Moloney, M.G.; Owen, K.L. *J Chem. Soc., Perkin Trans. I*, **1996**, 227-228.

65. Cossy, J.; Thellend, A. *Tetrahedron Lett.*, **1990**, *31*, 1427-1428.

66. Murray, A.; Proctor, G.R.; Murray, P.J. *Tetrahedron Lett.*, **1995**, *36*, 291-294.

67. Rando, R.R. *J. Am. Chem. Soc.*, **1972**, *94*, 1629-1631.

68. Doyle, M.P.; Oon, S.-M.; van der Heide, F.R.; Brown, C.B. *Bioorg. Med. Chem. Lett.*, **1993**, *3*, 2409-2414.

69. Padwa, A.; Austin, D.J.; Hornbuckle, S.F.; Semones, M.A. *J. Am. Chem. Soc.*, **1992**, *114*, 1874-1876.

70. Padwa, A.; Dean, D.C.; Fairfax, D.J.; Xu, S.L. *J. Org. Chem.*, **1993**, *58*, 4646-4655.

71. Doyle, M.P.; Protopopova, M.N.; Winchester, W.R.; Daniel, K.L. *Tetrahedron Lett.*, **1992**, *33*, 7819-7822.

72. Doyle, M.P.; Kalinin, A.V. *Tetrahedron Lett.*, **1996**, *37*, 1371-1374.

73. Minami, T.; Kamitamari, M.; Utsonomiya, T.; Tanaka, T.; Ichikawa, *J. Bull. Chem. Soc. Jpn*, **1993**, *66*, 1496-1500.

74. Padwa, A.; Austin, D.J.; Price, A.T. Tetrahedron Lett., 1994, 35, 7159-7162.

75. Padwa, A.; Austin, D.J.; Price, A.T.; Weingarten, M.D. *Tetrahedron*, **1996**, *52*, 3247-3260.

76. Yamanoi, K.; Ohfune, Y.; Watanabe, K.; Li, P.N.;Takeuchi, H. *Tetrahedron Lett.*, **1988**, *29*, 1181-1184.

77. Ohfune, Y.;Shimamoto, K.; Ishida, M.; Shinozaki, H. *Bioorg. Med. Chem. Lett.*, **1993**, *3*, 15-18.

78. Shimamoto, K.; Ohfune, Y. *Tetrahedron Lett.*, **1990**, *31*, 4049-4052.

79. Sturm, H.; Ongania, K.-H.; Daly, J.J.; Klotzer, W. *Chem. Ber.*, **1981**, *114*, 190-200.

80. Chiacchio, U.; Buemi, G.; Casuscelli, F.; Procopio, A.; Rescifina, A.; Romeo, R. *Tetrahedron*, **1994**, **50**, 5503-4414.

81. Chiacchio, U.; Corsaro, A.; Pistarà, V.; Rescifina, A.; Romeo, G.; Romeo, R. *Tetrahedron*, **1996**, 52, 7875-7884.

82. Chiacchio, U.; Casuscelli, F.; Corsaro, A.; Librando, V.; Rescifina, A.; Romeo, R.; Romeo, G. *Tetrahedron*, **1995**, *51*, 5689-5700.

83. Chiacchio, U.; Corsaro, A.; Librando, V.; Rescifina, A.; Romeo, R.; Romeo, *Tetrahedron*, **1996**, *52*, 14323-14334.

84. Mori, M.; Kanda, N.; Oda, I.; Ban, Y. *Tetrahedron*, **1985**, *41*, 5465-5474.

85. Yang, S.-C.; Shea, F.-R. *J. Chin. Chem. Soc.*, **1995**, *42*, 969-972.

86. Gill, G.B.; Pattenden, G.; Reynolds, S.J. *Tetrahedron Lett.*, **1989**, *30*, 3229-3232.

87. Gill, G.B.; Pattenden, G.; Reynolds, S. *J. Chem. Soc., Perkin Trans. 1*, **1994**, 369-378.

88. Marson, C.M.; Grabowska, U.; Walsgrove, T.; Eggeston, D.S.; Baures, P.W. *J. Org. Chem.*, **1994**, *59*, 284-290.

89. Marson, C.M.; Grabowska, U.; Walsgrove, T.; Eggleston, D.S.; Baures, P.W. *J. Org. Chem.*, **1991**, *56*, 2603-2605

90. Marson, C.M.; Grabowska, U.; Walsgrove, T. *J. Org. Chem.*, **1992**, *57*, 5045-.

91. Crocker, P.J.; Karlssonandreasson, U.; Lotz, B.T.; Miller, M.J. *Heterocycles*, **1995**, *40*, 691-716.

92. Martin, S.F.; Cheavens, T.H. *Tetrahedron Lett.*, **1989**, *30*, 7017-7020.

93. Fishwick, C.W.G.; Foster, R.J.; Carr, R.E. *Tetrahedron Lett.*, **1995**, 36, 9409-9412.

94. Fishwick, C.W.G.; Foster, R.J.; Carr, R.E. *Tetrahedron Lett.*, **1996**, *37*, 3915-3918.